Principles and Applications of
MILLIMETER-WAVE RADAR

Principles and Applications of
MILLIMETER-WAVE RADAR

Edited by
Nicholas C. Currie
Charles E. Brown

Artech House

Library of Congress Cataloging-in-Publication Data

Principles and applications of millimeter-wave radar.

Includes bibliographies and index.
1. Radar. 2. Millimeter waves. I. Currie, Nicholas C.
II. Brown, Charles E. (Charles Edward),
1947–
TK6580.P74   1987   621.3848   87-19460
ISBN 0-89006-202-1

ARTECH HOUSE, INC.
685 Canton Street
Norwood, MA 02062

International Standard Book Number: 0-89006-202-1
Library of Congress Catalog Card Number: 87-19460

10 9 8 7 6 5 4 3 2 1

Contents

Preface

This book had as its genesis a series of short courses in millimeter-wave radar and related topics offered through the Office of Continuing Education at the Georgia Institute of Technology. These courses have evolved over the past decade with continuing emphasis on radar to the extent that a four-day short course dedicated solely to millimeter-wave radar is now underway. This book was developed as a textbook for such a short course as well as a handbook for application-oriented engineers, scientists, and technical managers to aid in understanding the capabilities and limitations of millimeter-wave radars across a broad spectrum of systems applications.

The authors of the various chapters in this book were carefully chosen from among practicing application oriented research engineers and scientists. All of the authors work with millimeter-wave radar or related technology on a daily basis. Many authors are nationally recognized as technical experts in the fields addressed by their chapters.

Numerous persons have contributed to the development and publication of this book, and we wish to thank them all. Among these, we would like to acknowledge our contributing authors who worked very hard to prepare their respective chapters. Also, a very special thank you is due Dr. Edward K. Reedy, Director of the Radar and Instrumentation Laboratory of the Georgia Tech Research Institute, whose encouragement and support made the task of completing this book possible in light of consistently demanding research program schedules. Special thanks are due as well to Dr. James C. Wiltse, who initiated and nurtured many millimeter-wave short courses at Georgia Tech, which provided the momentum necessary to begin work on this book. Finally, we would like to thank Ms. Judy Truett and Ms. Phyllis Hinton for their hard work in helping with the manuscript typing, graphics, and printing.

Nicholas C. Currie
Charles E. Brown
June 1987

vii

The Authors

D. G. Bodnar

C. E. Brown (co-editor)

J. A. Bruder

J. C. Butterworth

C. H. Currie

N. C. Currie (co-editor)

J. D. Echard

J. A. Gagliano

W. A. Holm

M. M. Horst

E. B. Joy

R. W. McMillan

G. V. Morris

B. Perry, IV

S. O. Piper

E. K. Reedy

J. A. Scheer

R. N. Trebits

T. V. Wallace

J. C. Wiltse

In Memoriam

Harold L. Bassett served on the research faculty of the Georgia Institute of Technology for 18 years. During that time he rose to the rank of Principal Research Engineer and held the position of Chief of the Modeling and Simulation Division of the Georgia Tech Research Institute. He was internationally known for his research on high-temperature electronic materials, radar cross-section reduction (RCSR) techniques, and the computer modeling of electromagnetic signatures. He authored numerous reports, papers, and articles, as well as several chapters in major books on radomes and related subjects. He was in the process of developing a chapter for this text when he died on 25 November 1985 after a five-year battle with a respiratory disease. Mr. Bassett was a brilliant researcher and a good friend to all who knew him. This book is dedicated to his memory.

Part I
INTRODUCTION

Chapter 1
Overview and Background

C.E. Brown and J.C. Wiltse

Georgia Institute of Technology
Atlanta, Georgia

1.1 OVERVIEW

This book is both an introduction to and a study guide for the rapidly expanding field of *millimeter-wave* (MMW) *radar*. It is written for engineers, scientists, and managers, and addresses the fundamentals of MMW radar, and its capabilities and limitations, when used in various systems applications. Although there are numerous excellent textbooks in the broader field of radar, this book addresses the particular characteristics of MMW radar and contrasts its operation with that of conventional radar and other competing and complementary sensors. For both managers and researchers initially entering the MMW radar field, early chapters of this book will be useful as a primer and a tutorial; for the researcher involved in the design, development, and application of MMW radar to a particular systems problem, this book provides detailed treatment of the advantages of MMW radar and the pitfalls which should be avoided as they relate to specific system applications.

This book is structured around five major topical areas as follows:

- Part I—Introduction
- Part II—Theory and Phenomenology
- Part III—Components and Subsystems
- Part IV—Systems Applications
- Part V—Special Topics

Part I, "Introduction," consists of Chapters 1 and 2. Chapter 1, "Overview and Background," provides a brief introduction to the book and a short background and history of the progression of MMW radar to its current level of maturity. Chapter 2, "MMW Radar Fundamentals," provides an introduction to MMW radar and further serves as a primer on MMW radar. Chapter 2 covers the basics of MMW radar design and applications, and sets the stage for more detailed treatments of key topics in later chapters.

Part II, "Theory and Phenomenology," consists of Chapters 3 through 7. Chapter 3, "Target Detection in Noise and Clutter," addresses the fundamentals of target detection in the presence of competing noise and clutter; although this theory is the same as that applied to more conventional radars, target detection in clutter is emphasized, since many MMW radar applications are related to detecting military targets when immersed in ground clutter. Chapter 4, "MMW Propagation Phenomena," provides a detailed treatment of the propagation characteristics of MMW energy through the atmosphere. Because of differences in the interaction of the shorter wavelengths (as compared to microwaves) with various atmospheric constituents, a thorough understanding of this topic is necessary to assess the capabilities and limitations of MMW radars operating within the earth's atmosphere. In particular, the characteristics of the rather broad "windows" of relatively lower attenuation, along with the "walls" of rather high attenuation, are addressed to provide the reader with an understanding of how these atmospheric characteristics may be used to an advantage for the system developer. Particular caution is given to the significant variations in attenuation for adverse weather versus clear weather, especially as applied to effective use of these "windows" and "walls" in a particular system application.

Chapter 5, "MMW Clutter," provides a detailed discussion of MMW radar backscatter from various types of clutter, with particular emphasis on ground clutter. This chapter is of particular importance because, for short-range radar applications against small surface-targets, clutter backscatter is generally the dominant factor limiting system performance. Chapter 6, "MMW Signal Processing Techniques," describes signal processing techniques used to detect targets immersed in clutter, with emphasis on gaining additional subclutter visibility using advanced signal processing techniques.

Chapter 7, "MMW Modeling Techniques," describes the computer modeling techniques used to assess the performance of a MMW radar in a complex environment. Because of the complex nature of targets and clutter at MMW frequencies, it is not possible to assess the performance characteristics of various MMW radar configurations without the use of rather sophisticated computer models. These models and their use in MMW radar design, development, and evaluation are addressed in this chapter.

Part III, "Components and Subsystems," consists of Chapters 8 through 12. This section deals with the hardware associated with the development and fabrication of MMW radars. Although many of the components and subsystems associated with a MMW radar are similar to those of its microwave counterpart, the techniques and materials associated with such components and subsystems are sometimes drastically different. This section provides a thorough description of the key components of a MMW radar system and compares and contrasts these with components used in conventional microwave radars where applicable. Chapter 8, "MMW Solid State Sources," discusses the various types of solid state power sources, along with a description of solid-state modulator techniques. Chapter 9, "High Power MMW Transmitters," describes the currently available sources (primarily tubes) of high (100 W or greater) MMW RF power along with modulator techniques for several of the most common types. Chapter 10, "MMW Receivers," covers both design considerations and components. Chapter 11, "MMW Antennas," describes the characteristics of antennas currently utilized with MMW systems and describes some of the advantages and pitfalls of each type. Chapter 12, "MMW Radomes," gives a summary of the characteristics of radomes in general and then discusses particular problems characteristic to the MMW band.

Part IV, "Systems Applications," provides system application discussions and associated radar design examples for several specific system applications. Chapter 13, "Radar Design Considerations," introduces the reader to the primary design trades which must be addressed when developing a MMW radar for any system application. This chapter serves as a primer for the MMW radar designer and serves as a basis for understanding the specific system applications and radar design trades addressed in the subsequent chapters. Three specific system applications and associated design examples are then covered in Chapters 14 through 16 including Chapter 14, "MMW Seekers," Chapter 15, "MMW Airborne Mapping Radar," and Chapter 16, "MMW Low Angle Tracking Radar."

Part V, "Special Topics," covers two special topics which are very important to a thorough treatment of MMW radar which do not conveniently fit into one of the other sections. Chapter 17, "MMW Reflectivity Measurements," covers the techniques of making quantitative radar backscatter measurements at MMW frequencies and the particular pitfalls inherent at MMW frequencies as opposed to conventional microwave frequencies. Finally, Chapter 18, "MMW Radiometry," covers the theory and techniques associated with passive MMW radiometers in making effective use of *gray* emission and absorption at these frequencies. Some researchers refer to the MMW radiometric mode of operation as *passive MMW radar*, although it is not a form of radar at all, especially in its underlying theory. Although the basic principles underlying MMW radiometry are somewhat similar to those applicable to infrared radiometry, there are important and far-reaching differences due to the significantly longer wavelengths where material emissivity and reflectivity becomes dominant as compared to the physical temperature being dominant in the infrared region. MMW radiometry finds numerous applications in areas of covert surveillance, airborne imaging, target detection, and target tracking. Also, this passive sensing mode is often used synergistically with a radar mode to capitalize on inherent advantages of the preferred mode for various phases of operation in a given system application.

1.2 BACKGROUND

The MMW region of the electromagnetic spectrum has received increased interest in recent years due to significant advances in the development of transmitters, receivers, devices, and components, and their use in systems applications in such fields as radar, radiometry, remote sensing, missile guidance, radio astronomy, communications, and spectroscopy. This chapter provides an introduction and a brief history of activities in areas relating to MMW radar and radiometry and associated system applications.

The MMW region of the electromagnetic spectrum is generally defined as the frequency range from 30 to 300 GHz (or wavelengths between 1 cm and 1 mm). Other current terminology associated with the MMW region includes *near-millimeter waves* for frequencies from approximately 100 GHz (sometimes from 90 GHz) to 1000 GHz, and *submillimeter*, from about 150 to 3000 GHz (3 THz). Figure 1.1 shows commonly used terminology associated with MMW radar operation relative to adjacent bands. Complementing this pictorial view of the MMW spectrum, Table 1.1 shows specific frequency-band designations in common usage by various elements of the technical community. The reader can readily see that there is definitely room for both confusion and debate over the correct terminology to use. Extreme care must always be exercised to ensure that the reader understands which terminology is being used by a specific author.

One characteristic of the MMW frequency range is that for a given physical antenna size (aperture) the antenna beamwidth is smaller and the gain is higher than at microwave frequencies used for conventional radars; therefore, to obtain a specified gain or narrow beamwidth, a much smaller antenna may be used. This characteristic is important in many system

Table 1.1 Frequency-band designations.

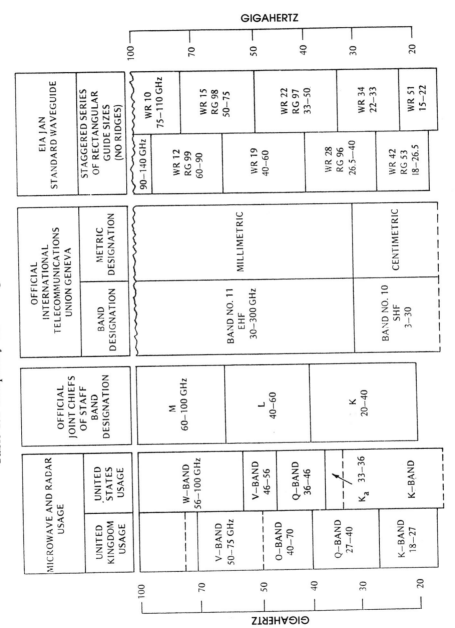

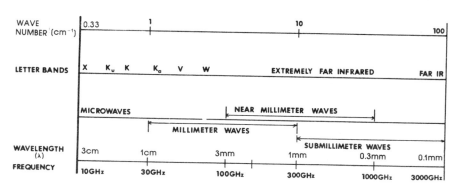

Fig. 1.1 Common terminology for millimeter and submillimeter waves and other frequency ranges.

applications where the size and weight of the hardware are constrained, such as for missile terminal guidance seekers and airborne surveillance sensors. As an example, for an air-to-surface missile seeker, a narrow antenna beamwidth is of paramount importance since it directly reduces the competing radar or radiometric return from the terrain "clutter patch" around a target.

In general, atmospheric propagation effects dominate design considerations relating to many MMW radar applications. This is true even for satellite-borne systems located outside the earth's atmosphere, since frequencies may be chosen for which the atmosphere is opaque, thus preventing detection of satellite-borne radars by ground-based receivers. Terrestrial systems desiring to prevent signal "overshoot" in range may similarly operate at a frequency of high atmospheric absorption to gain a specified degree of covertness. Typical values of atmospheric attenuation are shown in Figure 1.2 for propagation at sea level and 4-km altitude. Noteworthy are the absorption peaks or maxima regions, called "walls," due to atmospheric constituents such as oxygen and water vapor, and the minima regions called "windows." Also characteristic is the fact that the window minima increase monotonically with frequency. When reviewing data such as those presented in Figure 1.2, the reader must always be careful to note whether attenuation values are given for "one-way" or "two-way" propagation through the medium.

Additional signal attenuation (and backscatter) is produced by rainfall, clouds, and fog (see Figure 1.3), although fog losses are generally very low unless visibility is extremely limited (e.g., 100 m or less). Depending on total liquid water content, cloud attenuation rates may not be insignificant, but total attenuation is often low because of the limited range extent of the cloud. The attenuation of cirrus ice clouds is negligibly small, since the dielectric constant of ice is much lower than that of liquid water. Values of attenuation and backscatter due to atmospheric effects have been well established for frequencies up to 100 GHz and programs have recently been conducted to obtain better information about atmospheric effects above 100 GHz. Of particular interest are effects near the attenuation minima at 35, 94, 140, and 220 GHz.

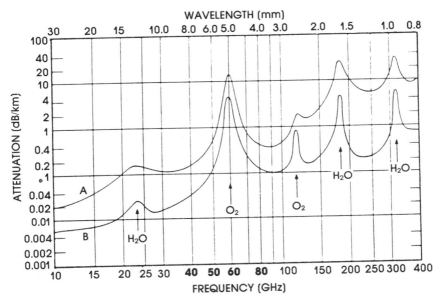

Fig. 1.2 Average atmospheric absorption of MMWs: (a) Sea level. T = ' 20°C, P = 760mm, ρ_{H_2O} = 7.5 g/m^3; (b) 4km. T = 0°C; ρ_{H_2O} = 1 g/m^3.

Atmospheric turbulence effects due to time-varying localized temperature and humidity variations in clear air can produce amplitude scintillations and angle-of-arrival changes. Amplitude fluctuations of several decibels can occur on near-earth paths of moderate length (up to a few kilometers) in clear air. The fluctuation rate is slow (spectral width of from a few cycles up to tens of cycles per second), so automatic gain control can compensate for this in many applications. However, angle of arrival effects may not be so easily compensated for. Since MMW systems are often designed to provide very good angular resolution, atmospheric angle-of-arrival fluctuations may contribute significant unwanted errors in some

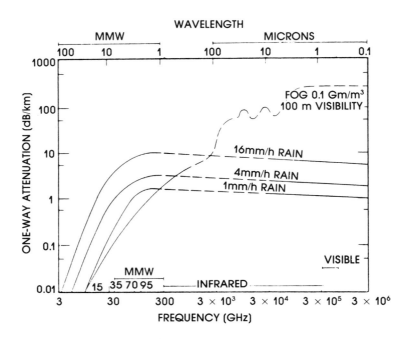

Fig. 1.3 Effects of rain and fog.

systems applications. Peak-to-peak fluctuations up to one-half milliradian may occur for clear-air path lengths of from 1 to 3 km.

The effects of smoke and dust on the propagation of MMW energy have also been well quantified. An example is a test (DIRT-I) conducted at White Sands Missile Range, New Mexico, by the Army Atmospheric Sciences Laboratory. Propagation through a dust and oil smoke cloud was simultaneously measured at 94 and 140 GHz and at several infrared wavelengths. In another test near Lake Havasu City, Arizona, the transmission through, and reflection from, a large dust cloud produced by 720 tons of high explosive were measured at several millimeter wave lengths. The results of such tests have provided systems engineers with specific information for design purposes. In other programs, the radar reflectivity and radiometric signatures of land and sea (i.e., surface clutter), as well as vehicles and other targets of interest, have been obtained, primarily at "window" frequencies.

The strong revival during the past several years of research and applications in the MMW region is attributed to the advent of new technology, the evolution of new requirements for sensors and communication links, and the superiority of MMW systems over optical and infrared systems for penetration of smoke, fog, haze, dust, clouds, and other adverse environments. The improved technology includes better sources (such as IMPATTs, Gunn oscillators, gyrotrons, extended interaction oscillators [EIOs], extended interaction klystron amplifiers [EIKAs], magnetrons, and traveling wave tubes [TWTs], which have higher-power outputs or operate at higher frequencies and, in some cases have longer lifetimes than earlier designs. Lower-noise mixers have also been developed, providing noise temperatures below 500 K (uncooled) or 100 K (cooled) for frequencies up to 115 GHz. Component development has progressed rapidly, particularly in the areas of integrated circuits, image lines, fin line waveguides, and quasi optical devices. Nonetheless, there is still great room for continued improvement in components and devices, particularly in extending ferrite and other nonlinear devices to frequencies above 100 GHz and in reducing losses in most components. For a variety of reasons the frequency range from 35 to 100 GHz (with a few special examples at 140 GHz) has seen the heaviest system development, while the range above 100 GHz (and on into the submillimeter range) is now seeing a concentration of research effort on components and techniques for numerous applications.

1.3 HISTORY OF MMW RADAR

The first reported activity in MMW technology began in the 1890s [1]. From then until the 1930s, researchers developed various components and sources, and measured the bulk absorption and spectroscopic properties of numerous materials. The principal source used to generate MMW energy during these early days was the spark-gap generator, and radiometers or thermopiles were used to detect MMW energy. In 1934 the first resonant spectral frequency measured in the microwave or MMW regions (the ammonia inversion at 27.3 GHz) was observed by Cleeton and Williams using an early type of magnetron, rather than a spark-gap generator. In 1936 they operated such a tube as high as 47 GHz (6.4-mm wavelength).

The late 1930s saw the development of several new radio-frequency sources based on various vacuum tube configurations. These included the klystron (invented during 1937 by the Varian brothers and reported by them in 1939) and the cavity magnetron (invented in 1939 by Boot and Randall). These sources were immediately used in the radars of World War II, and early applications occurred at microwave frequencies rather than MMW frequencies. Late in the war the microwave traveling wave tube also was developed. The packaged silicon crystal detector/mixer was developed in the early 1940s (primarily at Bell Laboratories), and by the latter part of 1944 the production rate for these devices at Western Electric was in excess of 50,000 units monthly.

The typical airborne radar in use by the Americans and British in the latter part of World War II operated in X band (near 10 GHz). However, as the war progressed, frequencies were pushed above 20 GHz to K band and beyond. The impetus to move to higher frequencies was prompted in part by the desire to obtain better angular resolution from apertures limited in size. Late in the war a new radar design operating near 24 GHz was produced. It was quickly found that this naive choice of frequency was very unfortunate because of the increased atmospheric attenuation (nearly an order of magnitude greater than that at X band), resulting from the fact that a broad absorption band due to water vapor is centered near 22.3 GHz. Because of a lack of knowledge of atmospheric absorption effects by the radar designers, these radars were rendered far less effective than originally anticipated.

Prior to World War II only one discrete microwave or MMW spectral transition had been measured. During the war, Beringer, using a crystal harmonic generator driven by a centimeter-wave klystron, obtained enough second-harmonic signal to measure the 5-mm wavelength (60 GHz) absorption of oxygen, but did not resolve its fine structure. Thus, the period through World War II closed out with the stage set for the extension of coherent radar techniques into the millimeter frequency region.

After the war the new sources were heavily used to develop the discipline of microwave and MMW spectroscopy. For the first time, molecular spectra could be studied with the use of coherent radiation. The resolving power was remarkable and permitted precise determination of molecule sizes and shapes. In addition, there was interest in MMW applications to radar, radio communication, and radiometric sensing. Gordy's group at Duke University was one of the first to take advantage of this capability. He obtained from Raytheon a few of the first klystrons developed for frequencies from 30 to 50 GHz and, starting in 1947 and continuing for many years, he and a succession of graduate students made many high-resolution spectral measurements.

Other very successful programs in MMW technology were undertaken at the University of Illinois, Lincoln Laboratory, the Johns Hopkins University, and Bell Laboratories, where an interest had developed even before World War II as an outgrowth of early waveguide work. While the members of the Duke group were developing their spectroscopy methods, a Columbia University group was experimenting with the filtered harmonic energy obtained directly from high-powered magnetrons. As early as 1949 they announced the detection of harmonic energy in the 1.5–3.0 mm wavelength region utilizing a Golay cell. Later, however, they discontinued the magnetron measurements and instead used the klystron harmonic-multiplier approach.

At the University of Texas, Straiton and Tolbert were leading a very active group which became well known for its research and measurements on atmospheric propagation and the development of radars and radiometers. Straiton's group produced a set of curves of atmospheric attenuation (for horizontal propagation) which have since become classics. Figure 1.2 shows these curves; they have been replicated in scores of articles, reports, and books during the past two decades. The curves, derived from theoretical analyses of absorption of O_2 and H_2O, as originally performed by Van Vleck, were supplemented with experimental data obtained at various frequencies by several groups, including the University of Texas and Bell Laboratories.

During the 1950s and early 1960s many other institutions were involved in MMW research. For instance, at the Georgia Institute of Technology, in Atlanta, there was considerable emphasis on antennas (particularly the geodesic lens) and on radar designs and, by 1959, a 70-GHz surveillance radar was developed to the point of being given U.S. Army nomenclature (AN/MPS-29) [2,3]. Research on high-power sources continued at the Columbia University Radiation Laboratory; by 1954, the magnetron was extended upward to 115 GHz with 3.3-kW output power, but its lifetime and duty cycle, and hence (average) power output, were very low. In 1960, the U.S. Army sponsored an extensive development program on electron tube sources, including switching, duplexing, and ferrite devices [4].

Throughout the 1970s, the MMW radar field continued to expand in level of effort, diversity, and technical progress. Although the advent of solid-state sources was a major boost to MMW system applications in the 1970s, recent progress in new vacuum tube sources such as extended interaction oscillators (EIOs), extended interaction klystron amplifiers (EIKAs), and gyrotrons have provided a similar boost to this technology field. By 1978 the U.S. Army's STARTLE (Surveillance and Target Acquisition Radar for Tank Location and Engagement) radars (developmental) were using all-solid-state transmitters consisting of a 94-GHz phase-locked Gunn oscillator chain whose frequency-controlled output was used to injection lock the IMPATT transmitter; this technique provided 4 W peak power in a pulse mode. Various modulation formats were employed: short-pulse/spread spectrum or coherent MTI (moving target indication); or pulse compression with frequency agility. Figure 1.4 further illustrates this type of progress.

Representative of the most advanced state-of-art MMW radars is the High Power Coherent Radar operating at 95 GHz (HIPCOR-95) developed by the Georgia Institute of Technology for the U.S. Army Missile

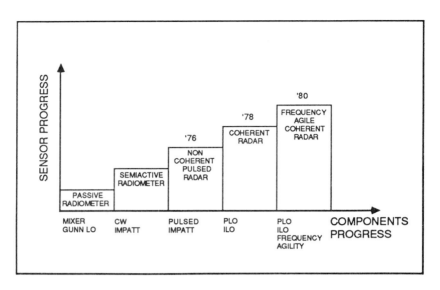

Fig. 1.4 MMW sensor technology development reflects progress of solid state components (courtesy of J. Kuno).

Command for research into advanced target detection and missile guidance techniques. The HIPCOR-95 radar (see Figure 1.5) employs two selectable output stages to support a broad range of research objectives. A high-power, moderate-bandwidth transmitter section uses a state-of-the-art extended interaction amplifier as its final output stage to achieve 2 kw of peak output power. A moderate-power, high-bandwidth mode utilizes a current technology traveling wave tube as its final output stage to achieve a transmitted bandwidth in excess of 2 GHz [6].

In each of the chapters of this book, specific references are provided to show some of the more recent work in advancing the development of various areas of MMW technology. However, some of the more distant historical perspectives may easily be lost due to the concentration on modern technology. For that reason, Table 1.2 is provided to summarize the early MMW technological milestones and Table 1.3 is included to provide examples of early MMW radars.

Fig. 1.5 High power coherent 95 GHz (HIPCOR-95) radar (courtesy of the Georgia Institute of Technology).

Table 1.2   Significant MMW technology milestones.

Date	Event
1936	First resonant spectral measurement (ammonia inversion at 27.3 GHz), by Cleeton and Williams
1937	Invention of the klystron oscillator by Russell and Sigurd Varian
1939	Invention of the cavity magnetron by Boot and Randall
1942	Development of the crystal mixer at Bell Laboratories
1946	First MMW magnetron developed
1947	First K_a-band radars (AN/SPS-7 surface radar developed by the US Navy, and AN/TPQ-6 cloud height measuring radar developed by Bendix for the US Air Force)
1960	Development of MMW klystrons and magnetrons by US Army Signal Corps Laboratories
1962	Gunn effect discovered in GaAs by J. B. Gunn
1964	IMPATT oscillator developed separately at Bell Laboratories and in the Soviet Union
1978	Commercial availability of the extended interaction klystron oscillator (EIKO or EIO)

Table 1.3 Examples of Early MMW Radars
(adapted from S. L. Johnston [5], © 1977 Microwave Journal).

Year	Identification	Application	Band	Sponsor	Source
1947	AN/SPS-7	Surface Search	K_a	Navy	—
1947	AN/TPQ-6	Cloud Height	K_a	Air Force	Bendix
1948	AN/APS-36	Search	K_a	Navy	Philco
1950	AN/SPN-10	Aircraft Landing	K_a	Navy	—
1952	Instrumentation Radar	Sea Clutter Measurements	K_a	Navy	Georgia Tech
1954	AN/APQ-58	Terrain Clearance	K_a	Air Force	—
1955	Unknown	Air Surveillance	K_a	United Kingdom	Decca
1956	AN/APQ-56	Mapping	K_a	Air Force	Westinghouse
1957	AN/TPQ-11	Cloud Height	K_a	Air Force	Bendix
1958	AN/BPS-8	Surface Search	K_a	Navy	Fairchild
1959	AN/MPS-29	Gnd Surveillance	W	Army	Georgia Tech
1959	Lacrosse	Missile Tracker	K_a	Army	Cal/Martin
1961	AN/APQ-97	Target Location	K_a	Army	Westinghouse
1962	AN/APQ-89	Terrain Clearance	K_a	Navy	NAFI
1964	AN/APN-161	Mapping	K_a	Air Force	Sperry
1965	MRL-1	Weather	K_a	USSR	—
1968	Lunar Radar	Moon Measurement	K_a	Air Force	MIT
1968	AN/APQ-137	Ground Targets	K_a	—	Emerson
1968	AN/APQ-144	Multimode	K_a	Air Force	GE

Table 1.3 (cont'd)

Year	Identification	Application	Band	Sponsor	Source
1968	River Radar	Ship Navigation	K_a	Fed. Rep. of Germany	—
1969	—	Meteorological Research	K_a	Rep. of South Africa	—
1970	AN/SPN-42	Aircraft Landing	K_a	Navy	Motorola
1973	AN/APQ-122	Dual-Band Multimode	K_a	Air Force	TI
1975	Instrumentation Radar	Land Clutter Measurements	W	Army	Georgia Tech
1976	TRAK X	Dual-Band Low Altitude Tracking	K_a	Navy	NRL
1976	Flycatcher	Dual-Band Air Defense	K_a	The Netherlands	HSA
1977	—	Automobile Braking	K_a	Dept. of Transportation	—

REFERENCES

1. J.C. Wiltse, "History of Millimeter and Submillimeter Waves," *IEEE Transactions on Microwave Theory and Techniques*, Vol. MTT-32, pp. 1118–1127, September 1984.

2. K.J. Button and J.C. Wiltse, eds., *Infrared and Millimeter Waves*, Vol. 4, *"Millimeter Systems,"* Academic Press, New York, 1981.

3. J.C. Wiltse, "Millimeter-Wave Radar Features Unique Characteristics and Designs," *Microwave Systems News*, Vol. 14, pp. 58–76, May 1984.

4. H.J. Hersch, E.J. Kaiser, F.E. Kavanaugh, and I. Reingold, "Electron Tubes and Devices in the 4.3-mm Frequency Range," *IRE Transactions*, Vol. MIL-4, No. 4, pp. 481–492, October 1960.

5. S.L. Johnston, "Millimeter Radar," *Microwave Journal*, Vol. 20, pp. 16–28, November 1977.

6. T.L. Lane, *et al.*, "Coherent 95 GHz High Power Radar," *SPIE Technical Symposium East Conference*, No. 544, April 1985.

Chapter 2
Fundamentals of MMW Radar Systems*

E.K. Reedy

Georgia Institute of Technology
Atlanta, Georgia

2.1 INTRODUCTION

Some of the operational fundamentals of MMW radar systems are presented in this chapter. In some ways, the chapter will serve as an overview of MMW radar, in general, and the rest of this book, in particular. Topics covered range from applications of MMW radar to considerations of specific component characteristics and limitations. Particular emphasis is placed on the propagation environment and its effect on MMW radar applications and performance because of the extreme importance of this area in understanding both the advantages of MMW radars and their limitations.

2.1.1 Key System Operational Considerations

Full understanding of the operational characteristics of MMW radar requires an appreciation for the location of the MMW band within the electromagnetic spectrum and the resulting effects on the propagation of MMW energy. Figure 2.1 illustrates the electromagnetic spectrum and identifies the region of conventional radar operation, along with standard radar operating band designations. Radars certainly can be designed and developed at any frequency at which electromagnetic energy can be generated and controlled; however, the primary operational region for radars

---

*Author's Note: Certain material in this chapter was excerpted from Reedy and Ewell ([1] © 1981 Academic Press).

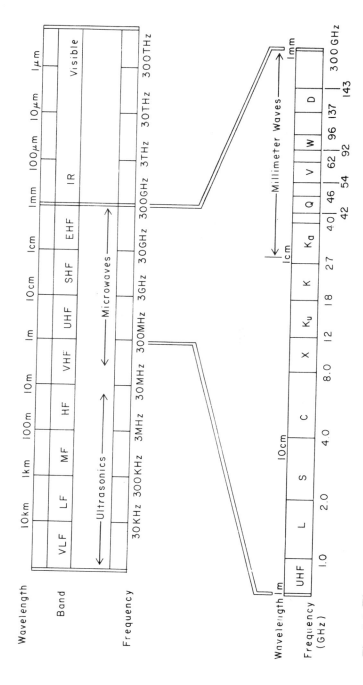

Fig. 2.1 Electromagnetic spectrum with radar band designations.

which must penetrate the atmosphere extends from below 300 MHz to perhaps 300 GHz or from 3×10^8 Hz to 3×10^{11} Hz, a three-order-of-magnitude frequency range. This region of the spectrum has been further subdivided into standard radar bands and given standardized letter designations which are also shown in Figure 2.1.

Differences of opinion exist among radar researchers about exactly what region constitutes millimeter waves. One obvious definition would be the region between 1 mm (300 GHz) and 10 mm (30 GHz). However, the IEEE Standard [2] defines millimeter waves as the region between 40 GHz and 300 GHz, excluding the operationally important and very active region around 35 GHz. This book will consider operation near 35 GHz in discussions of general MMW radar system performance and characteristics.

Other band designations have been proposed for regions of the MMW spectrum. For example, Kulpa and Brown [3] used the term *near millimeter* to refer to the region between 100 GHz and 1000 GHz. In some references, *submillimeter* refers to the region between 300 GHz and 3000 GHz. Other overlapping and sometimes confusing band designations have appeared in the literature. For example, see Table 1.1, p. 2, of Bhartia and Bahl, where US and United Kingdom frequency band designations are summarized [4]. For consistency, the standard band designations shown in Figure 2.1 will be used throughout this book.

Figure 2.2 provides a representation of the clear-air atmospheric attenuation between 3 cm and 0.3 μm wavelengths. Note that the atmospheric attenuation generally monotonically increases from 3 cm to 0.3 mm, after which it flattens and then decreases in the infrared region. Between 10 GHz and 500 GHz, where the attenuation strongly increases as a function of frequency, there is a series of peaks or regions of higher attenuation, and valleys or regions of lower attenuation, compared to that in the immediate vicinity. The peak frequencies are often referred to as "walls" and the valleys as "windows."

Most applications of radar involve achieving as long a detection range as possible; consequently, radar research and development has been concentrated in the areas of lower attenuation at the MMW window frequencies of 35 GHz, 95 GHz, 140 GHz, and 220 GHz. Very little radar development has occurred above 220 GHz. Water vapor and oxygen molecule electromagnetic interaction resonances provide regions of abnormally large absorption which, in turn, cause the relative maxima or walls occurring at approximately 60 GHz, 120 GHz, and 180 GHz. Radars have also been developed at these frequencies to take advantage of the covertness provided by the high atmospheric absorption, and to operate in exoatmospheric space where atmospheric absorption does not exist.

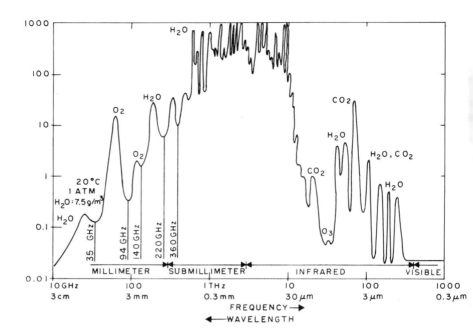

Fig. 2.2 Clear weather atmospheric attenuation (from Preissner, [10]).

Many of the key operational features of MMW radar are directly dictated by the atmospheric propagation characteristics discussed above. Obviously, not all the operational features of a MMW radar can be related to its frequency of operation. Table 2.1 presents some of the more important MMW radar system tradeoff considerations. This table compares the relative advantages and disadvantages or current limitations of MMW radars. The importance of the first two limitations should gradually be reduced as activity in the MMW area continues to increase and leads to production and large-scale manufacture and standardization of MMW components, along with the introduction of MMW circuit integration and increased use of microstrip and dielectric waveguide components. Additional discussion of other MMW radar operational characteristics is given in the following section.

2.1.2 MMW Radar Characteristics

As Table 2.1 indicates, MMW radar offers significant operational advantages when compared to microwave radar, especially in the area of high angular resolution resulting from smaller antenna beamwidths for a

Table 2.1 Millimeter-Wave Radar System Tradeoff Considerations

Advantages	*Limitations*
Physically Small Equipment	Component Cost High
Low Atmospheric Loss[a]	Component Reliability and Availability Low

High Resolution
  Angular
  Doppler
  Imaging Quality
  Classification
	Short Range (10–20 km)
	
	Weather Propagation[b]

Small Beamwidths
  High Accuracy
  Reduced ECM Vulnerability
  Low Multipath and Clutter
  High Antenna Gain	

Large Bandwidth
  High Range Resolution
  Spread Spectrum ECCM
  Doppler Processing	

[a]Compared to IR and Visual Wavelengths
[b]Compared to Microwave Frequencies

fixed antenna aperture size. Antenna half power or 3 dB beamwidth, θ, in the plane corresponding to the antenna dimension, is related to the operating frequency by the following expression for a diffraction limited antenna [5]:

$$\Theta = k\lambda/l \text{ (radians)} \tag{2.1}$$

where

    k = constant ($4/\pi$ for -25 dB sidelobes),
    λ = wavelength,
    l = aperture dimension.

Solving (2.1) for several aperture dimensions and frequencies yields Table 2.2 which provides antenna beamwidth (in degrees) as a function of radar operating frequency. Equation (2.1) and Table 2.2 also illustrate that Θ

is inversely proportional to frequency ($\Theta \propto 1/f$). Therefore, for a fixed antenna aperture, a radar operating at 9.5 GHz (X band) would radiate an antenna beam 10 times larger than a radar operating at 95 GHz. In other words,

$$\Theta_{95} = \frac{\Theta_{9.5}}{10} \tag{2.2}$$

The availability of significantly smaller antenna beamwidths results in several of the most important operational advantages attributable to MMW radar. Among the more important of these are:

(1) High Antenna Gain with Small Aperture

Skolnik [5] gives the gain, G, of an antenna relative to an isotropic radiator as

$$G = 4\pi A_e/\lambda^2 \tag{2.3}$$

where A_e = effective area of antenna aperture. Thus, the gain of an antenna increases in proportion to the frequency of operation squared, again for a fixed aperture. As shall be shown later, antenna gain is one of the most important radar parameters in determining target detection performance.

Table 2.2   Antenna Beamwidth versus Frequency.

Diffraction Limited Antenna

Beamwidth = Θ = $k\lambda/l$ (Radians)
k = Constant ($4/\pi$ for -25 dB Sidelobes)
λ = Wavelength
l = Aperture Dimension

Antenna Beamwidth (in Degrees) as a Function of Radar Frequency

Aperture Size (m)	Frequency (GHz)				
	10	35	95	140	220
0.01	—	62.5	23.1	15.6	9.9
0.1	21.9	6.3	2.3	1.6	1.0
1.0	2.2	0.63	0.23	0.16	0.1

(2) High Angular Tracking Accuracy

A radar's angular tracking accuracy is directly related to its antenna beamwidth, Θ, as shown below for a thermal noise limited case:

$$\sigma_t = (k_t\Theta)/[(S/N)n]^{1/2} \tag{2.4}$$

where

σ_t = *root-mean-square* (rms) angle tracking error,
k_t = constant depending on type of tracking,
(S/N) = signal-to-noise ratio at receiver output,
n = number of pulses integrated.

Thus, all other things equal, a MMW radar can be expected to achieve a reduction in tracking errors when compared to a lower frequency radar, as a direct result of smaller antenna beamwidth.

(3) Reduced Electronic Countermeasures (ECM) Vulnerability

Smaller radar antenna beamwidths provide less opportunity for a jammer to inject energy into the radar's main beam and thus reduce the radar's susceptibility to jamming. Also, higher antenna gains reduce vulnerability to jamming through the antenna sidelobes. The basic cross-range resolution of a radar, d_x, is defined in terms of the antenna half power beamwidth and distance or range to the target cell as follows:

$$d_x = \Theta_x R \tag{2.5}$$

where

d_x = cross-range resolution,
Θ_x = half-power beamwidth,
R = range to target cell.

The higher angular resolution available with MMW radars allows spatial isolation of a target in a volume clutter background such as might be present in a chaff cloud or rain background. This isolation provides enhanced target-to-clutter-plus-noise ratios and improved detection performance.

(4) Reduction in Multipath and Ground Clutter at Low Elevation Angles

A MMW radar with small beamwidth will typically have less ground intercept than a lower frequency radar with larger beamwidths. Since ground intercept is reduced, multipath propagation conditions and ground clutter effects are correspondingly reduced.

(5) Improved Multiple Target Discrimination and Target Identification

Again, a radar's ability to separate multiple, closely-spaced targets and to provide information for target identification is closely coupled to its resolution. Thus, MMW radar has inherent advantages in these areas.

(6) Mapping Quality Resolution Possible

Figure 2.3 illustrates the level of target resolution attainable from an airborne, real aperture 95 GHz radar. This figure shows a radar image of a Navy ship taken with a noncoherent 95 GHz radar at a range of 122–152 m. The radar beamwidth was approximately 10 mr.

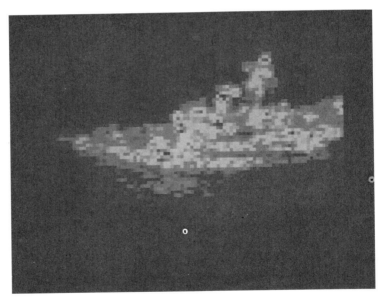

Fig. 2.3 Radar image of a ship taken with an airborne, noncoherent 95 GHz radar.

In addition to the high angular resolution of a MMW radar, another important characteristic is the radar's ability to measure and resolve target motion through the Doppler effect. Any target motion in the radar's beam will cause a shift in the received signal frequency in accordance with the following relationship:

$$f_d = 2V_r/\lambda \tag{2.6}$$

where

f_d = shift in transmitted frequency (Doppler frequency),
V_r = radial target velocity,
λ = wavelength of transmitted signal.

Table 2.3 presents the Doppler frequency shift (given in hertz per meter per second of target radial velocity) for several radar operating frequencies. For a target having a 30 m/s radial velocity with respect to the radar, a radar operating at 10 GHz (X band) would experience a 2.0 kHz frequency shift (i.e., the received signal frequency would be 10,000,002 kHz for a target radially approaching the radar). A 95 GHz radar observing the same target would experience a 19 kHz frequency shift. Such large Doppler shifts from relatively slow moving targets provide the capability for increased target detection and perhaps recognition of such target features as skin vibration, and second and higher order velocity signatures. These features aid in the automatic classification of targets based on radar signatures.

Table 2.3 MMW Radar Characteristics: Doppler Frequency Properties.

Doppler Frequency Shift:

$$f_d = 2 V_r/\lambda$$

V_r = Radial Target Velocity
λ = Wavelength of Transmitted Signal

Radar Frequency (GHz)	Doppler Shift (Hz/m/s)
10	66.7
35	233.3
95	633.3
140	933.3
220	1466.7

Even with the advantages given in Table 2.1 and discussed above notwithstanding, a MMW radar should not be considered a panacea for all radar surveillance and tracking problems. For terrestrial applications, MMW radar is basically a short-range, high-resolution sensor which may be severely degraded in some adverse weather environment. However, for many applications, it represents an excellent compromise between the performance characteristics of microwave and lower frequency radar on the one hand, and optical and infrared sensors operating at higher frequencies, on the other. Table 2.4, which is adapted from Bhartia and Bahl [4], compares the relative performance of MMW radar with its microwave and optical counterparts for several important radar operating characteristics. MMW radar performance usually falls between the extremes of the microwave and optical sensors.

Table 2.4   Radar System Performance Comparison
(Adapted From Bhartia and Bahl [4] © 1984 John Wiley and Sons).

Radar Characteristics	*Microwave*	*Millimeter Waves*	*Optical*
Tracking Accuracy	Fair	Fair	Good
Classification or Identification	Poor	Fair	Good
Covertness	Poor	Fair	Good
Volume Search	Good	Fair	Poor
Adverse Weather Performance	Good	Fair	Poor
Performance in Smoke, Dust	Good	Good	Poor

2.1.3   Typical Applications

Table 2.5 summarizes the known U.S. radars operating between 70 and 140 GHz prior to approximately 1975 [6]. Considerable activity at several different organizations is evident. Applications of these radars can be generally subdivided into the following three areas:

- Surveillance and target acquisition
- Instrumentation and measurement
- Fire control and tracking.

In addition to continuing development of MMW radars to satisfy requirements in these areas, emphasis in the last ten or so years has shifted to include development of radars to satisfy requirements in the additional area of guidance and seekers. Table 2.6 identifies four generic areas for application of MMW radar. Some specific examples of applications in these areas will be presented later in this chapter and other examples are covered in Part IV of this book.

2.2 FUNDAMENTAL CONSIDERATIONS

2.2.1 Functional Elements

The functional elements of a MMW radar are basically the same as those of the microwave radar, namely the transmitter, receiver, signal processor, and antenna. However, major differences in the operational and performance parameters of these functional elements and the passive components which connect these elements result in significant differences in the system level performance of a microwave and MMW radar.

Block Diagram

A basic block diagram for a typical MMW radar (or, for that matter, any other radar) is shown in Figure 2.4. In its simplest form, a basic radar consists of an antenna, which acts as a transducer between the radar and the space into which the radar is radiating energy, and the power source or transmitter and receiving or signal processing subsystems. Obviously, a functional radar must include several other components, not all of which are illustrated in Figure 2.4. Some of the most important include the modulator for forming the basic pulse shaping; a synchronizer to provide system timing; a duplexer to switch the antenna between the transmitter and the receiver; some form of receiver protection device; a display or indicator acting as the interface between the radar and the user of the information and data provided by the radar, antenna servo control (and tracking circuit or processor if required); various power supplies, conditioners, and converters. In addition, these elements must be linked together, usually by conventional means of connection (wires, printed circuits, *et cetera*), and some form of waveguide, microstrip, or MMW integrated circuit techniques.

Table 2.5 Early MMW Radar.
Known US Radars Operating Between 70 and 140 GHz
(Prior to 1975).

Identification	Application	Frequency	Sponsor and Source
AN/APQ-62	Side Looking Mapping Radar	70 GHz	—
JR-9	Search Mapping Radar	70 GHz	WPAFB and TRG and Raytheon
AN/BPS-8	Search Radar	70 GHz	—
AN/MPS-29	Search and Surveillance Radar	70 GHz	USAECOM and Georgia Institute of Technology
Experimental	Aircraft Obstacle Avoidance Aircraft Instrument Landing	70 GHz	USAECOM and Nordon Division, United Aircraft
Experimental	Obstacle Avoidance Sea Clutter Measurement	95 GHz	NADC, Warminster, PA
Experimental	Space Object identification	95 GHz	Aerospace Corp., El Segundo, CA
Experimental	Arctic Terrain Avoidance	95 GHz	Navy and Applied Physics Lab., Silver Springs, MD
Experimental	Airborne Applications Instrument Landing Short Range Weapon Delivery Sensor Cueing	95 GHz	WPAFB and Goodyear Aerospace Corp., Litchfield Park, AZ
Experimental	Low Altitude Aircraft Tracking Target Acquisition Basic Millimeter Wave Radar Studies	70 GHz	Ballistic Research Labs., Aberdeen Proving Grounds, MD

Table 2.5 (cont'd)

Identification	Application	Frequency	Sponsor and Source
Experimental	Noise Modulated Radar for Clutter Suppression and FM/CW for High Range Resolution	94 GHz	Ballistic Research Labs., Aberdeen Proving Grounds, MD
Experimental	Bistatic CW Radar for Cross Section Measurements	140 GHz	Ballistic Research Labs., Aberdeen Proving Grounds, MD
Experimental Rapid Scan	Ranging, Target Acquisition Command Fusing, Fire-control, and Navigation	70 GHz	Harry Diamond Laboratories and Georgia Institute of Technology
Experimental	Instrumentation for Basic Millimeter Radar Studies; Backscatter Studies, *et cetera*	95 GHz	Georgia Institute of Technology
Experimental	Instrumentation for Basic Millimeter Radar Studies; Backscatter Studies, *et cetera*	70 GHz	Georgia Institute of Technology
Experimental	Monopulse Tracking Investigations	70 GHz	Norden Division, United Aircraft

Table 2.6 Generic Applications for MMW Radar.

- Surveillance and Target Acquisition
- Instrumentation and Measurements
- Guidance and Seekers
- Fire Control and Tracking

Significant progress continues to be made in developing components and the basic technology to support radar programs in the MMW region. However, the performance characteristics of most of the critical millimeter components tend to decrease with increasing frequency. A wide range of components is currently available from commercial vendors up to, and including, the 95 GHz window. Above 95 GHz—at 140 GHz and, most especially 220 GHz—generally available components consist primarily of laboratory devices lacking the basic structural and dielectric integrity required for military and commercial use in a harsh field environment, although progress is currently underway to improve off-the-shelf and catalog devices at these frequencies. The cost of components remains high, however.

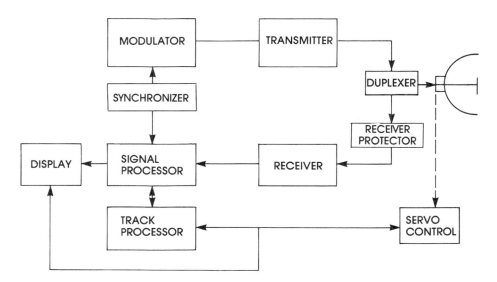

Fig. 2.4 Radar block diagram.

Transmitter

One of the primary performance-determining parameters in any radar design and development program is the amount of power available from the radar transmitter and the type of source available (i.e., amplifier, coherent or noncoherent oscillator, *et cetera*). This is particularly true for MMW systems, since they are invariably power limited. The transmitter normally consists of a high-power oscillator, perhaps a magnetron, or a lower power oscillator coupled to a high-power amplifier such as a *traveling wave tube* (TWT) or the newer *extended interaction klystron amplifier* (EIKA). In a pulsed radar, the transmitter generates a high-power MMW frequency during the time period that the modulator is producing a drive or control pulse—usually on the order of a few nanoseconds to hundreds of microseconds. In effect, the dc modulator pulse has been converted to a pulse of millimeter wavelength energy which is then passed through the duplexer to the antenna where it is shaped and radiated.

Sources of MMW transmitters can be either solid-state or thermionic devices. The solid-state sources consist primarily of the IMPATT and Gunn devices. Thermionic sources include *magnetrons, traveling wave tubes* (TWT), *klystrons, extended interaction klystron oscillators* (EIKO) and *amplifiers* (EIKA), *backward wave oscillators* (BWO), and *gyrotrons*. Shown in Figure 2.5 are several MMW source devices. Summarized in Table 2.7 are typical power levels for five devices that are extensively used as power sources in MMW radar—the magnetron, EIKO/EIKA, TWT, gyrotron, and solid-state devices (Gunn and IMPATT). All of these devices have improved power outputs and/or operate at higher frequencies and, in some cases, have longer lifetimes than previous versions of the device. The increasing availability of both TWTs and EIKAs supports the design and development of fully coherent *master oscillator power amplifier* (MOPA) transmitters at 95 GHz with sizable output powers to complement previously available coherent solid-state transmitters with much lower power.

Additional, more detailed consideration of both solid-state and thermionic, or high-power transmitters will be presented in Chapters 8 and 9 of this book.

Receiver

The receiver in a MMW radar converts the received signal, collected by the antenna, from a frequency corresponding to the approximate trans-

Fig. 2.5 Photograph of several commonly used high-power MMW transmitter sources. A 5-W Gunn oscillator is located in the center,
while surrounding it from left to right is an EIO with a samarium-
cobalt magnet, an EIO with an Alnico magnet, a 500-W 70-GHz
magnetron, a 6-kW 95-GHz magnetron, and a 1 kW 95-GHz
magnetron.

TABLE 2.7　MMW Radar Component Considerations.

	Typical Transmitter Source Power Levels (Peak) Frequency (GHz)			
Power Source	35	95	140	220
Magnetron	50–100 kW	1–6 kW	—	—
EIK/EIKA	2–3 kW	1–2 kW	200 W	60 W
TWT	3 kW	1–2 kW	—	—
(Water Cooled)	(30 kW)	(6 kW)		
Gyrotron[a]	340 kW	500 kW	180 kW	150 kW[b]
Solid State (Gunn, IMPATT)	28 W	22 W	5 W	1 W

[a]Higher power levels have appeared in the literature.
[b]Estimate

mitted frequency (depending on Doppler effects) to a lower or intermediate frequency (IF) where it can then be more conveniently filtered,
amplified, and processed. While direct detection of millimeter wavelength
signals is a possibility, the lack of sensitivity of such an approach effectively
argues against its use in most applications. The translation of the incoming
signal to an intermediate frequency and the subsequent processing of the
lower frequency signal is termed a heterodyne receiver. A low-noise mixer-
amplifer combination is normally used to accomplish the down-conversion
and first-level amplification. This device usually represents the key element
in the receiver chain in determining the radar's detection performance,
dynamic range, sensitivity, and noise properties.

Figure 2.6 shows the current status of MMW receiver technology.
Considerable progress has been made in recent years in developing improved performance for mixer-amplifier combinations. Generally the double sideband noise figure increases rather rapidly with frequency, ranging
from perhaps 3 dB at 35 GHz to 5 dB at 95 GHz, and 6.5 dB at 140 GHz.
Noise figures of approximately 13 dB have been reported at 220 GHz,
although work at this frequency has been somewhat limited. More detailed
treatment of MMW radar receivers and their properties will be presented
in Chapters 10 and 13 of this book.

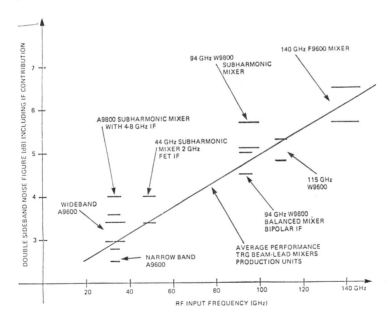

Fig. 2.6 Current status of MMW receiver technology.

Antenna

Most of the antenna technology and techniques currently in use at MMW involve direct extension of lower frequency approaches. In many applications, a microwave antenna design whose performance characteristics have been well established through design, theoretical prediction, and measurements at the lower frequencies will simply be "scaled" up in frequency to satisfy requirements at MMW frequencies. For this reason, research directed toward developing MMW antenna techniques and associated technology has been limited.

Extensive use has been made at MMW of the classical reflector antennas, such as the front-fed paraboloidal dish shown in Figure 2.7 and the Cassegrain-fed reflector shown in Figure 2.8. A Cassegrain-feed arrangement avoids the waveguide losses associated with the more conventional front-fed reflector which can become excessive at MMW frequencies.

Fig. 2.7 Front-fed paraboloidal dish MMW (35 GHz) antenna.

Fig. 2.8 Cassegrain 95-GHz antenna.

Horn and lens antennas are perhaps more popular at MMW than at microwave frequencies. These antennas avoid the aperture blockage and its associated effect on sidelobes common to the reflector antennas. Lens antennas, in particular, have seen limited application at microwave frequencies, due to size and weight limitations. However the size and weight of a MMW lens antenna is much less than those of its microwave counterpart. That advantage, when coupled with advantages of lens antennas over reflector antennas in angle scanning, spillover loss, and fabrication tolerances, makes the lens antenna much more capable of meeting requirements at MMW than lower frequencies. A zoned dielectric lens antenna for use at 95 GHz is shown in Figure 2.9.

Slotted waveguide antennas, which are composed of arrays of parallel waveguide runs having either broad-wall or narrow-wall slots as the radiating elements, have several characteristics which have resulted in their utilization at MMW. These include the capability for electronic scanning (by changing the transmitted frequency), ease of fabrication, well known and predictable radiation performance, and the elimination of a complex feed structure and phase shifters such as that required for a phase-scanned array antenna.

Other antenna techniques, such as the microstrip antenna, dielectric rod, and leaky-wave antenna combine small size, electronic scanning capability, compatibility with integrated circuit construction techniques, and

Fig. 2.9 Zoned dielectric lens.

other important characteristics in such a manner as to make them desirable for MMW applications. Table 2.8 summarizes the antenna techniques which have seen the most application at MMW. More complete discussions of MMW antenna technology are presented in Chapters 11 and 17 of this book and in Bhartia and Bahl [4].

Passive Components

As previously indicated, a MMW radar requires an array of components which are normally passive to transport, control, attenuate, filter, isolate, *et cetera*, the MMW energy. These passive components include such devices as waveguide, connectors, couplers, transitions, attenuators, hybrids, filters (preselectors), circulators, couplers, isolators, terminations, matching networks, and duplexers. Many of these devices are simply microwave designs of a particular component scaled up in frequency (and usually down in size). However, frequency scaling poses problems, particularly at the higher MMW frequencies of 140 GHz, 220 GHz and above, where manufacturing tolerances and increased losses can severely affect performance. In these situations, optical and quasi-optical techniques and structures (dielectric waveguide, slabs, gratings, lenses, *et cetera*) have

Table 2.8   MMW Antenna Techniques.

- Reflector Antennas
  Front Fed
  Dual Reflector (Cassegrain)
  Offset–Fed
  Shaped Reflector

- Lens Antennas
  Zoned Dielectric
  Luneburg

- Horn Antenna
  Flared
  Multimode
  Corrugated
  Lens-Corrected

- Dielectric Rod Antennas

  Slotted Waveguide Antennas
  Broad Wall
  Narrow Wall

- Leaky Waveguide Antennas
  Fast Wave Structure
  Slow Wave Structure

- Microstrip Antennas

provided practical solutions. Also, hybrid approaches, involving a combination of both waveguide and optical technologies, have been successfully developed which combine the advantages of the two techniques and circumvent many of their limitations.

At frequencies above 35 GHz, however, considerable work is still required to develop low cost, wide bandwidth, and high isolation components that are lightweight, low power, and complement the small size, light weight advantages of millimeter waves.

2.2.2 Environment

For radars operating at millimeter wavelengths, the environment has a significant effect on the overall radar system performance, directly influencing such performance parameters as target detection range, track accuracy, target discrimination, *et cetera*. Environment, as used in the discussion to follow in this section, refers to both the electromagnetic reflection and attenuation properties of the propagation path between the radar and the target (propagation effects), and the nature of the electromagnetic energy reflected back toward the radar by objects in the vicinity of the target (usually termed clutter if these returns interfere with detection of the desired target). Much of the discussion presented in this section has been extracted from Reedy and Ewell [1].

Propagation Effects

No attempt will be made in this section to review and present data that completely describe MMW propagation effects, since an extensive summary of propagation information is presented in Chapter 4. Only that information required to make this chapter reasonably self-contained or to introduce the reader to basic concepts is included here.

Attenuation and Reflectivity

Figure 2.10 illustrates, in block diagram form, the major factors affecting the signal received at any radar. Of particular interest are those attenuation and reflection or scattering influences present in the propagation path which, for terrestrial radar, is normally the atmosphere. The primary attenuation-producing factors for MMW radar are the molecular absorption of water vapor and oxygen in a relatively clear atmosphere, absorption of condensed or suspended water in the form of droplets in rain, fog, clouds, *et cetera*, and scattering from those water droplets. Backscatter of energy from the suspended water droplets in fog and clouds and from rain can severely limit radar performance, since the target return signal will be embedded in and corrupted by this noiselike reflected energy. Suspended particulate matter, such as dust particles and smoke, may also influence MMW propagation.

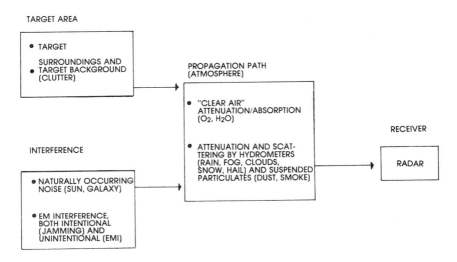

Fig. 2.10 Major factors affecting received radar signal.

Attenuation by Atmospheric Gases, Rain, and Fog

Although the transmission of electromagnetic energy through the atmosphere is a very complex phenomenon, certain aspects of this phenomenon have been extensively studied and documented [3, 7, 8]. However, much of the basic propagation data necessary to predict the performance of MMW sensors accurately either does not exist or the uncertainty associated with existing data is large enough to make accurate, analytical prediction of the sensor performance extremely difficult.

Figure 2.11, after Strom [9], illustrates this uncertainty by shading the window region two-way attenuation levels. Note that at 220 GHz the clear-air attenuation varies between approximately 1.5 dB/km and, perhaps, 11 dB/km, depending on such parameters as atmospheric pressure, temperature, and water density or humidity. Large regions of uncertainty are present at all of the window frequencies. To reduce this level of uncertainty and to predict radar performance more accurately, Kulpa and Brown [3] point out that a rather detailed specification of the conditions under which the sensor is expected to operate is required. For example, the geographic location of expected operation can be important since geographic as well as seasonal variations in humidity and temperature can affect clear air attenuation values by as much as 10 to 15 dB/km.

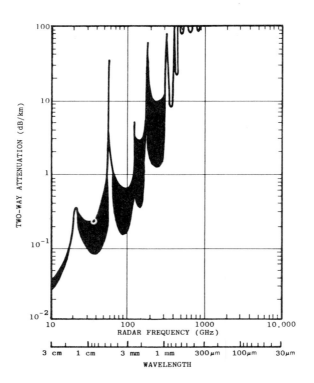

Fig. 2.11 Clear weather attenuation variation for horizontal, sea-level path (from Strom [9]).

Nevertheless, it is still worthwhile to document propagation effects on millimeter waves as completely as possible so that equipment performance can be estimated as accurately as possible using available data. Attenuation produced by basically clear air atmospheric gases at 20°C, one atmosphere pressure, and 7.5g/m^3 water vapor is shown in Figure 2.12 along with attenuation produced by certain hydrometers (rain and fog) [10]. Attenuation produced by rain increases as a function of frequency with maxima occurring in the 100–200 GHz region, after which it remains relatively constant. Ice clouds generally produce negligible attenuation at these frequencies [10]. We refer the reader to later chapters of this book or to Kulpa and Brown [3] for a more complete discussion of MMW propagation in both clear air and in the presence of naturally occurring hydrometers such as rain, snow, and hail.

The variation in currently available propagation data bases at millimeter waves cannot be overemphasized. For example, Table 2.9 (taken

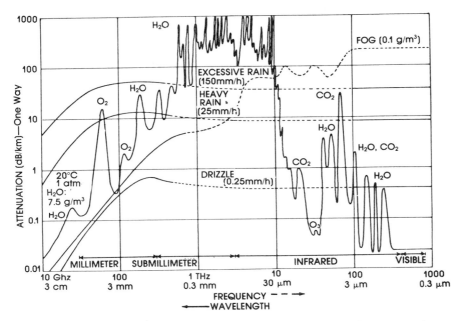

Fig. 2.12 MMW and near-MMW attenuation by atmospheric gases, rain, and fog (from Preissner [10]).**

from Kulpa and Brown [3]) summarizes the variation in total one-way attenuation at 220 GHz in a 100-m fog and 4-mm/hr rain using the best available data. Variations of 10–15 dB in attenuation are present in the data, illustrating the difficulty in analyzing performance of millimeter equipment and the necessity for specifying very accurately the operational environment in which the radar is expected to perform.

Table 2.9 Variation in Attenuation at 220 GHz*.

Condition	One-Way Attenuation (dB/km)
Clear Air	1.6 − 11.2
100-m Fog	0.4 − 4.7
TOTAL	2.0 − 15.9
Clear Air	1.6 − 11.2
4-mm/hr Rain	1.0 − 7.0
TOTAL	2.6 − 18.2

*From Kulpa and Brown [3]

**The original version of this material was first published by the Advisory Group for Aerospace Research and Development, North Atlantic Treaty Organisation (AGARD/NATO) in Conference Proceedings No. 249 published in 1978.

Rain Backscatter

Only a very limited number of theoretical and experimental investigations have attempted to quantify the radar backscattering properties of rain at millimeter wavelengths (Richard and Kammerer [11], Wilcox and Graziano [12], Currie et al. [13], Strom [9]). The basic source of experimental data relating rain radar backscatter characteristics to rain rates at the low millimeter wavelengths and X band is a series of measurements conducted at McCoy AFB, Florida, in August and September 1973 by the U.S. Army Ballistic Research Laboratories, Aberdeen, Maryland. The resulting data from these measurements are recorded and documented in Currie et al. [13] and Richard and Kammerer [11]. A summary of the data from these experiments is shown in Figure 2.13, where the measured average backscatter coefficient of rain is given as a function of rain rate for the operating frequencies of 9.375, 35, 70, and 95 GHz. Also, shown in Figure 2.13 are data for 45 GHz taken at a different time and location [14]. One interesting property of rain reflectivity is apparent from Figure 2.13. In the vicinity of 45 GHz, a significant change in the raindrop, electromagnetic wave interaction mechanism occurs, as evidenced by the change in the slope of the rain reflectivity *versus* frequency data and the decrease in MMW backscatter coefficient as the frequency increases from 45 GHz to 95 GHz. Data shown in Figure 2.13 with the exception of the 45 GHz data, were taken from Richard and Kammerer [11] since their method of data reduction resulted in a larger and, thus, somewhat more conservative value for the backscatter coefficient, especially at the lower rain rates. (See Chapter 5 for a discussion of this experiment.)

Dust, Smoke, and Other Obscurants

Even though one of the primary advantages of MMW sensors over their electro-optical counterparts is enhanced propagation characteristics through dust, smoke, and other battlefield and naturally occurring obscurants, very limited definitive data exist to specify attenuation and backscatter when propagation takes place under these conditions. Allen and Simonson [15] and Morgan *et al.* [16] presented some of the first comparative data for attenuation by smoke and dust. More recently, spurred on by the increased interest in, and usage of, the millimeter region, more investigators have examined the effects of smoke and dust on millimeter propagation [16, 17].

All of the above references confirm that suspended dust and smoke in the atmosphere in amounts typical of a battlefield or naturally occurring environment produce a practically imperceptible amount of attenuation at frequencies of 140 GHz and below. Knox [18] found that the only obscurant

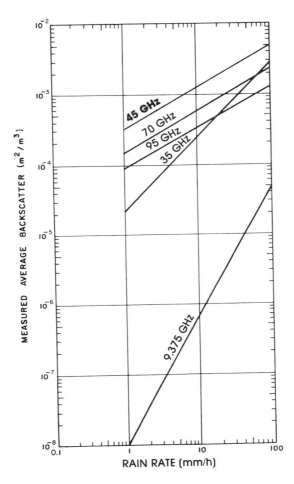

Fig. 2.13 Rain reflectivity coefficient versus rain rate, vertical polarization (from Richard and Kammerer [11]).

to affect transmissivity significantly during exercises conducted at Smoke Week II was dust and debris produced by exploding charges of C4 material that had been buried under several centimeters of dirt. Maximum attenuation measured during these trials was approximately 0.4 dB/km at 140 GHz.

Petito and Harris [19] observed that only during periods when exploding artillery rounds hurled debris (principally soil and metallic fragments) into the air was there significant attenuation and backscatter from artillery bursts when observed with a 95-GHz radar.

Turbulence Effects

Refractive index inhomogeneities in the propagation path of a MMW signal causes certain propagation phase shifts to result, which in turn result in propagation scintillation effects, angle of arrival fluctuations, and depolarization. Only scintillation and angle of arrival fluctuations are thought to be significant at millimeter wavelengths, however. Theoretical and experimental investigations of these effects at 94 and 140 GHz have shown that the scintillation effects are not likely to affect MMW systems except at the extreme limits of performance, but that atmospheric turbulence effects may result in angle of arrival variations of perhaps 0.35 mr, which is approximately the same level of accuracy required of many systems [20].

Multipath Effects

When an RF signal is incident on the surface of the earth, a portion of the incident electromagnetic energy is forward scattered; at the target, the energy that reaches the target by the direct path from the transmitter to the target and by reflection from the surface of the earth combines vectorally and can add either in or out of phase. This same phenomenon occurs on the return path from the target to the radar, where it affects tracking accuracy as well as signal strength. Effects produced by these reflected signals, collectively termed "multipath," can produce fluctuations in signal strength from a target, or in the case of a tracking system, can introduce considerable tracking errors. The magnitude of these multipath-related effects is related to the amount of energy incident on the surface, the reflection coefficient of the surface, the amount of energy that reaches the target by the direct path, and the relative phase of the direct and indirect components.

While the analysis of a general multipath situation may be quite complex, considerable insight into the process may be gained by a simplified analysis. In such a simplified analysis, the surface is considered to be a randomly rough surface having known dielectric properties. The voltage reflection coefficient for a smooth surface of the same dielectric material can be calculated, and this modified by a factor to compensate for roughness of the surface in order to obtain an approximate description of its forward scattering properties. The forward scattered reflection coefficient of a smooth uniform dielectric surface, ρ_0, can be calculated directly from the Fresnel equations.

The reflection coefficients are then modified by the specular scattering factor, ρ_s, where

$$\rho_s^2 = \exp\left[-(4\pi\sigma_h)\sin\gamma/\lambda^2\right] \qquad (2.7)$$

where σ_h is the rms deviation of the surface height, λ the wavelength and γ the grazing angle. Note that a surface is considered rough when ρ_s^2 is less than 0.5 or $\rho_s \leq 0.7$; this is approximately the Rayleigh roughness criterion.

The reflection coefficient for specular reflection is given by

$$\rho = \rho_0\rho_s. \qquad (2.8)$$

Equation (2.8) gives the specular reflection coefficient for rough surfaces; there will remain a diffuse component, not as significant in target fading but important in determining tracking error.

In general, use of MMW radar may offer the advantage of a reduction in multipath effects, primarily because for a given aperture size, the antenna beamwidth decreases with increasing frequency, and since narrow beamwidths can be used to illuminate the target without directing as much energy toward the reflecting earth surface. Also, since roughness is dependent on σ_h/λ and since λ is small for MMW radars, a given surface appears rougher as the frequency increases, thus decreasing the specular scattering factor and decreasing ρ.

Clutter Characteristics

When a radar system illuminates the earth's surface (land or sea), a portion of the energy is scattered forward, giving rise to multipath effects described above; in addition, a portion of the energy is reflected back toward the radar system. These unwanted signals are usually referred to as "clutter," and can seriously affect overall system performance for those situations where an appreciable amount of energy is backscattered toward the radar. Historically, such clutter has been described in terms of its radar cross section per unit area (σ°), a dimensionless quantity that, when multiplied by the resolution cell size of the radar system, gives the clutter radar cross section. Particularly when high resolution radar systems are utilized, clutter no longer appears to be homogeneous with a uniform spatial distribution, and the concept of σ° should be used with some care. However, it is still a quite useful quantity, and one almost universally utilized to report the magnitude of clutter backscatter.

Another problem in characterizing clutter returns in a radar system is that the return is a complex function of a number of parameters, including frequency, polarization, depression angle, clutter type, and system resolution. Analysis and reporting of clutter data are further complicated by the fact that clutter returns are not constant, but fluctuate with time; thus, in addition to the average value, some measure of both the amplitude fluctuations and the temporal behavior (or spectral width) of the clutter must also be provided if a useful description of clutter is to be obtained.

The properties of land clutter can vary widely. This variability is partially due to the fact that land clutter is present in a large number of forms, such as trees, fields, desert, snow, and cultured areas; even within such broad categories, different clutter regions can produce substantial differences in reflectivity. In addition, local environmental conditions such as wind velocity, wind direction, and moisture content can affect clutter returns.

Figure 2.14 presents a collection of MMW surface backscatter data taken from several sources and compiled by Dyer and Hayes [21]. In this figure, the surface backscatter coefficient, σ^o, is presented as a function of wavelength covering the frequency band from 10 GHz to approximately 100 GHz. Data for several different types of terrain backscatter are presented. Also, varying grazing angles are represented. The terrain backscatter data of Figure 2.14 illustrate a relatively weak dependence of σ^o on the radar frequency. In contrast, however, a relatively wide spread in the terrain backscatter or reflectivity coefficient values occurs for fixed frequency and terrain type. This last characteristic substantiates the general observation that the properties of land clutter tend to vary widely even within the same type of surface vegetation and for a constant frequency. Also, note in Figure 2.14 the tendency for the terrain backscatter coefficient to increase slowly as a function of frequency in the MMW region. This is especially true for a relatively smooth surface such as a concrete or asphalt runway or road surface.

A characteristic of surface backscatter which is extremely important in determining the performance of a particular radar over a range of operational scenarios is how the surface reflectivity varies as a function of grazing angle or depression angle. The terms grazing angle, depression angle, or, sometimes, incidence angle are normally used to refer to the angle of incidence or intercept of the radar's illumination pattern and the surface of reflection, where grazing angle is measured relative to the horizontal surface plane, and vertical incidence is measured with respect to the normal to the surface. Figure 2.15 illustrates the variation of σ^o as a function of grazing angle for several types of surface vegetation and a fixed radar frequency of 35 GHz. A rather strong dependence of σ^o on grazing angle is evident in this figure. This figure also illustrates one of the limitations of the current MMW surface reflectivity data bases—a lack of representative data for low ($< 10°$) grazing angles.

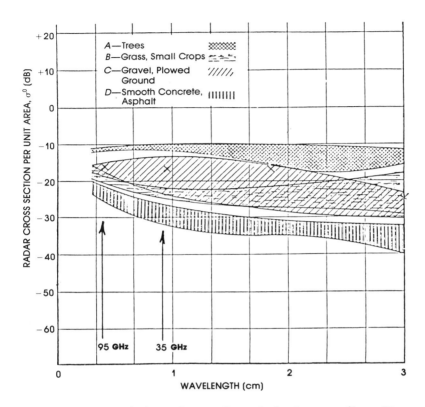

Fig. 2.14 Wavelength dependence of terrain backscatter (from Dyer and Hayes [21]).

Under certain conditions, the detection of targets against a snow covered background has been difficult to achieve with a consistently high probability using a MMW radar (typically a seeker on a small homing missile). Figure 2.15 provides one indication of why detection of targets in snow is a difficult problem. The radar reflectivity of snow at MMW is highly variable and dependent on atmospheric and surface conditions and, in some situations, is also extremely high, as shown in Figure 2.15 where the range of snow reflectivity values measured at fixed grazing angle of approximately 17° and a fixed frequency of 35 GHz is indicated. Under the conditions of these measurements, the snow reflectivity varied almost 30 dB.

While the reflectivity of snow is a function of a number of parameters; it appears to be primarily driven by the surface roughness, and by the free

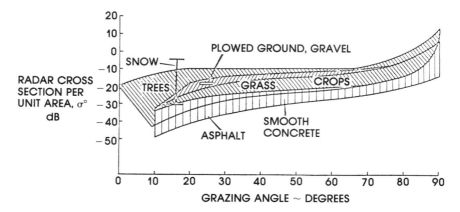

Fig. 2.15 Radar cross section versus grazing angle (35 GHz) (from Dyer and Hayes [21]).

water content of the snow. The effect of free water is clearly shown by the variations in reflectivity during the course of a day; Figure 2.16 presents σ° for both horizontal and vertical polarization as a function of time over a 24 hour diurnal cycle for several depression angles [22]. Also presented in Figure 2.16 is a plot of air temperature and surface snow free water content. The close correlation between free water content and radar reflectivity of the snow is clearly indicated in Figure 2.16. The properties of clutter at millimeter wavelength will be considered in considerably more detail later in Chapter 5.

2.2.3 Measures of Performance

Quantitative and tractable measures of a MMW radar's performance are required to provide a basis for comparison and evaluation of relative performance of several radars, as well as a basis on which the optimization of a particular radar design can be based. Different measures of performance are used for different radar applications (search, track, *et cetera*) and operational scenarios (benign or jamming environment, *et cetera*). However, several performance measures, such as maximum detection range, probability of detection, and probability of false alarm are almost invariably used in specifying a radar's performance.

MMW radar performance, like its microwave counterpart, is governed by, and can be predicted from, the basic radar range, beacon, and jamming equations. In the case of microwave radars, the atmospheric attenuation term in these equations can usually be neglected, whereas, for MMW radar, it may be the most important factor limiting the radar system's performance.

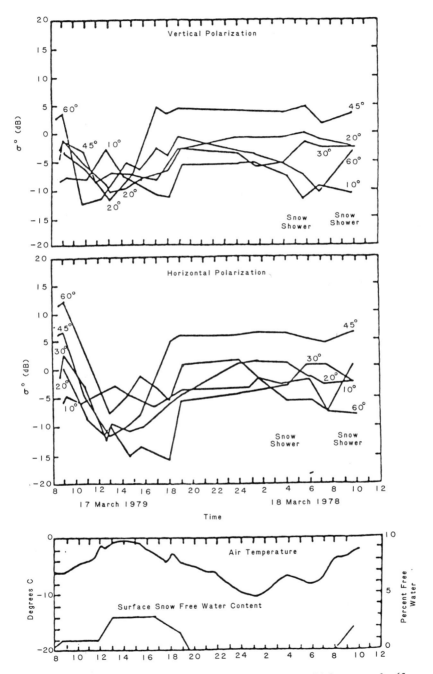

Fig. 2.16 MMW snow reflectivity versus time over a 24-hour cycle (from Hayes, Currie, and Scheer [22], © 1980 IEEE).

The performance of a radar system is governed by the complex interrelationships of several system parameters. Analytically, radar performance and tradeoff analysis is normally investigated by employing the radar range equation—a mathematical expression that can relate the radar maximum range performance to the complete set of system parameters:

$$R_m^4 = (P_t G^2 \lambda^2 \sigma 10^{-0.2\alpha R})/[(4\pi)^3 (kT_0 B)F_n(S/N)_1 L_s] \qquad (2.9)$$

where R_m is the maximum radar range corresponding to minimum single-pulse signal-to-noise ratio, $(S/N)_1$, P_t is the peak radar transmitted power, λ is the radar wavelength, B is the receiver bandwidth $\approx 1/\tau$ (for matched receiver), τ is the pulse width (integration time for a CW radar), G is the antenna gain, σ is the target radar cross section, k is the Boltzmann's constant (1.38×10^{23} joule/K), T_0 is the standard reference temperature (290 K), F_n is the receiver noise figure, α is the one-way atmospheric attenuation coefficient (dB/km), $(S/N)_1$ is the minimum equivalent receiver output single-pulse signal-to-noise ratio required for the radar's function (detection or tracking), and L_s is the system losses.

This form of the radar range equation incorporating a single-pulse signal-to-noise ratio does not measure the overall effectiveness of the radar since the effects of integration are not included. Integration can significantly enhance the performance of a radar for both detection and tracking. However, the single-pulse S/N represents a quantity that is widely used in the literature to compare and predict the performance of radars. Integration or signal processing gain can usually be included by adding a multiplicative factor in the numerator of Equation 2.9.

It is sometimes instructive to demonstrate certain basic concepts and limitations through the use of a system performance calculation for a simple radar. As an example of the use of the radar range equation for a simple, first-cut calculation of radar performance, consider the set of postulated radar parameters shown in Table 2.10 for a short range, MMW target acquisition radar. To avoid an iterative solution for determining maximum range (since the propagation loss is range dependent), for convenience a constant total loss factor L_T, independent of range is defined which includes both system and atmospheric losses. For this calculation, assume $L_T = 6$ dB. Also, assume no integration or signal processing gain; furthermore, a circular antenna aperture having a gain of 37 dB is assumed.

Using these parameters, the maximum expected radar range is plotted in Figure 2.17 as a function of target RCS. Indications of the general range of cross section of several classes of targets are also included at the top of this figure. The band of anticipated performance indicates the spread resulting from system compromises, atmospheric loss, range dependence, target RCS, and possible field degradation that might be reasonably ex-

Table 2.10 Postulated Radar Parameters for Radar Range Calculation.

Parameter	Value Assumed
Wavelength λ	3 mm
Transmitter Power P_t	4 kW
Antenna Gain G	37 dB
Pulse Length τ	50 ns
Bandwidth B	20 MHz
Noise Figure F_n	10 dB
Losses L_T	6 dB
Signal-to-Noise $(S/N)_1$	13 dB

pected. These performance curves are for clear weather and represent, for the parameters chosen, an upper boundary on performance, since such factors as clutter have not been included. The range performance achieved in this example is strongly influenced by the assumed constraints on the antenna aperture size; choice of a larger aperture would result in a corresponding increase in range.

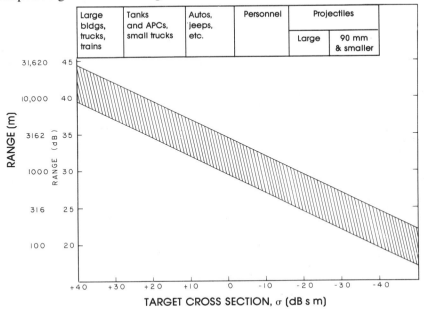

Fig. 2.17 Theoretical maximum target detection capability of a postulated MMW radar.

2.3   ELECTRONIC WARFARE

Increasing utilization of MMW sensors, such as radar, is resulting in the development of specific actions, equipment, and techniques to counter these sensors or limit their effectiveness. Because of the unique presentation of the example radar's performance, Figure 2.17 provides a good example of the effectiveness of one particular approach—reduction, suppression, or modification of the target's radar signature. Consider the performance of this radar against a tank with an RCS of 20 dBsm. The maximum detection range is predicted to be approximately 10 km from Figure 2.17. If, through the use of radar cross-section reduction techniques (such as radar camouflage using mesh, pads, blankets, and other radar-absorbing materials, shaping of the primary scattering surfaces of the tank, and other cross-section reduction techniques), the cross section of the tank were reduced by 20 dB from +20 dBsm to 0 dBsm, then the detection range of the radar would drop from 10 km to approximately 3 km—a 70% decrease in the range performance of this particular radar. Obviously, such a performance decrease would significantly decrease the operational effectiveness of this radar, perhaps to the point of rendering it operationally ineffective.

Table 2.11 summarizes most of the major countermeasure threats to MMW radar. In addition to the passive threats, such as the one discussed above, two other categories of countermeasures are presented. Active techniques include both intentional (enemy produced) and unintentional (usually naturally occurring) *electronic countermeasure* (ECM) or jamming. *Electronic support measures* (ESM) include such techniques as wide bandwidth reconnaissance surveillance-warning receivers and direction-finding systems.

However, MMW radar has inherent advantages to counter such threats. Several characteristics of MMW radar which help reduce the susceptability of these radars to electronic countermeasures threats are given in Table 2.12. Narrow mainbeams (high spatial resolution) and potentially low antenna sidelobes with well designed MMW radar antennas are probably the most important of the items listed in Table 2.12. Low sidelobes and narrow beamwidths make intercept and jamming extremely difficult; high spatial resolution reduces the effectiveness of chaff and the deleterious effects of volume clutter such as rain. Covert operation, with low propagation overshoot, can be achieved at some frequencies (near or on one of the high atmospheric attenuation bands, perhaps at 60 GHz as an example)

Table 2.11   Major MMW Radar Countermeasure Threats.

MMW RADAR THREATS		
ECM (*Active*)	*ECM* (*Passive*)	*ESM*

Intentional
  Noise
  Deception
Unintentional
  Mutual Interference
  EMI
	Chaff
	RAM/Signature
	  Modification
	Foliage/Natural Cover
	Camouflage Screens
	False Targets (Confusion)
	Decoys (Target-Like)
	Clutter
	  Rain, Snow, Hail
	  Ground
	  Sea
	Atmospheric
	  Contaminants
	    Fog
	    Smoke
	    Dust
		Direction Finding
		Elint Receivers
		Defense Suppression
		ARM

Table 2.12   MMW Radar Advantages in a Countermeasures
Environment.

- Narrow Beamwidths with Small Apertures
- Low Sidelobes
- Small, Lightweight Equipment (High Mobility)
- Large Bandwidths (Spread Spectrum; Frequency Agility)
- Covert Operation (Low Propagation "Overshoot")
- High Resolution

reducing the vulnerability of these systems to standoff intercept and limiting the use of long range surveillance or jamming. More conventional and proven electronic counter-countermeasures (ECCM) such as spread-spectrum and frequency agility, are potentially very effective at MMW because of the availability of large operational bandwidths thereby making MMW systems even more resistant to active noise and other bandwidth-limited countermeasures.

2.4 MMW RADAR APPLICATIONS

As an overview of the application of MMW technology to radar, and to summarize the discussion of MMW radar presented in this chapter, potential applications and earlier MMW radar developments will be reviewed in the four generic categories of application previously defined:

- Surveillance and Target Acquisition
- Instrumentation and Measurement
- Seekers and Terminal Missile Guidance
- Fire Control and Tracking

2.4.1 Surveillance and Target Acquisition

There are a number of applications where MMW radars operating at short-to-moderate ranges are quite attractive for surveillance and target acquisition. The advantages of millimeter waves for such applications include small size and weight coupled with high resolution in both azimuth and range, providing excellent resolution of the area under surveillance. Often, for such applications, a rapid scan over an extended angular sector is desirable; the unique capabilities of an electromechanically scanned geodesic lens have resulted in several prototype systems being fabricated for operation at millimeter wavelengths that incorporate such a scanning concept. The basic operation of the scanning geodesic lens is well-known and has been covered elsewhere [1]; characteristics of two 70 GHz radar systems, which employ such antennas, are summarized in Table 2.13. These radars are typical of early MMW radar development efforts.

The resolution achievable with a short transmitted pulse and narrow azimuthal beamwidth permits a considerable amount of detail to be displayed to the operators. In addition, the rapid scan rates achievable (up to 70 scans/second) permit a virtually flicker-free display to be realized without the necessity for digital or analog storage techniques. Geodesic antennas may also rapidly stop the scan for "searchlighting" on a given target in order to investigate the Doppler signatures that might be associated with specific targets of interest within the surveillance area. The

Table 2.13 Characteristics of Two 70-GHz Surveillance Radars.

	AN/MPS-29	Rapid Scan Radar
Frequency	70 GHz	70 GHz
Pulse Width	50 ns	20-45 ns
PRF	10 kHz	5-25 kHz
Azimuth Beamwidth	0.2°	0.55°
Elevation Beamwidth	0.3° (shaped)	3.5°
Peak Power	15 kW	500 W
Scan Sector	30°	45°
Scan Rate (scans/second)	Up to 40	Up to 70

fact that the system rapidly scans an extended area also permits the utilization of track-while-scan techniques, thus permitting accurate tracking of a number of targets simultaneously while maintaining area surveillance. Figure 2.18 is a photograph of the antenna and transmitter-receiver portion of one such surveillance radar, the AN/MPS-29 [23]. The degree of terrain mapping detail achievable with this radar permitted navigation of vehicles using only the radar generated information without use of optical information.

Fig. 2.18 Antenna and transmitter-receiver of a 70 GHz rapid scan system used for battlefield surveillance (from Long and Allen [23]).

2.4.2 Instrumentation and Measurement

Accurate and meaningful MMW radar performance prediction and analyses require an extensive data base of readily available data on signature and *radar cross section* (RCS) for targets and clutter of interest. This requirement, in turn, implies the need for instrumentation and measurement radars designed especially to perform calibrated reflectivity measurements over a wide variety of propagation conditions, target types, and clutter backgrounds. If this data set is to be complete, these radars must cover the complete frequency spectrum of interest. Instrumentation radars must maintain extremely high standards for overall system measurement accuracy (amplitude, time, phase, frequency, and polarization). Internal and external calibration must be included. As might be expected, highly accurate instrumentation-quality radars are relatively complicated and expensive to design, build, and operate.

Only limited propagation and radar reflectivity data are available for the potential radar operating band around 140 GHz. One of the reasons for this fact is that there are virtually no existing instrumentation radars capable of acquiring data at this frequency, characterizing target signatures and clutter reflectivity. To overcome this limitation, Georgia Tech has recently completed the development of a 140 GHz dual polarized instrumentation radar, shown in Figure 2.19. The radar incorporates a pulsed EIKO transmitter, dual channel superheterodyne receiver, and separate dual polarized transmit and receive lens antenna to improve isolation. The major radar operational characteristics are given in Table 2.14 and an overall system block diagram appears in Figure 2.20. The radar is intended to provide calibrated reflectivity measurements for radar cross section, parallel- and cross-polarization, relative phase, and polarization scattering matrix investigations. Potential applications for the radar and resulting data include MMW radar phenomenology studies, radar target reflectivity signatures, clutter and false target characterization, enviromental effects, and reflectivity model development and validation.

Another unique instrumentation radar has recently been developed at Georgia Tech. Called HIPCOR-95, for *high-power, coherent radar operating at 95 GHz,* this radar incorporates the heretofore difficult to achieve characteristics of combined high power with coherent operation in an integrated, MMW radar operating at 95 GHz. Furthermore, the radar is unique in that it combines both an EIKA and a TWT amplifier in one master oscillator, power amplifier coherent chain. The radar is also dual polarized, with two operational modes: high power (1–2 kW peak) and moderate power (100 W peak). Coherent frequency agility over 2 GHz bandwidth is available along with two pulsewidths of 50 and 10 ns.

Fig. 2.19 140 GHz dual polarized instrumentation radar.

Table 2.14   Georgia Tech Portable Instrumentation Grade
140-GHZ Radar.

(a) Pulsed EIO Transmitter	
Antenna	Dual Polarized Horn Lens
Gain	44 dB
Beamwidth	1.0°
Polarization	H, V, RHC, or LHC (Fixed or Pulse-to-Pulse Agile)
Peak Power	200 W
Pulse Width	50, 100, 400 ns (Selectable)
Pulse Repetition Frequency	10 kHz (Maximum)
RF Losses	3 dB
(b) Dual-Channel Superheterodyne Receiver	
Antenna	Dual Polarized Horn lens
Gain	44 dB
Beamwidth	1.0°
Polarization	H and V, or RHC and LHC
System Polarization Isolation	>20 dB
Noise Figure	7 dB
IF Bandwidth	50 MHz
Detection	
Amplitude	Logarithmic
Relative Phase	Quadrature
Dynamic Range	70 dB
Output Signals	
Video, Logarithmic Amplitude	Each Receiver Channel
Video, Relative Phase	Between Receiver Channels (I and Q)
Sample/Hold Boxcar	Each Log Video
Sample/Hold Boxcar	Each I and Q Video

    The salient features of the HIPCOR-95 radar are given in Table 2.15 and a complete block diagram in Figure 2.21. The radar incorporates a Varian VKB 2449 series EIKA (see Figure 2.22) and a Hughes 982H TWT (see Figure 2.23). HIPCOR-95 will be used by the U.S. Army Missile Command, Redstone Arsenal, Alabama, to evaluate a 95 GHz MOPA transmitter using the most recent EIKA and TWT devices, to evaluate high-range resolution performance parameters for a high-power MMW radar, to establish baseline MTI performance parameters for a high-power

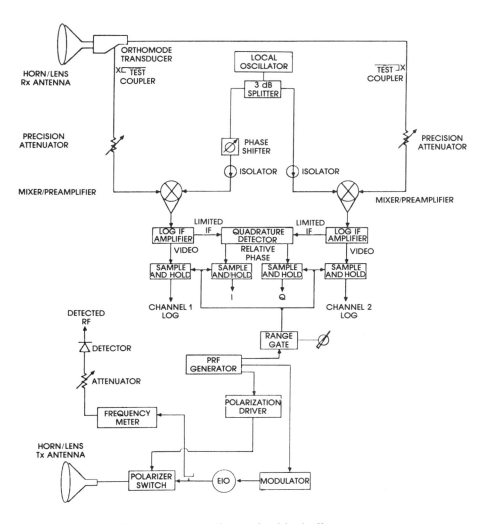

Fig. 2.20 140 GHz instrumentation radar block diagram.

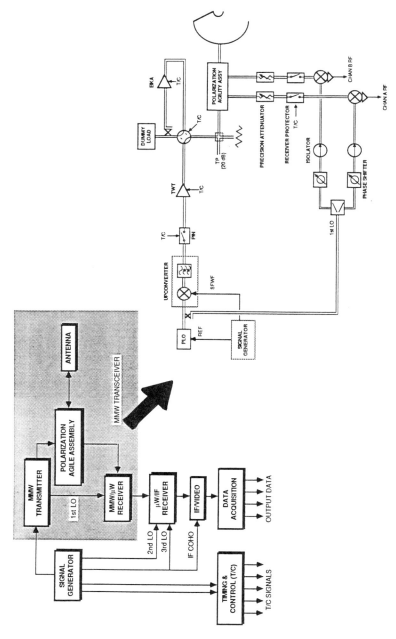

Fig. 2.21 HIPCOR-95 MMW transceiver block diagram.

MMW radar, and to provide a high technology test bed for evaluating advanced target detection, discrimination, and classification techniques, and obtaining target-clutter reflectivity data.

Table 2.15 HIPCOR-95 Salient Features.

Coherent	I and Q Outputs
Polarization Agile (Transmit)	H/V, RHC/LHC
Dual Polarized (Receive)	Co-Polarization
	Cross-Polarization
Operational Modes:	
High Power	1 kW (2 kW Goal)
	1 W (Average)
Moderate Power	100 W
	0.1 W (Average)
Coherently Frequency Agile	2 GHz BW at Moderate Power Level
Pulse Width	50, 10 ns
High Range Resolution	0.5 ft (Off-Line Processing) at Moderate Power Level
Multiple Range Samples	1 per Channel (6 Total)

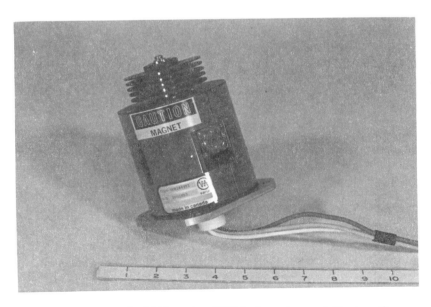

Fig. 2.22 Varian VKB 2449 series EIKA (courtesy of Varian Corporation).

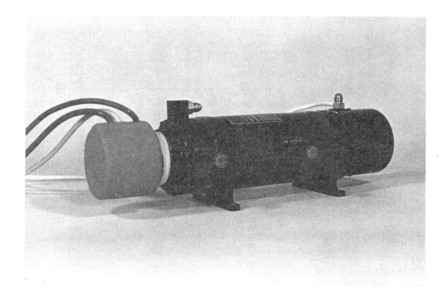

Fig. 2.23 Hughes 928H TWT amplifier (courtesy of the Hughes Aircraft Company).

2.4.3 Guidance and Seekers

The size and weight advantages inherent in MMW sensors make them ideal for applications involving missile seekers and terminal guidance. In fact, this appears to be the most active research area currently incorporating MMW sensors (both passive and active). Passive, fully active, semiactive, beamrider, and dual-mode sensors and guidance techniques are all being investigated and evaluated for MMW applications.

Shown in Figure 2.24 is a 94 GHz missile seeker for air-to-ground applications incorporating either an active monostatic pulsed radar mode or a passive or bistatic CW radar mode. The antenna is a 25.4 cm diameter Cassegrain with a conical scan feed arrangement. The radar is all solid state, utilizing an IMPATT transmitter (Bernues *et al.* [24]). Dual mode operation allows the seeker to circumvent active radar "aim point wander" and target-induced "glint" track errors in the terminal engagement phase by switching to the passive mode.

Figure 2.25 is a conceptual drawing of a MMW beamrider guidance concept, a technique receiving considerable attention for application at millimeter wavelengths. In the most simplified terms, a MMW beamrider guidance system can be defined as a system for guiding missiles to a target

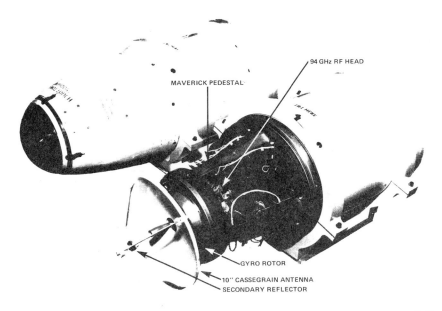

94 GHz RF HEAD

MAVERICK PEDESTAL

GYRO ROTOR

10" CASSEGRAIN ANTENNA
SECONDARY REFLECTOR

Fig. 2.24 A 95 GHz missile seeker (from Bernues, King, and Kaswen [24], © 1979 *Microwave Systems News*).

that incorporates a narrow pencil beam directed at the target along which a missile flies. Receiving and guidance equipment in the missile determines the missile's position with respect to the beam center and develops commands to control the missile's aerodynamic control surfaces in such a manner as to keep the missile trajectory in the middle of the beam.

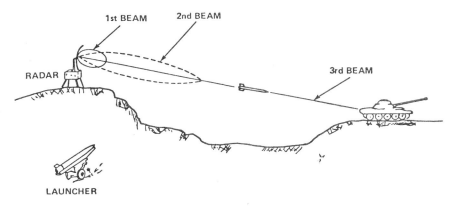

1st BEAM   2nd BEAM

RADAR

3rd BEAM

LAUNCHER

Fig. 2.25 MMW beamrider guidance concept.

2.4.4 Fire Control and Tracking

Short-range, closed-loop, fire control applications, such as that illustrated in Figure 2.26, where extremely high spatial accuracy, small size and weight, and high mobility are the prime system requirements, are well suited to MMW radar [25]. Such systems might serve as point defense antiaircraft weapons, ship defense systems, or for close-in battlefield air defense applications. Radars supporting these applications should provide some inherent target acquisition capability and, at least, first-order target classification capability.

Fig. 2.26 An artist's concept of a short-range antiaircraft weapons system incorporating a MMW radar-directed, rapid-fire cannon mounted on an armored personnel carrier.

Multipath propagation and ground main beam clutter are classical problems associated with low-angle tracking for air defense fire control systems—in particular, gunfire control systems. In this environment, MMW radar can provide significant, potential advantages due to its narrow beamwidths and high resolution properties. The U.S. Army has investigated and, in various research and development programs, has developed advanced fire control radar systems to test the feasibility of millimeter-wave radar to direct and control antiaircraft gun systems with high precision against low-flying aircraft and helicopters. Typical of these radar systems is a dual-band tracker, operating at Ku band and 94 GHz, utilizing a common monopulse antenna at both frequencies.

Millimeter-wave radar techniques have been investigated for such diverse applications as a target acquisition (detection and tracking) and engagement system (limited fire control) having capabilities commensurate with current main battle tank fire-control requirements, a fire-control radar operating near the absorption band frequencies for ship defense, and an advanced, modular gunfire-control system for use on fast patrol boats.

The tank target acquisition and engagement system referred to in the preceding paragraph is shown conceptually in Figure 2.27 (Backus [26]). This radar is intended to demonstrate the feasibility of MMW for tactical systems. Preliminary specifications are given below:

Frequency: 94 GHz (3.2 mm)
Bandwidth: 11 mr
Average Power: 0.1–0.5 W
Antenna Aperature: 35.6 cm
Field of View (Scan Sector): 15° × 7.5° in wide mode; 5° × 2.5° in narrow tracking mode
Frame Time: > 2 s
Target Detection: 3000 m with 100-m visibility
Target Tracking: 0.5 mr accuracy at 2 km with 100-m visibility

The radar is intended to complement, and be integrated with, the infrared tank thermal sight such that both systems share the same display. Spread spectrum waveforms and both coherent MTI and area MTI techniques are being investigated for detection of moving targets. The radar incorporates a solid-state IMPATT transmitter power source.

One of the simplest and most creative applications of MMW radar to a closed-loop fire control problem is discussed in Strom [9]. In this reference, a design is developed for a MMW radar fire control system for the Army infantryman and his rifle. The capability and performance parameters of this radar were examined and, and were found to compare favorably with the human eye.

Fig. 2.27 Surveillance and Target Acquisition Radar for Tank Location and Engagement (STARTLE) (from Backus [26], © 1979 *Journal of Electronic Defense*).

The millimeter-wave radar application examples discussed in this section are certainly not all-inclusive or complete, but were selected primarily to illustrate the variety of operational problems involving target acquisition, tracking, guidance, and instrumentation where MMW radar has some inherent advantages, and shows promise for providing improved performance over either higher- or lower-frequency radars.

2.5 FUTURE DIRECTIONS

As the discussion in the last section has indicated, extensive utilization of the MMW frequency spectrum for radar applications is currently taking place. The future is certainly bright also. Where is the future for MMW radars? What directions will radar applications take? Table 2.16 lists some of the expected directions for MMW radar applications and research. Fully coherent, high-power radars are currently being developed, as evidenced by HIPCOR-95, but these are still early in their research and development cycle. Dual frequency approaches, especially combining MMW radar and EO/IR sensors for missile guidance are currently being investigated. The increased emphasis on exoatmospheric (space) based defense systems, under the Strategic Defense Initiative (SDI) program, bodes well for MMW sensors because propagation limitations are virtually eliminated outside the earth's atmosphere.

Table 2.16 MMW Radar Future Directions

- Fully Coherent, High Power
- Dual Frequency Detection and Tracking Sensors (Dual Mode)
    Millimeter/Microwave
    Millimeter/EO (IR)
- Higher Power
- Packaging Improvements and Component Development
    Integrated Components
    Smaller Size, Lower Weight
    Solid State
- Electronic Scanning
- IFF, Target Recognition, and Classification
- ECM/ECCM/ESM
- Space Applications (SDI)
- Cost Reduction

The last item listed in Table 2.16 may well be the most important, however, in determining the extent of future MMW radar development and application. Fabrication costs must be significantly reduced for MMW radar to realize its full potential.

REFERENCES

1. Edward K. Reedy and George W. Ewell, "Millimeter Radar," Chapter 2 of *Infrared and Millimeter Waves,* Vol. 4, *Millimeter Systems*, Kenneth J. Button and James C. Wiltse, eds., Academic Press, New York, 1981.
2. IEEE Standard 521–1976, November 30, 1976.
3. Stanley M. Kulpa and Edward A. Brown, "Near-Millimeter Wave Technology Base Study, Volume I, Propagation and Target/Background Characteristics," Special Report Number HDL-SR-79-8 on Contract DA 1L161102AH44, Harry Diamond Laboratories, Adelphi, MD, November 1979.
4. P. Bhartia and I. Bahl, *Millimeter Wave Engineering and Application*, John Wiley and Sons, New York, 1984.

5. Merrill I. Skolnik, *Introduction to Radar Systems*, McGraw-Hill Book Company, New York, 1962.

6. F. B. Dyer, *et al.*, "Review of Millimeter Wave Radar Development at Georgia Tech," Technical Report 77-01, Georgia Institute of Technology, Atlanta, Georgia, 1977.

7. A. R. Downs, "A Review of Atmospheric Transmission Information in the Optical and Microwave Spectral Regions," Memorandum Report No. 2710, USA Ballistic Research Laboratories, December 1976.

8. E. J. McCartney, "Scattering: The Interaction of Light and Matter," Report No. AB-1272-0057, Sperry Rand Corporation, February 1966.

9. L. D. Strom, "Applications for Millimeter Radars," Report 108, System Planning Corporation, 1973.

10. J. Preissner, "The Influence of the Atmosphere on Passive Radiometric Measurements," AGARD Conference Reprint No. 245. *Millimeter and Submillimeter Wave Propagation and Circuits*, 1978.

11. V. W. Richard and J. E. Kammerer, "Rain Backscatter Measurements and Theory at Millimeter Wavelengths," Report No. 1838, USA Ballistic Research Laboratories, October 1975.

12. F. P. Wilcox, and R. S. Graziano, "Millimeter Wave Weather Performance," *1974 Millimeter Waves Techniques Conference*, Vol. 1, Naval Electronics Laboratory Center, San Diego, CA, March 1974.

13. N. C. Currie, F. B. Dyer, and R. D. Hayes, "Analysis of Radar Return at Frequencies of 9.375, 35, 70, and 95 GHz," Technical Report No. 2 on Contract DAAA 25-73-C-0256, Georgia Institute of Technology, February 1975.

14. J.D. Nespar, "Dual Frequency, Dual Polarization Backscatter Measurement of Rain," *Proceedings of the Tenth DARPA/Tri-Service Millimeter Wave Conference*, Harry Diamond Laboratory, Adelphi, MD, 1986.

15. G. Allen and B. Simonson, "Attenuation of Infrared Laser Radiation by HC, FS, WP, and Fog Oil Smokes," Technical Report EATR 4405, U.S. Army Edgewood Arsenal, May 1978.

16. R. L. Morgan, J. D. Stattler, and G. A. Tanton, "Results of MIRADCOM Workshop on Millimeter and Submillimeter Atmospheric Propagation Applied to Radar and Missile Systems," *Proceedings of the Eighth DARPA/Tri-Service Millimeter Wave Conference*, Eglin AFB, FL, April 1979.

17. E. E. Martin, "Radar Reflectivity of Dust Cloud Lofted by Misers Bluff II High Explosive Test," Final Technical Report, GIT Report A-2104, Georgia Institute of Technology, March 1979.

18. J. E. Knox, "Effects of Smoke Obscurants on Millimeter Waves," *Proceedings of the Eighth DARPA/Tri-Service Millimeter Wave Conference*, Eglin AFB, FL, April 1979.

19. F. C. Petito and R. L. Harris, "Millimeter Wave Radar Transmission through High Explosive Artillery Barrages," *Proceedings of the Eighth DARPA/Tri-Service Millimeter-Wave Conference*, Eglin AFB, FL, April 1979.

20. R. W. McMillian, "Atmospheric Turbulence Effects on Millimeter Wave Propagation," *IEEE EASCON-79 Conference Record*, Vol. 1, 1979.

21. F. B. Dyer and R. D. Hayes, "Computer Models for Fire Control Radar Systems," Technical Report No. 1 on Contract DAAA 73-C-0256, Georgia Institute of Technology, Atlanta, Georgia, 1973.

22. R. D. Hayes, N. C. Currie, and J. A. Scheer, "Reflectivity and Emissivity of Snow and Ground at MM Waves," *1980 IEEE International Radar Conference Record*, Washington, DC, April 1980.

23. M. W. Long and G. E. Allen, "Combat Surveillance Radar," Final Report on Contract DA36-039-SC-74870, Georgia Institute of Technology, 1960.

24. F. J. Bernues, R. S. King, and M. Kaswen, "Solid State Oscillators, Key to Millimeter Radar," *Microwave Systems News*, Vol. 9, No. 5, May 1979.

25. E. K. Reedy, and J. L. Eaves, "Millimeter Wave Instrumentation Radar Technology—Considerations and Applications," *Military Electronics Exposition '79*, Anaheim, CA, 1979.

26. P. H. Backus, "Electro Optics," *Electronic Defense*, Vol. 11, No. 3, March 1979.

Part II
THEORY AND PHENOMENOLOGY

Chapter 3
Target Detection in Noise and Clutter

J.D. Echard

Georgia Institute of Technology
Atlanta, Georgia

3.1 INTRODUCTION

Detection is the process by which the presence of a target is sensed in the presence of competing signals that arise from background echos, atmospheric noise, or noise generated within the radar receiver. In radar, the target is sensed by its reflection of radio-frequency (RF) energy originating in the radar's own transmitter. Interference always accompanies the target signal within the radar receiver. The interference usually consists of several types of "noise." First, some of the interference is *thermal noise* generated in the various stages of the radar receiver. In addition, the reflected target energy must compete with *atmospheric noise* generated externally to the receiver. However, in many MMW applications, the largest source of interference may be from ground or sea *clutter,* which is the reflection of radar energy from the ground, trees, vegetation, sea waves, or other background objects. Therefore, special emphasis is given to clutter in this chapter.

The properties of the interference as well as that of the target are important in determining detection performance. The ability of a radar receiver to detect a weak target echo signal is determined by the interference energy that occupies the same portion of the frequency spectrum as does the target energy. Target energy above a threshold level at the output of the receiver provides detection of the target. If the receiver output exceeds the threshold, a target is assumed to be present.

In this chapter, the radar range equation is first presented. Several forms of this equation are discussed including the so-called *clutter-limited* range equation. Then, the detection of targets in thermal noise is examined and several target models are presented, including Marcum, Swerling, and log-normal models. Next, the detection of targets in ground clutter is considered. The natural correlation properties of ground clutter and rain clutter are given, and the utilization of frequency agility to further decorrelate ground clutter is explained. Finally, the need for automatic detection of targets in a realistic environment is discussed. A summary at the end of this chapter describes the relations among these topics.

3.2 THE RADAR RANGE EQUATION

Several parameters are of major importance in determining the probability of detection of a target in background interference. One parameter is the amount of reflected power received by the radar receiver from the target, and another is the level of interference power within the receiver. In particular, the ratio of these two quantities is required to determine the probability of target detection. The mathematical equation that provides this ratio is called the *radar range equation*. It is derived by considering the power radiated by an isotropic antenna[1] in all directions. The power density at any point in space at a radial distance (or range) from this radar antenna is given by

$$\text{Power Density} = P_T/(4\pi r^2) \tag{3.1}$$

where P_T is the average power transmitted by the radar and $4\pi r^2$ represents the area of a sphere with radius r. However, most antennas are directive, and this characteristic is taken into account by including an antenna gain in the above equation. Antenna gain, G, is defined as the ratio of the antenna's maximum radiation intensity to that of an isotropic antenna. With this definition, the power density of the radar signal at a distance r from the radar in the direction of maximum antenna radiation intensity is given by

$$\text{Power Density} = P_T G/(4\pi r^2) \tag{3.2}$$

The target located a distance $r = R$ from the radar's antenna intercepts and reflects part of the transmitted energy. The power reflected from the target at distance r is given by

---

[1]An isotropic antenna is one that radiates in all directions with equal intensity.

Reflected Power $= P_T G\, \sigma_t/[(4\pi R^2)(4\pi r^2)]$ (3.3)

where σ_t is a quantitative measure of the intercepted and reflected power. It is referred to as the *target radar cross section* (RCS), and has units of area such as square feet or square meters.

The reflected power density at the radar's receiving antenna located at a radial distance $r = R$ from the target is therefore

Reflected Power Density at Receiving Antenna $= P_T G\sigma_t/(4\pi R^2)^2$ (3.4)

The electromagnetic capture area of an antenna is related to its gain by

$$G = 4\pi a A/\lambda^2$$ (3.5)

where

$A =$ the antenna area projected in the direction of the impinging wave,
$a =$ the antenna's efficiency,
$\lambda =$ the wavelength of the transmitted wave.

The power intercepted by the radar's receiving antenna is the product of the antenna's effective capture area, A, and the target reflected power impinging at the radar receiving antenna as indicated below:

Power Received at Antenna $= P_T G A \sigma_t/(4\pi R^2)^2$ (3.6)

Solving (3.5) for the *effective* antenna capture area, aA, and substituting it into (3.6) the power received from the target is

$$P_{rt} = \frac{P_T G^2\, \lambda^2 \sigma_t}{(4\pi)^3\, R^4}$$ (3.7)

Taking into account radar system loss, L_s, and propagation loss, $L_a(R)$, the amount of power, P_{rt}, received by the radar antenna due to reflection from a target is given by

$$P_{rt} = \frac{P_T\, G^2\, \lambda^2\, \sigma_t}{(4\pi)^3\, R^4 L_s\, L_a(R)}$$ (3.8)

where

P_{rt} = the power received from the target,
P_T = the average power transmitted by the radar,
G = the antenna gain (one way),
λ = the transmitting wavelength,
σ_t = the target RCS,
R = the range from the radar to the target,
L_s = the radar system loss (> 1),
$L_a(R)$ = the atmospheric loss (> 1).

The atmospheric loss can be expressed in terms of an attenuation factor, α, as follows:

$$L_a(R) = \exp[-0.2\alpha R] \tag{3.9}$$

where α is the one-way attenuation factor in dB/km. (See Chapter 4 for values of α.) The target RCS, σ_t, is a proportionality factor that is a function of the target's apparent size. The target RCS is a complicated function of many parameters, including the angle from which the target is viewed by the radar. In practice, the target RCS is often considered to be a random variable. Thus, the power received from the target, P_{rt} can also be considered a random variable. From (3.8) the *average* or mean power received by the radar receiver is determined to be

$$\overline{P_{rt}} = \frac{P_T\, G^2\, \lambda^2\, \overline{\sigma_t}}{(4\pi)^3\, R^4\, L_s\, L_a(R)} \tag{3.10}$$

where the overbar denotes a statistical average or mean. Therefore, $\overline{\sigma_t}$ is the mean target RCS, and $\overline{P_{rt}}$ is the mean receiver power. We can also calculate other statistical measures of P_{rt} such as variance. Histograms (probability density functions) of P_{rt} can also be obtained. Since all of the parameters in (3.8) except σ_t are deterministic, the histogram and statistical measures of P_{rt} are directly proportional to the histogram and statistical measures of σ_t. More will be said about this later.

The power received at the antenna output port due to radar echos from the background clutter for a system without range ambiguities can be expressed as

$$P_{rc} = \frac{P_T\, G^2\, \lambda^2\, \sigma_c}{(4\pi)^3\, R^4\, L_s\, L_a(R)} \tag{3.11}$$

where σ_c is the RCS of the clutter, and the remaining parameters are as defined for (3.8). Since the clutter return generally originates from the

same locality[2] as the target, the range equation contains the same parameter values for clutter as it does for target, except that the clutter RCS,
σ_c, is used instead of the target RCS, σ_t. As will be discussed later in more
detail, the clutter RCS, σ_c, is also considered a random variable. Thus,
the receiver power due to clutter, P_{rc}, is statistical in nature. It, too, can
be described in terms of a histogram, a mean, and other statistical moments. Often, (3.11) is expressed in terms of the mean clutter RCS, $\overline{\sigma_c}$,
as follows:

$$\overline{P_{rc}} = \frac{P_T G^2 \lambda^2 \overline{\sigma_c}}{(4\pi)^3 R^4 L_s L_a(R)} \tag{3.12}$$

where $\overline{P_{rc}}$ is the mean clutter power received at the antenna output terminals.

The mean noise spectral density, $\overline{N_0}$, at the input of the radar receiver
is given by

$$\overline{N_0} = k T_0 \overline{NF} \tag{3.13}$$

where

k = Boltzmann's contant (1.38×10^{-23} J/K)
T_0 = the standard temperature (defined as 290 K),
$\overline{NF}$ = the system noise figure.

The noise is that which effectively exists at the receiver's input terminals or antenna output port. Since atmospheric noise generally has the
same statistical characteristics as the receiver noise, it is sometimes included in the calculation of the system noise figure. In most MMW systems,
the receiver noise level is much larger than the atmospheric noise, and the
atmospheric noise contribution is therefore ignored (Barton [1]).

The receiver thermal noise power level has been carefully characterized statistically and found to have an exponential probability density
function or histogram. It is described only by statistical measures, such as
the mean spectral density level, $\overline{N_0}$, given in (3.13).

The ratio of mean target (signal) energy, $\overline{E_0}$, to the mean noise
spectral density is an important parameter in determining the probability
of detecting a target. The reflected target energy received at the radar is
given by

$$E_{rt} = P_{rt}\tau \tag{3.14}$$

---

[2]This is the same range-azimuth cell as that which contains the target.

where

E_{rt} = the energy received from the target,
P_{rt} = the power received from the target,
τ = the pulse width.

Combining (3.10) and (3.14), and forming a ratio of the target energy to the noise spectral density, yields the following:

$$\frac{\overline{E_{rt}}}{N_0} = \frac{P_T \tau \, G^2 \, \lambda^2 \, \overline{\sigma_t}}{(4\pi)^3 \, R^4 \, kT_0 \, \overline{NF} \, L_s \, L_a(R)} \tag{3.15}$$

where the terms are as defined previously. This ratio is commonly referred to as the *signal-to-noise ratio* (SNR). Equation (3.15) is the expression commonly called the "radar range equation." However, it only applies to the case where receiver (thermal) noise is the predominant source of interference: the so-called *noise-limited* case.

In MMW systems, the range to the target is sometimes quite short. For example, in an air-to-ground missile application, the target range may be just a few kilometers. In this circumstance, the receiver noise may not be a significant factor in the interference power. Instead, the ground clutter power received may be much larger than the thermal noise power. In this clutter-limited case, the ratio of the received target power to the received clutter power (the signal-to-clutter power ratio, SCR) will be the important factor in determining target detection performance. The SCR for a system without range ambiguities is obtained by taking the ratio of (3.10) and (3.13) as follows:

$$\text{SCR} = \frac{\overline{P_{rt}}}{\overline{P_{rc}}} = \frac{\overline{\sigma_t}}{\overline{\sigma_c}} \tag{3.16}$$

where the terms in the resulting equation have been previously defined. Since the radar echo from the ground clutter traverses the same path and radar circuitry as the target return, all the terms cancel, except for the target and clutter radar cross sections. However, there is another factor that must be included in (3.16): the constant false alarm rate (CFAR) loss, L_{CFAR}. In a clutter-limited case, the clutter power level must be estimated by a specially devised circuit or process that is discussed in Section 3.5. There is a loss associated with this clutter power level estimate which must be added. The resulting equation is

$$\text{SCR} = \frac{\overline{\sigma_t}}{\overline{\sigma_c} \cdot L_{\text{CFAR}}} \tag{3.17}$$

where L_{CFAR} is defined to be greater than unity. One variation of this equation is to treat the clutter RCS as a homogeneous medium and thus to define it as a normalized quantity, σ_c^o, such that

$$\overline{\sigma_c} = \overline{\sigma_c^o}\, A_c \tag{3.18}$$

where $\overline{\sigma_c^o}$ is the mean RCS of the clutter in a one square meter area expressed as m^2/m^2. The area on the ground that is illuminated by the radar antenna and included within a single range gate is denoted as A_c. See Chatper 5 for definitions of the area subtended by a radar. Combining (3.17) and (3.18), we obtain

$$\mathrm{SCR} = \frac{\overline{\sigma_t}}{\overline{\sigma_c^o}\, A_c\, L_{\mathrm{CFAR}}} \tag{3.19}$$

This equation is also sometimes expressed in decibels as follows:

$$\mathrm{SCR\ (dB)} = \overline{\sigma_t}(\mathrm{dBSM}) - \overline{\sigma_c^o}(\mathrm{dB}) - A_c(\mathrm{dBSM}) - L_{\mathrm{CFAR}}(\mathrm{dB}) \tag{3.20}$$

3.3 DETECTION OF TARGETS IN THERMAL NOISE

Target detection consists of differentiating a radar return of interest (target) from competing signals, which arise from background echos, atmospheric noise, and thermal noise generated in the radar receiver. The target is sensed by the reflection of RF energy originating in the radar. Noise is generated by the various electronic components in the radar receiver. In addition, the reflected target energy competes with atmospheric noise, even before entering the radar system.

The properties of the electromagnetic energy reflected by the target are important in calculating detection performance, and these will be discussed later in this chapter. The ability of a radar receiver to detect a weak target return is determined by the noise energy that occupies the same portion of the frequency spectrum as does the target energy. Detection is generally based on establishing a threshold level at the output of the IF amplifier in the radar receiver. If this output exceeds the threshold, a target is assumed to be present.

3.3.1 Thermal Noise as a Random Process [1–4]

Noise signals are generated by the random movement of free electrons. In the atmosphere, the electron movements generate an electromagnetic radiation that is intercepted by the radar. Inside the radar receiver, noise voltages are caused by random electron movement in the electronic components. Since the movement of free electrons is a random occurrence, the voltages resulting are described by statistical measures. Also, because the noise voltage varies with the passage of time, it is called a random *process*. The mathematical description of a narrowband random noise is given by

$$n(t) = a(t) \cos[w_0 t + \Theta(t)] \tag{3.21}$$

where

$n(t)$ = the noise *voltage* in the radar receiver,
$a(t)$ = the amplitude of the envelope modulation,
w_0 = the carrier frequency in radians per second,
$\Theta(t)$ = the phase modulation,
t = time.

In (3.21) it is assumed that the bandwidth of the random process, B_n, is much smaller than the carrier frequency, w_0:

$$B_n \ll w_0/2\pi \tag{3.22}$$

In conventional radars this assumption is almost always satisfied.

Both the amplitude and phase functions, $a(t)$ and $\Theta(t)$, of (3.21) are random processes, but the carrier frequency, w_0, is not random, it is deterministic. Also, t is deterministic and varies from zero to infinity in a linear fashion. For thermally generated noise, the amplitude of the noise *voltage* at any instant of time, t, is statistically described by the Rayleigh probability density function:

$$p[a(t)] = \frac{a(t)}{\sigma_n^2} \exp[-a^2(t)/2\sigma_n^2]; a(t) \geq 0 \tag{3.23}$$

where σ_n is the standard deviation. The phase of the noise voltage at any instant of time is described by the uniform density function:

$$p[\Theta(t)] = \frac{1}{2\pi}; (0 < \Theta(t) < 2\pi) \tag{3.24}$$

If the instantaneous value of the noise power is given by $b(t) = a^2(t)/2$, then (3.23) becomes

$$p[b(t)] = \frac{1}{\sigma_n^2} \exp(-b(t)/\sigma_n^2); b(t) \geq 0 \tag{3.25}$$

where σ_n^2 is now interpreted as the *mean* noise power and $p[b(t)]$ is an exponential distribution.

When a target return is present at the input of the radar receiver, this target signal is added to the noise voltage in the receiver, and the combined *voltages* are represented as

$$s(t) = r(t) + n(t) \tag{3.26}$$

where $r(t)$ is the target return voltage as a function of time, and $n(t)$ is the noise voltage as a function of time. For some types of targets, $r(t)$ may vary randomly with time. The nature of the target return, $r(t)$, will be discussed in Sections 3.3.2 and 3.3.3. The ratio of the target return power (signal power) to the noise power (σ_n^2 in (3.25)), is called the signal-to-noise ratio.

The *probability density functions* (pdf) discussed in the preceeding do not alone adequately describe a random process. In addition, the values of the process must be related from one time instant to another. For example, the random noise process, $n(t)$, at time instants, t_1, t_2, *et cetera*, must be related in some way. The formulation that provides this time relationship is called *autocorrelation*, ϕ_n, and is defined as

$$\phi_n(t_1, t_2) = E\left[n(t_1) \cdot n(t_2)\right] \tag{3.27}$$

where t_1 and t_2 are arbitrary times, and $E(\cdot)$ is the statistical expectation. However, this relationship is seldom used because it is difficult to evaluate. An easier autocorrelation formulation to use is the time averaged function:

$$R_n(\tau) = \lim_{T \to \infty} \frac{1}{2T} \int_{-T}^{T} n(t)\, n(t - \tau)\, dt \tag{3.28}$$

where $\tau = t_2 - t_1$. If the random process is ergodic[3], then

$$\phi_n(t_1, t_2) = \phi_n(\tau) = R_n(\tau) \tag{3.29}$$

In most common radar applications, (3.28) and (3.29) apply if the random process is stationary. This means that the statistical moments of the random process must not vary with time. This condition is satisfied for receiver generated noise which is the subject under discussion in this section.

---

[3]For a definition and discussion of ergodicity, See Papoulis [4]. For a random process to be ergodic, it must be stationary. However, the converse is not true. In many radar situations, the random process is ergodic.

The noise that exists at the electrical component, where it is generated in the radar receiver, is very wideband and is therefore uncorrelated over very short time increments. The most significant source of thermal noise in the receiver is at its front end, near the antenna. The amplifiers used in the receiver are limited in bandwidth and determine the noise bandwidth at the detector. Often, the IF amplifier has the smallest bandwidth in the radar and determines the noise bandwidth. Thus, the noise out of the IF amplifier is correlated over a longer time duration than it is at the receiver front end.

The filter response of the radar receiver can be designed to maximize the signal energy out of the IF amplifier while at the same time minimizing the noise energy at that point. A filter that has a frequency-response function which maximizes the output signal-to-noise ratio is called a *matched filter* (Skolnik [2]). If the bandwidth of the receiver is wide compared with that occupied by the signal energy, extraneous noise is introduced by the excess bandwidth, which lowers the output signal-to-noise ratio. On the other hand, if the receiver bandwidth is narrower than the bandwidth occupied by the signal, the signal energy is reduced more than the noise energy. This is because for $B < 1/\tau$ the signal energy varies as B^2 and the noise energy only varies as B (Barton [1]). The result is again a lowered signal-to-noise ratio. Thus, there is a receiver bandwidth at which the signal-to-noise ratio is a maximum. A rule often used is that the receiver bandwidth, B, should be approximately equal to the reciprocal of the pulse width, τ, for pulsed radars.

The frequency response of the radar receiver is given by the response from the antenna terminals to the output of the IF amplifier. The overall receiver bandwidth is most often determined by the IF amplifier. The bandwidths of the RF and mixer stages of the radar receiver are usually large compared with the IF bandwidth. Therefore, the frequency-response function of the receiver is usually taken to be that of the IF amplifier. The IF amplifier may be considered as a filter with gain. The response of this filter as a function of frequency is of particular interest.

For a radar return, $r(t)$, it can be shown that the frequency-response function of a linear, time-invariant filter which maximizes the output signal-to-noise ratio for a fixed input signal-to-noise ratio is

$$H(f) = R^*(f) \exp\left(-j2\pi f t_1\right) \tag{3.30}$$

where

$$R(f) = \int_{-\infty}^{\infty} r(t) \exp(-j2\pi ft) \, dt,$$

$R^*(f)$ = complex conjugate of $R(f)$,

t_1 = fixed value of time at which the signal is observed to be maximum.

The noise that accompanies the signal is assumed to be stationary and to have a uniform spectrum (white noise).[4] The filter that has a frequency response function as given by (3.30) is called a matched filter.

The frequency response of the matched filter is the conjugate of the spectrum of the received waveform, except for the phase shift $\exp(-j2\pi ft_1)$. This phase shift causes a constant time delay so that the filter is physically realizable. The frequency spectrum of the received signal may be expressed as an amplitude spectrum $|R(f)|$ and a phase spectrum $\exp[-j\phi_R(f)]$. The matched filter frequency response may be similarly written in terms of its amplitude and phase spectrum $|H(f)|$ and $\exp[-j\phi_H(f)]$. Equation (3.30) for the matched filter may be written as

$$|H(f)| = |R(f)| \tag{3.31}$$

and

$$\phi_H(f) = -\phi_R(f) + 2\pi ft_1 \tag{3.32}$$

Thus, the amplitude spectrum of the matched filter is the same as the amplitude spectrum of the signal, but the phase spectrum of the matched filter is the negative of the phase spectrum of the signal plus a phase shift that is proportional to frequency.

The output of the matched filter is not a replica of the input signal. However, from the point of view of detecting signals in noise, preserving the shape of the signal is of no importance.

3.3.2 Nonfluctuating Targets

As mentioned earlier, the temporal characteristic of the target return, $r(t)$, depends upon the nature of the target being illuminated by the radar. Several standard models of $r(t)$ have been developed over the years and are common to the radar community. The first target model to be discussed is the so-called nonfluctuating or "Marcum" target (Marcum [5, 6]). Examples of targets which match this model are a metalized balloon or a

---

[4]In practice, the noise only has to be uniform over the frequency passband of the filter.

corner reflector when viewed from the front. The radar cross section of this target model is constant over time periods that are long compared to the integration time of the radar. For example, the target return may be described by

$$r(t) = c \cos [wt + \alpha] \tag{3.33}$$

where both c and α are constants. When added to thermal noise, described by (3.21), the total signal appears as

$$s(t) = c \cos (wt + \alpha) + a(t) \cos [wt + \Theta(t)] \tag{3.34}$$

Note that $s(t)$ is a random process with a deterministic component, i.e., $c \cos (wt + \alpha)$. Thus, $s(t)$ is mathematically described by a probability density function. The envelope of $s(t)$ is defined by the Ricean density function (Davenport and Root [3]):

$$p(\hat{s}) = \frac{\hat{s}}{\sigma_n^2} \exp [-(\hat{s}^2 + c^2)/2\sigma_n^2)] I_0\left(\frac{c\hat{s}}{\sigma_n^2}\right); (\hat{s} > 0) \tag{3.35}$$

where

$\hat{s}$ = the envelope of the target-plus-noise signal $s(t)$,
σ_n^2 = the variance (power) of the noise voltage,
c = the peak magnitude of the target return,
$I_0(\cdot)$ = the modified Bessel function of order zero.

The term $c^2/(2\sigma_n^2)$ is called the signal-to-noise ratio, which is sometimes denoted as SNR or S/N. It is the ratio of average signal power $c^2/2$, to noise variance or noise power, σ_n^2. When the signal power is zero (i.e., there is no target return present), then (3.35) is the density function for noise alone and is identical to (3.23). The probability density function describing the envelope of the "target-plus-noise" signal is shown in Figure 3.1 for various SNRs.

The detection of signals in the presence of noise is equivalent to deciding whether the receiver output is due to noise alone or to signal-plus-noise. When the detection process is carried out automatically by electronic means, the detection process cannot be left to chance and must be specified and built into the decision-making device by the radar designer.

The detection process is described in terms of threshold detection. If the envelope of the receiver output exceeds a pre-established threshold, a target is said to be present. The purpose of the threshold is to divide the output into regions: detection and no detection. In other words the

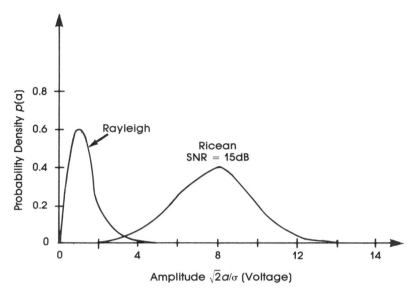

Fig. 3.1 Probability density function of noise envelope (Rayleigh) and signal-plus-noise envelope (Ricean).

threshold detector allows a choice between two hypotheses. One hypothesis is that the receiver output is due to noise alone; the dividing line between these two regions depends upon the probability of false alarm, which is related to the average time between false alarms. This situation is illustrated in Figure 3.2. Here, the envelope of the noise is shown with its rms value and threshold denoted by horizontal lines. Note that occasionally the threshold is exceeded by the noise voltage, and a "false alarm" hence occurs.

The threshold level is usually selected so as not to exceed a specified false alarm probability; that is, the probability of detection is maximized for a fixed probability of false alarm. This approach to selecting the threshold level is illustrated in Figure 3.3, where the probability density functions for noise alone and signal plus noise are shown.

The device that extracts the modulation from the carrier is called the *detector*. As used here, "detector" implies more than simply a rectifying element. It includes that portion of the radar receiver from the output of the IF amplifier to the input of the data processor. The effect of the amplifier is not of importance to our present discussion. The major concern is the effect of the detector on the desired signal and the noise.

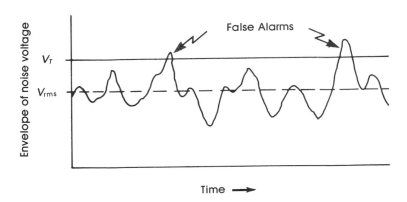

Fig. 3.2 Envelope of noise *versus* time with threshold voltage shown and two false alarms illustrated.

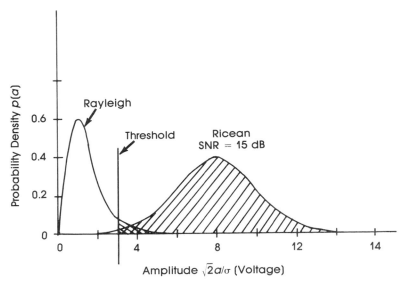

Fig. 3.3 Probability density functions of noise and signal-plus-noise with threshold shown for $P_{FA} = 10^{-2}$.

One type of detector is the envelope detector, which only removes the signal represented by the amplitude of the carrier envelope. It is also possible to design a detector that utilizes only phase information. If the exact phase of the echo carrier were known, it would be possible to design a detector that makes use of both the phase information and the amplitude information contained in the radar return. It would perform more efficiently than a detector that used either amplitude information only or phase information only. The synchronous or *I-Q* detector is an example of one which uses both phase and amplitude information. Two of these detectors will be discussed in the following subsections.

Envelope Detector [1, 2, 7]

The function of the envelope detector is to extract the amplitude modulation, and reject the carrier and the phase modulation. Thus, the detection decision is based solely on the envelope amplitude. The detector characteristic relates the output signal to the input signal and is called the detector law. Most detector laws approximate either a linear or a square-law characteristic. In the so-called "linear" detector, the output signal is directly proportional to the input signal envelope. Similarly, in the square-law detector, the output signal is proportional to the square of the input signal envelope. In general, the difference in performance between the two is small.

The probability of detection (P_d) *versus* SNR for a envelope detector is given by the curves in Figure 3.4 with constant false alarm probabilities (P_n) as a parameter. The target is assumed to be nonfluctuating and the number of pulses summed is one (no pulse-to-pulse integration).

Suppose one desired to determine the single-pulse SNR that is required to provide a probability of detection of 50% with a false alarm probability of 10^{-8}, assuming a nonfluctuating target. The procedure for using the detection curves of Figure 3.4 is as follows. First, select the appropriate false alarm curve. Second, locate a point on this curve corresponding to a detection probability of 50% (horizontal dashed line in Figure 3.4). Next, follow a vertical line from this point to the abscissa to determine the required SNR: + 12.5 dB. This example is applicable only to single-pulse detection, but will be extended later to include multiple pulses.

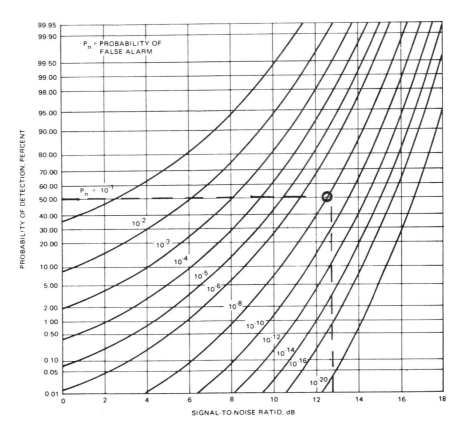

Fig. 3.4 Probability of detection *versus* signal-to-noise ratio for a non-fluctuating target (single pulse detection). (From Hovanessian [7], © 1973 Artech House.)

Logarithmic Detector

If the output of the receiver is proportional to the logarithm of the input signal envelope, the receiver is called a *logarithmic receiver*. This receiver is another type of envelope detector and finds application where large variations of input signals are expected. It might be used to prevent receiver saturation, or to reduce the effects of unwanted clutter in certain types of radar receivers.

Synchronous Detectors [8]

A detector known as a synchronous detector, shown in Figure 3.5, is used to extract both the modulation envelope and phase information from the IF signal. In Figure 3.5, the reference oscillator for each mixer is assumed to have the same frequency as the transmitted signal, but the two reference signals are out of phase with respect to each other by 90°. Usually, the reference signal is obtained from the same source as the transmitted signal so that it is synchronous with or follows the transmitter frequency. With this type of detector, the phase of the received signal may have any value and the amplitude and phase modulation information will be fully represented by the in-phase (I) and quadrature (Q) signals. If the IF signal is represented by the mathematical expression given by

$$s(t) = a(t) \cos[w_0 t + \Theta(t)] \tag{3.36}$$

then the I and Q signals out of the synchronous detector will be given by

$$I = a(t) \cos[\Theta(t)] \tag{3.37}$$

and

$$Q = a(t) \sin[\Theta(t)] \tag{3.38}$$

where

$a(t)$ = the envelope amplitude voltage,
$\Theta(t)$ = the phase voltage.

If the IF modulation is represented by complex valued mathematics, the detector output, $y(t)$, can be written as

$$y(t) = I + jQ \tag{3.39}$$

and

$$y(t) = a(t) \exp[j\Theta(t)] \tag{3.40}$$

where $a(t)$ and $\Theta(t)$ are the amplitude and phase modulation, respectively. Thus, the detector signal, $y(t)$, is proportional to the complex valued modulation.

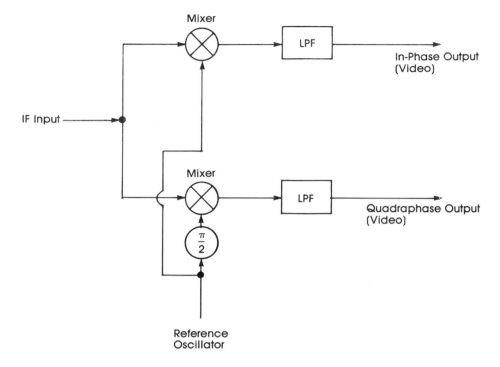

Fig. 3.5 Synchronous or *I-Q* detector. (Reference frequency oscillator
tracks transmitter frequency.)

Pulse Integration

Only in unusual cases will a radar detection decision be based on a
single received pulse. Instead, a train of several to several hundred pulses
will be received from each target, and the entire train will be processed
before it is decided if a target is present. Acceptable detection probability
can then be obtained, even when the single-pulse SNR is near or below
unity. The process by which the pulses are combined is known as *pulse
integration*.

In an ideal processing arrangement, the energy from all of the radar
returns is added directly in an integrator prior to envelope detection. The
result is to achieve an effective, or integrated, value of signal-to-noise
ratio, $(SNR)_i$, equal to the single-pulse SNR multiplied by N, the total
number of pulses in the train. This process is referred to an *coherent
integration*. This type of integration requires a highly stable transmitter.

When a synchronous detector is used, coherent integration can also be accomplished *after* the detector because the detector reference signal is derived from the same source as the transmitted signal and signal phase information is available. If the target reflected signal is Doppler shifted, then the phase shift from pulse-to-pulse must be removed before coherent integration.

Many radars use *noncoherent integration* to improve target detection capability. The noncoherent integrator can achieve a gain that approaches N for cases where only a few pulses are processed, and can be approximated by $N^{0.8}$ over a wide range of conditions. In general, noncoherent integration is not as efficient as coherent integration. However, this may be offset by the fact that noncoherent detection and integration are more easily implemented.

The integration gain is defined for a specific detection and false alarm probability. The gain is calculated by determining the SNR required before integration and after integration, and taking the ratio (or difference in dB). The probability density function of noise and target-plus-noise before integration may be as shown in Figure 3.6(a). Normalized integration [5] of radar returns causes the variance of the pdf values to become smaller according to the amount of statistical independence between radar returns. If the interference is only thermal noise, the variance will decrease by N, the number of samples integrated. The mean of the pdf remains unchanged. This is illustrated in Figure 3.6(b). The net effect is to increase the separation (in terms of variance) between the two pdf values. Thus, a smaller SNR is required to achieve a specific probability of detection and false alarm.

Detection curves similar to the ones shown in Figure 3.4 can be generated for various numbers of pulses integrated. In Figure 3.4, probability of detection (in percent) is plotted *versus* SNR (in dB) with false alarm probability shown as a parameter, for a single-pulse detection. When the number of pulses is added as a parameter, multiple plots are required. An example is shown in Figure 3.7, where the probability of detection (in percent) is plotted *versus* the number of pulses integrated with the SNR (in dB) as a parameter. These curves apply for only one value of false alarm probability or *false alarm number*, n'. In reference (9) from which these "Meyer" plots are taken, the probability of false alarm (P_{FA}) is related to false alarm number n' as follows: $P_{FA} = 0.693/n'$. In Figure 3.7, the false alarm number is given as 6×10^7, yielding a false alarm probability of $P_{FA} \approx 1.155 \times 10^{-8}$, which is very close to the 10^{-8} curve in

---

[5]Normalized integration consists of summing N of the radar returns from a given range-azimuth cell and dividing by the number, N.

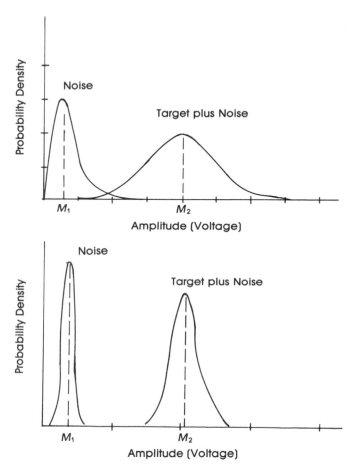

Fig. 3.6 Probability density functions of noise and target-plus-noise (a) before integration and (b) after integration.

Figure 3.7. In the curves of Figure 3.7, the target model assumed is non-fluctuating or Case 0.

Continuing the detection example given earlier, if the probability of detection of 50% is desired with a false alarm probability of about 10^{-8}, then the single-pulse SNR required is approximately +12.3 dB as shown in Figure 3.7 by the circle. If 10 pulses are transmitted and are reflected by a nonfluctuating target and are *noncoherently* integrated, the required *single-pulse SNR* (before integration) is found to be approximately +4.6 dB as indicated by the square in Figure 3.7. In other words, 10 radar returns with a single-pulse SNR of +4.6 dB when noncoherently integrated yields the same detection probability as a single radar return with SNR of

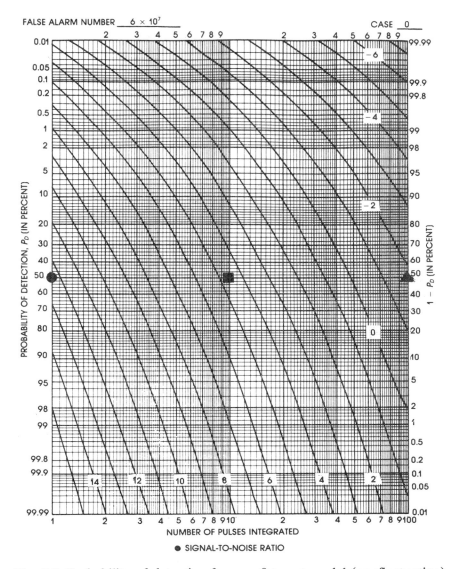

Fig. 3.7 Probability of detection for case 0 target model (nonfluctuating) in Rayleigh noise *versus* number of pulses integrated. Signal-to-noise (SNR) is shown as a parameter. False alarm number = 6×10^7. Probability of false alarm = 1.15×10^{-8}. (From Meyer and Mayer [9], © 1973 Academic Press.)

+ 12.3 dB. Referring to Figure 3.7, we note that 100 pulses integrated requires a single-pulse SNR of approximately -1.7 dB to give an integrated 50% probability of detection (shown by the triangle).

If *coherent* integration is used, then the required single-pulse SNR in the above example is calculated as follows for 10 and 100 pulses, respectively:

$$SNR = +12.3 \text{ dB} - 10 \log_{10}(10) = +2.3 \text{ dB} \qquad (3.41)$$

and

$$SNR = +12.3 \text{ dB} - 10 \log_{10}(100) = -7.7 \text{ dB} \qquad (3.42)$$

Note that coherent integration provides a SNR *gain* of 2.3 dB and 6.0 dB, respectively, as compared to noncoherent integration. That is, smaller single-pulse SNRs are required with coherent integration to achieve a 50% detection probability.

If the reader does not have sources (e.g., [9]) to use in determining detection performance, there are ways to estimate the performance by using Figure 3.7 and a couple of additional graphs (Barton [1], Skolnik [2]).

3.3.3 Fluctuating Targets [9, 10, 11]

Most often the radar echoes from targets usually encountered vary over very short periods of time. This is because the target and scatterers on the target move. A good example is a slow moving, propeller-driven aircraft. The movement of the propellers and the yawing motion of the aircraft changes the viewing angle of the target from the radar's perspective and the RCS varies with time. On many targets, the RCS varies rapidly with very small changes of target viewing angle. This is true for ground targets as well as for air targets.

Several standard target models have been established for fluctuating targets (Schwartz [11]). They are listed in Table 3.1. The first target model is the nonfluctuating or Marcum target model which has already been discussed in Section 3.2.2. This model is followed by four Swerling models and the log-normal and Weibull target models. In the table, several characteristics of the target models are listed. First, the pdf used to describe the target RCS is listed in the second column. Swerling models 1 and 2 are described by the same pdf as thermal noise voltage (i.e., Rayleigh). Swerling models 3 and 4 are modeled by a chi-squared pdf with four degrees of freedom, which has not yet been described. The chi-squared pdf with four degrees of freedom is given by

$$p(r) = \frac{4r}{\bar{r}^2} \exp\left(\frac{-2r}{\bar{r}}\right); \, r > 0 \tag{3.43}$$

where

r = the signal proportional to the target RCS,
$\bar{r}$ = the mean signal proportional to the mean target RCS.

Table 3.1 Target Models.

Target Model	Amplitude Density Function	Description of Radar Return Fluctuation
Marcum	Nonfluctuating	No fluctuation (perfect correlation pulse-to-pulse)
Swerling 1	Rayleigh	Slow fluctuation (scan-to-scan variation; perfect correlation pulse-to-pulse within scan)
Swerling 2	Rayleigh	Fast fluctuation (uncorrelated pulse-to-pulse)
Swerling 3	Chi-squared (4 d.f.)	Slow functuation (scan-to-scan variation; perfect correlation pulse-to-pulse within scan)
Swerling 4	Chi-squared (4 d.f.)	Fast fluctuation (uncorrelated pulse-to-pulse)
Rayleigh	Rayleigh	As determined or assumed
Weibull	Weibull	As determined or assumed
Log-normal	Log-normal	As determined or assumed

Another characteristic listed in Table 3.1 is a verbal description of the time variation of the target RCS. The pulse-to-pulse fluctuation refers to the RCS variation of the target when it is subjected to the transmission of a series of radar pulses. Some target models assume the RCS from one pulse to another is uncorrelated as indicated in the third column. This fluctuation characteristic is assumed for Swerling target models 2 and 4.

During one sweep of the antenna beam a series of pulses illuminate the target. This is called a single scan. When the beam illuminates the target a second time, the second scan has occurred, and so on. Swerling target models 1 and 3 assume that, during a single scan, the target RCS does not vary or change. However, between scans the target moves. So, there is scan-to-scan variation in the target RCS. This model might be applicable to a metallic covered balloon or some other large, slowly moving object.

Detection curves for Swerling target models 1, 2, 3, and 4 in a thermal noise interference environment (Rayleigh pdf) are given in Figures 3.8 through 3.11. The Marcum and Swerling models just described have served the radar community for many years, but they have shortcomings when applied to the MMW band. In the MMW band where the wavelength is very short, almost all targets are fluctuating types over some time periods. Even stationary ground targets such as vehicles usually provide a fluctuating RCS because of the radar platform motion relative to the stationary target. In many scenarios, the aircraft or missile on which the radar is mounted has a changing viewing angle of the ground vehicle, and thus the vehicle RCS fluctuates with time.

The description of the RCS variation with time given by Swerling models are usually inadequate for MMW wave situations. Instead, the autocorrelation function of the RCS fluctuation with time is utilized. This function relates the value of target RCS at one time to it's value at another time. For example, suppose the autocorrelation function of a land vehicle is described by the exponential function in Figure 3.12. In this case, the autocorrelation is a function of the target aspect angle and 90% decorrelation occurs at one degree of azimuth angle change. If the angular rate at which the target is viewed by the radar is known, the autocorrelation function *versus* time can be determined. For example, if in an air-to-ground system the moving platform on which the radar is mounted is moving at 100 m/s (237 m/h) and the stationary ground target is 1000 m away (slant range between radar and target) and offset 20° from the platform movement direction, the rate of azimuth angle change is

$$\frac{d\theta_a}{dt} = 2.0°/s \qquad\qquad (3.44)$$

Therefore, one degree of azimuth is tranversed in 0.5 s and the corresponding autocorrelation function is given in Figure 3.13. In this example, radar returns that are about 500 ms apart in time are almost totally uncorrelated. For some radar systems, the pulse integration time may be much shorter than 500 ms, and only partial decorrelation may occur due to target RCS fluctuation. Nonetheless, the amount of decorrelation that takes place must be considered in determining the radar detection performance.

If the airborne radar platform is headed directly toward the ground target, target RCS fluctuation will probably not occur during pulse integration period. Thus, the pulses may be assumed to be highly correlated.

One way in which the partial decorrelation may be taken into account in determining detection performance is to simulate the radar returns and mathematically impart the appropriate correlation to the signal samples.

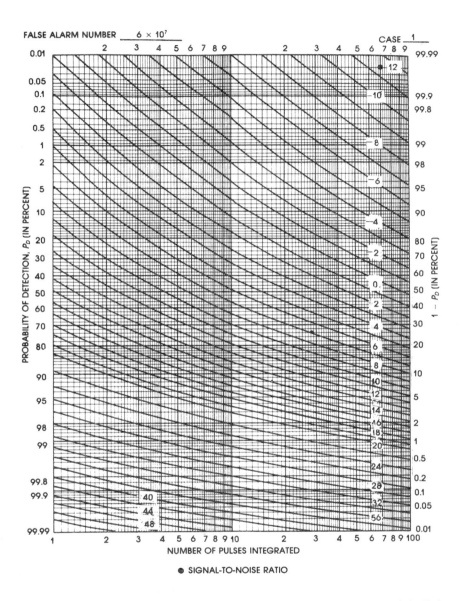

FALSE ALARM NUMBER   6 × 10⁷   CASE 1

NUMBER OF PULSES INTEGRATED

● SIGNAL-TO-NOISE RATIO

Fig. 3.8 Probability of detection for Swerling case 1 target model. False alarm probability = 1.15×10^{-8}. (From Meyer and Mayer, [9], © 1973 Academic Press.)

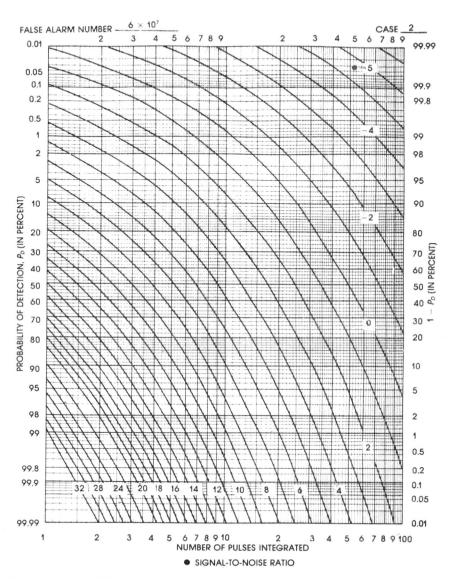

Fig. 3.9 Probability of detection for Swerling case 2 target model. False alarm probability $= 1.15 \times 10^{-8}$. (From Meyer and Mayer, [9], © 1973 Academic Press.)

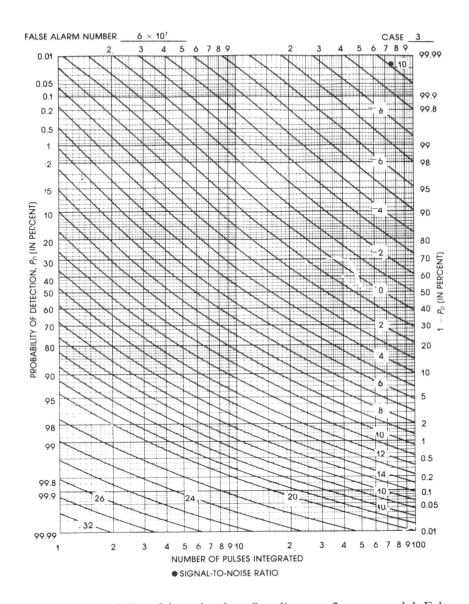

Fig. 3.10 Probability of detection for a Swerling case 3 target model. False alarm probability = 1.15×10^{-8}. (From Meyer and Mayer, [9], © 1973 Academic Press.)

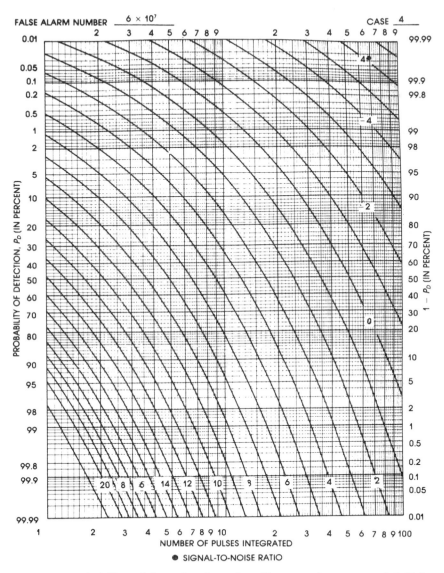

Fig. 3.11 Probability of detection for a Swerling case 4 target model. False alarm probability $= 1.15 \times 10^{-8}$. (From Meyer and Mayer, [9], © 1973 Academic Press.)

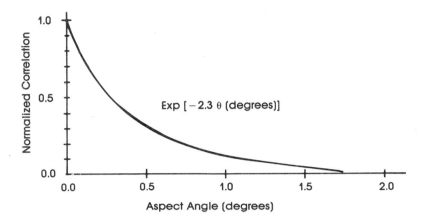

Fig. 3.12 Hypothetical normalized correlation function for land vehicles *versus* aspect angle.

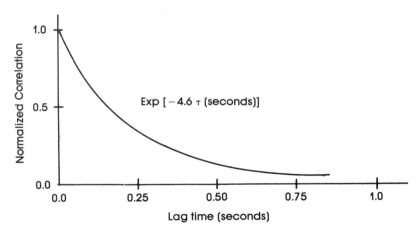

Fig. 3.13 Hypothetical normalized correlation function for land vehicles *versus* lag time for the example in the text.

The signal samples may then be summed to determine the integration gain achieved. Another approach is to use the appropriate Swerling models to bound the integration gain expected. In the example cited, if the integration period is 10 ms, for example, then for all practical purposes, all the pulses being integrated can be assumed to the correlated, and either the Swerling 1 or 3 target models may be used if the probability density functions are appropriate. However, if the integration period is 500 ms, for example, then a few of the many pulses integrated can be assumed to be uncorrelated; however, most of them are still highly correlated. Pulses separated by 250 ms are only 35% correlated and those separated by 100 ms are about 60% correlated. Thus, some integration gain will be achieved in these cases, but how much is difficult to say without simulating and summing signal samples with the appropriate correlation properties.

Another way in which the Swerling models may be inadequate for MMW applications is in the target RCS probability density function assumed. Experimental results in the MMW band indicate that ground vehicles exhibit a log-normal probability distribution of RCS values when all target aspect angles are considered (Einstein [12]).

The lognormal probability density function is defined as (Hines and Montgomery [13]):

$$p(x) = \frac{1}{x\sqrt{2\pi\ \text{SD}_y^2}} \exp\left[-\ln^2\left(x/m_x\right)/(2\ \text{SD}_y^2)\right] \tag{3.45}$$

where

x = the log-normal variate,
y = the natural logarithm of x,
m_x = median of x,
SD_y = the standard deviation of y.

There are several equivalent variations of (3.46). The one which appears to be most convenient is where the logarithmic base 10 of x is utilized:

$$P(x_{\text{dB}}) = \frac{1}{\sqrt{2\pi\ \text{SD}_{x\text{dB}}^2}} \exp\left[-(x_{\text{dB}} - \mu_{x\text{dB}})^2/2\ \text{SD}_{x\text{dB}}^2\right] \tag{3.46}$$

where

x_{dB} = $10\log_{10}(x)$
$\mu_{x\text{dB}}$ = $10\log_{10}(m_x)$
$\text{SD}_{x\text{dB}}$ = $4.343\ \text{SD}_y$

Sometimes, the relationships between the statistical moments of x and x_{dB} are useful. They are given below for convenience [13]:

$$\mu_x = \exp\left[\mu_{xdB}/c + SD^2_{xdB}/2c^2\right] \tag{3.47}$$

$$SD^2_x = \exp\left[2\mu_{xdB}/c + 2SD^2_{xdB}/c^2\right] - \exp\left[2\mu_{xdB}/c + SD^2_{xdB}/c^2\right] \tag{3.48}$$

and

$$\mu_{xdB} = 2c \ln(\mu_x) - c \ln(\mu_x^2 + SD_x^2)/2 \tag{3.49}$$

$$SD^2_{xdB} = c^2 \ln(\mu_x^2 + SD_x^2) - 2c^2 \ln(\mu_x) \tag{3.50}$$

where $c = 10 \log_{10}(e) = 4.343$

One of the nice characteristics of this target RCS model is that the variable x_{dB} is Gaussian (normally) distributed with mean μ_{xdB} and standard deviation SD_{xdB} as indicated by (3.46). More will be said about this target model in the next section on detection of targets in clutter.

3.4 DETECTION OF TARGETS IN CLUTTER

As previously indicated, radar target detection in an interference background is a statistical process, rather than a deterministic one. Thus, it must be described and evaluated in terms of probabilities and autocorrelation functions. The total composite signal that is processed in the radar receiver is composed of several parts: target, thermal noise, and background clutter. The purpose of the radar signal processor is to decide correctly whether a target signal is present. The noise and clutter signals interfere with this decision.

In many MMW applications, the range to the target is relatively short, for example, several kilometers. Therefore, the major source of interference is usually ground or sea clutter. When rain is present, it also may be a substantial source of interference. Thermal noise interference under these conditions is likely to be much smaller than the ground, sea, or rain clutter, and the radar's performance is said to be *clutter-limited*. This section concentrates on ground, sea, and rain clutter as sources of interference.

3.4.1 Clutter Return As a Random Process

Clutter may be defined as radar return signals that are unwanted in the radar situation being considered. In most common MMW applications, radar returns from the ground or falling rain are unwanted, and they are thus clutter.

The unwanted clutter returns detected by the radar usually come from a number of spatially distributed scatterers. This is illustrated in Figure 3.14, where N scatterers are indicated. The scatterers are illuminated by the radar transmitter, and each scatterer reradiates part of the incident energy back to the radar.

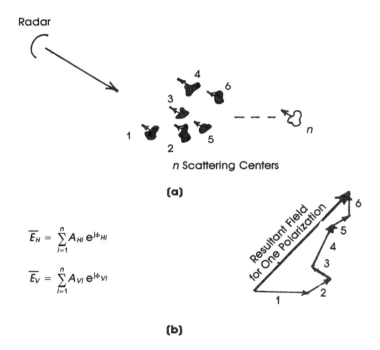

$$\overline{E_H} = \sum_{i=1}^{n} A_{Hi} e^{j\phi_{Hi}}$$

$$\overline{E_V} = \sum_{i=1}^{n} A_{Vi} e^{j\phi_{Vi}}$$

(b)

Fig. 3.14 Reflectivity from N scatterers: (a) Vector-phasor summation of individual field contributions; (b) clutter return from an assembly of scattering centers (From Corriher, *et al.*, [27]).

The total effect of the reflected (or reradiated) energy is described by the vector sum of the electric fields reflected from each scatterer that has been illuminated by the radar. The electric field reflected by the ith scatterer may be described as

$$E_i = A_i \cos (wt + \phi_i) \tag{3.51}$$

or

$$E_i = R_e [A_i \, e^{j\phi i} \, e^{jwt}] \tag{3.52}$$

where

A_i = the amplitude (voltage) of the reflected field,
ϕ_i = the phase of the reflected field,
w = the RF corner frequency in radians.

For mathematical convenience, the ith reflected electric field will be represented as

$$\overline{E_i} = A_i e^{j\phi i} \tag{3.53}$$

where it is understood that

$$E_i = R_e (\overline{E_i} e^{jwt}) \tag{3.54}$$

The total electric field scattered from the N scatterers is the vector (complex) sum of the individual fields as follows:

$$\overline{E} = \sum_{i=1}^{N} \overline{E_i} = \sum_{i=1}^{N} A_i e^{j\phi i} \tag{3.55}$$

Because electric fields have polarimetric properties, a more accurate description of the total electric field would be

$$\overline{E_V} = \sum_{i=1}^{N} A_{Vi} e^{j\phi_{Vi}} \tag{3.56}$$

and

$$\overline{E_H} = \sum_{i=1}^{N} A_{Hi} e^{j\phi_{Hi}} \tag{3.57}$$

if a vertical and horizontal polarization basis is being used. The electric field E_v is the vertically polarized field reflected by the scatterers, while E_H is the corresponding quantity when a horizontally polarized field is reflected by the scatterers.

For any given situation where the clutter scatterers are quantified as to size, shape, location, and composition, the electric fields described by (3.56) and (3.57) are deterministic, and can be completely quantified. However, in practice, very little is known about *each* scatterer. Usually, only a general description of the clutter being illuminated is available. Also, the description varies from one location to another. Therefore, the radar analyst and designer resorts to a statistical description of the reflected electric fields described by (3.56) and (3.57). In other words, the quantities A_{Vi}, A_{Hi}, ϕ_{Vi}, and ϕ_{Hi} are considered random variables, and therefore the total reflected fields E_V and E_H are also random variables.

When the radar illumination (antenna beam) is swept over an area in space, the reflected electric fields, E_V and E_H, fluctuate with time because the number and type of clutter scatterers change from one area to another. Thus, E_V and E_H become random processes that can be characterized by a probability density function and an autocorrelation function in a manner similiar to that of thermal noise, a fluctuating target, or any other random process.

The probability density and autocorrelation functions which best describe a certain type of clutter are determined by careful and accurate radar measurements of that clutter type. The measured data are usually matched to a mathematical model, which is easily expressed and manipulated. In the remainder of this section, several common models will be presented for the clutter probability density and the autocorrelation functions. See Chapter 7 for a discussion of clutter models.

As mentioned earlier, the autocorrelation function and the power density spectrum have a Fourier transform relationship. Some of the models are most conveniently expressed in the autocorrelation domain, while other are best expressed in the frequency domain as a power density spectrum.

Clutter Probability Density Functions

Three general types of clutter will be considered in this section: ground, sea, and rain. Each of the clutter types has its own special characteristics, and these will be briefly introduced. However, the emphasis in this section will be on how to use these probability density functions to obtain the probability of target detection in a clutter background. Details of clutter characteristics and statistical descriptions are presented in Chapter 5.

Classical radar analysis assumes that the clutter radar returns (reflected electromagnetic fields measured in volts per meter) are Rayleigh distributed both within a single range-azimuth resolution cell as a function of time, and between such cells when the antenna illumination beam is swept in angle. This model is the natural result obtained when the received signal is the composite of radar returns from a large number of similarly sized scatterers. However, while this model may be adequate for micro-wave radars (carrier frequency of less than 30 GHz), it is not totally adequate at the MMW band (carrier frequency greater than 30 GHz). Since the wavelength in the MMW band is shorter than in the microwave band, individual scatterers sometimes appear larger in RCS. The result is that a composite of clutter scatterers more often have large RCS values in the MMW band than at microwaves. Statistically, this fact manifests itself in longer "tails" on the clutter probability density function. Thus, the Rayleigh pdf is sometimes not adequate for description of MMW clutter.

Current practice tends to support the Rayleigh pdf for intracell temporal fluctuations for many clutter types, and either the log-normal or the Weibull pdf family for many *intercell* clutter types. The intracell pdf model is applicable to stationary (nonscanning) radar systems, while the intercell pdf models apply to scanning radar systems. In practice, almost all deployed radar systems are of the antenna-beam-scanning type. However, if the pulse integration period is very short relative to the beam motion, the Rayleigh clutter pdf model may still be appropriate.

The likelihood of a large RCS scatterer being present in the illuminated clutter is greater for scanning systems, and the longer pdf tails characteristic of the log-normal and Weibull pdf results better describe the clutter statistics (Currie [14]). The log-normal pdf was already introduced in Section 3.3.3 as a model for ground vehicles in the MMW band. The Weibull pdf family is mathematically described as follows:

$$p(x) = \frac{ax^{(a-1)}}{\alpha} \exp\left[-\frac{x^a}{\alpha}\right] \qquad (3.58)$$

where

$x =$ the clutter variate (RCS),
$a =$ the spread parameter or shape factor of the distribution,
$\alpha =$ the scale factor.

The mean value of x is given by $x = \alpha^{1/a}\Gamma(1 + 1/a)$, where Γ is the Gamma function. The median of x is given by $m_x = \alpha \ln 2$. When a is unity, the Weibull distribution is identical to the exponential pdf. As a increases in value beyond unity the tail of the pdf becomes longer. As compared with the log-normal distribution, the Weibull pdf has a moderately long tail. The Rayleigh pdf has a shorter tail, and the log-normal pdf has a longer tail than the Weibull. Thus, the log-normal pdf represents the worst case.

Clutter Autocorrelation Functions and Power Density Spectra

Autocorrelation functions of clutter returns can be calculated utilizing the clutter's temporal radar return. For ground clutter, the movement of vegatation with the breeze or wind provides the clutter RCS variation with time. In sea clutter, the wave motion, which is due primarily to the wind and tides, causes the temporal variation of the clutter RCS return. With rain clutter, the movement of the falling raindrops, modulated by wind currents, causes the clutter fluctuations. In addition to these factors, movement of the radar platform may also cause temporal variations due to changing viewing angle, the Doppler effect, or both. In Section 3.4.3, the additional factor of carrier frequency stepping from pulse-to-pulse may also cause clutter RCS fluctuations.

The time variations of the clutter RCS can be characterized by either the autocorrelation function $R(\tau)$ or, equivalently, the *power density spectrum* (PDS). Methods exist whereby $R(\tau)$ or PDS can be estimated by carefully utilizing measured radar returns from various clutter types.

Convenient mathematical models are used to describe the measured variations. For autocorrelation functions, the exponential function is often used:

$$R(\tau) = A \exp{(-\tau/\delta_0)} \tag{3.59}$$

where

τ = the lag time,
A = the peak power in the reflected power,
δ_0 = the decay constant.

In the frequency domain, the corresponding power density spectrum is given by

$$W(f) = \frac{W_0}{1 + \left(\dfrac{f}{f_c}\right)^2} \tag{3.60}$$

where

f = frequency in Hz,
f_c = the "cut-off" frequency or a measure of bandwidth,
W_0 = the peak value of PDS.

Equations (3.59) and (3.60) are Fourier transform pairs. An extension of this PDS model is as follows:

$$W(f) = \frac{W_0}{1 + \left(\dfrac{f}{f_c}\right)^n} \tag{3.61}$$

where n is 1 or greater and is chosen to give the best match to measured data. Some clutter data measurements have a PDS that is best modeled with an n of 2.5 or 3.0 (see Chapter 5).

Another PDS model in use is the Gaussian model, which is described as

$$W(f) = W_0 \exp\left[-f^2/2\ \sigma_f^2\right] \tag{3.62}$$

where

    f = frequency in Hz,
    σ_f = the "one-standard-deviation" frequency or a measure of bandwidth,
    W_0 = peak PDS value.

The autocorrelation function corresponding to the Gaussian PDS model is also Gaussian in form.

In some cases, the clutter autocorrelation function or PDS may not match any of these conventional models. The use of digital versions of the autocorrelation function and the Fourier transform allows accurate calculation of nonconventional correlation factors or power density spectra.

Autocorrelation lag times that are typical of rain, ground, and sea clutter fall in the mid-to-low millisecond range. More detailed information on actual measured values are presented in Chapter 5.

3.4.2 Single Pulse Detection

In determining single pulse probability of detection, the appropriate threshold level must first be selected. This is accomplished by finding the threshold setting (voltage level) that provides the desired probability of false alarm. In general, the probability of false alarm, P_{FA}, is given by

$$P_{FA} = \int_{V_T}^{\infty} p(x)\ dx \tag{3.63}$$

where

    x = the clutter variate,

$p(x)$ = the clutter pdf,
V_T = the voltage threshold.

For some of the clutter pdf's just described, this integral may be very difficult to solve in closed form. In those cases, either an infinite series or a computerized solution may be found. However, the control of the error in this calculation is very important because the P_{FA} is typically quite small (e.g., 10^{-6}).

The calculation of the probability of detection, P_D, can be even more difficult than P_{FA}. The first step is to determine the pdf of the target-plus-clutter radar return. This is sometimes not easy in closed form. For example, suppose the target model is log-normal and the clutter model is also log-normal. What is the target-plus-clutter pdf model? In order to determine the answer, the statistical relationship between the target and clutter returns must be known or assumed. Usually the returns are assumed to be statistically uncorrelated. Because both the target and clutter radar returns are vector quantities (each described by amplitude and phase), the target-plus-clutter return (reflected electric fields) is a vector sum of the individual target and clutter electric field vectors (voltages). The target-plus-clutter voltage in the radar receiver would be described vectorally by

$$\overline{V}_{T+c} = \overline{V}_T + \overline{V}_c \tag{3.64}$$

where

$$\begin{aligned} \overline{V}_T &= k_T \overline{E}_T \\ \overline{V}_c &= k_c \overline{E}_c \end{aligned} \tag{3.65}$$

and where

$\overline{E}_T$ = the electric field reflected from the target,
$\overline{E}_c$ = the electric field reflected from the clutter,
k_T and k_c are constants of proportionality.

If the target and clutter returns are assumed to be statistically uncorrelated, then a good approximation to the target-plus-clutter voltage magnitude is

$$|\overline{V}_{T+c}|^2 = |\overline{V}_T|^2 + |\overline{V}_c|^2 + 2|\overline{V}_T| |\overline{V}_c| \cos\phi \tag{3.66}$$

where ϕ is a random variable, uniformly distributed between 0 and 2π.

Finding the pdf of $|V_{T+c}|$ or $|V_{T+c}|^2$ is difficult for most of the pdf types described in this section. A computerized Monte Carlo technique will generally be required to obtain the pdf. As with the calculation of P_{FA}, the error in the P_D calculation will be important.

Although log-normal target and clutter models have been cited as an example, the proposed techniques for calculating the P_{FA} and P_D also apply to other target and clutter models.

If the radar system is operating at a range or transmitter power level such that the thermal noise level is comparable to the clutter return level, then the pdf of the target-plus-noise-plus-clutter must be determined to calculate both the P_{FA} threshold level and the P_D.

For single pulse detection, the autocorrelation or PDS is of no importance because only one time instant (one pulse) is involved.

3.4.3 Pulse Integration

When pulse integration is employed, the calculation of P_D and P_{FA} generally becomes even more difficult. If the number of pulses integrated is large and the interference is known or assumed to be uncorrelated from pulse-to-pulse, these calculations are simplified by employing the well known *central limit theorem* [3, 4]. If the number of pulses summed (integrated) is large enough, the resulting target and clutter pdf's become Gaussian. Therefore, the target-plus-noise-plus-clutter pdf is also Gaussian, and the P_{FA} and P_D calculations are easy and straightforward. Unfortunately, some radar systems are limited in the number of pulses integrated, and so the central limit theorem cannot be used.

If, for a limited number, the pulses can be assumed as statistically uncorrelated, then the Monte Carlo technique can be used to obtain the pdf of the integrated (summed) variates. Uncorrelated random numbers can be generated by a computer, the appropriate number of them are summed, and a histogram or estimated pdf of the summed variates is formed. For example, if the number of pulses integrated is N, then N uncorrelated random numbers are summed and one integrated random number is obtained. This is repeated M times and a histogram of M uncorrelated numbers is formed to provide the estimated pdf. Because computer computations are becoming less expensive, this technique is increasingly popular for difficult cases.

Even when the pulses to be integrated are partially correlated, the Monte Carlo technique can sometimes be used. The major challange is to determine how to impose the correct amount of correlation on the N random numbers to be summed.[6]

At MMW frequencies, all types of clutter are correlated over some finite time period if a single transmit carrier is used. Quite often, the pulse integration period is much shorter than the natural decorrelation time of

---

[6]A discussion of this topic is beyond the scope of the present chapter.

the clutter. Thus, the integration does not reduce the clutter RCS distribution or spread about the RCS mean as much as desired. This same situation exists for most target situations. Thus, the probability of detection is not significantly enhanced by pulse-to-pulse integration.

Because of this fact, pulse-to-pulse frequency *stepping,* or *hopping,* is sometimes used in MMW systems to help decorrelate both clutter and target radar return pulses (Finn and Johnson [17], Goldstein [18]). If the frequency step size is large enough, the returning pulses from either clutter or targets are decorrelated and pulse integration enhances the detection of targets in clutter of various types. The changing of frequency causes the phase relationship between individual scatterers in clutter or on a target to change from pulse-to-pulse, thus decorrelating the radar returns. If a CW radar is utilized, a similar decorrelation of target and clutter can be achieved by sweeping the frequency over a finite bandwidth. This type of system is usually called a *frequency modulated CW* (FMCW) radar. The hardware implications of using frequency stepping or sweeping instead of a single frequency carrier are explained in Chapter 5.

3.5 AUTOMATIC DETECTION

The rate of information production in a typical MMW radar signal is much greater than can be utilized by a human operator. An operator can only accept information at the rate of 10 to 20 samples per second, but the data rate out of a radar system can be many orders of magnitude greater than this. Thus, there is a tremendous mismatch between the information content of a radar and the information-handling capability of a human operator.

When radar signal detection is performed by electronic circuitry without the intervention of a human operator, the process is called *automatic detection.* The main reason for using automatic detection is to overcome the human limitations of the operator and to aid the operator in interpreting the radar output information.

One of the important elements of automatic detection is the *constant false alarm rate* (CFAR) processor. Its function is to maintain a *constant* and *low* rate of false alarms, and thus reduce the amount of information that must be considered by the operator.

In MMW radar systems, which are often clutter limited,[7] the false alarm rate can be quite high unless the CFAR processor is carefully designed. The CFAR processor estimates the interference levels in the vicinity of the range-azimuth cell being tested for the presence of a target.

---

[7]Clutter signals comprise the predominant source of interference as compared with receiver noise.

On the basis of the estimated interference level, the detection threshold is set to provide a prescribed false alarm probability or rate. Thus, the CFAR processor is an adaptive device, which adjusts the threshold level according to nearby interference levels. The CFAR processor is a vital part of a radar system.

Since there are a great variety of clutter types, each of which has its special characteristics, it is difficult to design one CFAR processor that is suitable for all clutter situations. Therefore, there has been a proliferation of CFAR processor designs. Each design tends to excell in some type of interference situation, but has weaknesses in others. The author will briefly review some of the major CFAR types with their strengths and weaknesses, but an exhaustive treatment is beyond the scope of this chapter. More information may be obtained by referring to the many references on this subject, listed at the end of the chapter [16–26].

There are generally two classes of CFAR processors: *parametric* and *distribution-free*. A parametric CFAR is one that is specifically designed for an assumed interference probability distribution and which only performs well with said type of interference. However, a nonparametric or distribution-free CFAR processor, which is not designed for a specific probability distribution, works fairly well for a wide variety of interference probability distributions. If the interference probability distribution is known and not expected to change, then a parametric CFAR is best. However, if this is not the case, then a distribution-free CFAR may be the best choice.

3.5.1 Parametric CFAR Processors

Most parametric CFAR processors can be categorized in several different ways. Parametric processors can be single- or multiple-parameter processors, and they can also be range-only, angle-only, or area CFARs. For example, a one-parameter area CFAR processor would be one that utilizes radar information found in range-azimuth cells in *both* the range and angle extent in the vicinity of the cell being tested for the presence of a target. This situation is illustrated in Figure 3.15(a), where the cross-hatched cells are those used in estimating one statistical parameter (e.g., the RCS mean). This is accomplished by simply summing the RCS values in each range-azimuth cell in the cross-hatched area, then dividing by the number of cells. This represents the average interference RCS level surrounding the cell being tested for a target. For each range-azimuth cell being tested, this calculation is repeated by utilizing the N cells surrounding the test cell.

The one-parameter CFAR is useful for interference characterized by a one-parameter probability density function, such as the Rayleigh pdf. If the interference is described by a two-parameter pdf, such as log-normal or Weibull clutter, then both the mean and variance of the interference must be estimated by using the RCS values in range-azimuth cells surrounding the test cell.

Other variations in the type of parametric processors are the range-only and angle-only CFARs. For the first CFAR type, only range-azimuth cells in the range dimension are utilized in the mean or the mean and variance estimates. Likewise, for the angle-only CFAR, only the range-azimuth cells in the angular or azimuthal direction are used. These two CFAR situations are illustrated in Figure 3.15(b and c).

In some MMW radar applications, there may be too few range-azimuth cells in the range dimension available (perhaps only one) to provide a good parameter estimate and only cells in the angle dimension can be used.

The accuracy of the interference RCS mean and variance estimates will depend upon the number of range-azimuth cells used. Thus, the tendency will be to use as many as possible. However, the more cells that are used in the estimate, the more likely it is that the interference included in the estimate will be spatially nonhomogeneous,[8] and thus violate the underlying assumption made in applying parametric CFAR processors.

An example of a one-parameter parametric CFAR processor is shown in Figure 3.16. This CFAR concept is taken from [17]. As indicated in this figure, only one parameter is estimated as a function of range and azimuth angle.[9] The one-parameter CFAR is usually used where the interference voltage level is assumed to have a Rayleigh probability amplitude distribution or, equivalently, the interference power is assumed to have an exponential probability amplitude distribution. If the envelope detector used in this example is a square-law detector, then its output will be proportional to the interference power level (clutter RCS). The mean interference level in the N range-azimuth cells surrounding the target test cell will be given by

---

[8]The probability density function describing the RCS values in the CFAR cells will change with spatial position.

[9]In [17], the two CFAR variables used are range and Doppler. However, Doppler is not generally used in MMW systems for CFAR purposes. Therefore, azimuth angle is used instead in Figure 3.16.

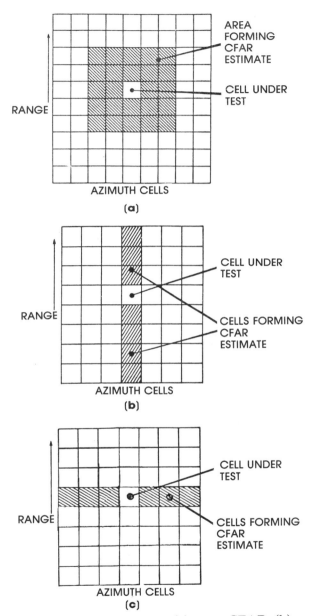

Fig. 3.15 Range-azimuth cells used for (a) area CFAR (b) range-only CFAR, and (c) azimuth angle-only CFAR.

$$\hat{m}_c = \frac{1}{N} \sum_{i=1}^{N} V_i^2 \qquad\qquad (3.67)$$

where

V_i = the envelope amplitude (voltage) in the ith range-azimuth cell,
N = the number of range-azimuth cells,
$\hat{m}_c$ = the estimated mean interference level.

As indicated in Figure 3.16, the threshold level estimate is linearly proportional to the mean interference level as follows:

$$\hat{T} = k\hat{m}_c \qquad\qquad (3.68)$$

where the selected value of k determines the average probability of false alarm that is achieved.

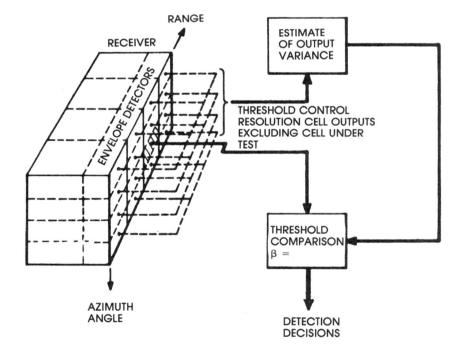

Fig. 3.16 Automatic threshold control for constant false alarm rate (CFAR) processor (From Finn and Johnson [17]).

The signal in the test cell, which is proportional to the RCS is declared to be from a target if the signal level is greater than the threshold level, and the signal is declared to be interference (clutter) if it is less than the threshold level.

This CFAR processor can be converted to a two-parameter processor simply by additionally calculating the variance of the *interference power* in the CFAR cells as follows:

$$\hat{SD}_c^2 = \frac{1}{N} \sum_{i=1}^{N} (V_i^2 - \hat{m}_c)^2 \tag{3.69}$$

and, using both the mean and standard deviation calculations given by (3.67) and (3.69), it is *estimating* the threshold level by

$$\hat{T} = \hat{m}_c + k\hat{SD}_c \tag{3.70}$$

where the selected value of k determines the average probability of false alarm that is achieved.

Since the threshold level is an estimate, not the exact level required to obtain a given probability of false alarm (P_{FA}), we would expect the detection performance to be a function of how well the needed threshold setting has been estimated. The accuracy of the threshold setting is dependent on the number of range-azimuth cells used in the estimate. In (Finn and Johnson [17]), the P_D *versus* SNR is calculated for a Swerling case 3 target in Rayleigh voltage interference (exponential power interference), with the number of CFAR cells, N, as a parameter. The resulting detection performance curves are shown in Figures 3.17 and 3.18 for two different probabilities of false alarm.

As indicated in these figures, the fewer CFAR cells that are used, the larger the SNR is required to achieve a given P_D. For example, with four CFAR cells, a SNR of about 32 dB is required to yield a P_D of 0.5 and a P_{FA} of 10^{-10}. If a much larger number of CFAR cells are used (≥ 100), the required SNR is only about 15 dB—a difference of 17 dB! If this difference is defined as a CFAR processing *loss*, then the CFAR loss *versus* N and P_{FA} for a P_D of 0.5 can be determined as that illustrated in Figure 3.19.

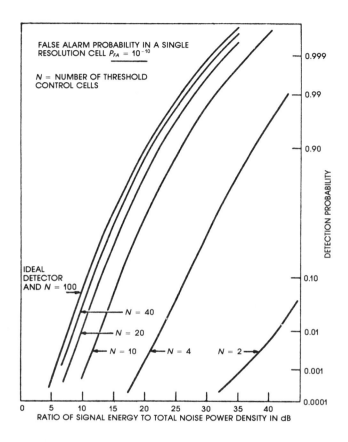

Fig. 3.17 Detection probability for Swerling case 3 target, $P_{FA} = 10^{-10}$, *versus* SNR with number of CFAR cells as a parameter (From Finn and Johnson [17]).

As expected, as N increases, the CFAR loss decreases and better detection performance is obtained. The performance presented is based on the assumption that the interference statistics are the same in each cell being used to estimate the interference level. This assumption is violated if the mean interference level or the probability density function, is different in some of the CFAR cells; therefore, the CFAR processor does not perform well. For the mean level CFAR processor illustrated in Figure 3.16, these perturbations result in a variable false alarm rate, and thus the processor is no longer CFAR!

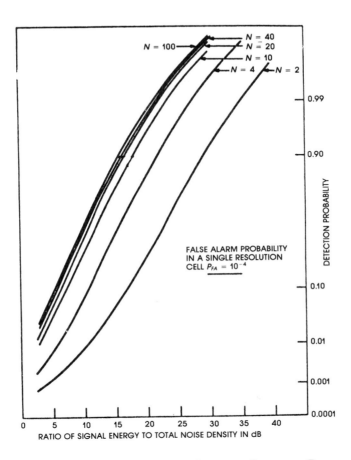

Fig. 3.18 Detection probability for Swerling case 3 target, $P_{FA} = 10^{-4}$, *versus* SNR with number of CFAR cells as a parameter (From Finn and Johnson [17]).

False alarm regulation for log-normal and Weibull clutter is discussed in (Goldstein [18]). The optimum threshold setting in the presence of log-normal clutter is found to be

$$T = \hat{m}_c + k\hat{S}D_c \qquad (3.71)$$

and, in this case,

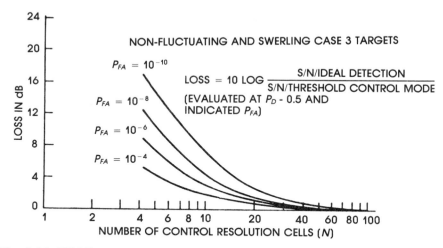

Fig. 3.19 CFAR processor loss *versus* number of control resolution (CFAR) cells (From Finn and Johnson [17]).

$$\hat{m}_c = \frac{1}{N} \sum_{i=1}^{N} \ln (V_i) \tag{3.72}$$

$$\hat{SD}_c^2 = \frac{1}{N} \sum_{i=1}^{N} [\ln(V_i) - \hat{m}_c]^2 \tag{3.73}$$

where ln ( ) is the natural logarithm.

The detection performance of a Rayleigh target model in log-normal interference is derived in [18] and will not be repeated here because it is fairly complicated and lengthy. The CFAR loss associated with the threshold given by (3.71), (3.72), and (3.73) is also given in [18].

The expression for the threshold estimate given by (3.71), (3.72), and (3.73) also provides for CFAR operation with Weibull-distributed clutter. However, in the Weibull case, this threshold estimate is not necessarily optimum.

Variations of the parametric CFAR processor shown in Figure 3.16 have also been proposed to overcome some of the difficulties that arise when the interference statistics being estimated vary from cell to cell, or when multiple targets occur a few resolution cells apart.

In general, the CFAR processor shown in Figure 3.16 provides an incorrect estimate of threshold level when either of these conditions is encountered. Reference [19] proposes modifications to the cell averaging CFAR to help reduce or eliminate these unwanted effects. The CFAR processor shown in Figure 3.20 is an example of one such modified CFAR processor. The selection logic shown in the figure allows various combinations of the estimated clutter levels "in front of" and "behind" the cell under test such as the *sum*, the *maximum*, and the *minimum*. A discussion of the detection performance of this CFAR processor is beyond the scope of this chapter and is not presented. However, it is discussed in some detail in [19].

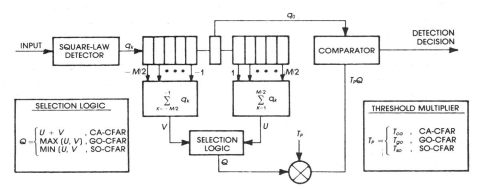

Fig. 3.20 A multiple-cell averaging CFAR processor (From Weiss [19] © 1982 IEEE).

3.5.2 Nonparametric or Distribution-Free CFAR Processors

Distribution-free CFAR implies that the probability of false alarm is maintained at a constant value, independently of the statistical distribution of the interference, even if drastic and unknown changes in the underlying distribution of the interference occur. There have been a number of distribution-free CFAR processors proposed [20-26]. However, the detection performance of each is generally decreased as compared with that of the parametric CFAR processors, and some are not entirely distribution-free, but tend to be so for certain classes of distributions.

A practical distribution-free CFAR processor for a multiple-range-bin radar, referred to as the *modified sign test processor*, is shown in Figure 3.21 (Dillard and Antoniak [21]). As indicated, the range samples out of an envelope detector are denoted as

$$X_i, X_{i+1}, X_{i+2}, \ldots, X_{i+j} \tag{3.74}$$

where j is the number of range-gate samples used in the CFAR process. In this figure, the ith sample is the radar return from the "test" cell. All the other j samples are from adjacent range cells, and are used to represent the surrounding interference. The CFAR procedure is to compare the ith range sample (the test cell sample) to each of the other j range samples. If

$$X_i, > X_i + j; j = 1, 2, \ldots, j \qquad (3.75)$$

for all values of j, then a one is assigned to that test cell range bin; otherwise, a zero is assigned. This process is repeated for each range cell. After *each* transmitted pulse, the output of this CFAR process is a series of ones and zeros, a one or zero for each range bin. After N pulses are transmitted, the resulting binary numbers for *each* range bin are summed (integrated) and compared to a threshold. A detection decision is made for each range bin. This CFAR and integration process can be repeated for each subsequent transmitted pulse[10] as long as the radar antenna illumination dwells on a potential target.

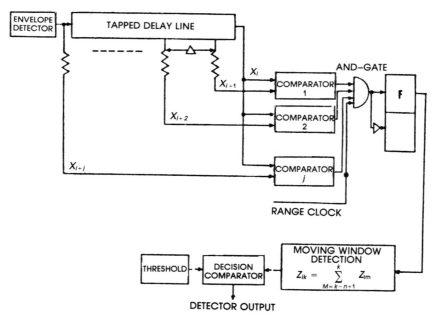

Fig. 3.21 A distribution-free CFAR processor for a multiple-range-bin radar (From Dilliard and Antoniak [21], © 1970 IEEE).

---

[10]This is called moving window integration.

Detection probability as a function of the number of samples used in the CFAR processor is shown in Figure 3.22 for three target cases, a 10^{-6} probability of false alarm, and 50 pulses integrated per range bin. Note that about 10 to 15 CFAR cells provide the maximum probability of detection that can be achieved.

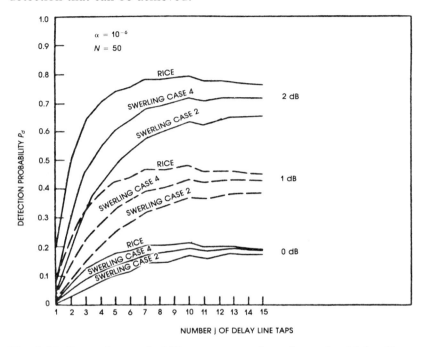

Fig. 3.22 Detection probability *versus* number of samples (delay line taps) used in a distribution-free processor (From Dillard and Antoniak [21], © 1970 IEEE).

In Figure 3.23, the detection probability *versus* single-pulse SNR is shown for three different types of integration. The video and binary integration procedures referenced in this figure are parametric-dependent processes. For all three processes (including the modified sign test), the target model assumed was nonfluctuating (Ricean), the probability of false alarm is 10^{-6}, and two values of N (number of pulses integrated) are assumed.

Note that the distribution-free CFAR process (modified sign test) requires about 1 or 2 dB more in SNR than do the binary and video integration schemes, respectively. This loss in SNR is the price paid for a distribution-free CFAR process as compared with two common, distribution-dependent, CFAR integration processes.

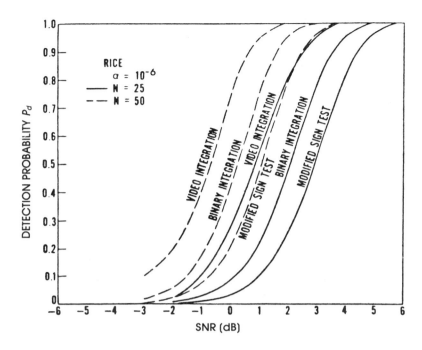

Fig. 3.23 Detection probability as a function of SNR, comparing the distribution-free processor (modified sign test) with parametric processors for linear envelope detector and nonfluctuating signal (From Dillard and Antoniak [21], © 1970 IEEE).

There are certain problems encountered with the modified sign test CFAR processor in some applications. One is the problem of small-signal suppression. With the processor shown in Figure 3.21 the occurance of a strong signal return within the j range bins along with a weak signal (possibly a target) will lower the probability that the weak signal is detected. Another problem with this CFAR processor is its inability to distinguish range and pulse-to-pulse correlated interference from true signals (targets). However, there are techniques discussed earlier in this chapter for decorrelating many types of interference.

3.6 SUMMARY

In this chapter, the following topics have been discussed:
- radar range equation,
- detection of targets in receiver noise,

- detection of targets in clutter,
- mathematical models of targets,
- types of signal detectors,
- statistical and temporal descriptions of radar clutter,
- pulse integration,
- constant false alarm rate (CFAR) processing.

We have discussed various mathematical models used in describing the detection of targets in thermally generated noise and in clutter. Basically, the detection process involves random processes. These processes are described by both probability density functions and by autocorrelation or power density functions. Once we have adequately described the target-plus-noise-plus-clutter signals in these terms, we can determine the effects of various types of detection and intergration processes.

The success with which target signals in noise and clutter are sensed or detected is measured by two parameters: the probability of detection, P_D, and the probability of false alarm, P_{FA}. Knowing the required P_D and P_{FA}, the target, clutter and noise statistics and correlation properties, the needed *signal-to-clutter ratio* (SCR) and *signal-to-noise ratio* (SNR) can be calculated. Then, using the radar range equations described in the first part of this chapter, we can determine the range at which the target can be detected.

Of course, the detection performance determination can be reversed. Given the required detection range and the other necessary parameters, we can obtain the necessary SCR and SNR, and determine the P_D for a specified P_{FA}.

REFERENCES

1. D.K. Barton, *Radar Systems Analysis*, Artech House, Dedham, MA, 1973.
2. M.I. Skolnik, *Introduction to Radar Systems*, McGraw-Hill, New York, 1980.
3. W.B. Davenport, Jr., and W.L. Root, *An Introduction to the Theory of Random Signals and Noise*, McGraw-Hill, New York, 1958.
4. A. Papoulis, *Probability, Random Variables, and Stochastic Processes*, McGraw-Hill, New York, 1965.
5. Marcum, J.I., "A Statistical Theory of Target Detection by Pulsed Radar," Rand Corp. Research Memo. RM-754, December 1947. Reprinted in *Trans. IRE*, Vol. IT-6, No. 2, April 1960.
6. J.I. Marcum, "Statistical Theory of Target Detection by Pulsed Radar: Mathematical Appendix," Rand Corp. Research Memo. RM-753, July 1948. Reprinted in *Trans. IRE*, Vol. IT-6, No. 2, April 1960.

7. S.A. Hovanessian, *Radar System Design and Analysis*, Artech House, Dedham, MA, 1984.

8. J.W. Taylor and J. Mattern, "Receivers, Chapter 5 in M.I. Skolnik, ed., *Radar Handbook*, McGraw-Hill, New York, 1970.

9. D.P. Meyer and H.A. Mayer, *Radar Target Detection*, Academic Press, New York, 1973.

10. P. Swerling, "Probability of Detection for Fluctuating Targets," Rand Corp. Research Memo. RM-1217, March 1954.

11. M. Schwartz, "Effects of Signal Fluctuation on the Detection of Pulsed Signals in Noise," *Trans. IRE*, Vol. IT-2, No. 2, June 1956.

12. T.H. Einstein, "Effect of Frequency Averaging on Fixed-Target Detection Performance Using an Amplitude CFAR Threshold Detector," Project Report TT-58, MIT, Lincoln Laboratory, Lexington, MA, 16 December 1982.

13. W.W. Hines and D.C. Montgomery, *Probability and Statistics in Engineering and Management Science*, Ronald Press, 1980, Chapter 7.

14. N.C. Currie, ed., *Techniques of Radar Reflectivity Measurement*, Artech House, Dedham, MA, 1984.

15. R.D. Hayes, "95 GHz Pulsed Radar Returns from Trees," *Proceedings of IEEE Eascon*, October 1980.

16. T.H. Einstein, "Effect of Frequency-Averaging on Estimation of Clutter Statistics Used in Setting CFAR Decision Thresholds," Project Report TT-60, MIT, Lincoln Laboratory, Lexington, MA, 9 November 1982.

17. H.M. Finn and R.S. Johnson, "Adaptive Detection Made With Threshold Control as a Function of Spatially Sampled Clutter-Level Estimates," *RCS Review*, September 1968, pp. 415–464.

18. G.B. Goldstein, "False-Alarm Regulation in Log-Normal and Weibull Clutter," *IEEE Transactions on Aerospace and Electronic Systems*, Vol. AES-9, No. 1, January 1973, pp. 84–92.

19. M. Weiss, "Analysis of Some Modified Cell-Averaging CFAR Processors in Multiple-Target Situations," *IEEE Transactions on Aerospace and Electronic Systems*, Vol. AES-18, No. 1, January 1982, pp. 102–114.

20. R.E. Lefferts, "Adaptive False Alarm Regulation in Double Threshold Radar Detection," *IEEE Transactions on Aerospace and Electronic Systems*, Vol. AES-17, No. 5, September 1981, pp. 666–675.

21. G.M. Dillard and C.E. Antoniak, "A Practical Distribution-Free Detection Procedure for Multiple-Range-Bin Radars," *IEEE Transactions on Aerospace and Electronic Systems*, Vol. AES-6, No. 5, September 1970, pp. 629–636.

22. G.V. Trunk, B.H. Cantrell, and F.O. Queen, "Modified Generalized Sign Test Processor for 2-D Radar," *IEEE Transactions on Aerospace and Electronics Systems*, Vol. AES-10, No. 5, September 1974, pp. 574–582.

23. R. Nitzberg, "Application of Invariant Hypothesis Testing Techniques to Signal Processing," Ph.D. Dissertation, Syracuse University, Syracuse, NY, 1970.

24. R. Nitzberg, "Constant-False-Alarm Rate Processors for Locally Nonstationary Clutter," *IEEE Transactions on Aerospace and Electronic Systems*, Vol. AES-9, pp. 399–405, May 1973.

25. L.E. Vogel, *et al.,* "An Examination of Radar Signal Processing Via Non-Parametric Techniques," *Proceedings of the 1975 IEEE International Radar Conference*, Washington, DC, April 1975.

26. L.D. Davisson, *et al.*, "The Effects of Dependence on Nonparametric Detection," *IEEE Transaction on Information Theory*, Vol. IT-16, No. 1, January 1970.

27. H.A. Corriher, *et al.*, "Elements of Radar Clutter," Chapter XVII of *Principles of Modern Radar*, H.A. Ecker, ed., Georgia Institute of Technology, Atlanta, GA, 1972.

Chapter 4
MMW Propagation Phenomena

R.N. Trebits

Georgia Institute of Technology
Atlanta, Georgia

4.1 INTRODUCTION

The operational performance of a radar system during surveillance, tracking, and target detection or discrimination ultimately is a direct function of the received signal-to-interference power ratio. When backscatter from radar clutter is not a major factor, then the interference is simply that of receiver noise. For a given radar receiver configuration, any factors external to the radar that affect the received radar signal power level will then likewise affect the radar's performance.

This chapter describes the atmospheric effects on electromagnetic propagation in the MMW region, including (1) attenuation due to atmospheric gases, hydrometeors, and particulates; (2) refraction due to atmospheric density nonhomogeneity with respect to altitude. In addition, near-ground propagation effects are discussed, including foliage penetration and multipath interference.

Attenuation of MMW energy along the two-way propagation path between a radar and a target cell naturally results in a degraded signal-to-noise ratio, and therefore degraded radar performance. However, multipath interference can have the effect of actually enhancing the received signal-to-noise ratio under certain geometric and reflectivity conditions (or greatly attenuating the signal at a different time). Furthermore, refractive effects can extend the apparent radar range beyond a straight line of sight by causing the electromagnetic energy to follow a curved path.

131

4.2 ATTENUATION EFFECTS

The absorption of electromagnetic energy by the intervening medium between the radar transmitter or receiver and the desired target degrades all aspects of radar performance by decreasing the received signal-to-noise level. This section discusses energy absorption due to (1) a clear atmosphere; (2) hydrometeors such as rain, fog, and snow; (3) particulate matter such as smoke and dust; (4) tree foliage. Empirical or analytical attenuation models will be included for many of these cases if justified by collaborating experimental data. Tabular or graphical attenuation data sets are included from representative experiments.

4.2.1 Clear Atmosphere Attenuation

The so-called standard dry (no water vapor) atmosphere consists of approximately 78.088% nitrogen (N_2), 20.949% oxygen (O_2), and 0.93% argon (Ar). The remaining 0.03% of the standard atmosphere includes fixed percentages of carbon dioxide (CO_2), neon (Ne), helium (He), methane (CH_4), krypton (Kr), nitrous oxide (NO), carbon monoxide (CO), xenon (Xe), and hydrogen (H_2). Variable amounts of ozone (O_3) and water vapor (H_2O) are included in a typical atmospheric composition, with the percentage of concentration dependent on local environmental conditions.

The amount of water vapor in the atmosphere is typically measured in terms of density (g/m^3). The popular measure, however, has historically been in terms of relative humidity, a measurement of the percentage of actual water vapor in the air relative to the saturation level. Water vapor density may be related to relative humidity values, pressures, and temperatures, straightforwardly as shown in Figure 4.1. Water vapor density varies with time of year, altitude, and geographic location. Figure 4.2 represents plots of water vapor density *versus* altitude for the 1962 US Standard Atmosphere and two cases representative of extreme conditions (Kulpa and Brown [1]). A frequently used value for water vapor density is 7.5 g/m^3. Figure 4.3 shows atmospheric absorption values *versus* frequency and relative humidity, calculated by R. Rogers, using equations developed by R.D. Hayes at Georgia Tech (Bohlander and McMillan [2]).

Atmospheric attenuation at MMW frequencies is dominated by molecular absorption by the water vapor and oxygen components of the atmosphere. Ozone has been demonstrated to have absorption effects in the MMW region also, but this effect will only arise in high-altitude situations. For conventional, tactical radar applications (i.e., air-to-ground

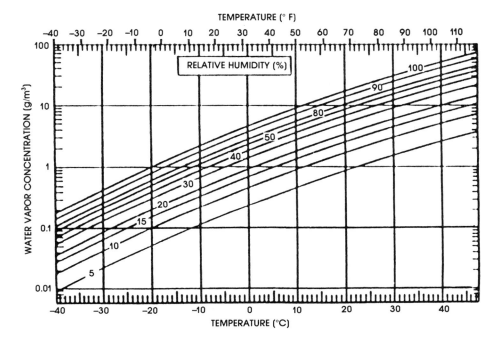

Fig. 4.1 Plots of H_2O concentration *versus* temperature for various percentages of relative humidity (from Kulpa and Brown [1]).

and air-to-air seekers), electromagnetic energy absorption due to ozone will not be a significant factor. Ozone effects would be relevant, however, for an application such as air-to-satellite detection and tracking.

The water molecule, with an oxygen atom nonsymmetrically located between two hydrogen atoms, comprises a rotor having three unequal moments of inertia. The water molecule is polar and has a strong dipole electric moment. The nature of this electromagnetic absorption is due to transitions between water molecule rotational states, in closely spaced absorption resonances, in three specific regions of the MMW spectrum: around 22, 183, and 323 GHz [2].

Free oxygen is a diatomic molecule, with no electric dipole moment. Magnetic dipole transfers are responsible for oxygen absorption in the MMW region. Several resonance lines occur due to rotational state transitions, but these are weak and they are dominated by the much stronger water vapor resonances, which are nearby in frequency. The major oxygen absorption resonances around 60 and 118 GHz (single line) are due to

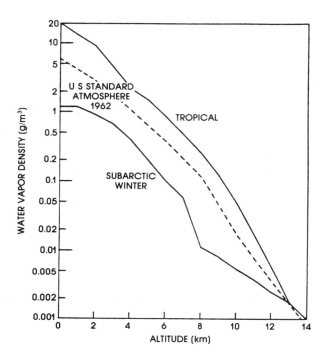

Fig. 4.2 Plots of water vapor density *versus* altitude for three atmospheric models (from Kulpa and Brown [1]).

transitions between triplet components of the oxygen rotational ground energy state (Wiltse [3]). Figure 4.4 shows attenuation values for the atmosphere as a function of radar frequency, highlighting locations of water vapor and oxygen resonance peaks [3].

The attenuation theory of Van Vleck [4] is based on the approximations of collision broadening theory. Oxygen attenuation, γ_1, in decibels per kilometer, at $T = 293K$ and standard atmospheric pressure is predicted by

$$\gamma_1 = \frac{0.34}{\lambda} \left[\frac{\Delta\nu_1}{1/\lambda^2 + \Delta\nu_1^2} + \frac{\Delta\nu_2}{(2 + 1/\lambda)^2 + \Delta\nu_2^2} \right.$$

$$\left. + \frac{\Delta\nu_2}{(2 - 1/\lambda)^2 + \Delta\nu_2^2} \right] \tag{4.1}$$

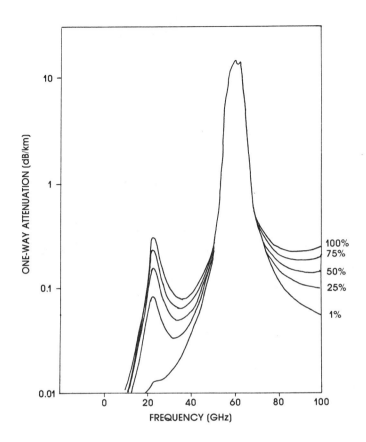

Fig. 4.3 Atmospheric absorption as a function of relative humidity (from Bohlander and McMillan [2], © 1985 IEEE).

where

λ = radar wavelength;

$\Delta v_1, \Delta v_2$ = line-width factors (from Bean and Dutton [5]).

The analogous analytic expression for water vapor attenuation γ_2, for $T = 293$ K and 22 GHz (1.35 cm wavelength line) is

$$\gamma_2 = \frac{3.5 \, \rho \times 10^{-3}}{\lambda^2} \left[\frac{\Delta v_3}{(1/\lambda - 1/1.35)^2 + \Delta v_3^2} \right.$$

$$\left. + \frac{\Delta v_3}{(1/\lambda + 1/1.35)^2 + \Delta v_3^2} \right] \tag{4.2}$$

where

ρ = absolute humidity (water vapor density, gm/m^3);

Δv_3 = line-width factor at 22 GHz, from Bean and Dutton [5].

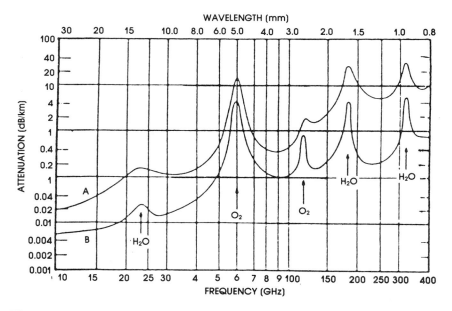

Fig. 4.4 Average atmospheric absorption of millimeter waves. A—Sea level: $T = 20°C$, $P = 760$ mm, $P_{H2O} = 7.5$ g/m^3; B—4 km: $T = 0°C$: $P_{H2O} = 1$ g/m^3 (from Wiltse [3], © 1981 Academic Press).

The water vapor attenuation γ_3, for $T = 293$ K and frequencies greater than 22 GHz, is described by

$$\gamma_3 = \frac{0.05 \, \rho \Delta v_4}{\lambda^2} \tag{4.3}$$

where Δv_4 is the effective line width for frequencies greater than 22 GHz [5]. Note that γ_3 has been increased by a factor of 4 over the original formulation of Van Vleck, a modification justified by fitting this analytical expression to experimentally recorded attenuation data (Becka and Autler [6]).

The $\Delta\nu$ values used in both of these oxygen and water vapor attenuation expressions have been derived from fitting these expressions to experimentally recorded attenuation data. In addition, other sets of $\Delta\nu$ values proposed by later researchers may be more accurate. Accommodation may also be made to the values for atmospheric pressure and temperature dependencies (Bean, Dutton, and Warner [7]).

4.2.2 Hydrometeor Attenuation

Among the perceived advantages of MMW radar systems over optical sensors is the ability to penetrate more effectively through suspended water in the atmosphere (i.e., through fog, haze, and clouds). This penetration capability is of obvious benefit in areas such as Western Europe, where, in approximately one of three mornings during the fall and winter, visibility due to ground fog is reduced to less than 1 km. Furthermore, the total cloud cover in the North Atlantic exceeds 50% approximately two-thirds of the time (Wicker and Webb [8]).

True all-weather sensor capability must also include sufficient penetration through falling hydrometeors: rain, sleet, hail, and snow. Thus, the attenuation of electromagnetic energy at candidate radar frequencies is of utmost concern to determine effective range in fulfilling sensor mission requirements. At MMW frequencies, signal attenuation due to hydrometeor absorption may be one of the most limiting system performance factor because transmitter sources, especially solid-state ones, tend to be power-limited.

The meteorological conditions commonly called *fog* and *haze* generally represent liquid water suspended in the atmosphere near the ground. Ice fog can form in the range $-30°$ to $-40°C$ for water droplets having radii in the 1 to 10 μm range (Kulpa and Brown [1]). Fog is qualitatively differentiated from haze simply on the basis of one's ability to distinguish objects at a distance. Fog is simply that set of suspended water vapor conditions limiting visibility to 1 km or less. This definition of visibility is defined by Middleton to be the distance at which "large dark objects can be discerned against the horizon by human observers" (Middleton [9]).

Two types of fogs are identified in the literature: advection fogs and radiation fogs. An advection fog forms when warm, moist air overrides a cooler surface. A radiation fog forms when the ground cools after sunset, thereby cooling the adjacent air mass until the air becomes supersaturated with water (Altschuler [10]). Figure 4.5 depicts visibility *versus* liquid water

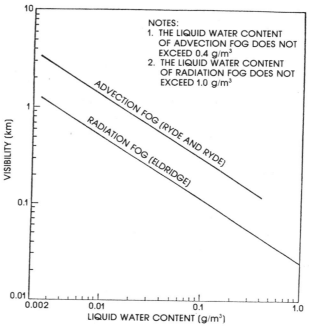

Fig. 4.5 Visibility *versus* liquid water content (from Altschuler, [10] ©
1984 IEEE).

density for advection and radiation fogs. Figure 4.6 depicts MMW attenuation *versus* liquid water density, overlaid with visibility in fog (Eldridge
[11]).

Eldridge's empirical relationship between visibility V in fog and liquid
water density M is

$$V = 0.024 \, M^{-0.65} \tag{4.4}$$

where V is in units of km and M is in units of g/m^3. Eldridge believes this
expression to be valid for radiation fogs, where drop diameters are usually
less than 10 μm. Advection fogs may contain drops with diameters that
exceed 100 μm, and Eldridge recommends for this case that the coefficient
0.024 be replaced by 0.017 [11, 12].

Altschuler's analysis of fog attenuation data at MMW wavelengths
indicates a straightforward expression relating normalized fog attenuation
A, radar wavelength λ, and temperature T:

$$A = -1.347 + 0.372\lambda + \frac{18.0}{\lambda} - 0.022 \, T \tag{4.5}$$

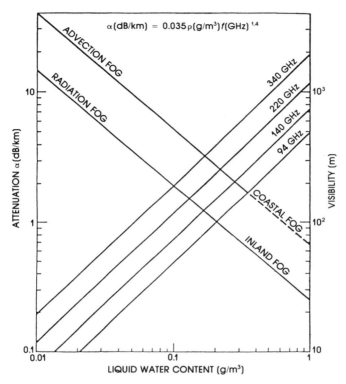

$$\alpha(\mathrm{dB/km}) = 0.035\,\rho(\mathrm{g/m^3})\,f(\mathrm{GHz})^{1.4}$$

Fig. 4.6 MMW attenuation by fog (from Altschuler [10], © 1984 IEEE).

where

A = attenuation, in dB/km/g/m^3;
λ = radar wavelength, in mm;
T = temperature, in °C.

The validity of this relationship should be restricted to a radar wavelength range 3 mm < λ < 3 cm (frequency range 100 GHz > f > 10 GHz) and a temperature range -8°C < T < $+25$°C (Altschuler [10]).

Figure 4.7 shows experimentally recorded attenuation data *versus* visibility at 140 GHz (2.14 mm) [13], while Figure 4.8 shows similar data at 35 GHz (8.57 mm) [14]. Calculated attenuation data *versus* temperature and radar wavelength are shown in Figure 4.9. In this calculation process, the water vapor contribution was linearly extrapolated from absolute humidity without temperature corrections. Fog contributions were calculated in the Rayleigh limit for liquid water at a temperature of 24°C. Thus, fog components may be as much as 50% in error (Kulpa and Brown [1]).

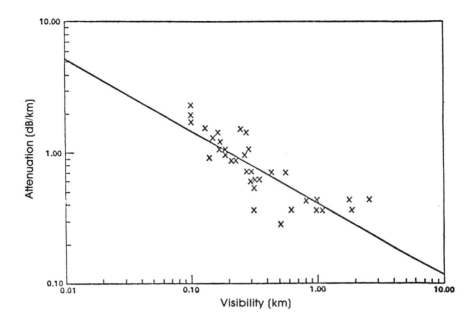

Fig. 4.7 Experimental data at 140 GHz (from Bauer, *et al.* [13]).

Rain Attenuation

Millimeter-wave energy propagating through rain undergoes two processes that result in signal attenuation. The first effect is defined by an absorption cross section Q_a that is equivalent to the power absorbed by the hydrometeor content of the intervening medium when it is radiated by a plane wave of unit power per unit area. The second effect is defined by a scattering cross section Q_s that is equivalent to the power scattered in all directions from the hydrometeor content when it is radiated by the same unit plane wave (Oguchi [15]). The total cross section Q_t is the sum of Q_a and Q_s, and represents the total power removed from the plane wave. Q_t is directly related to the attenuation of the electromagnetic energy.

The signal attenuation analysis is typically treated in the manner of Mie [16]. The absorption cross section Q_a and the scattering cross section Q_s are functions of raindrop diameter D, dielectric constant, and radar frequency. Raindrop diameters, in turn, are represented by distribution functions that depend on type of storm, rain rate, time of year, geographic location, and position within the storm cell. If there are $n(D)dD$ rain drops

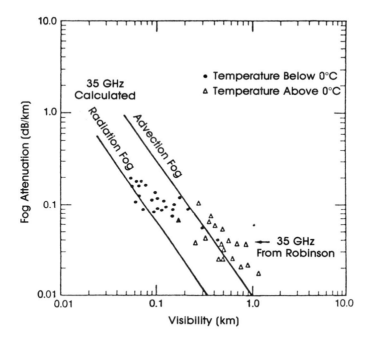

Fig. 4.8 Measured and calculated fog attenuation at 35 GHz (from Robinson [14], © 1955 IEEE).

per unit volume, having diameter between D and $D + dD$, then the total number of drops N per unit volume is given by

$$N = \int_0^\infty n(D)dD \tag{4.6}$$

The absorption cross section Q_a is, then,

$$Q_a = \int_0^\infty n(D)A_a dD \tag{4.7}$$

and the scattering cross section Q_s is

$$Q_s = \int_0^\infty n(D) A_s dD \tag{4.8}$$

where A_a and A_s are the absorption and scattering cross sectional areas, respectively, of each spherical rain drop (Dutton and Steele [17]). The

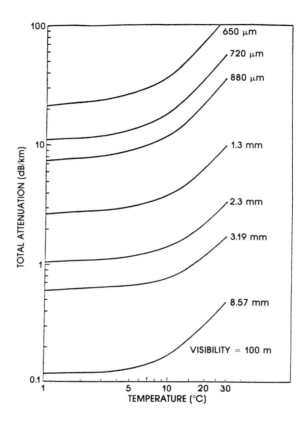

Fig. 4.9 Fog attenuation as a function of temperature and wavelength (from Kulpa and Brown [1]).

attenuation A is related to the total extinction cross section Q_t by the expression:

$$A = 4.343 \times 10^3 \, Q_t \tag{4.9}$$

where A is in units of dB/km. The reader should note, however, that nonspherical rain drops in the medium result in an attenuation value that also depends on the polarization of the electromagnetic signal (Oguchi [15]).

Some further discussion of drop size distributions is in order. The two widely cited distribution models are those of Laws and Parsons [18] and Marshall and Palmer [19]. Laws and Parsons recorded the percent of the total volume reaching the ground contributed by drops of different size ranges. This volume percentage may be written as $m(a)da$, where a is the drop size radius and da is the incremental size interval around the value a. The desired size distribution, however, is $n(a)da$, which may be analytically related to $m(a)da$ through the fall velocity $v(a)$ of the rain drops by

$$n(a)da = \frac{10^3}{4.8\pi} \frac{Rm(a)da}{a^3v(a)} \tag{4.10}$$

where

R = rain rate, in mm/hr;
$v(a)$ = fall velocity, in m/s;
a = drop radius, in mm.

Marshall and Palmer proposed an exponential relationship:

$$n(a)da = N_0 \, e^{-\Lambda a} da \tag{4.11}$$

where

$N_0 = 1.6 \times 10^4 \text{m}^{-3}\text{mm}^{-1}$;
$\Lambda = 8.2R^{-0.21} \text{ mm}^{-1}$;
a = drop radius, in mm;
R = rain rate, in mm/hr.

Figure 4.10 depicts both the drop size distribution proposed by Laws and Parsons and by Marshall and Palmer. These functions characterize typical drop size distributions for both widespread rain (in the lower rain rate range) and for convective rain (in the higher rain rate range) (Oguchi [15]). Table 4.1 lists the Laws and Parsons drop size distributions for various rainfall rates, and Table 4.2 lists the terminal velocity for various size rain drops (Setzer [20]).

A given rainfall rate does not imply a specific drop size distribution, although general trends appear to be true. For example, heavier rainfall rates tend to indicate drop size distributions weighted toward larger drops. Rainfall rates that are equal in value, but occur at different times of the year or at different locations, may have significantly different drop size distributions.

Table 4.3 lists rain attenuation values calculated for several rainfall rates and radar wavelengths, based on a study of drop size distributions. Table 4.4 lists rain attenuation values for measured drop size distributions [21].

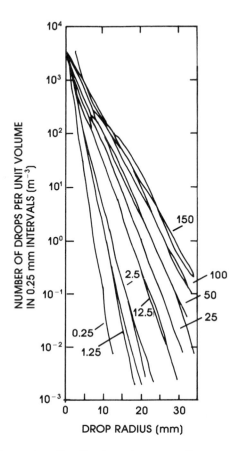

Fig. 4.10 Raindrop size distributions in space. Parameters are rain rates. Thick solid curves are from measurement by Laws and Parsons and fine solid lines are the negative exponential relation of Marshall and Palmer. Rainfall rates are in mm/hr (from Oguchi [15], © 1983 IEEE).

Oguchi and Hosoya calculated the effect of rain drop distortion on signal attenuation, and determined that the attenuation for a horizontally polarized signal always exceeds that for a vertically polarized signal. These researchers assumed in their calculation that all oblately distorted drops were oriented at the same canting angle and had the same angle of incidence to the radar's line of sight. The attenuation values A_h and A_v, for horizontally and vertically polarized signals, respectively, are listed in

Table 4.1 Various Precipitation Rates
(From Setzer [20] © 1970 Bell System Technical Journal).

Drop Diameter (cm)	*Rain Rate* (mm/hr)								
	Percent of Total Volume								
	0.25	1.25	2.5	5	12.5	25	50	100	150
0.05	28.0	10.9	7.3	4.7	2.6	1.7	1.2	1.0	1.0
0.1	50.1	37.1	27.8	20.3	11.5	7.6	5.4	4.6	4.1
0.15	18.2	31.3	32.8	31.0	24.5	18.4	12.5	8.8	7.6
0.2	3.0	13.5	19.0	22.2	25.4	23.9	19.9	13.9	11.7
0.25	0.7	4.9	7.9	11.8	17.3	19.9	20.9	17.1	13.9
0.3		1.5	3.3	5.7	10.1	12.8	15.6	18.4	17.7
0.35		0.6	1.1	2.5	4.3	8.2	10.9	15.0	16.1
0.4		0.2	0.6	1.0	2.3	3.5	6.7	9.0	11.9
0.45			0.2	0.5	1.2	2.1	3.3	5.8	7.7
0.5				0.3	0.6	1.1	1.8	3.0	3.6
0.55					0.2	0.5	1.1	1.7	2.2
0.6						0.3	0.5	1.0	1.2
0.65							0.2	0.7	1.0
0.7									0.3

Table 4.2 Raindrop Terminal Velocity
(From Setzer [20] © 1970 Bell System Technical Journal).

Radius (cm)	*Velocity* (m/s)
0.025	2.1
0.05	3.9
0.075	5.3
0.10	6.4
0.125	7.3
0.15	7.9
0.175	8.35
0.2	8.70
0.225	9.0
0.25	9.2
0.275	9.35
0.30	9.5
0.325	9.6

Table 4.3 Attenuation in Decibels per Kilometer for Different Rates of Rain Precipitation at Temperature 18° C. (From Burrows and Atwood [21], © 1949 Academic Press.)

Precipitation Rate, R (mm/hr)	Wavelength, λ (cm)								
	$\lambda = 0.3$	$\lambda = 0.4$	$\lambda = 0.5$	$\lambda = 0.6$	$\lambda = 1.0$	$\lambda = 1.25$	$\lambda = 3.0$	$\lambda = 3.2$	$\lambda = 10$
0.25	0.305	0.23	0.16	0.16	0.037	0.0215	0.00224	0.0019	0.0000997
1.25	1.15	0.929	0.72	0.549	0.228	0.136	0.0161	0.0117	0.000416
2.5	1.98	1.66	1.34	1.08	0.492	0.298	0.0388	0.0317	0.000785
12.5	6.72	6.04	5.36	4.72	2.73	1.77	0.285	0.238	0.00364
25.0	11.3	10.4	9.49	8.59	5.47	3.72	0.656	0.555	0.00728
50.0	19.2	17.9	16.6	15.3	10.7	7.67	1.46	1.26	0.0149
100.0	33.3	31.1	29.0	27.0	20.0	15.3	3.24	2.8	0.0311
150.0	46.0	43.7	40.5	37.9	28.8	22.8	4.97	4.39	0.0481

Table 4.4 Attenuation in Rain of Known Drop Size Distribution and Rate of Fall (db/km) (From Burrows and Atwood [21]).

Precipitation Rate, R (mm/hr)	Wavelength, λ (cm)					
	λ = 1.25	λ = 3	λ = 5	λ = 8	λ = 10	λ = 15
2.45	1.93×10^{-1}	4.92×10^{-2}	4.24×10^{-3}	1.23×10^{-3}	7.34×10^{-4}	2.80×10^{-4}
4.0	3.18×10^{-1}	8.63×10^{-2}	7.11×10^{-3}	2.04×10^{-3}	1.19×10^{-3}	4.69×10^{-4}
6.0	6.15×10^{-1}	1.92×10^{-1}	1.25×10^{-2}	3.02×10^{-3}	1.67×10^{-3}	5.84×10^{-4}
15.2	2.12	6.13×10^{-1}	5.91×10^{-2}	1.17×10^{-2}	5.68×10^{-3}	1.69×10^{-3}
18.8	2.37	8.01×10^{-1}	5.13×10^{-2}	1.10×10^{-2}	6.46×10^{-3}	1.85×10^{-3}
22.6	2.40	7.28×10^{-1}	5.29×10^{-2}	1.21×10^{-2}	6.96×10^{-3}	2.27×10^{-3}
34.3	4.51	1.28	1.12×10^{-1}	2.32×10^{-2}	1.17×10^{-2}	3.64×10^{-3}
43.1	6.17	1.64	1.65×10^{-1}	3.33×10^{-2}	1.62×10^{-2}	4.96×10^{-3}

Table 4.5 for various radar frequencies and rainfall rates. Figure 4.11 depicts the normalized differential attenuation $(A_h - A_v)/A_v$ *versus* frequency, from the data of Table 4.5. Note that the relative differential attenuation indicates a major relative extremum at 5 GHz and a minor relative extremum at 20 GHz [15].

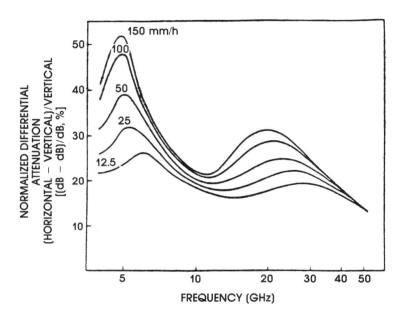

Fig. 4.11 Normalized differential attenuation *versus* frequency. The angle of incidence $\alpha = 90°$ (from Oguchi [15], © 1983 IEEE).

Numerous investigators have empirically fit measured rain attenuation data to the relationship between attenuation and rainfall rate proposed by Gunn and East [2]:

$$\alpha = AR^B \qquad (4.12)$$

where

α = attenuation, in dB per km;
R = rainfall rate, in mm/hr.

Currie, Dyer, and Hayes of Georgia Tech have proposed a set of A and B parameters for this empirical relationship between rainfall rate and signal attenuation, including both microwave and MMW frequencies. A listing of A and B parameters and signal attenuation values α for selected

Table 4.5 Attenuation Values for Various Frequencies and Rainfall Rates (from Oguchi [15] © 1983 IEEE).

Frequency (GHz)	Rain Rate (mm/hr)	Specific Attenuation (dB/km)					
		Spherical Drop	Spheroidal Drop		Pruppacher-Pitter Drop		
			Vertical Polarization	Horizontal Polarization	Vertical Polarization	Horizontal Polarization	
4	12.5	0.008992	0.008440	0.01028			
	50	0.04380	0.03954	0.05220			
	100	0.1028	0.09050	0.1259			
5	12.5	0.1855	0.01730	0.02144			
	50	0.1106	0.01730	0.1343			
	100	0.2921	0.2450	0.3420			
6	12.5	0.03726	0.03418	0.04316			
	50	0.02471	0.02175	0.2911			
	100	0.6558	0.5704	0.7760			
11	12.5	0.3110	0.2852	0.3349	0.2902	0.33568	
	50	1.717	1.554	1.855	1.577	1.884	
	100	3.898	3.504	4.230	3.533	4.332	
15	12.5	0.6609	0.6023	0.6989			
	50	3.136	2.791	3.376			
	100	6.562	5.745	7.169			
19.3	12.5	1.089	0.9822	1.152	0.9966	1.148	
	50	4.835	4.222	5.231	4.258	5.279	
	100	9.938	8.483	10.89	8.523	11.08	

Table 4.5 (cont'd)

Frequency (GHz)	Rain Rate (mm/hr)	Specific Attenuation (dB/km)					
		Spherical Drop	Spheroidal Drop		Pruppacher-Pitter Drop		
			Vertical Polarization	Horizontal Polarization	Vertical Polarization	Horizontal Polarization	
30	12.5	2.426	2.166	2.578			
	50	9.664	8.429	10.34			
	100	18.62	16.06	19.93			
34.8	12.5	3.093	2.774	3.275	2.847	3.236	
	50	11.62	10.25	12.33	10.46	12.26	
	100	21.57	18.91	22.85	19.22	22.82	
40	12.5	3.787	3.424	3.987			
	50	13.37	11.95	14.07			
	100	24.08	21.45	25.29			
50	12.5	4.915	4.519	5.121			
	50	15.75	14.41	16.37			
	100	27.52	25.13	28.59			

rainfall rates is shown in Table 4.6 [23]. A comparison of the Georgia Tech attenuation model is shown in Figure 4.12 against 94 GHz attenuation data measured by Keizer, *et al.* [24].

Table 4.6 Predicted Rainfall Attenuation as a Function
of Rainfall Rate
(From Currie, Dyer, and Hayes [24]).

	$\alpha = AR^B$			
	$f = 10$ GHz $\lambda = 3.2$ cm $A = .00919$ $B = 1.16$	$f = 35$ GHz $\lambda = 0.86$ cm $A = .273$ $B = .985$	$f = 70$ GHz $\lambda = 0.43$ cm $A = .634$ $B = .868$	$f = 95$ GHz $\lambda = 0.32$ cm $A = 1.6$ $B = .64$
R (mm/hr)	α(dB/km)			
5	0.0509	1.33	2.56	4.48
10	0.133	2.64	4.68	5.984
15	0.213	3.93	6.65	9.05
20	0.297	5.22	8.54	10.88
25	0.385	6.50	10.36	12.55
30	0.475	7.78	12.14	14.11
35	0.568	9.06	13.88	15.57
40	0.663	10.33	15.18	16.96
45	0.760	11.60	17.26	18.29
50	0.859	12.87	18.91	19.56
55	0.960	14.14	20.55	20.79
60	1.062	15.40	22.16	21.99
65	1.165	16.67	23.75	23.14
70	1.270	17.93	25.32	24.27
75	1.375	19.19	26.89	25.36
80	1.482	20.45	28.44	26.43
85	1.590	21.71	29.98	27.48
90	1.700	22.97	31.50	28.50
95	1.809	24.22	33.02	29.50
100	1.920	25.48	34.52	30.49

Brinks [25] has attempted to reconcile several measured 35 GHz attenuation values that greatly exceed recognized model predictions, incorporating a Laws-Parsons drop size distribution, as modified by Krasyuk, *et al.* [27]. Figure 4.13 shows attenuation *versus* rainfall rate for both models that bound these measured data points [26–28]. Brinks concluded that

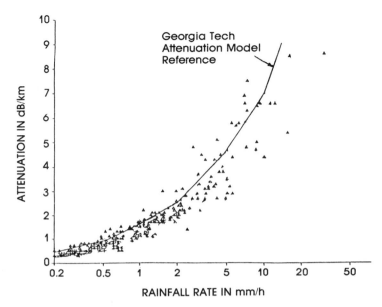

Fig. 4.12 Comparison of Georgia Tech attenuation model at 95 GHz to measured data (adapted from Keizer, *et al.* [31]*; see Currie, Dyer, and Hayes [24]).

either the Soviet researchers used a more accurate value of the index of refraction for water, or some of their measured rain may have been contaminated by natural or artificial pollutants.

Several experimenters have made relative attenuation measurements over a rain-filled path without characterizing the actual rain rate or drop size distribution. Ihare and Furuhama [29] made relative attenuation measurements at 81.8, 34.5, and 11.5 GHz to investigate the applicability of frequency scaling techniques. Applying a Laws and Parsons drop size distribution results in an underestimation of attenuation levels for frequencies above 10 GHz by using their techniques. Figure 4.14 shows the ratio of 81.8 to 34.5 GHz attenuation values as a function of the 34.5 GHz values [29].

Similarly formatted data presentations are shown in Figure 4.15. Relative attenuation data are plotted for attenuation measurements at 36 and 110 GHz. Rain attenuation data were recorded during prolonged, widespread rainfall to ensure uniformity over the entire propagation path.

---

*The original version of this material was first published by the Advisory Group for Aerospace Research and Development, North Atlantic Treaty Organisation (AGARD/NATO) in Conference Proceedings No. 249 published in 1978.

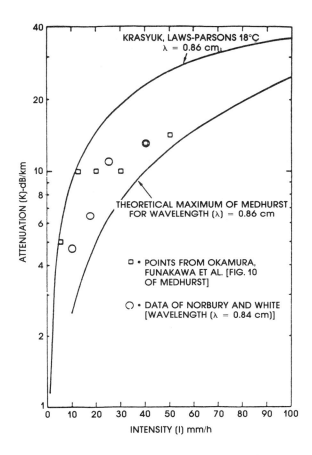

Fig. 4.13 Extreme data points of Okamura, *et al.*, and of Norbury and White (near λ = 0.86 cm) (from Brinks [25]).

The graphed data are consistent with computations based on the Laws and Parson rain drop distribution model (Ho, *et al.* [30]).

Figure 4.16 depicts measured 95 GHz rain attenuation values against five popular drop size distribution models. Considerable spread is evident in these data (Keizer, Sneider, and de Haan [31]). Figures 4.17 and 4.18 depict 140 GHz rain attenuation *versus* rain rate data measured by two different experimenters (Bauer [13], Richard, *et al.* [32]). Relatively good agreement between these two sets of data is evident.

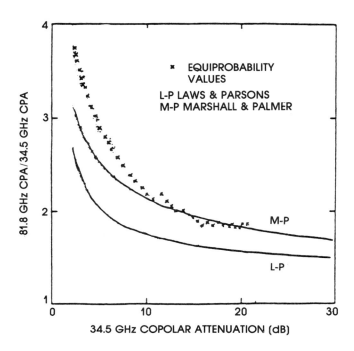

Fig. 4.14 Ratio of 81.8 GHz to 34.5 GHz CPA (from Ihara and Furuhama [29]).

Nemarich, *et al.* recorded attenuation of rainfall at 96, 140, and 225 GHz, and found the attenuation to be relatively insensitive to radar frequency. The AR^B relationship for the Laws and Parsons rain drop size distribution fit the recorded data better than did the Marshall-Palmer distribution. Figures 4.19, 4.20, and 4.21 depict rain attenuation values at 96, 140, and 225 GHz, respectively. Attenuation curves on each plot were calculated by using the AR^B relationship for a Laws and Parsons rain drop size distribution (for high and low rainfall rates) and for a Marshall-Palmer distribution as a comparison (Nemarich [33]).

Recent developments of radar techniques to exploit the vector or matrix nature of the backscattered electromagnetic energy require that the polarization effects of the intervening medium also be well understood. For example, several stationary target discrimination techniques that have

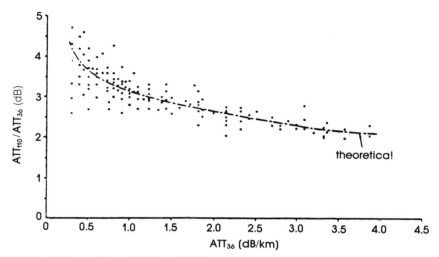

(a) Ratio of Attenuations at 110 and 36 GHz (att$_{110}$/att$_{36}$) *versus* Attenuation at 36 GHz (att$_{36}$)

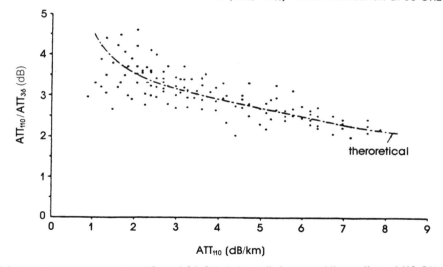

(b) Ratio of Attenuations at 110 and 36 GHz (att$_{110}$/att$_{36}$) *versus* Attenuation at 110 GHz (att$_{110}$)

Fig. 4.15 Relative attenuation due to rain at (a) 36 GHz and (b) 110 GHz (from Ho, *et al.* [30], © 1978 IEEE).

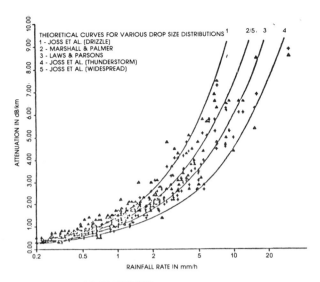

Fig. 4.16 Measured and calculated rain attenuation at 94 GHz (from Keizer, Sneider, and de Haan [31]).*

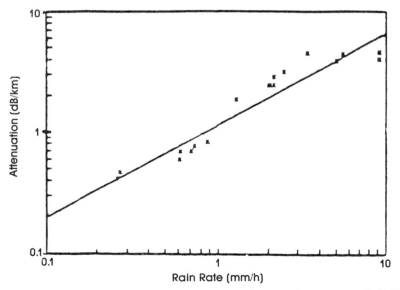

Fig. 4.17 140 GHz rain attenuation *versus* rain rate (Bauer, *et al.* [13]).

*The original version of this material was first published by the Advisory Group for Aerospace Research and Development, North Atlantic Treaty Organisation (AGARD/NATO) in Conference Proceedings No. 249 published in 1978.

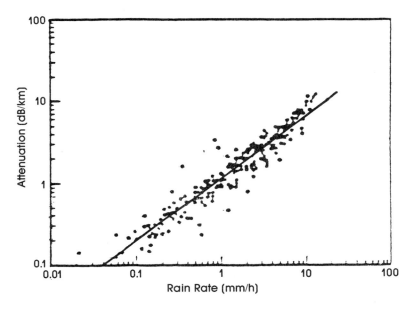

Fig. 4.18 Measured attenuation at 140 GHz *versus* rainfall rate (from Bauer, *et al.* [13]).

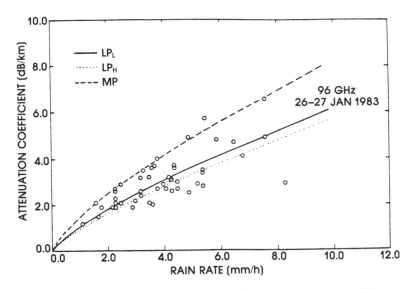

Fig. 4.19 Dependence of attenuation at 96 GHz on rain rate. The curves were calculated using an AR^B relationship for Laws-Parsons distributions for low and high rain rates and a Marshall-Palmer distribution (from Nemarich, *et al.* [33]).

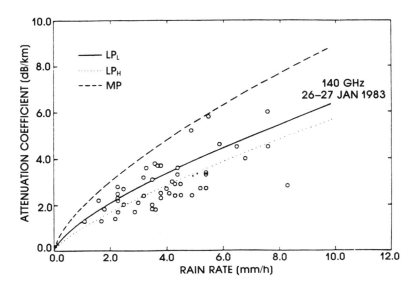

Fig. 4.20 Dependence of attenuation at 140 GHz on rain rate. The curves were calculated using an AR^B relationship for Laws-Parsons distributions for low and high rain rates and a Marshall-Palmer distribution (from Nemarich, *et al.* [33]).

been developed utilize polarization agility, where the transmitted polarization is switched on a pulse-to-pulse or batch-to-batch basis. Techniques that utilize all of the information content of the backscattered electromagnetic signal (i.e., the full polarization matrix content) have been named *matrix techniques.* Those techniques that utilize only part of this content and require polarization agility or diversity, or require dual-polarization receivers, have been named *vector techniques.* Conventional detection and discrimination techniques that assume a constant polarization state are called *scalar techniques.* Chapter 6 contains more detail with regard to the use of polarization parameters for radar signal processing.

The exploitation of polarization effects for radar and communication has included applications in rain return rejection and frequency re-use techniques (the use of two orthogonal polarizations for communication of two channels at the same carrier frequency). Circular polarization, for example, is often used in air surveillance applications to minimize radar returns from rain. This phenomenon is highly dependent on rainfall rate, as greater rainfall rates are associated with larger drop sizes. Larger rain drops, in turn, become nonspherical, undergo vibration, and assume a canted axial orientation in the presence of sheer winds. All of these physical effects result in depolarization of the energy backscattered from rain drops.

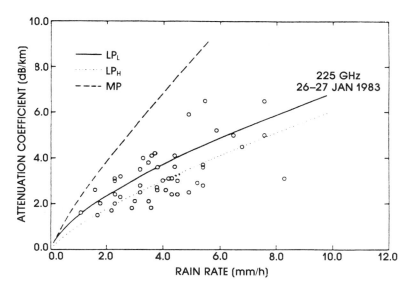

Fig. 4.21 Dependence of attenuation at 225 GHz on rain rate. The curves were calculated using an AR^B relationship for Laws-Parsons distributions for low and high rain rates and a Marshall-Palmer distribution (from Nemarich, *et al.* [33]).

In the ideal case, the rain backscatter has the opposite-sense circular polarization to that which was transmitted. In reality, depolarization of the received signal occurs, due to the backscatter from the nonspherical shape of the rain drops themselves as well as differential attenuation and phase shifts between orthogonal polarization components of the radar signal. The latter effect is well known in long-distance communication, where these differential characteristics increase the cross-talk between orthogonally polarized information channels.

The typical rain drop model that is used to evaluate this phenomenology is that of a canted, oblate spheroid, as shown in Figure 4.22. Cross-coupling, for example, between horizontally and vertically polarized radar channels, results from differential attenuation and phase shift between the linear polarization components, which are parallel and perpendicular to the major axis of the rain drops (Chu [34]). Several experimenters have measured canting angles for rain, and showed a somewhat consistent finding that about 40% of the drops had positive canting angles exceeding 15°, while a much smaller percentage had negative canting angles of less than $-15°$. Mean canting angles tend to be in the $+10°$ range, and are independent of drop size (Oguchi [15]).

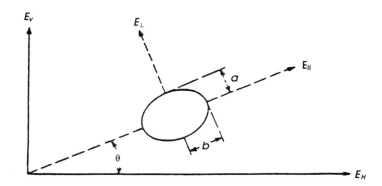

Fig. 4.22 Canted oblate spheroidal raindrop (from Chu [34], © 1974 Bell System Technical Journal).

Chu compared calculated differential attenuation $A_H - A_V$ with measured data at several radar bands. Figure 4.23 shows plots of differential attenuation *versus* sum attenuation at three sets of canting angles for data collected by Semplak at 30 GHz. The data denote the scatter of the original data, and the crosses are median values. Calculated attenuation coefficients for horizontally and vertically polarized radar signals are plotted over the MMW band in Figure 4.24 [34]. Oguchi's calculations of differential attenuation shift are shown in Figure 4.25 for comparison. Figure 4.26 shows calculated differential phase shift normalized by attenuation [34], while Figure 4.27 shows actual calculated differential phase shift.

Frozen Hydrometeor Attenuation

In comparison with the quantity of attenuation data for rain at MMW frequencies, the quantity of available data for snow and hail is scant. Snow data are especially difficult to compare because of the many different forms of snow and snow-sleet mixtures. Snowflakes are typically between 2 and 5 mm, but they can be as large as 15 mm. Relatively dry snow falls at speeds between 1.0 and 1.5 m/s, but snow with a high water content can fall as fast as 5 to 6 m/s. Hail, however, is usually spherical, with diameters from several millimeters to a few centimeters, and it is almost entirely composed of ice.

Oguchi points out that attenuation due to dry snow is 10 dB less than that due to rain at the same rate of precipitation, at microwave frequencies.

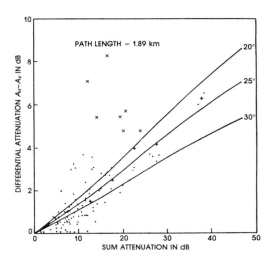

Fig. 4.23 Comparison between calculated and measured differential attenuation at 30 GHz (from Chu [34], © 1974 Bell System Technical Journal).

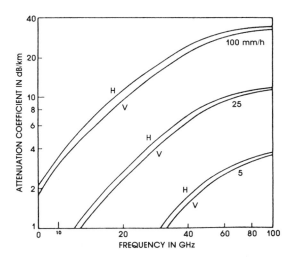

Fig. 4.24 Attenuation coefficients of vertically and horizontally polarized waves (from Chu [34], © 1974 Bell System Technical Journal).

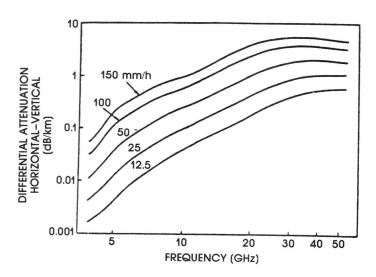

Fig. 4.25 Frequency characteristics of differential attenuation for spheroidal raindrops (from Oguchi [15], © 1983 IEEE).

Wet snow attenuation is comparable to equivalent rain in both the microwave and MMW regions (Oguchi [15]). Gunn and East propose the following relationship for snowfall attenuation at 0°C:

$$\alpha = 0.00349 \frac{R^{1.6}}{\lambda^4} + 0.00224 \frac{R}{\lambda} \tag{4.13}$$

where R is the snowfall rate, in millimeters of water per hour, and λ is the radar wavelength, in centimeters (Gunn and East [35]). The authors caution that this expression may be valid for dry snow only, precipitation rates of less than 10 mm/hr, and wavelengths of less than 1.5 m.

Nishitsuji, *et al.* made snow attenuation measurements at 35 GHz, and compared their data with calculations based on Mie-Stratton cross-section expressions. Figure 4.28 shows these measured data and computed attenuation *versus* precipitation rates (Nishitsuji and Matsumato [36]). Additional snow attenuation data are shown in Figure 4.29 for 35, 54, and 312 GHz (Richard [37]).

Nemarich, *et al.* recorded snowfall attenuation at 96, 140, and 225 GHz. Figures 4.30, 4.31, and 4.32 depict snow attenuation for this set of MMW frequencies as a function of equivalent water density. The solid line in each case is a least-squares fit to individual data points that were additionally forced to intercept the attenuation line of a clear atmosphere

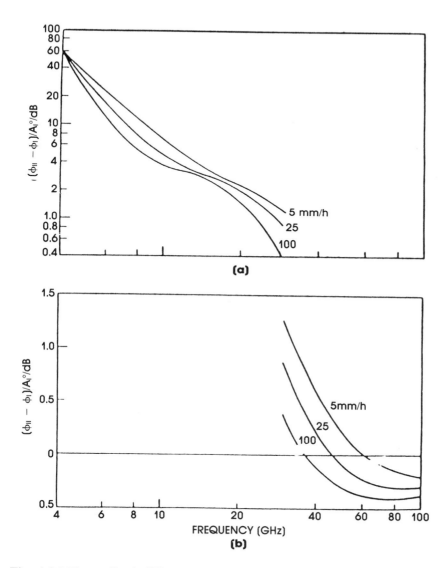

Fig. 4.26 Normalized differential phase shift with respect to *A*: 4 to 30 GHz; *B*: 30 to 100 GHz (from Chu Technical [34], © 1974 Bell System Technical Journal).

[33]. Figure 4.33 depicts snow attenuation data at 140 and 225 GHz, while Figure 4.34 depicts attenuation at these same radar frequencies for mixed snow and rain. Attenuation at 96 GHz was also measured in snowfall for

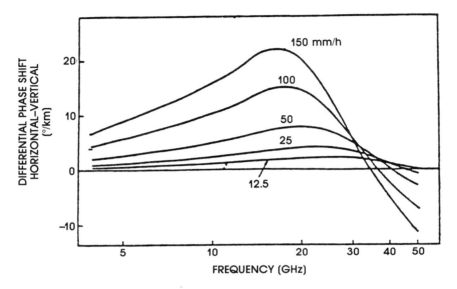

Fig. 4.27 Frequency characteristics of differential phase shift for spheroidal raindrops (from Oguchi [15], © 1983 IEEE).

various transmitter-receiver polarizations, but no significant polarization-
dependent effect was noted [38].

Hail attenuation data are even rarer than snow, a result due in part
to the infrequency and short duration of this phenomenon. Table 4.7 lists
hail attenuation calculations for several sizes of hailstorms [39]. Significant
hail attenuation has also been recorded by Antar and Hendry at 28.56
GHz during measurements along an earth-satellite path [40].

Table 4.7   Attenuation in Hail with Intensity of Hail = 10 mm/hr
(Equivalent Water)
(From Aganbekyan, *et al.* [39]).

Hail Particle Diameter (mm)	Attenuation (dB/km)	
	$\lambda = 1.0$ mm	$\lambda = 0.1$ mm
2.5	2.7	2.0
5.0	0.9	0.8
10.0	0.4	0.4

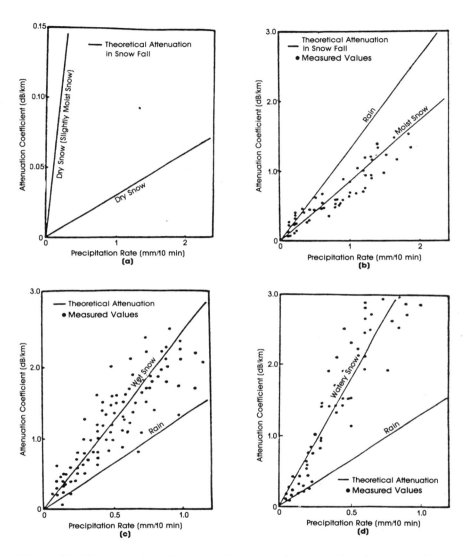

Fig. 4.28 Theoretical and measured values of attenuation due to various types of snowfall at 35 GHz (from Nishitsuji and Matsumato [36], © Hokkardo University 1971).

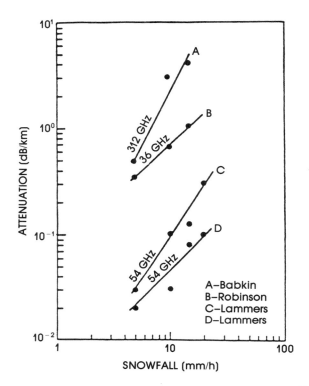

Fig. 4.29 Measured attenuation in snow *versus* snowfall rate (from Richard [37]).

4.2.3 Particulate Attenuation

The radar signal may be attenuated by particulate matter thrown into the air from bomb bursts, movement of ground vehicles over dry terrain, sand and dust storms, smoke from vehicle engines and burning objects, or intentionally placed smoke screens. These particulants will seriously degrade visibility of optical sensors, and such particulate matter may also degrade the performance of infrared sensors if it is hot. The degree of radar performance under these conditions is highly relevant in a land battlefield scenario of armored vehicles, burning debris, and optical or infrared countermeasures, where development of air-to-ground "smart" munitions is being pursued.

Knox reported smoke and obscurant attenuation data during the *Smoke Week I* and *Smoke Week II* test programs. Various military ordnance was exploded within the line of sight of MMW propagation paths.

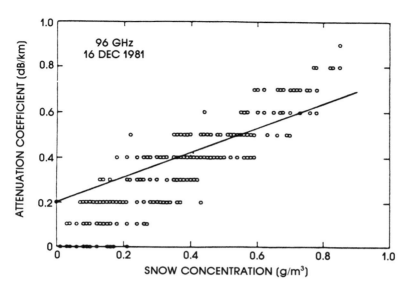

Fig. 4.30 Dependence of attenuation at 96 GHz on snow mass concentration. The solid line is a least-squares linear fit that was forced to intercept the clear air value (from Nemarich, *et al.* [33]).

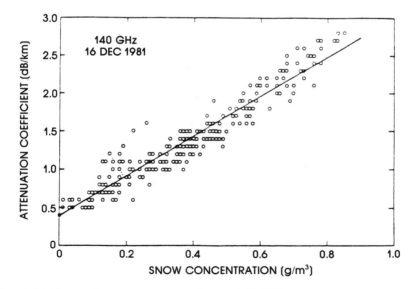

Fig. 4.31 Dependence on attenuation at 140 GHz on snow mass concentration. The solid line is a least-squares linear fit that was forced to intercept the clear air value (from Nemarich [33]).

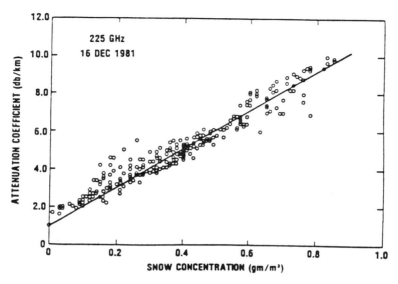

Fig. 4.32 Dependence of attenuation at 225 GHz on snow mass concentration. The solid line is a least-squares linear fit that was forced to intercept the clear air value (from Nemarich [33]).

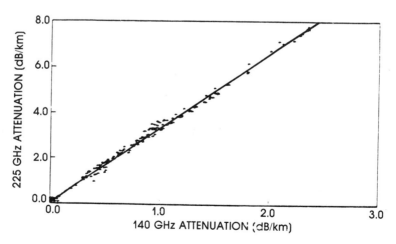

Fig. 4.33 Plot of the linear correlation between 225 and 140 GHz attenuation for 5 January 1984 snowstorm (from Nemarich, *et al.* [38].

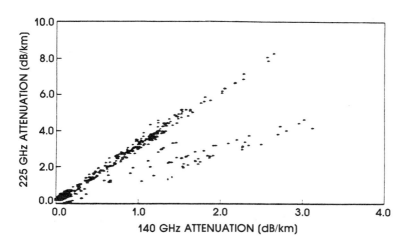

Fig. 4.34 Plot of correlation between 225 and 140 GHz attenuation for the mixed conditions observed on 14 December 1983 (from Nemarich, *et al.* [38]).

Obscurants included hexachloroethene, white and red phosphorus, dust, and fog oil. Of these materials, only dust dispersed by the detonation of high explosives resulted in appreciable attenuation. Figure 4.35 shows attenuation at 35, 94, and 140 GHz due to red phosphorus over the 600 m one-way propagation path. Since the physical extent of the obscurant cloud was not measured, attenuation data are presented only as a two-way total decrease in radar signal. Figure 4.36 depicts attenuation due to dust as a function of time for the same three MMW frequencies. A 0.4 dB worst-case attenuation was experienced, which lasted for approximately 10 s (Knox [41]).

Petito and Wentworth measured MMW radar transmission and backscatter during the *Dusty Infrared Test II* (DIRT II). High explosive projectiles of 105 mm and 155 mm, and C4 explosives were detonated singly, lying on or below the ground surface, along the radar's line of sight. Figure 4.37 depicts transmission and backscatter along the line of site (upper graph) and 10 m perpendicular to the line of sight (lower graph). The degree of attenuation depended on the type of soil and its moisture content, while the duration of attenuation depended upon the wind velocity (Petito and Wentworth [42]).

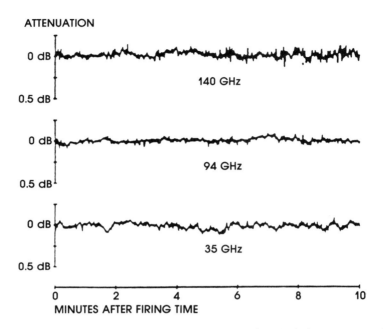

Fig. 4.35 Attenuation due to red phosphorous (wedge) (from Knox, [41], © 1979 IEEE).

Schwerling, *et al.* [43] summarized attenuation data from smoke and dust at Grafenwohr, Germany, during 1979 DIRT III experiments. Their data, at 38 and 60 GHz, indicated that attenuation due to smoke and dust was minimal, but that artillery shell detonations could temporarily interrupt signal transmission, usually for less than a second.

Several researchers have measured MMW attenuation through the dust and debris resulting from a (simulated) nuclear burst. Altschuler calculated a sensitivity analysis of attenuation, and determined that the attenuation depends heavily on the maximum particle radius, the number of large particles, and the complex-valued index of refraction. Losses accounted for fireball ionization, dust, and atmospheric oxygen and water vapor. Figure 4.38 depicts calculated attenuation of sand (upper graph) and clay (lower graph) as a function maximum particle radius, at frequencies between 10 and 95 GHz [44].

4.2.4 Foliage Penetration

Currie, Martin, and Dyer [45] made a series of attenuation measurements for microwave and MMW frequencies through foliage. Figure 4.39 shows dry foliage attenuation distributions that were measured for

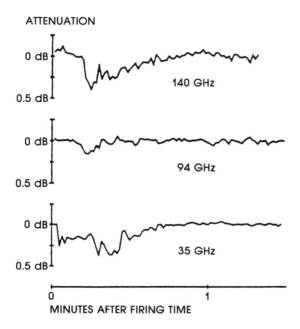

ATTENUATION

Fig. 4.36 Attenuation due to H.E. dust (from Knox [41], © 1979 IEEE).

35 and 95 GHz, while Table 4.8 lists mean and median values at these frequencies as well as polarization dependency at 35 GHz. Note that the median values are less than the mean values, and there is little statistical difference between horizontal and vertical polarization data. A least-squares fit to the dry foliage attenuation data results in the expression for attenuation α:

$$\alpha \; (\text{dB/m}) \; = \; 1.102 \; + \; 1.48 \; \log_{10} \; (F) \qquad (4.14)$$

where F is the radar frequency in GHz. Wet foliage attenuation tended to be higher than dry foliage attenuation and followed a power-law relationship with frequency rather than a logarithmic pattern [45].

Schwerling, *et al.* [43] reported that MMW attenuation through foliage greatly exceeds values extrapolated from data below 9 GHz. Polarization effects due to tree and bush attenuation were slight, but horizontally polarized signals were less attenuated through tall grass. Attenuation increased during rain, when water covered, or droplets formed on, leaves and branches of a willow tree. The increase was noted several minutes after the rain began, and indicated a delay of effect until the leaves and branches were wet [42, 46].

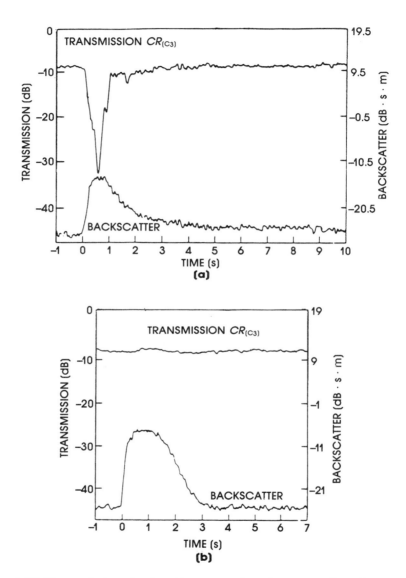

Fig. 4.37 Transmission, backscatter *versus* time for two surface-detonated C4 explosives at locations: (a) CL, and (b) 10 m east of CL (from Petito and Wentworth, [42]).

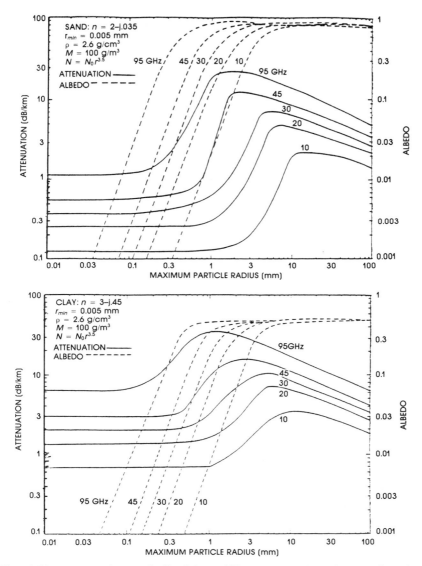

Fig. 4.38 Attenuation and albedo at millimeter wavelengths as a function of maximum particle radius (from Altschuler [44]).*

*The original version of this material was first published by the Advisory Group for Aerospace Research and Development, North Atlantic Treaty Organisation (AGARD/NATO) in Conference Proceedings, published in 1983.

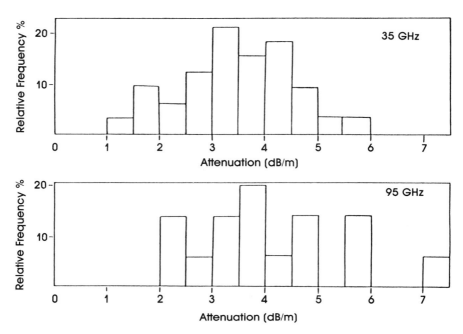

Fig. 4.39 Relative frequency of occurrence of the attenuation coefficient for two-way penetration measurements at 35 GHz and 95 GHz (from Currie, Martin, and Dyer, [45]).

Table 4.8 Summary of Two–Way Attenuation Measurements

Frequency (GHz)	9.4		16.2		35		95
Polarization	V	H	V	H	V	H	V
Median (dB/m)	1.9	2.3	2.7	3.1	3.4	3.3	3.9
Mean (dB/m)	2.0	2.5	2.7	3.5	3.7	3.0	4.1
Standard Deviation (dB)	0.84	0.61	0.76	1.10	0.30	1.56	1.30

4.3 ANOMALOUS PROPAGATION EFFECTS

Atmospheric refraction and multipath interference are additional phenomona that may alter or degrade radar system performance. Bending or trapping of microwave radiation is a well known effect that can extend the range of a radar beyond a straight line of sight. Reflection of the radar signal off the sea, ground, or buildings can cause interference effects at the receive antenna when backscatter converges from simultaneous, but distinctly different, paths. See Chapters 16 and 17 for more information on multipath interference effects on radar performance.

4.3.1 Refractive Propagation Effects

The path of an electromagnetic wave may be affected by the non-homogeneities of the medium. We have already noted the attenuation and relative phase shift effects due to the atmospheric content itself, plus those due to hydrometeors and particulates in the air. Additional homogeneity of the medium is represented by changes in the electrical properties of the medium due to thermal, gravitational, and local weather causes. These characteristics may result in actual bending of the propagating radar energy (refraction) or trapping of the energy within an "atmospheric waveguide" (ducting).

The complex-valued index of refraction for air is conventionally represented as

$$n = n' - jn'' \tag{4.15}$$

where n' and n'' are the real and imaginary components, respectively, of the index of refraction. For radar applications it proves useful to define the refractivity N:

$$N = (n - 1) \, 10^6 = N' - jN'' \tag{4.16}$$

where N is in parts per million (ppm) (Yu and Hodge [47]).

Over the 10–300 GHz frequency range, Omoura and Hodge [48] express N as the sum of the refractivities due to each atmospheric component absorption line, principally those of oxygen and water vapor. The sum, in turn, may be separated into two major components: a frequency-independent term and a frequency-dependent term.

The frequency independent refractivity N_0 can be expressed as the sum of the low-frequency limits of all the air component absorption line refractivities:

$$N_0 = \sum_i N_{0i}, \quad i \text{ for all lines} \tag{4.17}$$

In practice, this summation is impractical. Bean and Dutton propose an empirical equation for N_0:

$$N_0 = 77.6 \frac{P_d}{T} + 72 \frac{e}{T} + (3.75 \times 10^5) \frac{e}{T^2} \qquad (4.18)$$

where

P_d = dry air pressure, in millibars;
e = partial pressure of water vapor, in millibars;
T = temperature, in K.

An approximate equation for N, to within 0.02% of the previous equation, is

$$N_0 = \frac{77.6}{T} \left(P + \frac{4810e}{T} \right) \qquad (4.19)$$

where P is the total pressure, in millibars (Liebe and Layton [49]).

For higher radar frequencies, the effects of individual absorption lines are observed (e.g., water vapor at 22 GHz, oxygen around 60 GHz), and the refractivity value itself becomes frequency dependent. The imaginary component of the refractivity is shown in Figure 4.40 as a function of radar frequency. Figure 4.41 depicts the real component (attenuation) of the refractivity. These data were calculated for summertime atmospheric conditions in the Columbus, Ohio area (Yu and Hodge [47]).

It is well known that the refractive index of the atmosphere varies as a function of altitude in an exponential manner. The decrease ΔN in refractivity N over the first kilometer can be empirically expressed in terms of the refractivity at the surface N_s as:

$$\Delta N = 7.32 \exp (0.005577 \, N_s) \qquad (4.20)$$

A simplified model then predicts that the refractivity N at radius r can be expressed as

$$N = N_s \exp \left[- \frac{\Delta N}{N_s} (r - r_s) \right] \qquad (4.21)$$

where $N = N_s$ at radius $r = r_s$ (the surface).

The angle τ through which a radar beam is bent has been shown to be approximated by the following expression, for a high initial elevation angle θ_i:

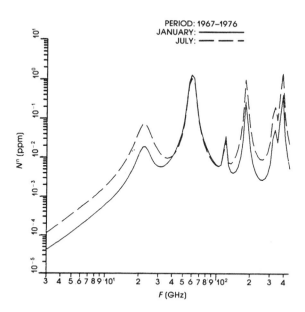

Fig. 4.40 Average imaginary part of refractivity (from Yu and Hodge [47]).

$$\tau = N_s \cdot 10^{-6} \cos\theta_i \exp\left[-\frac{\Delta N}{N_s} (r_i - r_s) \right]$$

$$\times \left[\frac{1}{\sin\theta_i} - \frac{\exp\left(-H\Delta N/N_s\right)}{\sqrt{(2H/r_i) + \sin^2\theta_i}} \right] \quad (4.22)$$

for $\theta_i \gg 3°$, where

r_i = initial beam altitude,
H = difference in altitude between initial and final points [47].

The essential result of this phenomenon is an incorrect (high) elevation angle and an incorrect (long) range to the target. At microwave frequencies it has been common practice to compensate for these effects by using a 4/3 earth radius (standard atmosphere) to model the beam trajectory simply as a straight line.

Liebe and Layton measured attenuation and refraction due to mixtures of water vapor and nitrogen at 138 GHz. Attenuation data behaved in accordance to accepted water vapor line broadening theory (Liebe and Layton [49]).

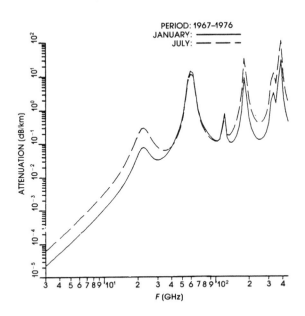

Fig. 4.41 Average attenuation (from Yu and Hodge [47]).

Patton, *et al.* surveyed MMW propagation effects, and concluded that degradation due to lens focusing loss and elevation and range errors would be significant for a shipboard fire control application. Figure 4.42 shows that the elevation angle and range will be in error by less than 2 mrad and 80 feet, respectively [50, 51, 54].

McMillan, Wiltse, and Snider surveyed atmospheric turbulence effects for the MMW region. They also extended calculations of angle-of-arrival fluctuations in the atmospheric windows at 94 and 140 GHz, and compared their results with Russian experimental data (McMillan, Wiltse, and Snider [52]).

An atmospheric duct (called *superrefraction*) will occur when there exists a rapid decrease in refractive index with altitude. This condition may be created by a temperature inversion or a high humidity gradient. The radar beam becomes trapped in the duct in a manner not unlike an electromagnetic wave confined within a waveguide structure.

Atmospheric ducts can extend from 10 meters to as great as 200 m in height. As a waveguide structure, the duct size also implies a long wavelength cut-off.

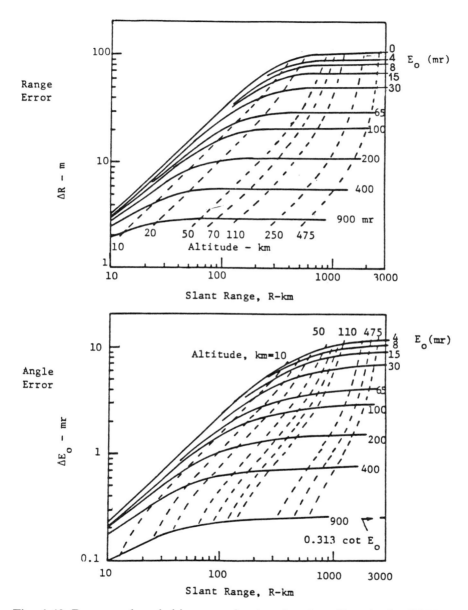

Fig. 4.42 Range and angle bias error due to refractive effects in the CRPL
exponential reference atmosphere ($N_0 = 313$) (from Hitrey [51],
© 1975 IEEE).

4.3.2 Multipath Interference

When the main beam of a radar antenna is pointed at a low incidence angle toward a target such that part of the beam intersects the ground, the radar signal can follow four different two-way paths. As shown in Figure 4.43, the energy can follow a direct-direct, direct-indirect, indirect-direct, or indirect-indirect path from transmission to reception. These various and simultaneous paths can result in interference effects at the receiving antenna, which acts as an electromagnetic vector-summing device. This electromagnetic vector summation is called *multipath interference*.

It is illustrative to model this effect on a flat earth. Figure 4.44 shows a monostatic radar configuration in which the antenna is at height h_a, and the target is at slant range R and height H_t. The field strength (after Skolnik [53]) at the target due to this multipath interference may be expressed in terms of the ratio F, or propagation factor, where

$$F = \frac{\text{Electric field strength at target in multipath field}}{\text{Electric field strength at target if in free space}} \qquad (4.23)$$

It is assumed that the angle θ is very shallow, the indirect and direct path lengths are nearly the same, and the antenna gains in both directions are also nearly the same. Let the complex-valued reflection coefficient on the ground be described by the quantity $\Gamma = \rho\, e^{i\psi}$, where ρ defines an amplitude change and ψ defines a phase change on reflection. With the previous assumption, it is then obvious that the field strength of the direct or indirect paths will be due essentially to path length differences plus reflection phase changes.

When the ground reflection coefficient $\Gamma = -1$, a phase change of 180° results, with no amplitude change. It is straightforward from the geometry defined by Figure 4.44 and the definition of Γ to show that the ratio of the power incident on the target in a multipath field to one in free space is

$$F^2 = 4\sin^2\left[\frac{2\pi h_a h_t}{\lambda R}\right] \qquad (4.24)$$

where λ is the radar wavelength. By reciprocity, the analogous power ratio at the receiving antenna will be

$$F^4 = 16\sin^4\left[\frac{2\pi h_a h_t}{\lambda R}\right] \qquad (4.25)$$

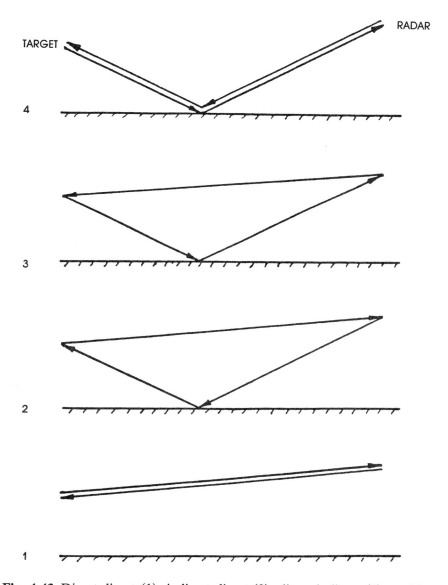

Fig. 4.43 Direct-direct (1), indirect-direct (2), direct-indirect (3), and indirect-indirect (4) paths.

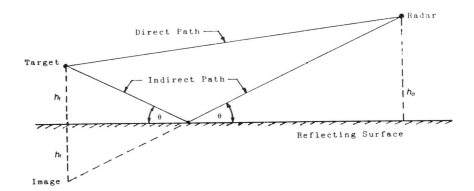

Fig. 4.44 Multipath interference geometry.

The integrated power will be a maximum when the argument of the sine function is an odd multiple of $\pi/2$ and a minimum when it is a multiple of π. Thus, the extrema for F occur when

$$\frac{4\,h_a\,h_t}{\lambda R} = 2n + 1: \text{maximum} \tag{4.26}$$

$$\frac{4\,h_a\,h_t}{\lambda R} = n: \text{minimum} \tag{4.27}$$

for $n = 1, 2, 3, \ldots$. Figure 4.45 is a plot of F^4. For $h_t \gg h_a$ the first interference lobe occurs at a radian elevation angle of approximately $\lambda/4h_a$. When the propagation factor F is included in the radar range equation, the expression for the received power P_r becomes

$$P_r = \frac{P_t G^2 \lambda^2 \sigma}{(4\pi)^3 R^4} \cdot 16 \sin^4 \left[\frac{2\pi h_a h_t}{\lambda R}\right] \tag{4.28}$$

which, for small incidence angles, simplifies to

$$P_r = \frac{4\pi P_t G^2 \sigma (h_a h_t)^4}{\lambda^2 R^8} \tag{4.29}$$

Skolnik [53] notes that care must be exercised in the interpretation of this equation. In reality the reflection point is rarely smooth and the reflection coefficient for land clutter will vary from -1. Furthermore Γ is a strong function of the electromagnetic polarization. Interference nulls

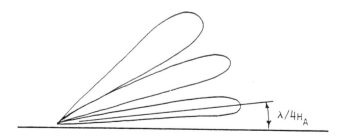

Fig. 4.45 Multipath interference lobing structure.

will tend to be "filled in" and peaks lowered, relative to the ideal case previously described. Surface roughness tends to be more important than reflection coefficient values in determining the specularity of the surface reflection.

More recent studies separate the reflected energy from the sea surface into a coherent and an incoherent component. The coherent component is that which has been described previously. The incoherent component is characterized by a random phase and amplitude [53]. Barton [54] proposed that specular reflection will apply if the root-mean-square surface roughness σ in the first Fresnel zone of reflection is

$$\sigma < \frac{1}{8}\frac{\lambda}{\sin\xi} \tag{4.30}$$

where

ξ = incidence angle to the ground,
λ = radar wavelength.

Wallace measured multipath interference effects at 140 GHz in a forward scattering experiment. He recorded received power over three types of terrain: high weeds, mowed weeds, and asphalt. Vegetative cover resulted in both an attenuation of the electromagnetic energy and a noisy modulation of the reflected signal due to wind. The reflection coefficient for the weeds was less than -0.1, while for asphalt it was approximately -0.5 (Wallace [55]).

4.4 SUMMARY OF ATTENUATION EFFECTS

Table 4.9 summarizes one-way attenuation values for MMW electromagnetic energy propagating through the indicated media. Obviously, there exists a significant variance in experimental and analytical data values, and no attempt has been made to weigh individual contributor's data. Each entry simply represents an attempt to provide the radar engineer a reasonable attenuation value for each radar frequency and each atmospheric condition. Where an entry is omitted, there was insufficient data, or none, with which to make a judgement.

Table 4.9   One-Way Attenuation Coefficients in (dB/km).

	Frequency (GHz)				
Condition	35	60	95	140	220
Clean Air	0.05	10	0.3	0.5	1
Fog (100 m visibility)	0.2	—	1.5	2	4
Rain 1 mm/hr	0.3	—	1	1	—
4 mm/hr	1	—	4.5	5	—
10 mm/hr	2.6	—	6	8	—
Snow (1 mm/hr rain equivalent)	0.3	—	3	3	—
Dust*	0	0	0	0	0

*Low dust concentrations.

REFERENCES

1. S.M. Kulpa and E.A. Brown, Co-chairmen, *Near-Millimeter Wave Technology Base Study*, US Army Electronics Research and Development Command, Harry Diamond Laboratories, Report No. HDL-SR-79-8, November 1979, p. 37.
2. R.A. Bohlander and R.W. McMillan, "Atmospheric Effects on Near-Millimeter-Wave Propagation," *Proceedings of the IEEE,* Vol. 73, No. 1, January 1985, pp. 49-60.
3. J.C. Wiltse, "Introduction and Overview of Millimeter Waves," Chapter 1 in K.J. Button and J.C. Wiltse, eds., *Infrared and Millimeter Waves*, Volume 4, Millimeter Systems, Academic Press, New York, 1981, p. 4.
4. J.H. Van Vleck, "The Absorption of Microwaves by Oxygen," *Physical Review*, Vol. 71, April 1947, pp. 413-424.
5. B.R. Bean, and E.J. Dutton, *Radio Meterology*, Dover Publications, New York, 1968.
6. G.B. Becka and S.H. Autler, "Water Vapor Absorption of Electromagnetic Radiation in the Centimeter Wavelength Range," *Physical Review*, Vol. 70, September 1 and 15, 1946, pp. 300-307.

7. B.R. Bean, E.J. Dutton, and B.D. Warner, "Weather Effects on Radar," M.I. Skolnik, ed., *Radar Handbook*, McGraw-Hill, New York, Chapter 24.

8. L.R. Whicker and D.C. Webb, "The Potential Military Applications of Millimeter Waves," AGARD Conference Proceedings, CP245, 1978, pp. 1-1 to 1-6.

9. W.E.K. Middleton, *Vision Through the Atmosphere*, University of Toronto Press, Toronto, 1963.

10. E.A. Altschuler, "A Simple Expression for Estimating Attenuation by Fog at Millimeter Wavelengths," *IEEE Transactions on Antennas and Propagation*, Vol. AP-32, No. 7, July 1984, pp. 757-758.

11. R.G. Eldridge, "Haze and Fog Aerosol Distributions," *Journal of Atmospheric Science*, Vol. 23, September 1966, pp. 601-613.

12. H.H. Jenkins, T.L. Lane, and J.A. Scheer, "Millimeter Wave DF Antenna/Converter/Processor," Volume II, Georgia Tech, July 1982.

13. D.G. Bauer, *et al.*, "140 GHz Beamrider Feasibility Experiment," Ballistic Research Laboratories, Interim Memorandum Report No. 538, January 1977.

14. N.P. Robinson, "Measurements on the Effect of Rain, Snow, and Fog on 8-6 mm Radar Echoes," *Proceedings of the IEE*, Vol. 203-B pp. 709-714, September 1955.

15. T. Oguchi, "Electromagnetic Wave Propagation and Scattering in Rain and Other Hydrometers," *Proceedings of the IEEE*, Vol. 71. No. 9, September 1983, pp. 1029-1078.

16. G. Mie, *"Berträge zur Optik trüber Median, spegiell kolloidaler Metallösungen,"Ann. Physik*, Vol. 4, No. 25, 1908, pp. 377-445.

17. E.J. Dutton and F.K. Steele, "Some Further Aspects of the Influence of Raindrop-Size Distributions on Millimeter Wave Propagation," National Telecommunications and Information Administration, Report No. 84-169, December 1984.

18. J.O. Laws and D.A. Parsons, "The Relation of Raindrop Size to Intensity," *Transactions of the American Geophysical Union*, Vol. 24, 1943, pp. 452-460.

19. J.S. Marshall and W.M. Palmer, "The Distribution of Raindrops with Size," *Journal of Meteorology*, Vol. 5, August 1948, pp. 165-166.

20. D.E. Setzer, "Computer Transmission Through Rain at Microwave and Visible Frequencies," *Bell System Technical Journal*, October 1970, pp. 1873-1893.

21. C.R. Burrows and S.S. Atwood, "Radio Wave Propagation, Consolidated Summary Technical Report of the Committee on Propagation, NDRC," Academic Press, New York, 1949, p. 219.

22. K.L.S. Gunn and T.W.R. East, "The Microwave Properties of Precipitation Particles," *Journal of the Royal Meteorological Society*, Vol. 80, 1954, pp. 522-545.

23. N.C. Currie, F.B. Dyer, and R.D. Hayes, "Analysis of Radar Rain Returns at Frequencies of 10, 35, 70, and 95 GHz," Technical Report No. 2 on Contract DAAA 25-73-C-0256, Georgia Institute of Technology, 1975.

24. N.C. Currie, F.B. Dyer, and R.D. Hayes, "MMW Short Course Notes," Georgia Institute of Technology, Atlanta, GA, 1983.

25. W.J. Brinks, "A Discussion of Excessive Rainfall Attenuation at Millimeter Wavelengths," Harry Diamond Laboratories Report No. TM-73-14, July 1973.

26. R.G. Medhurst, "Rainfall Attenuation of Centimeter Waves: Comparison of Theory and Measurements," *IEEE Transactions on Antennas and Propagation*, Vol. 13, July 1985, p. 550.

27. N.P. Krasyuk, V.I. Rozenberg, and D.A. Christyakov, "Radar Characteristics of Precipitation of Different Nature, Spectra, Intensity and Temperatures in the Centimeter and Millimeter Ranges of Radio Waves," prepared by Foreign Technology Division, Wright-Patterson AFB, OH, Report No. FTD-MT-24-246-69.

28. J.R. Norbury and W.J.K. White, "Microwave Attenuation at 35.8 GHz due to Rainfall," *Electronic Letters*, Vol. 8, No. 4, 24 February 1972.

29. T. Ihara and Y. Furuhama, "Frequency Scaling of Rain Attenuation at Centimeter and Millimeter Waves Using a Path-Averaged Drop Size Distribution," *Radio Science*, Vol. 16, No. 6, November–December 1981, pp. 1365-1372.

30. K.L. Ho, N.D. Mavrokoukoulakis, and R.S. Cole, "Rain-Induced Attenuation at 36 GHz and 110 GHz," *IEEE Transactions on Antennas and Propagation*, Vol. AP-26, No. 6, November 1978, pp. 873-875.

31. W.P.M.N. Keizer, J. Sneider, and C.D. de Haan, "Rain Attenuation Measurements at 94 GHz: Comparison of Theory and Experiment," *AGARD Conference Proceedings No. 245, Millimeter and Submillimeter Wave Propagation and Circuits*, AGARD-CP-245, France, September 1978, pp. 44-1 to 44-9.

32. V.W. Richard, J.E. Kammerer, and R.G. Reitz, "140 GHz Attenuation and Optical Visibility Measurements of Fog, Rain, and Snow," US Army Ballistic Research Laboratories, Memorandum Report ARBRL-MR-2000, December 1977.

33. J. Nemarich, *et al.*, "Comparative Near-Millimeter Wave Propagation Properties of Snow and Rain," *Proceedings of Snow Symposium III*, Hanover NH, August 1983, pp. 115-129.

34. T.S. Chu, "Rain-Induced Cross-Polarization at Centimeter and Millimeter Wavelengths," *Bell System Technical Journal*, Vol. 53, No. 8, October 1974, pp. 1557-1665.

35. K.L.S. Gunn and T.W.R. East, "The Microwave Properties of Precipitation Particles," *Quarterly Journal of the Royal Meteorological Society*, Vol. 80, October 1954, pp. 522-545.

36. A. Nishitsuji and A. Matsumato, "Calculation of Radio Wave Attenuation Due to Snowfall," SHF and EHF Propagation in Snowy Districts, Monograph of the Research Institute of Applied Electricity, Hokkardo University, No. 19, 1971, pp. 63-78.

37. V.W. Richard, "Low Angle Tracking at Millimeter Wavelengths," TTCP Ad-Hoc Study Group 102, Electro-Optical Low Angle Tracking, December 1976.

38. J. Nemarich, *et al.*, "Attenuation and Backscatter for Snow and Sleet at 96, 140, and 225 GHz," *Proceedings of Snow Symposium IV*, Hanover, NH, August 1984, pp. 41-52.

39. K.A. Aganbekyan, *et al.*, "The Propagation of Submillimeter, Infrared, and Visible Waves in the Earth's Atmosphere," *Rasprostraneniye Radiovoln*, Institut Radioteckhniki i Electroniki, published by Nauka, 1975, pp. 187-227.

40. Y.M.M. Antar and A. Hendry, "Attenuation of Radio Waves by Atmospheric Wet Ice and Mixed Phase Hydrometers," *Proceedings of the URSI Commission F*, 1983 Symposium, Louvain, Belgium, pp. 455-461.

41. J.E. Knox, "Millimeter Wave Propagation in Smoke," *IEEE EASTCON-79 Conference Record*, Vol. 2, 1979, pp. 357-361.

42. F.C. Petito and E.W. Wentworth, "Measurements of Millimeter Wave Radar Transmission and Backscatter During Dusty Infrared Test II (DIRT II)," Night Vision and Electro-Optics Laboratory, Report No. DELNV-TR-0011, May 1980.

43. F.K. Schwerling, *et al.*, "Effects of Vegetation and Battlefield Obscurants on Point-to-Point Transmission in the Lower Millimeter Wave Region (30–60 GHz)," June 1982.

44. E.E. Altschuler, "The Effects of Low-Altitude Nuclear Burst on Millimeter Wave Propagation," *AGARD Conference Proceedings*, Spatind, Norway, October 1983.

45. N.C. Currie, E.E. Martin, and F.B. Dyer, "Radar Foliage Penetration Measurements at Millimeter Wavelengths," Georgia Institute of Technology, Final Technical Report on Contract DAAA25-73-C-0256, December 1975.

46. E.J. Violette, R.H. Espeland, A.R. Mitz, and F.A. Goodnight, "SHF-EHF Propagation through Vegetation on Colorado East

Slope," Research and Development Technical Report CECOM-81-C-CS020-F, June 1981.

47. E. Yu and D.B Hodge, "Atmospheric Microwave Refractivity and Refraction," Technical Report 712759-1, Contract No. NASW-3393, December 1980.

48. A.I. Omoura and D.B. Hodge, "Microwave Dispersion and Absorption Due to Atmospheric Gases," Ohio State University, Electro-Science Laboratory, Technical Note No. 10, August 1979.

49. H.J. Liebe and D.H. Layton, "Experimental and Analytical Aspects of Atmospheric EHF Refractivity," *Proceedings of the URSI Commission F,* 1983 Symposium, Louvain Belgium, pp. 477-486.

50. T.N. Patton, J.J. Petronic, and J. Teti, "Propagation Effects for MM Wave Fire Control System," 6th DARPA-Tri-Service Millimeter Wave Conference, 1977.

51. H.V. Hitney, "Radar Detection Range Under Atmospheric Ducting Conditions," *Record of the IEEE International Radar Conference,* April 1975, pp. 241-243.

52. R.W. McMillan, J.C. Wiltse, and D.E. Snider, "Atmospheric Turbulence Effects on Millimeter Wave Propagation," *IEEE EASCON-79 Conference Record,* Vol. 1, 1979, pp. 42-47.

53. M.I. Skolnik, *Introduction to Radar Systems,* McGraw-Hill, New York, 1980, pp. 442-446.

54. D.K. Barton, *Radar System Analysis,* Artech House, Dedham, MA, 1979.

55. H.B. Wallace, "140 GHz Multipath Measurements Over Varied Ground Covers," *IEEE EASCON-79 Conference Record,* Vol. 2, 1979, pp. 256-260.

Chapter 5
MMW Clutter Characteristics

N.C. Currie

Georgia Institute of Technology
Atlanta, Georgia

5.1 INTRODUCTION AND DEFINITIONS

5.1.1 Definitions

Radar Clutter

The subject of radar clutter is one of the subject areas of greatest interest to radar scientists and engineers, and the study of clutter dates back to the earliest days of radar. One complicating factor in the study of radar clutter is the fact that it may be different things to different people. For example, to an engineer trying to develop a missile seeker to detect and track a tank, the return from vegetation and other natural objects would be considered "clutter." However, a remote sensing scientist would consider the return from natural vegetation as the primary target, and the return from the tank or other man-made objects would be treated as clutter. Thus, it can be seen that clutter can be defined as follows:

Radar Clutter is the RF return from a physical object or a group of objects that is undesired for a specific radar application.

For the majority of radar applications the desired targets are man-made, and clutter, thus, consists of returns from natural objects. The goal of the radar designer is to eliminate these returns without affecting the return from the desired target. Examples include detecting a small boat in the presence of sea clutter, a tank surrounded by trees, or an airplane in a rain storm. For the remainder of this chapter, clutter returns will be assumed to be primarily returns from natural objects.

Millimeter-wave clutter, as might be expected, describes unwanted radar returns for the frequency region of 30 to 300 GHz. In practice, however, most data exist only within the commonly used radar frequency bands near 35, 95, 140, and 220 GHz. Generally, it is tacitly assumed that the properties of returns between these commonly used bands are similar to those of the bands where data exist.

The properties of MMW clutter are of particular interest since millimeter waves span the frequency region between conventional microwave frequencies and the near infrared region. As such, a combination of the good and bad properties of scattering for the microwave and infrared regions are exhibited by millimeter wave scatterers. Currently, system designers are attempting to exploit the helpful properties while minimizing the bad characteristics. In order to do this, detailed knowledge of millimeter wave scattering is required.

Radar Scattering Matrix

The reflectivity from clutter or any target can be completely described by the *radar scattering matrix* (sometimes called the *polarization scattering matrix*). This matrix describes the radar reflectivity characteristics in terms of two orthogonal transmitted polarizations and two orthogonal received polarizations. Thus, for vertical and horizontal polarizations the received signals from a target can be expressed as

$$E_{rH} = a_{HH}E_{tH} + a_{VH}E_{tV} \tag{5.1}$$

$$E_{rV} = a_{HV}E_{tH} + a_{VV}E_{tV} \tag{5.2}$$

or in matrix form:

$$\begin{vmatrix} E_{rH} \\ E_{rV} \end{vmatrix} = \begin{vmatrix} a_{HH} & a_{VH} \\ a_{HV} & a_{VV} \end{vmatrix} \cdot \begin{vmatrix} E_{tH} \\ E_{tV} \end{vmatrix} \tag{5.3}$$

Each of the a_{ij} components of the matrix is complex, that is, they contain both an amplitude and a phase term. The terms of the matrix can be related to radar cross section by the equation:

$$a_{ij} = \sqrt{\sigma_{ij}}\ e^{j\phi_{ij}} \tag{5.4}$$

For the monostatic radar case, generally, a_{ij} equals a_{ji} so that there are only three independent coefficients in the scattering matrix.

The scattering components for circular polarization can be related to those for linear polarization by the following relationships:

$$|C_{RR}| = |\tfrac{1}{2}(a_{HH} - a_{VV}) + ja_{HV}| \tag{5.5}$$

$$|C_{LL}| = |\tfrac{1}{2}(a_{HH} - a_{VV}) - ja_{HV}| \tag{5.6}$$

$$|C_{RL}| = |\tfrac{1}{2}(a_{HH} + a_{VV})| \tag{5.7}$$

For the case of circular transmission and linear polarization reception (one form of noncoherent polarimetric processing), the relationships are as follows:

$$|a_{LH}| = \tfrac{1}{2}|a_{HH} + ja_{VH}| \tag{5.8}$$

$$|a_{LV}| = \tfrac{1}{2}|a_{HV} + ja_{VV}| \tag{5.9}$$

$$|a_{RH}| = \tfrac{1}{2}|a_{HH} - ja_{VH}| \tag{5.10}$$

$$|a_{RV}| = \tfrac{1}{2}|a_{HV} - ja_{VV}| \tag{5.11}$$

The phase between the horizontally and vertically polarized components of the received electric field is called the *polarimetric phase* Φ. This parameter has been used in some MMW systems as a descriminant for separating targets from clutter (see Chapters 2 and 6). The polarimetric phase can be computed by performing a dot-product vector operation between the horizontal and vertical received components of the electric field:

$$\overline{E}_{rH} \cdot \overline{E}_{rV} = |\overline{E}_{rH}|\,|\overline{E}_{rV}|\cos\Phi \tag{5.12}$$

or

$$\cos\Phi = \frac{a_{HH}\,a_{HV} + a_{VH}\,a_{VV}}{\sqrt{a_{HH}^2 + a_{VH}^2}\,\sqrt{a_{HV}^2 + a_{VV}^2}} \tag{5.13}$$

Area and Volume Scattering

In general, natural clutter consists of many scatterers located within the radar resolution cell of a radar defined by its antenna beam shape, grazing angle to the ground, and pulse length. Since the number of scatterers illuminated by the radar is a function of the resolution cell size, the backscattered energy also varies with the cell size. To account for this fact, the concepts of *sigma zero* ($\sigma°$) and *eta* (η) were introduced in the late 1940s. Sigma zero is defined for surface scattering by the equation:

$$\sigma° = \sigma/A \tag{5.14}$$

where

$\sigma°$ = radar reflectivity per unit area,
σ = radar cross section of the illuminated area on the surface,
A = illuminated surface area.

The units of $\sigma°$ are m^2/m^2, and they are often expressed as decibels, for example, -10 dB. Reflectivity data will be sometimes expressed in terms of the parameter *gamma* (γ), where γ is related to $\sigma°$ by the equation:

$$\gamma = \sigma°/\sin\Theta \tag{5.15}$$

where

γ = radar scattering coefficient,
$\sigma°$ = radar cross section per unit area,
Θ = grazing angle between the radar beam and the surface.

The area subtended by the radar illumination on the surface can be computed by considering the geometry shown in Figures 5.1 and 5.2. The figures give the geometry for two cases: case 1, where the grazing angle to the surface is large, and the area illuminated by the radar is determined by the azimuth and elevation beamwidths; case 2, where the grazing angle is small, and the area illuminated by the radar is determined by the azimuth angle and the transmitted pulse width.

From Figure 5.1, for the case of large grazing angle, the area illuminated is an ellipse in which the width is equal to $2R \tan (\Phi_{AZ}/2)$, and the length is $2R \tan (\Phi_{EL}/2) \csc(\Theta)$. Thus, the area of the ellipse is π times the product of the two radii, or

$$A = (\pi R^2/\alpha^2) \tan(\phi_{AZ}/2) \tan(\Phi_{EL}/2) \csc(\Theta) \tag{5.16}$$

where

A = the area illuminated on the surface,
R = the slant range to the surface,
α = the beam shape factor* = 1.33 (for a Gaussian shaped beam),
Φ_{AZ} = the azimuth 3 dB two-way beamwidth,
Φ_{EL} = the elevation 3 dB two-way beamwidth,
Θ = the grazing angle. (Note that for a flat earth the grazing angle (measured relative to the ground plane) is equal to the grazing angle (measured relative to the horizon).)

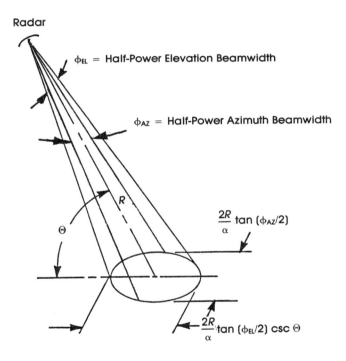

$$\tan \Theta > \frac{2\pi R \tan (\phi_{EL}/2)}{\alpha(c\tau/2)}$$

Fig. 5.1 Beam-limited resolution cell (adapted from Trebits [1], © 1984 Artech House).

---

*The beam shape factor accounts for the fact that the radar beam is not rectangular, and therefore it varies in gain between the 3 dB points.

For small beamwidths of less than 10°, the small angle approximation of $\tan(\phi) \approx \phi$ can be invoked, and (5.16) becomes:

$$A = \left(\frac{\pi R^2}{4\alpha^2}\right)\Phi_{AZ}\Phi_{EL}\csc(\Theta) \tag{5.17}$$

For the case of low grazing angle, the length of the illuminated area is determined by the transmitted pulsewidth rather than the elevation beamwidth, as shown in Figure 5.2. Thus, the length equals $c\tau/2 \sec(\Theta)$ (i.e., the pulse length mapped onto the surface), and the width remains the same as for the first case. If the pulse length mapped onto the ground is small relative to the length of the ellipse defined by the elevation beamwidth, then the area is approximately a rectangle, and its area is the product of the length times the width, or

$$A = (Rc\tau/\alpha) \tan(\Phi_{AZ}/2) \sec \Theta \tag{5.18}$$

where

    c = speed of light (3×10^8 m/s);
    τ = transmitted pulse length, and the other parameters are as defined above.

Radar

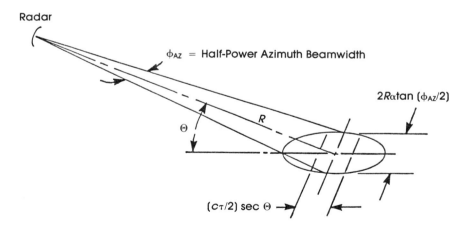

$$\tan \Theta < \frac{2\pi R \tan (\phi_{EL}/2)}{\alpha(c\tau/2)}$$

Fig. 5.2 Pulse-limited resolution cell (from Trebits [1], © 1984 Artech House).

Once again for small beamwidths, A is given by

$$A \approx \frac{Rc\tau\Phi_{AZ}}{2\alpha}\sec(\Theta) \tag{5.19}$$

The transition from the beam-limited case to the pulse-limited case can be found by setting (5.16) equal to (5.18). The resulting value for Θ is

$$\tan\Theta = \frac{2\pi R \tan(\Phi_{EL}/2)}{\alpha(c\tau/2)} \tag{5.20}$$

An analogous normalization parameter to $\sigma°$ exists for volume scatterers such as hydrometeors in the atmosphere. This parameter is called η or σ^v, and it is given by

$$\eta = \text{RCS of illuminated volume/illuminated volume}$$

The illuminated volume can be computed from a consideration of the geometry shown in Figure 5.3. The illuminated volume is an elliptical cylinder in which the faces are perpindicur to the radar line of sight, and the height is defined by the transmitted pulse length. If it is assumed that the volume is far from the radar, then the two diameters of the faces are $R\Phi_{AZ}$ and $R\Phi_{EL}$, and the height is $c\tau/2$. The volume of an elliptical cylinder is π times the the product of the two radii times the height. Thus, the volume is

$$V = \frac{\pi R^2 \Phi_{AZ}\Phi_{EL}c\tau}{8\alpha^2} \tag{5.21}$$

5.1.2 Differences between MMW and Microwave Reflective Properties

Obviously, the primary difference between microwave and MMW scattering is the wavelength of the energy. Since the difference in frequency between the upper part of the microwave region (K_u band) and the lower part of the MMW region (K_a band) is only a factor of two, large differences in reflectivity properties of scatterers at the two bands would not be expected, and reflectivity properties at the two bands are similar to a certain extent. However, there are differences that would not be expected for such a small change in frequency. These differences can be explained in terms of resonance effects. For example, if we consider the scattering of

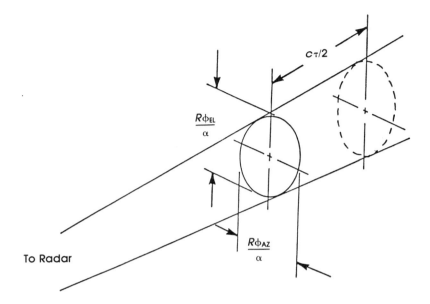

Fig. 5.3 Volumetric resolution cell (from Trebits [1], © 1984 Artech House).

electromagnetic energy by symmetrical bodies, such as cylinders and spheres, then the scattering properties are determined by the ratio of the normalized body diameter to the wavelength (Trebits [1]). In particular, for spherical bodies, the scattered waves can be described by Bessel equations. When the argument of the Bessel equation is small (i.e., if the radius of curvature is much smaller than the wavelength), then we use only the first term of the series representation of the Bessel function. The resulting solutions of Maxwell's equations take the same form as used by Rayleigh to describe the scattering of light from small particles. An effective cross section is defined in terms of the ratio of the reflected power to incident power, and the equation takes the form:

$$\sigma = 4\pi R^2 \frac{S_{\text{refl}}}{S_{\text{inc}}} = \pi^5 |K|^2 \frac{D^6}{\lambda^4} \tag{5.22}$$

where

K = function of the material of the scattering object,
D = diameter of the scattering object,
λ = wavelength.

When the argument of the Bessel function is large (i.e., the radius of curvature is large compared to the wavelength), then Maxwell's equations reduce to the same form employed in geometrical optics. The resulting effective cross section is independent of wavelength, and takes the form:

$$\sigma = \frac{\pi D^2}{4} \tag{5.23}$$

where D = the diameter of the sphere.

The region represented by ratios of diameter to wavelength between the Rayleigh and optical regions has generally been called the *Mie scattering region*. The Bessel functions must be applied in detail, and the typical decaying oscillatory characteristic is evident [2]. The three regions of scattering are generally described in terms of the object diameter D and the radar signal's wavelength λ as

Rayleigh scattering–$0 < \pi D/\lambda < 1$
Mie scattering–$1 < \pi D/\lambda < 10$
Optical scattering–$10 < \pi D/\lambda < 100$

Figure 5.4 shows the effective normalized cross section of a sphere for the three scattering regions.

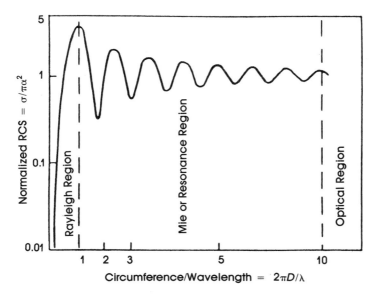

Fig. 5.4 Radar cross section of a conducting sphere showing three scattering regions (Rayleigh, Mie, optical).

At millimeter wavelengths, many scattering objects range from the Rayleigh region through the Mie region and into the optical region. Table 5.1 gives the typical sizes of many scattering objects and the ratios of D/λ at several millimeter wavelengths.

Table 5.1   Ratio of Typical Diameter to Wavelength of Common Scatterers at Millimeter Wave Frequencies.

Scatterer	Diameter (mm)	Ratio of Diameter to Wavelength (D/λ)			
		35 GHz	95 GHz	140 GHz	300 GHz
Raindrops	(0.2–6)	0.02–0.7	0.6–2	0.09–2.8	0.2–6
Sea Spray	(0.2–10)	0.02–1.15	0.6–3.3	0.09–4.7	0.2–6
Pine Needles	(0.5–1.5)	0.057–0.171	0.167–0.5	0.23–0.7	0.5–1.5
Screw Heads	(1.5–25)	0.17–2.9	0.5–8.3	0.7–11.7	1.5–25
Rivets	(10)	1.15	3.3	4.7	10
Grass Blades	(2–8)	0.23–0.92	0.7–2.7	0.93–3.7	2–8
Deciduous Leaves	(6–20)	0.7–2.31	2.0–6.7	2.8–9.4	6–20
Branches	(5–76)	0.7–8.78	2.0–25	2.8–35.5	6–76
Snow Crystals	(5–50)	0.58–5.77	1.7–16	2.3–23.3	5–50
Hail	(1–10)	0.12–1.15	0.3–3.3	0.47–4.7	1–10

A second cause for differences between MMW and microwave reflective properties involves the effect of surface roughness on scattering. Generally, a surface appears rough at a specific wavelength when the rms surface roughness is greater than $\lambda/8$ for perpendicular incidence. For other incidence angles, the effective surface roughness is defined by the *Rayleigh roughness criterion*, which states that a rough surface no longer appears rough to microwave energy when

$$\delta h \, \sin\Theta \leqslant \lambda/8 \tag{5.24}$$

where

   δh = rms surface roughness of the surface,
   Θ = grazing angle to the surface,
   λ = the wavelength of the incident energy.

Since millimeter wavelengths are up to a factor of 10 smaller than corresponding microwave wavelengths, surface roughness is much more important. The primary effects appear to be the disappearance of the smoothness of large surfaces, and thus disappearance as well of the expected, corresponding, large specular returns and deep nulls in the patterns of complex scatterers. Clutter almost always appears rough at these frequencies (even relatively smooth pavement can appear rough).

The third major cause for differences between microwave and MMW scattering properties has to do with the effects of atmospheric gasses and water in the two regions. In addition to rain and other hydrometeors spanning the Mie-resonance scattering region, oxygen and water in its various forms exhibit different properties as the frequency increases, including absorption regions of high loss, which occur periodically at various frequencies starting at 20 GHz (see Chapter 4). There is also a gradual increase of the attenuation in the "window" frequency bands as the frequency increases.

5.2 MMW ATMOSPHERIC CLUTTER CHARACTERISTICS

5.2.1 Overview

Scattering from particles in the atmosphere is generally less of a problem at microwaves than at MMWs. At the MMW bands, even very light precipitation can cause appreciable backscattered return. Thus, the atmosphere can interfere with the signal from a target by both attenuating it and returning an interfering signal. This section will discuss atmospheric scattering data that are currently available in the unclassified literature, and will summarize the results.

5.2.2 Scattering from Dust, Debris, and Smoke

Although considerable effort has been expended in recent years to measure the return from airborne aerosols, such as debris or smokes, most of the results are restricted and cannot be presented here. However, the return from dust and debris produced by a large explosion at 35 GHz has been reported by Martin [3], and compared to the return at 10 GHz as shown by Figure 5.5(a and b). As can be seen from the figure, the value of η for dust is very similar between 10 and 35 GHz, except that the 35 GHz return is almost 10 dB higher. Note that after 20 s have passed from the initial explosion, there is still considerable return at 35 GHz. However, the explosion utilized for this test was equilivent to several hundred tons of TNT, and one should not expect to encounter reflectivity values that would be this high for dust storms or smaller explosions.

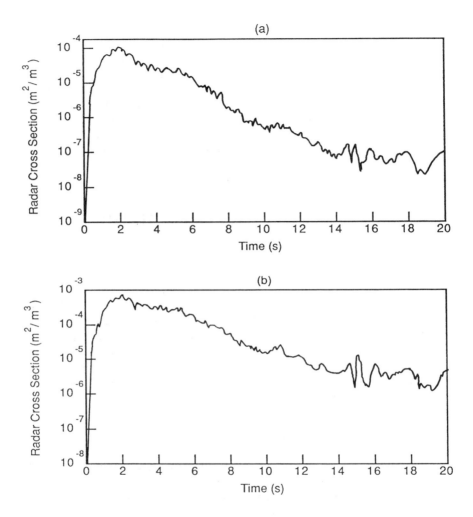

Fig. 5.5 Maximum backscatter as a function of time from a large dust explosion (a) at 10 GHz and (b) at 35 GHz (from Martin [3]).

5.2.3 Scattering from Rain

Average Values

The backscatter from rain is appreciable at millimeter wavelengths. According to theory, as discussed in section 5.1.2, in the Rayleigh scattering region the backscatter from rain increases rapidly with decreasing wavelength until the ratio of the drop diameters to the RF wavelength of the rain approaches one, whereupon the backscatter increase becomes much less as the wavelength approaches the optical region. For a typical rain storm, the number of drops does not increase appreciably as the rain rate increases, but rather the size of the drops increases. Thus, for Rayleigh scattering, the backscatter increases rapidly with increasing rain rate, while for higher frequencies this strong dependence moderates. This is illustrated by Figure 5.6, which gives the theoretically calculated reflectivity of rain, assuming a Marshall-Palmer drop size distribution. (The Marshall-Palmer distribution is often considered to be a good model for frontal rain storms.) This theoretical model shows a strong dependence of the backscatter on rain rate up through 30 GHz, with a weaker dependence on rain rate at 100 GHz, and almost no dependence on rain rate at 300 GHz.

Although few data have been obtained on the reflectivity of rain at MMWs, one major experiment was performed by the US Army Ballistic Research Laboratory (BRL) in 1973. In that experiment, four radars at 10, 35, 70, and 95 GHz with matching antenna sizes and pulsewidths were utilized to obtain simultaneous radar backscatter data from rain. Data were collected in two ways: by making A-scope photographs of the rain returns and by recording the rain returns on magnetic tape. BRL reduced the A-scope data and Georgia Tech reduced the magnetic tape data. The results of the two different analyses are presented in Figure 5.7. Plotted on the figure are least-square fits to the data resulting from each analysis. As we can see, the 10 GHz least-square-fit lines are identical, and the 70 GHz and 95 GHz lines are similar, although not quite identical. This is not too unlikely, since the data spread around each least-squares-fit line was 10 to 15 dB. However, the 35 GHz least-square-fit lines are quite different and have quite different slopes. Comparison of the two 35 GHz lines to the theoretical curve shows the BRL data have about the right value but the wrong slope, while the Georgia Tech data have the right slope but appear to be 10 dB low. Both analyses have potential problems. The BRL data were read by hand from an A-scope presentation. The 35 GHz radar had a "lin-log" receiver characteristic, which may have caused an overestimation of small values, thus resulting in a flatter slope. Also, the peak values of the data were read from the scope as opposed to the average data, which would tend to make the BRL data somewhat high. The Georgia Tech data utilized an estimate of the attenuation because the actual attenuation was not recorded on tape, and thus may have an error in the attenuation used to calculate the reflectivity. Also, the Georgia Tech approach averaged log data, whereas the theoretical data assume the average is of linear data. This difference would make the Georgia Tech data appear low.

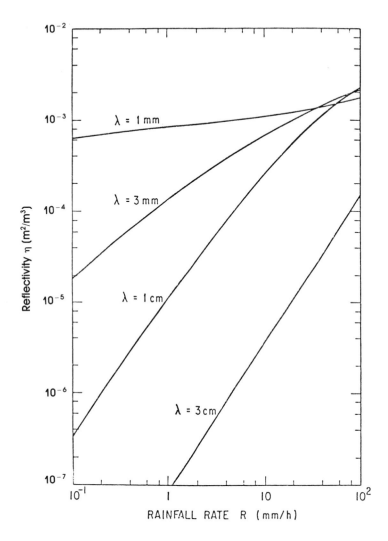

Fig. 5.6 Theoretical rain reflectivity *versus* rainfall rate with Marshall-Palmer drop size distribution (adapted from Mitchell [4]).

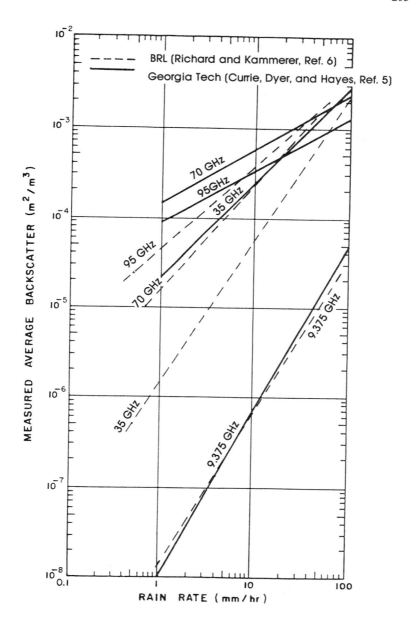

Fig. 5.7 Comparison of the measured backscatter coefficient of rain *versus* rain rate for BRL and Georgia Tech data.

In summary, the 10, 70, and 95 GHz data in Figure 5.7 agree reasonably well, but a large discrepancy exists at 35 GHz for low rain rates. Barring a new experiment, this problem cannot be resolved at present. Thus, the radar designer is "on his or her own" to decide which data set is more reasonable.

Amplitude Distributions

Traditionally, the return from rain has been considered to be Rayleigh distributed (see Chapter 3 for a definition of the Rayleigh distribution). However, at MMW frequencies, the returns have been measured, and the distributions appear to be better approximated by log-normal distributions, as shown in Figure 5.8. For the cumulative distribution shown, log-normally distributed data would appear as a straight line, which closely approximates the actual data shown. The apparent non-Rayleigh statistics could be due to unusual drop size distributions generated by the thunderstorms, or because of the fact that logarithmic receivers were used in the experiment. Figures 5.9 and 5.10 give measured standard deviations for the rain returns at 35 and 95 GHz. The standard deviations typically range between 2.5 and 6 dB. For a Rayleigh distribution, one would expect a standard deviation of approximately 3.7 dB.

Frequency Spectra

Since rain drops are falling and often blown by the wind, a spectral broadening of the return signal is created by the phase changes of the signal from each scatterer as the round-trip path length varies. Such spectral broadening interferes with the ability of a Doppler or moving-target-indication (MTI) radar to detect a moving target. Traditionally, a Gaussian spectrum has been assumed for clutter spectra based on randomly sized scatterers moving at random velocities, but in recent years non-Gaussian spectra have been measured by several experimenters. Data measured by Fishbein, *et al.* [6] on the returns from foliage indicated that the spectrum is of the form:

$$W(f) = \frac{A}{1 + (f/f_c)^n} \tag{5.25}$$

where

A = zero-frequency spectral density, in w/Hz;
f_c = the "corner frequency";
n = a positive number (equal to 3, according to Fishbein).

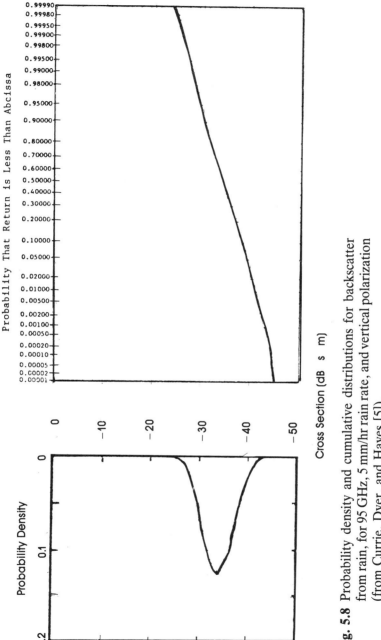

Fig. 5.8 Probability density and cumulative distributions for backscatter from rain, for 95 GHz, 5 mm/hr rain rate, and vertical polarization (from Currie, Dyer, and Hayes [5]).

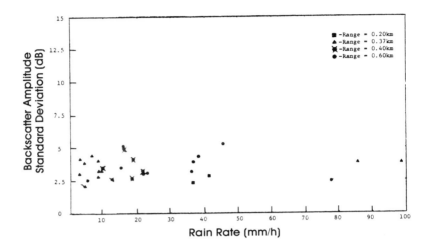

Fig. 5.9 Standard deviation *versus* rain rate for 35 GHz rain data (from Currie, Dyer, and Hayes [5]).

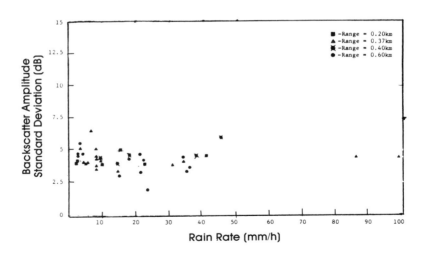

Fig. 5.10 Standard deviation *versus* rain rate for 95 GHz rain data (from Currie, Dyer, and Hayes [5]).

Analysis by Georgia Tech of the spectra of rain returns from the BRL experiment described above yielded a similar non-Gaussian spectral roll-off, except that measured values for n varied from 3 at 10 GHz to 2 for 35 and 95 GHz. This variance is illustrated in Figures 5.11 and 5.12. The Georgia Tech analysis also indicated that the corner frequency f_c increases with increasing frequency and rain rate. The cause of these non-Gaussian spectra are subject to debate at the present time. One physical explanation is the effect of turbulence in thunderstorms (since the data analyzed were from thunderstorms), which might give rise to higher frequencies than would normally be predicted.

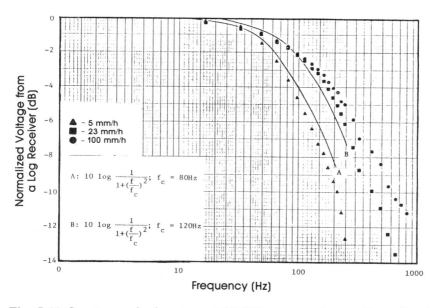

Fig. 5.11 Spectrum of rain return at 35 GHz *versus* rain rate (from Currie, Dyer, and Hayes [5]).

Decorrelation Times

Another way to describe the frequency spectrum of the return from clutter is to use the autocorrelation function because the power spectral density and autocorrelation function are Fourier transform pairs. However, the autocorrelation function is often used for analysis in a different manner

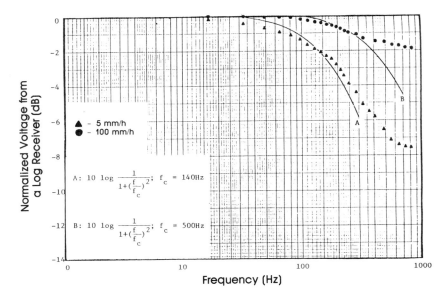

Fig. 5.12 Spectrum of rain return at 95 GHz *versus* rain rate (from Currie, Dyer, and Hayes [5]).

than the frequency spectra. The autocorrelation function indicates the time required for a given signal to become uncorrelated. Thus, if the decorrelation time (the time for the autocorrelation function to decay to some value, often $1/e$) is known, then the maximum sample rate which will yield uncorrelated (noise like) samples can be determined. If the sample rate (i.e., the pulse repetition frequency (PRF)) exceeds this value, then no additional integration gain can be achieved by averaging the additional samples. Figure 5.13 gives the measured decorrelation times (and the corresponding maximum sample rate) for rain as a function of frequency and rain rate. As we can see, the decorrelation time decreases with increasing frequency and rain rate. The decorrelation time is a function of the wind shear and the beamwidth, as discussed by Nathanson [28]. For these data, the beamwidths were constant for each frequency band (1°), and a number of data samples from different storms were averaged in order to remove wind shear effects. Thus, these data should represent the internal turbulence of the rain cell and give the relative decorrelation times as a function of frequency.

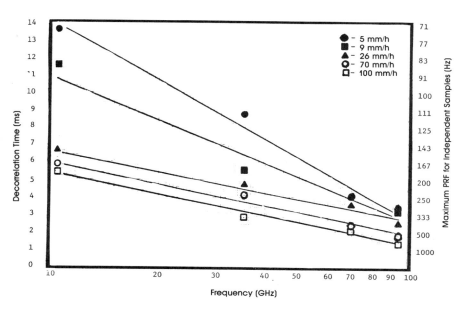

Fig. 5.13 Decorrelation time *versus* rain rate and frequency for rain backscatter (from Currie, Dyer, and Hayes [5]).

5.2.4 Scattering from Snow

In recent years, considerable effort has been expended by the US Army to measure the MMW backscatter from falling snow. As a result, data currently exist from 35 GHz up through 220 GHz, although the 35 GHz data are not as prevalent as the higher frequency data. Nemarich, *et al.* [8, 9] have been leaders in the collection of this type of data. Figure 5.14 compares the radar reflectivity of falling snow at 95 and 225 GHz as a function of the time of day and the snow concentration (the average weight of snow within a given cubic meter of air at a given time). As one can see, the snow reflectivity at 225 GHz is consistently higher than at 96 GHz. Figures 5.15 through 5.17 give the dependence of the snow backscatter on snow concentration at 96, 140, and 225 GHz for the same day. As can be seen, the reflectivity is strongly dependent on snow concentration up to 0.2 gm/m^3, while less dependence is exhibited above this value.

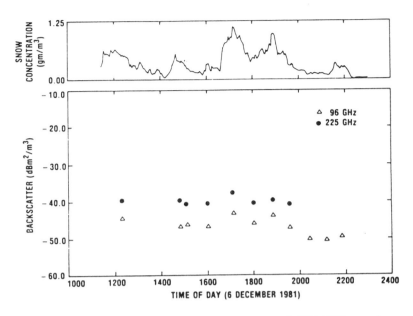

Fig. 5.14 Backscatter coefficients for 96 GHz and 225 GHz, and corresponding snow mass concentration (from Nemarich, *et al.* [8]).

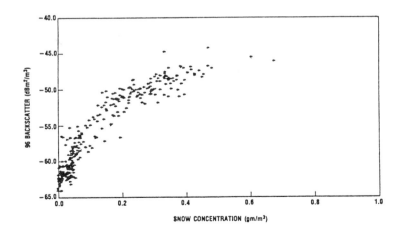

Fig. 5.15 Dependence of 96 GHz backscatter levels on snow mass concentration for 14 December 1983 (from Nemarich, *et al.* [9]).

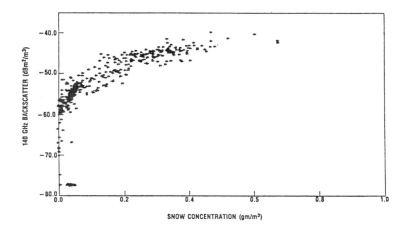

Fig. 5.16 Dependence of 140 GHz backscatter levels on snow mass concentration for 14 December 1983 (from Nemarich, *et al.* [9]).

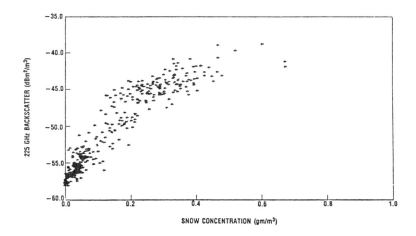

Fig. 5.17 Dependence of 225 GHz backscatter levels on snow mass concentration for 14 December 1983 (from Nemarich, *et al.* [9]).

The dependence of the snow reflectivity on frequency is further illustrated by Figures 5.18 and 5.19, which compare 96 and 140 GHz backscatter with 140 and 225 GHz backscatter measured at the same time. From Figure 5.18, 140 GHz is clearly higher than 96 GHz. (The line indicates equality between the values.) However, Figure 5.19 shows little difference between 225 and 140 GHz reflectivity. Other conclusions made by Nemarich, *et al.* [8, 9] include: snow backscatter at 96 GHz was somewhat higher for rain than for snow; snow backscatter was higher on days in which mixed precipitation fell as opposed to snow only; no dependence on polarization for either vertical, horizontal, or circular was observed for snow at 96 GHz.

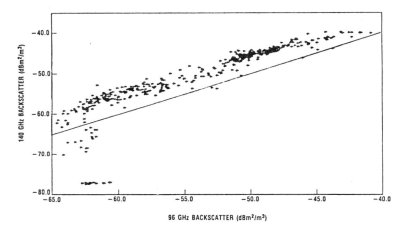

Fig. 5.18 Comparison of simultaneously measured snow backscatter levels at 96 and 140 GHz for 14 December 1983 (from Nemarich, *et al.* [9]).

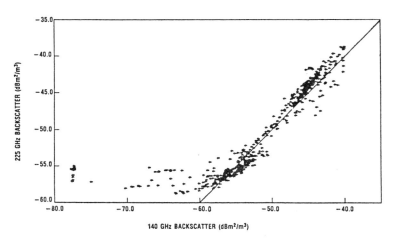

Fig. 5.19 Comparison of simultaneously measured snow backscatter levels at 140 and 225 GHz for 14 December 1983 (from Nemarich, *et al.* [9]).

5.3 MMW SURFACE SCATTERING CHARACTERISTICS

5.3.1 General Comments

At microwave frequencies, scattering from the surface of the earth can generally be described by the classical theory for scattering from rough surfaces. The radar cross section per unit area, $\sigma°$, has the general dependence on grazing angle as shown in Figure 5.20. At large grazing angles near 90°, the value of $\sigma°$ increases rapidly with increasing angle and reaches a maximum at 90°. At low grazing angles below the Rayleigh critical angle (see Section 5.1.2), the surface appears "smooth," and the reflectivity rapidly decreases with decreasing angle. Between these two extremes is the plateau region in which the reflectivity is a weaker function of angle.

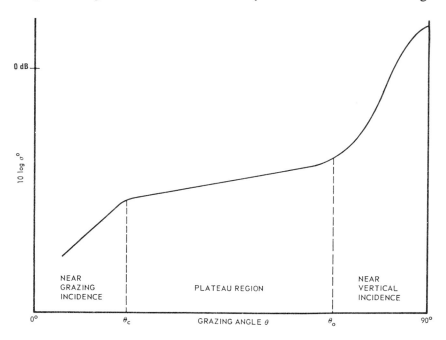

Fig. 5.20 General dependence of $\sigma°$ on depression angle (from Long [10], © 1984 Artech House).

At millimeter wavelengths, the reflective properties differ in two important ways: the critical angle is smaller because of the smaller wavelengths and resonant scatterers are often present. Thus, although the general dependence on depression angle shown in Figure 5.20 is still exhibited at MMWs, there are some surprises. The following sections will discuss the specific characteristics of MMW surface scattering.

5.3.2 MMW Land Clutter

Average Values

In general, average clutter values for millimeter wavelengths are higher than those for microwaves. Figure 5.21 gives the range of reflectivity values measured for several types of clutter from 3 mm to 3 cm in wavelength. The reflectivity increase with decreasing wavelength is approximately $1/\lambda^n$, where $0 \leq n \leq 1$.

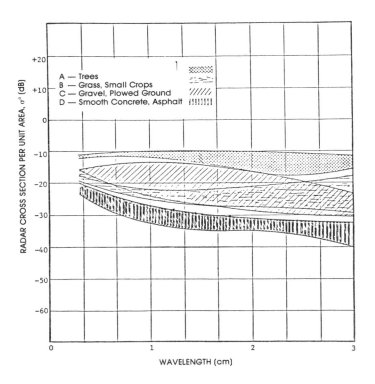

Fig. 5.21 Clutter σ° values between 20° and 70° depression angle as a function of wavelength (from Dyer and Hayes [11]).

Most of the currently available MMW land clutter data have been measured at 35 and 95 GHz. Some programs are underway to measure clutter at higher frequencies, but the results are not available as of this writing. Thus, the data presented here will be for 35 and 95 GHz only.

Figures 5.22 and 5.23 present data from several sources for the reflectivity of grass and crops at 35 and 95 GHz, respectively. Chapter 7 (Table 7.5) presents a MMW land clutter model of the form $A(\Theta + C)^B$, where Θ is the grazing angle, and A, B, and C are arbitrary constants that have been empirically fit to available MMW data.

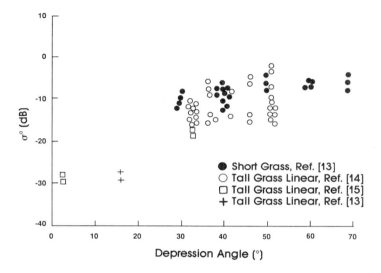

Fig. 5.22 Summary of available data on grass and crops reflectivity at 35 GHz (from Currie, Echard, and Efurd [12]).

For the 35 GHz data, it was convenient to divide the data into two groups, representing short grass and tall grass. No such distinction could be made for the 95 GHz data. There is little difference in the reflectivity between the 35 GHz data and the 95 GHz data, except that the critical angle seems to be lower for the 95 GHz data, as would be predicted by theory.

Figures 5.24 and 5.25 present data on the reflectivity of trees at 35 and 95 GHz, respectively. For this case, the 95 GHz data are several dB higher than the 35 GHz data, and neither data set shows the decrease in reflectivity that would be expected below the critical angle. However, this is not too surprising because trees should be the roughest terrain of all. Interestingly enough, although the tree data exhibit more variations in the measured values than do the grass data, the average values are lower in magnitude.

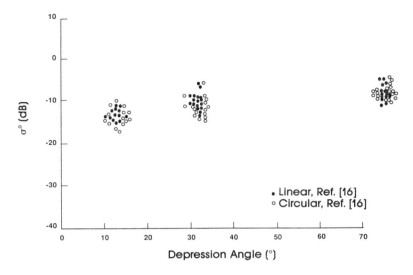

Fig. 5.23 Summary of available data on grass and crops reflectivity at 95 GHz (from Currie, Echard, and Efurd [12]).

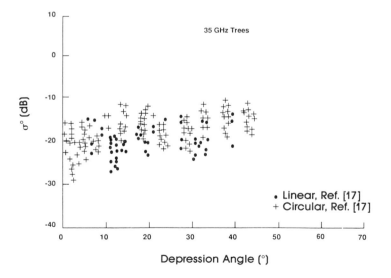

Fig. 5.24 Summary of available data on deciduous trees reflectivity at 35 GHz (from Currie, Echard, and Efurd [12]).

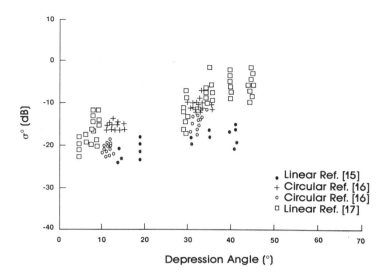

Fig. 5.25 Summary of available data on deciduous trees reflectivity at 95 GHz (from Currie, Echard, and Efurd [12]).

Figures 5.26 and 5.27 give the reflectivity of snow-covered ground as a function of depression angle at 35 and 95 GHz, respectively. The data are divided into two groups: wet snow and dry (or refrozen) snow. The reason for this is that measurements have determined the reflectivity of snow to be a function of the snow state. Thus, wet snow has relatively low reflectivity, while refrozen, dry snow can be much larger in magnitude. As can be seen from Figures 5.26 and 5.27, there is a difference of almost 10 dB between the wet snow and the refrozen snow, on the average. Also, the reflectivity is higher at 95 GHz, with σ° being greater than 0 dB on occasion. The reflectivity of the snow can change rapidly with time, as shown by Figure 5.28, which gives the return from a snow covered field at 35 GHz as a function of time. For these data, the sun began illuminating the snow-covered ground at 8:30 a.m. causing rapid melting, as indicated by the rapid increase of free water in the snow. Almost immediately, a corresponding decrease in reflectivity occurred. Attempts have been made to develop a parametric relationship between free water and snow reflectivity, but these efforts have met with limited success. The notable change in reflectivity with snow state appears to be greatest at MMWs, as shown by Figure 5.29 which compares the return from snow-covered ground as a function of time of day at 9, 17, and 35 GHz. As we can see, the 35 GHz data show the largest backscatter and the greatest variation with time. Apparently, some sort of resonance phenomenon is involved in the scattering mechanism that leads to the largest effects occurring at MMWs.

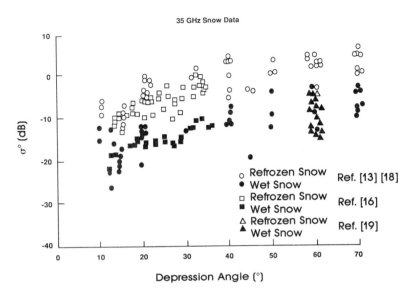

Fig. 5.26 Summary of available data on snow reflectivity at 35 GHz (from Currie, Echard, and Efurd [12]).

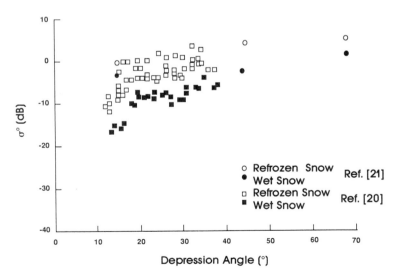

Fig. 5.27 Summary of available data on snow reflectivity at 95 GHz (from Currie, Echard, and Efurd [12]).

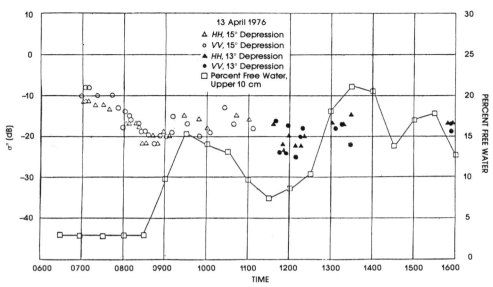

Fig. 5.28 Radar backscatter per unit area from two snow-covered fields as a function of time and the percent of free water present in the top snow layer for 35 GHz (from Currie, Dyer, and Ewell [22]).

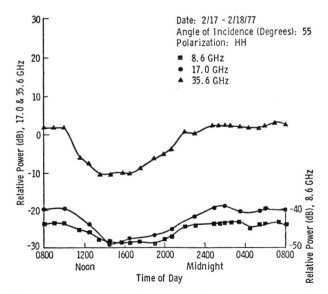

Fig. 5.29 Diurnal variation of the received power measured by the radar at 55° angle of incidence (from Stiles and Ulaby [18]).

Spatial Variations

Spatial variations of clutter are important whenever a radar scans the land surface to look for specific targets of interest. Since the terrain is not generally uniform, variations occur in the backscattered signal to the radar. The typical terrain consists of areas with relatively uniform properties that are adjacent to other areas with different features, but also are relatively uniform, such as a field next to a forest. Thus, one must not only consider variations within a given terrain type, but one must also consider what happens when the radar beam scans across different types of terrain. Generally, at microwave frequencies, the spatial distributions within uniform areas are Rayleigh distributed, except for elevation angles of less than 5° (Boothe [23]). (See Chapter 4 for a definition of the Rayleigh distribution.) Some data on the spatial distributions of various types of terrain have been measured at 95 GHz and some of the results are listed in Table 5.2, which gives the measured standard deviations of uniform clutter areas at three depression angles, from Lane [24]. When the statistics of the radar return from a number of terrain types are considered, the distribution appears more nearly log-normal, and the standard deviation is greater. Figure 5.30 gives the spatial distribution of a number of terrain backgrounds for 95 GHz, from Lane [24]. The scale that is used results in a straight line if the data are log-normally distributed. As can be seen, a straight line nearly fits the data.

Table 5.2   95 GHz Standard Deviations for Spatial Distributions (HH Polarization) (from Lane [16]).

Clutter Type	Depression Angle (°)	Standard Deviation (dB)
Barley	12	1.1
	30	2.1
Coniferous Trees	12	1.4
	30	1.7
	75	2.25
Corn	12	1.0
	30	1.7
	75	2.45
Deciduous Trees	12	2.0
	30	2.0
Plowed Field	12	1.0
	30	1.5
	75	2.5
Grass	12	0.9
	30	1.9
	75	3.3

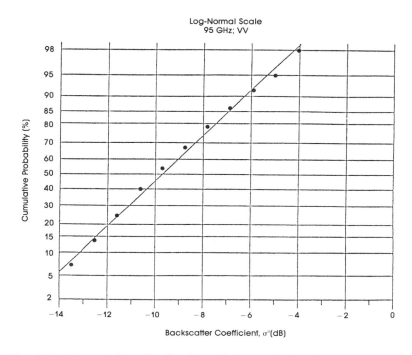

Fig. 5.30 Cumulative distribution of the spatial variations for a number of clutter types at 95 GHz (from Lane [16]).

The measured standard deviations in Table 5.2 are relatively small because the radar that was used to collect the data averaged them over a 250 MHz bandwidth, and thus greatly reduced the variations in the returns. If wideband averaging were not used, then the variations could be much greater. Figure 5.31 gives the measured standard deviation of the spatial variation of snow-covered ground at 35 GHz as a function of averaging bandwidth. For 5 MHz bandwidth, the standard deviation is almost 6 dB, while for 640 MHz the standard deviation is approaching 1 dB. Thus, wideband frequency averaging can be seen to be very helpful in reducing variations in uniform terrain.

Temporal Fluctuations

As was indicated in Section 5.1, temporal fluctuations are those variations that occur within a single radar cell with passing time, and generally can be described by amplitude statistics and frequency spectra or decorrelation times. Some data have been obtained on the amplitude variations of trees and grass at MMWs, and are given in Table 5.3, which summarizes

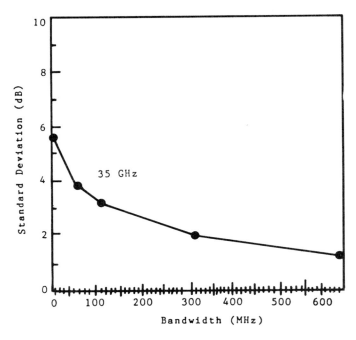

Fig. 5.31 Standard deviation *versus* bandwidth for dry snow at 35 GHz (from Trostel, *et al.* [24]).

the measured standard deviations of several types of trees and grass at 10, 16, 35, and 95 GHz. At 95 GHz, the standard deviation is larger and the distributions more closely approximate a log-normal as opposed to a Rayleigh distribution. Figure 5.32 gives a temporal amplitude distribution for deciduous trees at 35 GHz, while Figure 5.33 gives the distribution for the same clutter cell for 95 GHz. As we can see, the 35 GHz data are more closely approximated by a Rayleigh distribution, while the 95 GHz data more closely resemble a log-normal distribution. The data in Table 5.3 and Figures 5.32 and 5.33 are narrowband data. Wideband processing would be expected to narrow the variations, as in the case of the spatial distributions.

Frequency spectra have been measured for deciduous trees as a function of wind speed and found to have the general appearance of those presented in Figure 5.34, which gives the averaged spectra of deciduous trees at 95 GHz for two wind speed ranges. The spectra appear to have a

Table 5.3 Summary of the Standard Deviation for Various Classes of Clutter [from Currie, Dyer, and Hayes [15]).

Clutter Type	Polarization	*Average Value of Standard Deviation* (GHz)			
		9.5	16.5	35	95
Deciduous Trees	Vertical	3.9	—	4.7	—
Summer	Horizontal	4.0	—	4.0	5.4
	Average	4.0	—	4.4	6.6
Deciduous Trees	Vertical	3.9	4.2	4.4	6.4
Fall	Horizontal	3.9	4.3	4.3	5.3
	Average	3.9	4.2	4.3	5.0
Pine Trees	Vertical	3.5	3.7	3.7	6.8
	Horizontal	3.3	3.8	4.2	6.3
	Average	3.4	3.7	3.9	6.5
Mixed Trees	Vertical	4.3	—	4.0	—
Summer	Horizontal	4.6	—	4.2	—
	Average	4.4	—	4.1	—
Mixed Trees	Vertical	4.1	4.1	4.7	6.3
Fall	Horizontal	4.5	4.3	4.6	5.0
	Average	4.4	4.2	4.6	5.4
Tall Grassy Field	Vertical	1.5	—	1.7	2.0
	Horizontal	1.0	1.2	1.3	—
	Average	1.3	1.2	1.4	2.0
Rocky Area	Vertical	1.1	2.2	1.8	1.6
	Horizontal	1.2	1.7	1.7	1.7
	Average	1.1	1.9	1.8	1.7
10″ Corner Reflector (located in grassy field)		1.0	1.0	1.2	1.2

power function roll-off characteristic, as reported by Fishbein [7] for X-band data in 1967. Table 5.4 gives the measured exponents, N, and the corner frequencies, f_c, of the power functions in the form as given by (5.26) for deciduous trees at 35 and 95 GHz, measured by Georgia Tech. A different roll-off function was measured, depending on whether the receiver characteristic was linear or log.

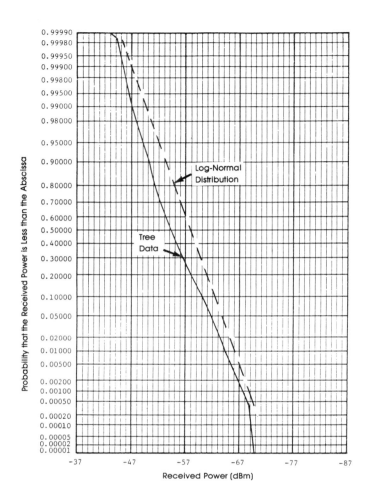

Fig. 5.32 Cumulative probability distribution of the received power from deciduous trees at 35 GHz (from Currie, Dyer, and Hayes [15]).

The causes for non-Gaussian spectra are not known at the present time. Perhaps they are an artifact of the measuring system, or perhaps they have a physical explanation. One possible physical explanation for non-Gaussian spectra from tree clutter is that trees are not composed of uniformly distributed scatterers, but rather represent a complex rotational mechanism. For example, the tree truck is a rotational body as a whole.

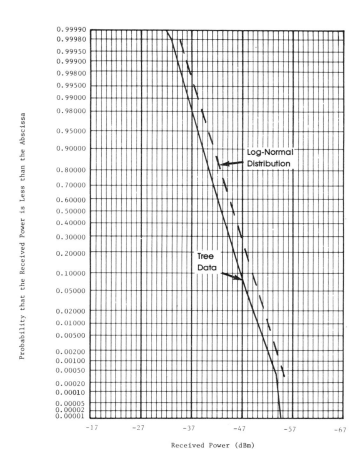

Fig. 5.33 Cumulative probability distribution of the received power from deciduous trees at 95 GHz (from Currie, Dyer, and Hayes [15]).

Each large limb is a separate rotational body; each small twig can rotate separately; each leaf or needle can flutter. Thus, the spectrum from a tree comprises the vector sum at a given time of the motion of the tree as a whole, the large limbs, the twigs, and the leaf flutter. Also, not all components are visible at the same time. At low frequencies, the twig and leaf movement that are likely to provide the highest frequencies are not important because of the large wavelength. As the frequency is increased,

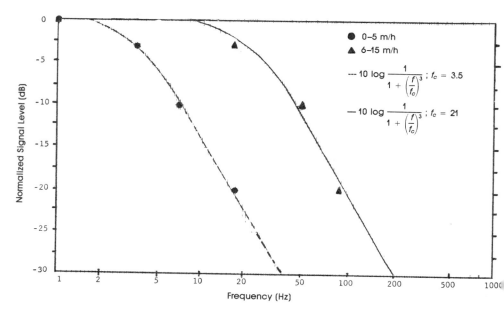

Fig. 5.34 Normalized frequency spectrum of a logarithmic receiver for the return from deciduous trees for two ranges of wind speed at 95 GHz (from Currie, Dyer, and Hayes [15]).

Table 5.4 Foliage Spectral Shapes (from Currie, Dyer, and Hayes [15]).

Frequency (GHz)	Wind Speed (m/h)	f_c		N	
		Linear	Log	Linear	Log
35	0–5	7	4	3	3
35	6–15	16	11	2.5	3
95	0–5	6	3.5	2	3
95	6–15	35	21	2	3

$$W(f) = \frac{A}{1 + \left(\dfrac{f}{f_c}\right)^N}$$

where A is a dc value, and f_c and N are as given herein.

scattering from these rapidly moving scatterers becomes more important, thus giving rise to higher frequencies in the spectrum. It should be noted that curve-fitting is an inexact science at best, and where one analyst will fit a power function to a set of measured data, another will fit a Gaussian with a wider corner frequency. Thus, the important implication of Figure 5.34 and Table 5.4 is not the shape of the curves, but rather that the spectral width is wider than might normally be expected from standard theory.

Decorrelation times were computed for the deciduous tree data in [15] as a function of frequency and wind speed. Figure 5.35 summarizes these data and indicates that wideband frequency processing is required to reduce the decorrelation times to a value sufficient for detection. If the radar PRF exceeds a value of one divided by the decorrelation time, then the extra returns from the clutter are strongly correlated, and no signal-to-clutter gain is achieved by averaging the clutter cell returns. Frequency agility can be used to decorrelate the clutter returns in a shorter time, thus enhancing the effectiveness of clutter averaging.

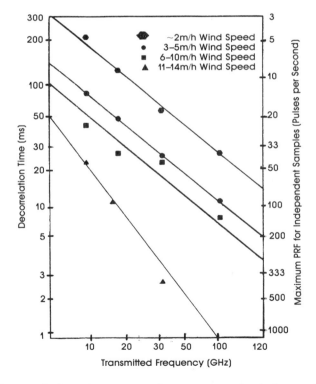

Fig. 5.35 Decorrelation time *versus* frequency and wind speed for the return from deciduous trees (from Currie, Dyer, and Hayes [15]).

5.3.3   Sea Clutter Characteristics

Definitions

The surface of the sea is capable of moving with time, and is strongly affected by interaction with the wind and tides. Thus, some definitions need to be discussed, which are unique to the problem of describing the sea surface. Table 5.5 summarizes these definitions, which include: sea state, average wave height, significant wave height, peak wave height, wind-wave direction, and fetch. Each of these terms is associated with attempts to describe the surface of the sea at a given time. The terms are important because radar reflectivity data are often described in relationship to one or more of these parameters, and many sea clutter models require wave height as an input variable. More detailed information on the definition of these parameters is available in Long [10].

General Characteristics

Only a small number of measurements have been performed to date on the reflectivity of the sea at millimeter wavelengths, while a large body of data exists at microwave frequencies. Thus, the best way to discuss MMW reflectivity characteristics is to compare the few data that exist at MMWs to those at lower frequencies. This is done in Table 5.6. For microwave frequencies, the reflectivity at low to moderate elevation angles is dependent on the wind's speed and direction and the waves' height and direction. At microwave wavelengths for a "fully developed" sea, where the waves have reached their maximum height for the wind speed and are in the same direction as the wind, the reflectivity is proportional to the wave height, it is a maximum in the upwind-upwave direction, and it is a minimum in the downwind-downwave direction. This trend appears to hold also at millimeter wavelengths. The average value of HH polarization is lower than VV polarization at microwaves for low wavelengths, while this effect appears to be reversed at millimeter wavelengths. Finally, the average value of the return at 10 GHz is less than that at 35 GHz, but greater than that at 95 GHz.

The standard deviations measured for MMW sea return are also different from those measured at microwaves. At microwave wavelengths, the standard deviation for HH polarized returns are almost always higher than VV returns, except for the case of very rough seas. This trend is reversed at 95 GHz, where the standard deviations of VV polarized returns are sometimes larger than HH polarized returns.

Table 5.5 Sea Clutter-Related Definitions.

Parameter	Definition
Sea State	A descriptive scale for the roughness of the sea, which is based on a given trough-to-crest height of the highest $\frac{1}{3}$ of the waves. (See Long [10] for more information.)
Average Wave Height	The true mean peak-to-trough height of ocean waves at a given time.
Significant Wave Height	The peak-to-trough height of the highest $\frac{1}{3}$ of the waves. It is thought to be the height normally given by an observer.
Peak Wave Height	The peak-to-trough height of the highest $\frac{1}{10}$ of the waves. It is approximately twice the average wave height.
Wind-Wave Direction	The direction of the wind and waves relative to the radar's line of sight.
Fetch	The interaction distance between the wind and water, which leads to a given wave height for a given wind speed.
Fully Developed Sea	A sea for which enough time has passed for the waves to reach their maximum value for a given fetch and wind speed.

The dependence of the reflectivity on wind-wave direction seems to be independent of the wavelength up through 95 GHz. Hence, the return from the sea is always greatest in the upwind-upwave direction, and the least in the downwind-downwave direction, except at depression angles near nadir, where the return is approximately independent of wind-wave direction. When the wind and wave directions are not the same, the backscatter dependence is variable and difficult to predict.

Table 5.6   Sea Clutter Radar Reflectivity Characteristics

Characteristic	Value	
	Polarization Dependence	
	Microwave	Millimeter Wave
Average Values		
Low Grazing Angles	$\sigma^{\circ}_{HH} < \sigma^{\circ}_{VV}$	$\sigma^{\circ}_{HH} > \sigma^{\circ}_{VV}$
Higher Grazing Angles	$\sigma^{\circ}_{HH} \approx \sigma^{\circ}_{VV}$	$\sigma^{\circ}_{HH} \approx \sigma^{\circ}_{VV}$
Standard Deviations		
All Grazing Angles	$\sigma^{\circ}_{HH} > \sigma^{\circ}_{VV}$	$\sigma^{\circ}_{HH} \approx \sigma^{\circ}_{VV}$
Wind-Wave Dependence		
High Grazing Angles	$\sigma^{\circ}_{UPWIND} \approx \sigma^{\circ}_{CROSSWIND} \approx \sigma^{\circ}_{DOWNWIND}$	
All Other Grazing Angles	$\sigma^{\circ}_{UPWIND} > \sigma^{\circ}_{CROSSWIND} > \sigma^{\circ}_{DOWNWIND}$	
Frequency Dependence		
Average Values	$\sigma^{\circ}_{10\ GHz} < \sigma^{\circ}_{35\ GHz}$; $\sigma^{\circ}_{10\ GHz} > \sigma^{\circ}_{95\ GHz}$	
Standard Deviations	$\sigma^{\circ}_{10\ GHz} < \sigma^{\circ}_{35\ GHz} < \sigma^{\circ}_{95\ GHz}$	

Average Values

Figure 5.36 gives some measured data on sea return at 35 GHz as compared to the Georgia Tech MMW sea clutter model, from Trebits, Currie, and Dyer [25], while Figure 5.37 gives data measured at the same time for 95 GHz. (See Chapter 7 for a discussion of the Georgia Tech sea clutter model.) These data exhibit the trends discussed above in that the upwind data are the highest and the downwind data are the lowest, the HH data are higher than the VV at 95 GHz, while they are approximately equal at 35 GHz, and the 95 GHz data are slightly lower than the 35 GHz data.

Figure 5.38 is a scatter diagram that compares the reflectivity of the sea as measured at 3 cm *versus* the reflectivity measured simultaneously at 3 mm. As one can see from the figure, the return at 3 cm is larger than that at 3 mm most of the time. Figure 5.39 illustrates the change in polarization dependence as a function of frequency. The figure is another scatter diagram that compares σ°_{VV} with σ°_{HH} at 35 and 95 GHz, respectively. For the 35 GHz data, σ°_{VV} is generally larger than σ°_{HH}, while at 95 GHz, the two returns are approximately equal, with a slight advantage of σ°_{HH} over σ°_{VV}.

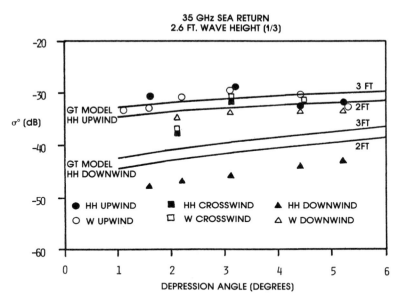

Fig. 5.36 Sea return at 35 GHz (from Trebits, Currie, and Dyer [25]).

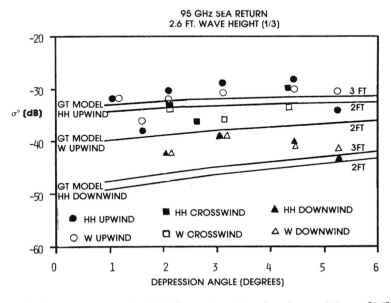

Fig. 5.37 Sea return at 95 GHz (from Trebits, Currie, and Dyer [25]).

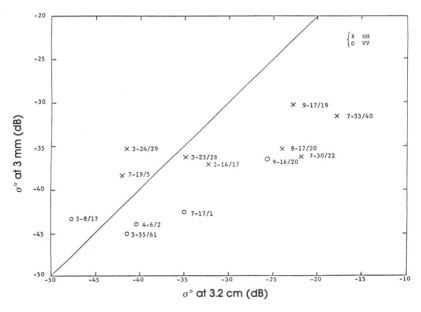

Fig. 5.38 Scatter diagram of σ° at 9.5 GHz *versus* 95 GHz (from Rivers [26]).

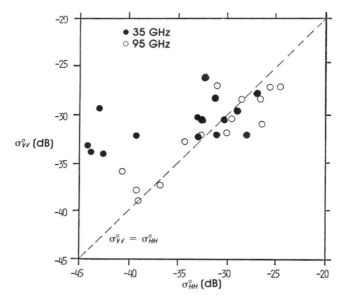

Fig. 5.39 Comparison of σ° measured simultaneously for VV and HH polarizations at 35 GHz and 95 GHz (from Trebits, Currie, and Dyer [26]).

No data have been published to date for cross-polarized returns from the sea at MMWs (i.e., HV or VH), but it is expected that these data would generally be a few dB less than the parallel-polarized returns, as is the case for microwave returns. Also, no known circularly polarized data are available, but little difference would be expected from the linearly polarized data.

Temporal and Spatial Properties of Sea Return

The return from the sea differs in two important ways from land clutter: the sea is generally much more uniform, and the temporal and spatial statistics are essentially the same. In fact, because gravity waves move across the surface, over periods of time equal to several seconds, the primary temporal characteristics are caused by the spatially distributed features moving through a given fixed point on the surface. As in the case for atmospheric and land clutter, the variations in time of the return from the sea can be described in terms of amplitude statistics and frequency spectra. Figures 5.40 and 5.41 give measured histograms and cumulative distributions of sea returns at 35 and 95 GHz, respectively, which were recorded simultaneously. As we can see, the 95 GHz distribution is wider than the 35 GHz distribution, and the horizontal is wider than the vertical at 35 GHz, while the two distributions for horizontal and vertical polarized returns are equal at 95 GHz. Figures 5.42 and 5.43 give uncalibrated frequency spectra for sea return at 35 and 95 GHz. Although the vertical axes are uncalibrated, they are linear, so we can see that the 95 GHz spectral width is approximately three times that of the 35 GHz data, which is about the ratio of the wavelengths. The data in Figures 5.42 and 5.43 are noncoherent, and thus represent only the amplitude variations from the various scatterers on the sea. Were the data coherent, a Doppler component would be present from the continuous motion of the sea surface. The frequency of the moving component can be expressed by the following equations, as given by Long [10].
The classical hydrodynamic equation for gravity wave velocity:

$$v = \sqrt{gL/2\pi} \tag{5.26}$$

where

g = the accelleration due to gravity,
L = the average spacing of the wave fronts,
v = the wave velocity.

The familiar Doppler equation for the frequency shift from a moving target:

$$f_D = 2v/\lambda \tag{5.27}$$

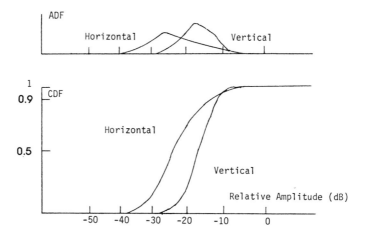

Fig. 5.40 35 GHz sea return amplitude distributions.

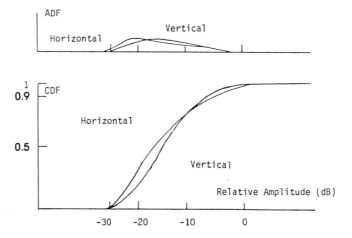

Fig. 5.41 95 GHz sea return amplitude distributions.

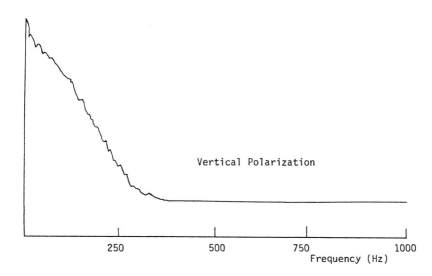

Fig. 5.42 Power spectral density of 35 GHz sea return.

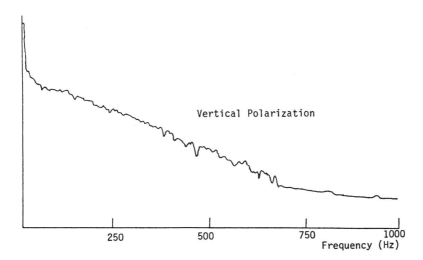

Fig. 5.43 Power spectral density of 95 GHz sea return.

If $L = \lambda/2$, then,

$$f_D = \sqrt{g/\pi\lambda} \tag{5.28}$$

Furthermore, Wright [27] showed that the first-order Bragg equation best fit the spectra measured at X band, and derived the equation:

$$f_D = \sqrt{\frac{g \cos\Theta}{\lambda\pi} + \frac{16\pi s \cos^3\Theta}{\lambda^3}} \tag{5.29}$$

where s is the ratio of surface tension to water density. Presumably, a similar effect would be seen at MMWs, although no current data exist to confirm this relationship.

5.4 SUMMARY

This chapter has attempted to summarize the primary features of MMW clutter and to describe the primary differences between microwave and MMW clutter reflectivity properties. As a summary of the available data for MMW clutter, Table 5.7 provides typical values for various types of clutter. However, one should not use the data in Table 5.7 as the values for predicting radar detection because clutter returns are statistical in nature. The data can only be utilized for detection calculation, as discussed in Chapter 3, where clutter is considered as a random process with a specific distribution. Table 5.7 can be useful for comparing relative performance as a function of frequency in the MMW region of the spectrum.

This chapter has only touched on the very complicated subject of radar clutter. Other clutter references of value include: Long [10], Nathanson [28], Skolnik [29], and Eaves and Reedy [30].

Table 5.7 MMW Reflectivity Data Summary.

ATMOSPHERIC CLUTTER

Rain (RCS per Unit Volume)

Frequency (GHz)		*Rain Rate* (mm/hr)			
		4	10	25	100
35	η (dBm^{-1})	-50	-43	-38.5	-27
95		-40	-37	-31	-25.2

Snow (RCS per Unit Volume)

Frequency (GHz)		*Snow Concentration* (gm/m^3)		
		0.1	0.2	0.4
96	η (dBm^{-1})	-58	-54	-40
140		-53	-50	-45
225		-54	-48	-44

SURFACE SCATTERING

Land Clutter

Frequency (GHz)	*Type*		*Depression Angle* (°)			
			5	10	30	60
35	Grass/Crops	$\sigma°$(dB)	-28	-25	-14	-8
95			—	-16	-12	-8
35	Trees		-22	-18	-16	—
95			-22	-18	-10	—
35	Wet Snow		—	-25	-15	-12
95			—	-20	-8	-4
35	Dry Snow		—	-15	-4	0
95			—	-12	-2	$+2$

Sea Return (0.6 m Wave Height)

Frequency (GHz)	*Wind Direction*		*Depression Angle* (°)		
			1	3	6
35	Upwind	$\sigma°$(dB)	-32	-30	-30
95			-32	-30	-29
35	Crosswind		—	-37	—
95			—	-36	—
35	Downwind		-48	-47	-45
95			—	-42	-42

REFERENCES

1. Robert N. Trebits, "Radar Cross Section," Chapter 2 of *Techniques of Radar Reflectivity Measurement*, N.C. Currie, ed., Artech House, Dedham, MA, 1984, pp. 27-53.
2. N.C. Currie, F.B. Dyer, and R.D. Hayes, "Radar Return from Rain at 10, 35, 70, and 95 GHz," *1980 IEEE International Radar Conference Digest*, Washington, DC, April 1980.
3. E.E. Martin, "Radar Propagation Through Dust Clouds Lofted by High Explosive Tests, MISERS BLUFF Phase II," Final Technical Report on Project A-2465 for SRI International, Georgia Institute of Technology, Atlanta, GA, October 1980.
4. R.L. Mitchell, "Radar Meteorology at Millimeter Wavelengths," Report No. SSD-TR-66-117 on Contract AF 04(695)-669, Aerospace Corporation, Los Angeles, CA, June 1966.
5. N.C. Currie, F.B. Dyer, and R.D. Hayes, "Analysis of Radar Rain Return at Frequencies of 9.375, 35, 70, and 95 GHz," Technical Report No. 2 on Contract DAAA 25-73-0256, Georgia Institute of Technology, Atlanta, GA, February 1975.
6. Victor W. Richard and John E. Kammerer, "Rain Backscatter Measurements and Theory at Millimeter Wavelengths," Report No. 1838, US Army Ballistic Research Laboratories, Aberdeen Proving Ground, MD, October 1975.
7. W. Fishbein, *et al.*, "Clutter Attenuation Analysis," Technical Report No. ECOM-2808, USAECOM, March 1967, AD665352.
8. J. Nemarich, *et al.*, "Comparative Near-Millimeter Wave Propagation Properties of Snow and Rain," *Proceedings of Snow Symposium III*, US Army Cold Regions Research and Engineering Laboratory, Hanover, NH, August 1983, pp. 115–129.
9. J. Nemarich, *et al.*, "Attenuation and Backscatter for Snow and Sleet at 96, 140, and 225 GHz," *Proceedings of Snow Symposium IV*, US Army Cold Regions Research and Engineering Laboratory, Hanover, NH, August 1984, pp. 41-52.
10. Maurice W. Long, *Radar Reflectivity of Land and Sea*, Artech House, Dedham, MA, 1984.
11. F.B. Dyer and R.D. Hayes, "Computer Modeling of Fire Control Radar Systems," Technical Report No. 1 on Contract DAAA 25-73-0256, Georgia Institute of Technology, Atlanta, GA, July 1974.
12. N.C. Currie, J.D. Echard, and R.B. Efurd, "Millimeter Wave Radar Performance Study," Final Technical Report on Hughes Aircraft Company Purchase Order No. 04-435812-UJX, Georgia Institute of Technology, Atlanta, GA, October 1985.

13. R.D. Hayes, N.C. Currie, and J.A. Scheer, "Backscatter and Emissivity of Snow at MM Wavelengths," *IEEE 1980 International Radar Conference*, Washington, DC, April 1980.

14. W.H. Stiles, *et al.*, "Backscatter Response of Roads and Roadside Surfaces," Technical Report No. SAND78-7069 on Contract No. DE-AC04-76DP00789, University of Kansas Center for Research, Lawrence, KS, 1979.

15. N.C. Currie, F.B. Dyer, and R.D. Hayes, "Backscatter from Land Clutter at 9.5, 16, 35, and 95 GHz," Technical Report No. 3 on Contract DAA 25-73-0256, Georgia Institute of Technology, Atlanta, GA, April 1975.

16. T.L. Lane, "95 GHz Clutter Backscatter TABILS V Data Base Summary," Final Technical Report on Boeing Airplane Company Contract No. BB 3041, Georgia Institute of Technology, Atlanta, GA, December 1982.

17. N.C. Currie, *et al.*, "MMW Signature Measurements: Vol I, Land Clutter," Technical Report No. AFATL-TR-76-xx, Georgia Institute of Technology, Atlanta, GA, June 1976.

18. W.H. Stiles and F. Ulaby, "Snow Measurements and Models," Technical Report, University of Kansas Center for Research, Lawrence, KS, 1980.

19. J.E. Knox, "Radar Backscatter Studies at SNOW II," *CRREL Special Report 84-35 on SNOW Symposium IV*, Hanover, NH, August 1984.

20. T.L. Lane, *et al.*, "Field Report No. 3 on SNOWMAN I, Covering the Period 11 March–25 March 1984," Interim Report on Contract DAAAH01-83-A013, Georgia Institute of Technology, Atlanta, GA, March 1984.

21. D.T. Hayes and U.V. Lammers, "MMW Reflectivity of Snow," *1978 IEEE APS International Radar Conference Digest*, Seattle, WS, June 1978.

22. N.C. Currie, F.B. Dyer, and G.W. Ewell, "MMW Radar Reflectivity Measurements From Snow," Technical Report No. AFATL-TR-77-4, Georgia Institute of Technology, Atlanta, GA April 1977.

23. R.R. Boothe, "The Weibull Distribution Applied to Ground Clutter Backscatter Coefficient," Report No. RE-TR-69-15, US Army Missile Command, Huntsville, AL, June 1969.

24. J.M. Trostel, *et al.*, "MM Wave Tower Snow Data Collection Program," *SNOW SYMPOSIUM V Proceedings*, US Army CRREL, Hanover, NH, August 1985, pp. 291–300.

25. R.N. Trebits, N.C. Currie, and F.B. Dyer, "Multifrequency Radar Sea Return," *URSI Commission F Symposium*, Quebec, June 1980.

26. W.K. Rivers, "Low Angle Sea Return at 3 mm Wavelength," Final Technical Report on Contract N62269-70-C-0489, Georgia Institute of Technology, Atlanta, GA, 1970.
27. J.W. Wright, "A New Model for Sea Clutter," *IEEE Transactions on Antennas and Propagation*, Vol. AP-16, No. 3 March 1968, pp. 217-223.
28. F.E. Nathanson, *Radar Design Principles*, McGraw-Hill, New York, 1969.
29. M.I. Skolnik, ed., *Radar Handbook*, McGraw-Hill, New York, 1975.
30. J.L. Eaves and E.K. Reedy, *Principles of Modern Radar*, Van Nostrand-Reinhold, New York, 1987.

Chapter 6
MMW Radar Signal Processing Techniques

W.A. Holm

Georgia Institute of Technology
Atlanta, Georgia

6.1 INTRODUCTION

A radar operating in the millimeter-wavelength portion of the electromagnetic spectrum (35–220 GHz) is similar in many respects to conventional microwave (<35 GHz) radars. This is especially true in terms of the radar signal processing techniques employed in the two classes of radars. Once the received radar signal has been down-converted in frequency to video (base) band, it makes little difference from a signal processing point of view whether that radar signal was originally at millimeter or microwave wavelengths. Many of the signal processing techniques used in microwave radars (e.g., *constant false alarm rate* (CFAR) processing, Doppler processing, pulse compression) are used in MMW radars.

The signal processing difference between conventional radars and MMW radars is in the radar signal processing philosophy adopted for MMW radars, which is to maximize the target information that can be processed out of the MMW radar signal. MMW radars have made it possible for the first time to have an imaging radar sensor for vehicle-sized objects at frequencies below infrared without using synthetic aperture techniques. Thus, the possibility of doing more that just target "detection and ranging" becomes evident. Object or target identification by radar becomes a realistic goal. Techniques are being developed to obtain and process high-range-resolution and full-polarimetric target information to complement the high cross-range resolution and scatterer information inherent in the MMW radar.

One of the principal applications of MMW radar systems is in the area of smart sensors, which are sensors capable of interpreting what they sense and then making a decision based on that information. One application of a smart sensor is the seeker system for a terminally guided smart munition. Here, the dual requirements of reasonable cross-range resolution coupled with sensor space and weight limitations often dictate the use of a MMW radar system.

In this chapter, those signal processing techniques that have come to be associated with MMW radar systems designed for smart sensor applications are discussed in detail. These techniques include coherent and noncoherent Doppler processing techniques for achieving moving target identification and for identifying stationary objects with moving components (e.g., a scanning antenna). Other stationary target identification techniques discussed in this chapter include *scalar techniques* (such as CFAR processing, high-range-resolution processing, clutter decorrelation techniques and spatial feature algorithms) and *polarimetric techniques*. Other processing techniques associated with MMW radar systems designed for other applications, such as synthetic aperture techniques for a ground-mapping MMW radar and monopulse techniques for target tracking are discussed in Chapters 15 and 16.

After a discussion of standard Doppler processing techniques is given, noncoherent pulsed Doppler (PD) and moving target identification (MTI) techniques are discussed. This is followed by a discussion of the nature of the *stationary target identification* (STI) problem, including a general discussion of various STI techniques. One STI technique is one-dimensional imaging by high-range-resolution profiles. These profiles are often obtained in MMW radar systems by using a frequency-stepped waveform. This pulse compression technique is discussed in detail. Finally, we cover polarimetric processing techniques to extract target geometrical features to achieve STI.

6.2 DOPPLER PROCESSING TECHNIQUES

The Doppler processing techniques discussed here are those radar signal processing techniques that utilize the Doppler effect to distinguish moving targets (or moving components on targets) from stationary targets or clutter. These techniques can be divided into two classes: MTI techniques and PD techniques. MTI processing techniques reject the backscattered return from stationary targets and clutter. PD processing techniques also reject the backscattered return from stationary objects and clutter, and, in addition, determine the range-rate (radial speed) of moving targets via the appropriate Doppler filtering. Historically (Skolnik [1]), in the airborne radar community, the difference between MTI and PD radars was not based on range-rate determination capability, but on *pulse repetition frequency* (PRF) rates: MTI radars had low PRFs, which implies no range ambiguities; PD radars had high PRFs, which implies no Doppler frequency ambiguities. This distinction between the two types of Doppler processing techniques based on PRF rates is no longer applicable, since, for example, PD radars exist at low, medium, and high PRF rates.

In this section, we cover some fundamental concepts of the pulsed waveform's spectral characteristics, coherent radar detection, and MTI and PD processing techniques. This is followed by a discussion of noncoherent radar detection and noncoherent pulsed Doppler techniques.

6.2.1 Spectral Characteristics of a Pulsed Waveform

A brief discussion of the Fourier transform and the pulsed waveform spectra of stationary and moving targets is given here. These concepts are necessary before a complete understanding of Doppler processing techniques can be obtained.

The signal received by a pulsed radar is a time sequence of pulses for which the amplitude (in a single-polarization-channel radar) and phase (in a coherent radar) are measured. Doppler processing techniques are based on measuring the spectral (frequency) content of this signal. The frequency content of this *time-domain* signal is obtained by taking its Fourier transform, thus resulting in a *frequency-domain* signal, or spectrum of the time-domain signal. Shown in Figure 6.1 is a plot of amplitude *versus* time of an arbitrary time-domain signal, and the resulting plot of the amplitude *versus* frequency of the frequency-domain signal obtained by taking the Fourier transform of the time-domain signal. If the time-domain signal is described by the complex function $f(t)$, then the Fourier transform of $f(t)$ is given by

$$F(\omega) = \int_{-\infty}^{\infty} f(t) \, e^{-j\omega t} dt \qquad (6.1)$$

The time-domain signal, $f(t)$, can be regained by taking the inverse Fourier transform of $F(\omega)$, which is

$$f(t) = \frac{1}{2\pi} \int_{-\infty}^{\infty} F(\omega) \, e^{j\omega t} d\omega \qquad (6.2)$$

Time-domain pulsed radar signals can be (partially) described in terms of four time scales: wave period, pulsewidth, *pulse repetition interval* (PRI), and dwell time. The frequency-domain pulsed radar signal can be

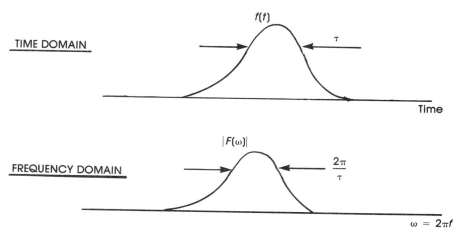

Fig. 6.1 The Fourier transform.

described in terms of four corresponding frequency scales: wave frequency, pulse bandwidth, PRF, and spectral line bandwidth. A fundamental property of Fourier analysis is that the time and frequency scales have an inverse relationship to each other: the larger is the wave period, the smaller is the wave frequency; the larger is the pulsewidth, the smaller is the pulse bandwidth, *et cetera*. This relationship and its resulting consequences is crucial to the understanding of Doppler processing techniques. The pulsed waveform will now be described by considering these four time and frequency scales.

The simplest radar waveform is the infinite continuous wave (CW) signal of frequency f_0. Only one time scale is required in the description of this waveform: wave period, t_0. Shown in Figure 6.2 is the time-domain plot of this waveform as a function of time, and the resulting frequency-domain amplitude *versus* frequency plot of this waveform. The frequency-domain plot consists of two impulse, or delta (Dirac), functions, one at f_0 and the other at $-f_0$. Shown in Figure 6.3 is a time-domain plot of a single pulse of pulsewidth, τ, and its corresponding frequency domain plot. The frequency-domain plot has a $\sin (x)/x$ form centered at dc with a bandwidth of approximately $1/\tau$. Notice that the nulls occur at integer multiples of $1/\tau$.

The two waveforms given above are combined and shown in Figure 6.4. Two time scales, t_0 and τ, are required in the description of this waveform. The plot of the amplitude of the Fourier transform of this waveform has the form of two $\sin (x)/x$ curves of bandwidth τ centered at $\pm f_0$. The time-domain and frequency-domain plots of an infinite sequence of these pulses is shown in Figure 6.5. Here, the pulses are separated in

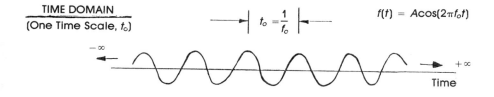

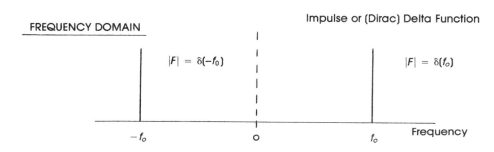

Fig. 6.2 Infinite continuous wave (CW) signal of frequency f_0.

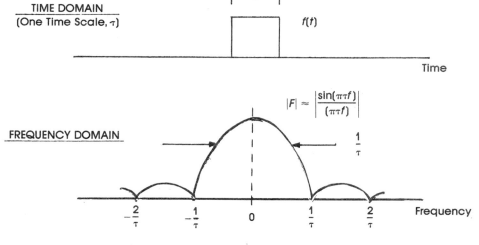

Fig. 6.3 Pulse envelope of pulsewidth τ.

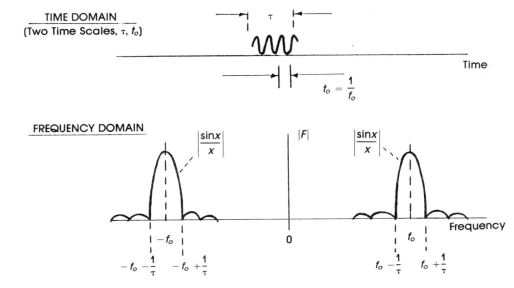

Fig. 6.4 Single pulse (τ) signal of frequency f_0.

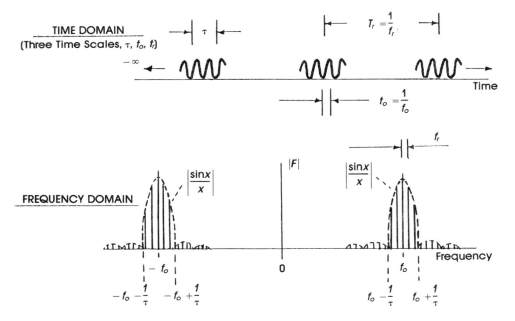

Fig. 6.5 Infinite pulse (τ) train signal of PRF f_r and frequency f_0.

time by T_r, the PRI. The resulting frequency-domain plot shows that the sin $(x)/x$ curves have been resolved into PRF spectral lines within the two sin $(x)/x$ envelopes. These spectral lines have zero bandwidth, and are separated by $f_r = 1/T_r$ (i.e., the PRF).

Finally, to obtain the spectral characteristics of a realistic pulsed waveform, the infinite series of pulses is truncated into a finite series of duration T. The resulting time-domain and frequency-domain plots are shown in Figure 6.6. In the frequency-domain plot, the PRF spectral lines now have a non-zero bandwidth given by $1/T$. Thus, four frequency scales are required in the description of the spectrum of a pulsed waveform: bandwidth of spectral lines ($1/T$), spacing of spectral lines (PRF $= 1/T_r$), bandwidth of sin $(x)/x$ envelopes ($1/\tau$), and center frequencies of sin $(x)/x$ envelopes ($\pm f_0 = \pm 1/t_0$).

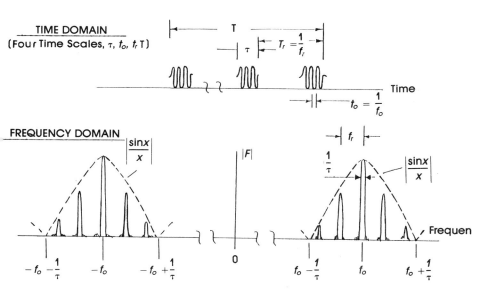

Fig. 6.6 Finite (T) pulse (τ) train signal of PRF f_r and frequency f_0.

If the pulsed radar return is from a cell that contains both a stationary target (clutter) and a moving target, then the return will consist of a superposition of signals. The signal from the clutter will be at the transmitted frequency, and the signal from the moving target be at the transmitted frequency plus the Doppler frequency shift. The Doppler frequency shift, f_d, has the magnitude of (Holm [2]):

$$f_d = 2v/\lambda \tag{6.3}$$

where v is the component of the moving target's velocity along the line of sight between the radar and the target, and λ is the wavelength of the transmitted wave. The Doppler frequency shift is positive for approaching targets, and negative for receding targets. Doppler frequency shifts for various radar frequency bands and target speeds are shown in Table 6.1. The resulting spectrum of the received signal from a moving target plus stationary clutter is shown in Figure 6.7.

Table 6.1 Doppler Frequency Shifts (Hz) for Various Radar Frequency Bands and Target Speeds.

Radar Frequency Band	Radial Target Speed		
	1 m/s	1 knot	1 mi/hr
L (1 GHz)	6.67	3.43	2.98
S (3 GHz)	20.0	10.3	8.94
C (5 GHz)	33.3	17.1	14.9
X (10 GHz)	66.7	34.3	29.8
K_u (16 GHz)	107	54.9	47.7
K_a (35 GHz)	233	120	104
(95 GHz)	633	326	283
(140 GHz)	933	480	417

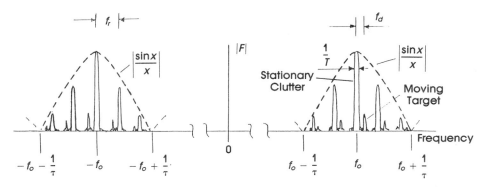

Fig. 6.7 Spectrum of the received signal from a moving target pulse stationary clutter.

The basic moving target discrimination approach inherent in Doppler processing techniques is now obvious. In the spectrum of the received signal, the moving and stationary targets are separated. Thus, by applying the appropriate filtering, the stationary target can be removed from the spectrum, thus leaving only the signature of the moving target. Before this can occur, however, the received signal must be detected. Coherent radar detection of a pulsed waveform is discussed next.

6.2.2 Coherent Detection of a Pulsed Waveform

Coherent radar systems extract not only amplitude, but also phase information from the signal reflected from the target. This coherency is usually required because the bandwidth of a single pulse is usually several orders of magnitude greater than the expected Doppler frequency shift (i.e., $1/\tau \gg f_d$). Thus, the Doppler frequency shift is lost in the single pulse spectrum. To extract the Doppler frequency shift, the returns from many pulses (over an observation time T) must be frequency analyzed so that the single pulse spectrum will separate into individual (PRF) lines with bandwidths approximately given by $1/T$. Now, as was shown in Figure 6.7, the Doppler frequency shift of the moving object becomes apparent and separates out from any stationary object return. For this process to work, however, a deterministic phase relationship from pulse to pulse must be maintained (and measured) over the observation time T. Radars capable of maintaining and measuring such a deterministic phase are called *coherent*.

In some cases, the Doppler shift is greater that the single-pulse bandwidth, making coherent detection unnecessary. Shown in Figure 6.8 is the single-pulse spectra for signals received from stationary clutter, and from a target moving at 2.24 m/s and a target moving at 6.7 km/s (15,000 mi/hr), received with a K_a-band radar transmitting a 10 μs pulse. (For convenience, the target and clutter powers are taken to be equal.) In the slow-moving target case, the target and clutter spectra, which both have bandwidths of 100 kHz, are separated by only 520 Hz, whereas in the fast moving target case they are separated by 1.6 MHz. Thus, in the fast moving target case, the target and clutter are clearly separable by using single pulse detection, making pulse-to-pulse phase coherency unnecessary. For the slow-moving target case, multiple pulse processing is required to separate the moving target and clutter. This is dramatically illustrated in Figure 6.9, which shows the high-pass-filtered video time signal for the two cases. For the fast-moving-target case, the moving target Doppler frequency is clearly visible in a single pulse, whereas in the slow-moving-target case the Doppler frequency is only established over many coherently detected pulses. Only multiple pulse processing will be considered further.

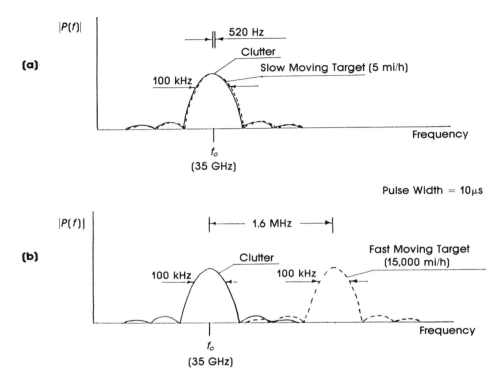

Fig. 6.8 Single pulse spectra of received signals containing clutter for (a) a slow-moving target and (b) a fast-moving target.

Radar coherency is obtained by deriving the transmitted radio-frequency (RF) signal from very stable oscillator sources that exhibit very little phase drift. Any undesired phase drift in these oscillators will impart a false-Doppler signature to all objects including stationary objects, thus making the measurement of the moving object's spectrum impossible. These same stable oscillators are used in the linear detection process that superheterodynes (beats) the received signal from RF to intermediate frequency (IF) and finally to video frequency. To realize full coherent integration gain and to discriminate between positive and negative Doppler frequency shifts, the video detector must be an in-phase, quadrature (I/Q) detector, consisting of a pair of synchronous detectors. This I/Q video detection is sometimes referred to as *synchronous detection,* since the phase information in the original IF signal is totally preserved in the video signal (Stimson [3]). Before discussing I/Q detection further, however, a coherent single-channel (I) radar is discussed in detail.

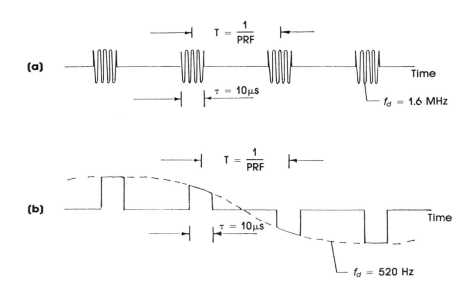

Fig. 6.9 High-pass filtered video signals of return from (a) a fast moving target, and (b) a slow moving target.

A block diagram of a typical single-channel coherent radar is shown in Figure 6.10 (Skolnik [4]). This configuration if often referred to as a *master oscillator, power amplifier* (MOPA) configuration. Here, the transmitted signal is obtained by mixing an RF signal from a *stable local oscillator* (STALO) with an IF, f_{IF}, signal (usually on the order of tens or hundreds of megahertz) from a *coherent oscillator* (COHO), which is also very stable. The resulting CW signal at frequency f_0 is then pulse-modulated to produce the pulsed waveform. The reflected signal (from a single stationary specular scatterer) has the spectrum shown in Figure 6.6. The first step in the detection process is to down-convert this signal to IF by mixing it with the STALO signal. The spectrum of the resulting signal is the same as shown in Figure 6.6, except now the $\sin(x)/x$ envelopes are centered at plus and minus the intermediate frequency, i.e., this the down-conversion has had the effect of "sliding" the center positions of the two $\sin(x)/x$ envelopes from $\pm f_0$ toward each other to $\pm f_{IF}$. This resulting IF spectrum is shown in Figure 6.11 for a signal reflected from stationary clutter and an approaching target. Amplification is now done at IF (as opposed to being done later at a lower frequency) to enhance the sensitivity of the receiver.

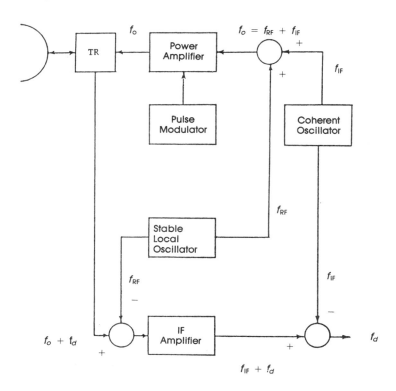

Fig. 6.10 Block diagram of a simple coherent pulsed radar.

The signal is next down-converted to baseband (video) by mixing it with the COHO signal. The $\sin(x)/x$ envelopes now overlap each other as both are now centered at dc. This is shown in Figure 6.12 for the same IF spectrum of clutter and an approaching target shown in Figure 6.11. One consequence of this single-channel down conversion, is that the moving target spectral lines now appear on either side of the stationary target spectral lines. That is, the sense of the direction of motion of the moving target is lost, since a target receding from the radar at the same rate would produce exactly the same video spectrum. As mentioned above, this ambiguity can be avoided by employing an I/Q detector, as shown in the block diagram in Figure 6.13.

In an I/Q detector, the IF signal is now divided into two channels: one channel (the quadrature, or Q channel) is phase shifted 90° with respect to the other channel (the in-phase, or I channel). The signals in the I and

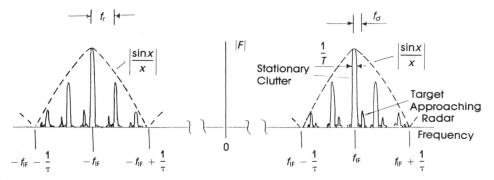

Fig. 6.11 IF Spectrum of the received signal from a target approaching the radar and stationary clutter.

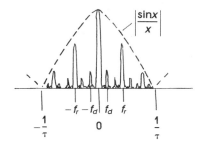

Fig. 6.12 Video spectrum of the received signal from a target approaching the radar and stationary clutter.

Q channels can be thought of as the respective real and imaginary components of a complex IF signal. If the I/Q power spectrum of the complex IF signal in a receiver employing I/Q detection and the power spectrum of the real IF signal from a single-channel receiver were obtained, they would conceptually appear like those shown in Figure 6.14. For the complex IF signal, all the power is contained in the positive frequency portion of the spectrum. Thus, compared to the positive IF spectrum of the single-channel receiver, the target, clutter, and noise spectral components have doubled in their power content (although power ratios such as the target-to-noise ratio remain unchanged).

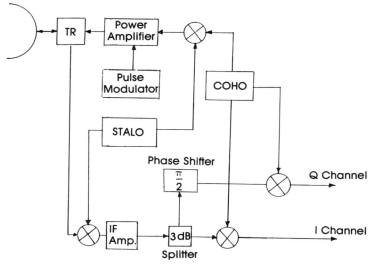

Fig. 6.13 Block diagram of a coherent pulsed radar with a synchronous (I/Q) detector.

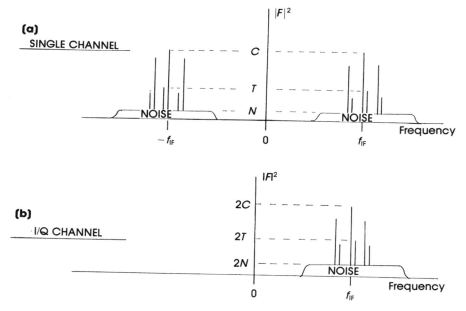

Fig. 6.14 IF power spectrum of a coherent radar with (a) a single channel receiver and (b) I/Q channel receiver. (Spectral linewidths are set equal to zero for clarity.)

Both IF channels in the I/Q detector are now separately down-converted to baseband. The power spectrum of the resulting complex video signal is shown in Figure 6.15 along with the spectrum of the real signal from a single-channel receiver. The spectral lines of the moving target now appear only on the right-hand side of the clutter spectral lines, indicating an approaching target. Another benefit of I/Q detection over single-channel detection is an average 3 dB gain in integrated target-to-noise ratio as shown in Figure 6.15.

The video signal from the coherent radar can then be filtered to achieve moving target discrimination. Two types of filtering are now discussed: MTI and PD filtering.

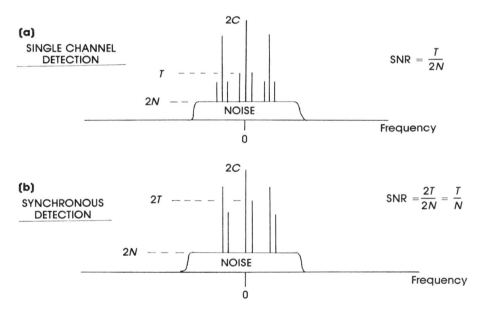

Fig. 6.15 Video power spectrum of a coherent radar with (a) a single channel receiver and (b) I/Q channel receiver. (Spectral linewidths are set equal to zero for clarity.)

6.2.3 MTI Filtering [5, 6, 7, 8]

Moving-target-indication (MTI) filtering attempts to remove the stationary target (clutter) signatures from the detected video spectrum, while passing the moving target signatures. No attempt is made to measure the Doppler frequency of the moving target. Two basic MTI filtering techniques have been employed in radar processors: the *range-gated filter* and the *delay-line canceller*.

A block diagram of a range-gated filter is shown in Figure 6.16. Each range gate is sequentially activated once every PRI, and the video signal is sampled just long enough (one pulsewidth in duration) to correspond to the range resolution of the radar. This sampling preserves the range resolution of the radar that would have been lost in the narrow-band filtering which follows. The sample-and-hold (boxcar) circuit holds the amplitude of each sample for one PRI until the next sample is received. This *pulse-stretching* essentially makes the pulsewidth equal to the PRI, and the fundamental of the Doppler frequency thus is emphasized and harmonics of the PRF are eliminated. The Doppler filter is a narrow bandpass filter that rejects all low-frequency clutter returns. The rectifier converts the bipolar video to unipolar video, and the low-pass filter (integrator) provides an essentially unmodulated video signal to be compared with threshold. The bandwidth of the Doppler filter will depend on the frequency extent of the expected clutter return. The frequency response of a four-pole Butterworth Doppler filter is shown in Figure 6.17.

A block diagram of a two-pulse (single-stage) delay-line canceller is shown in Figure 6.18. A delay-line canceller is essentially a comb-filter applied to the detected video signal, which filters out all dc signals and the PRF spectral lines. This is accomplished (in the simple two-pulse canceller shown) by simply subtracting the returns from two contiguous PRIs. All return pulses that have a zero-Doppler shift (as will be the case for stationary clutter), or Doppler shifts that are equal to an integer multiple of the PRF, will be unmodulated when down-converted to video. Thus, these pulses will cancel perfectly in the delay-line canceller. The frequency response of the two-pulse delay-line canceller is shown in Figure 6.19. The relatively narrow dc clutter notch can be widened and shaped by employing a multistage canceller with the appropriate feedback or feed-forward loops.

Delay-line cancellers are often implemented in digital form. A block diagram of a digital delay-line canceller is shown in Figure 6.20. The detected video is sampled at a rate commensurate with obtaining one sample within each range resolution cell, and then digitized by the analog-to-digital (A/D) converter. The delay-line canceller is a series of shift registers with the appropriate logic gates for the feedback (feed-forward) loops. The bipolar output is rectified and compared to a threshold to determine the presence of a moving target.

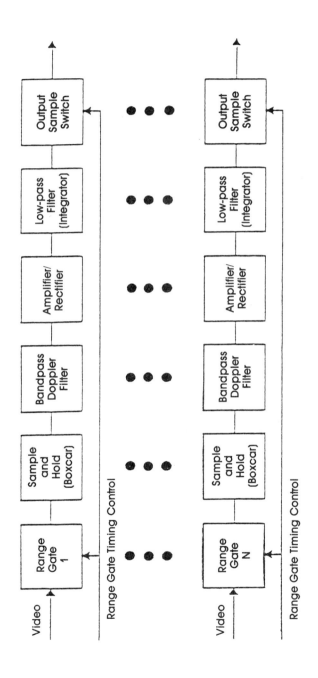

Fig. 6.16 Range-gated-filter (RGF) MTI signal processor.

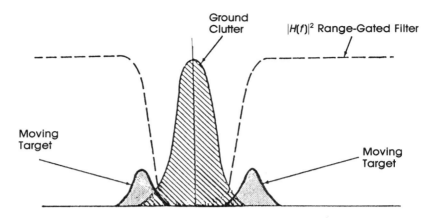

Fig. 6.17 Filter response of a four-pole Butterworth Doppler filter.

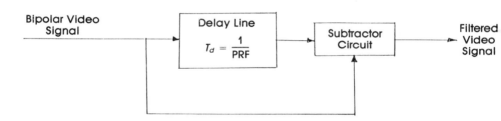

Fig. 6.18 Block diagram of a two-pulse delay line canceller.

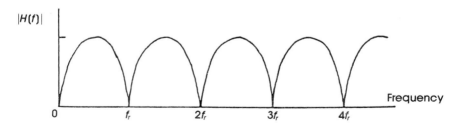

Fig. 6.19 Frequency response of a two-pulse delay line canceller.

Fig. 6.20 Digital delay line canceller and signal processor.

Both the range-gated filter and delay-line canceller techniques described above will eliminate returns from targets moving at speeds that produce Doppler shifts equal to an integer multiple of the radar's PRF. These speeds are called *blind speeds*. Blind speeds (i.e., blind frequencies), are inherent to all sampling systems (e.g., pulsed radars), and are not related to the particular Doppler filtering technique employed. Blind speeds can be eliminated by using multiple or staggered PRFs (PRF switching on a pulse-to-pulse basis). In this way, a target moving at a blind speed for one PRF will not be moving at a blind speed for the other PRFs.

6.2.4 Pulsed Doppler Filtering [9, 10, 11]

Pulsed Doppler filtering is accomplished by performing a Fourier transform on the detected video signal. This can be done by using a contiguous (in frequency) set of Doppler bandpass filters, usually by digitizing the video signal and employing digital filters, or a *discrete Fourier transform* (DFT), such as the *fast Fourier transform* (FFT) (see Figure 6.21). A block diagram of an FFT processor is shown in Figure 6.22. This type of PD processor will be discussed next in detail.

As shown in Figure 6.22, the video outputs of the I and Q channels are sent through a sample-and-hold circuit and an A/D converter, and then introduced into the N-point FFT processor. The sample-and-hold circuit samples at a rate commensurate with obtaining one sample from each of the M range bins contained within one PRI. After these samples are digitized by the A/D converter, they are stored in the buffer memory of the N-point FFT processor. This process is repeated N times (for N PRIs, which is usually equal to the target dwell time T) so that $N \times M$ digital complex numbers are stored in buffer memory. The FFT processor then begins computing an N-point FFT for each of the M range bins, while the next set of $N \times M$ digital numbers are being stored.

The N-point FFT processor will take N samples of the complex video signal ($2N$ real numbers), sampled at rate R (the PRF) for time $T = N/R$, perform a discrete Fourier transform on these samples, and arrange the results at N discrete frequencies of magnitude k/T (where k is an

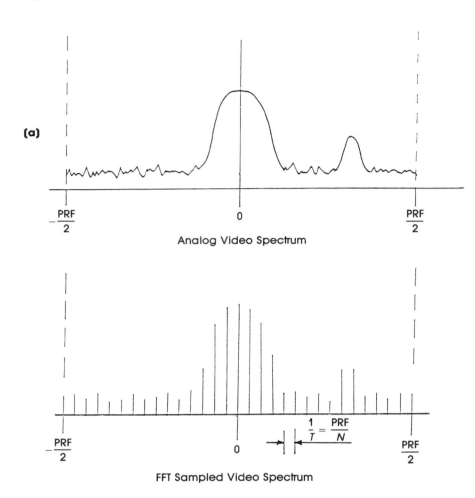

Fig. 6.21 (a) Analog and (b) FFT sampled video spectra.

integer) that extends in frequency from $-R/2$ to $R/2$. The output of the N-point FFT processor are N complex-valued numbers, i.e., an I and Q channel for each of the N discrete frequencies. The frequency-domain magnitude and phase diagrams of the video signal can now be constructed from these N complex-valued numbers. Of interest here is only the frequency-domain magnitude diagram, which is determined in the magnitude detector by taking the square root of $I^2 + Q^2$ for each discrete frequency.

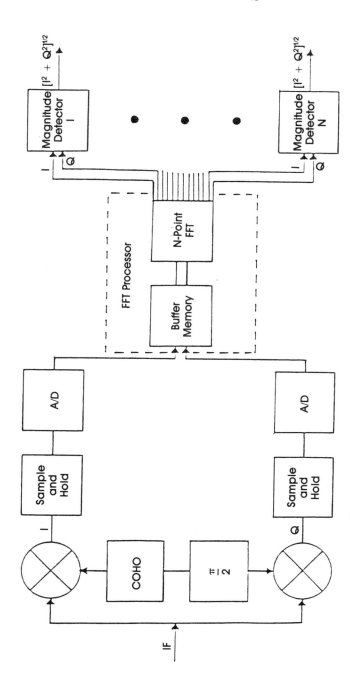

Fig. 6.22 FFT processor for a coherent radar.

The frequency resolution of the FFT processor is $1/T$; hence, any monochromatic signal of frequency f, where f has the range $(n-0.5)/T \leq f \leq (n + 0.5)/T$ (n is an integer and $|n| \leq N/2$), will appear in the output of the FFT processor at frequency n/T. This signal will also appear in other frequency bins due to the finite sampling time T, which causes the filter response of any given bin to have frequency sidelobes. This effect can be minimized with appropriate window weighting. The maximum unambiguous frequency that can be determined by the FFT processor is $R/2$, and higher frequencies present will "fold over" into the FFT processor's frequency interval $[-R/2, R/2]$; thus, an arbitrary frequency,

$$f = kR + f' \tag{6.4}$$

where k is an integer and $-R/2 \leq f' \leq R/2$, will appear in the frequency bin containing f'.

The output of each of the N channels from the N magnitude detectors is then sent to a thresholding circuit. The output from the dc channel (and perhaps several channels around dc) is assumed to be the return from stationary clutter, and is therefore discarded. A threshold detection in any of the other channels is assumed to be from a moving target. The particular channel that gives the moving target detection also gives an indication of the target's radial speed.

Perfect MTI or pulsed Doppler processing performance requires a total elimination of the clutter signal. Any clutter residue present at the output of the processor results in a degradation of moving target performance. Clutter residue originates from two general sources: *clutter spectral broadening* and *radar system imperfections* (instabilities). System instabilities include variations in transmitted power, pulsewidth, or frequency, and pulse-to-pulse timing errors. The main contributions to clutter spectral broadening come from scanning antenna motion, radar platform motion, and internal clutter motion.

Many applications for MMW radar systems require that they be attached to a moving platform; therefore, an apparent Doppler shift is imparted to the return from stationary objects as well as moving objects. The utilization of coherent Doppler processing techniques in these systems, therefore, requires that platform motion (translation, rotational, vibrational) be quantified, and a corresponding compensation should be made to the received signal to eliminate all moving platform effects. This *motion compensation* processing can be significant, and it can extract a high price, especially when platform payload volume and weight are limited. Motion compensation problems can be eliminated by use of noncoherent radar detection techniques.

6.2.5 Noncoherent Pulsed Doppler Radars [12, 13]

Often, a stationary target of interest will have a moving component, such as a scanning antenna. The spectral characteristics (spectrum) of the backscattered MMW radar signal from these moving components are often very useful in identifying the target. In this section, we discuss simple MMW radar systems that are capable of measuring the Doppler signatures of stationary targets with moving components, i.e., noncoherent pulsed Doppler MMW radars.

Noncoherent pulsed MMW radars have transmitters in which there exists a random phase relationship between transmitted pulses; thus, all targets including stationary targets are given false Doppler signatures. The IF spectrum of a signal reflected from stationary clutter and an approaching target as detected by a coherent radar is shown in Figure 6.23. The same spectrum as detected by a noncoherent radar (e.g., magnetron-based radar), is also shown in the same figure. Since there is no deterministic pulse-to-pulse phase relationship in the transmitted signal of the noncoherent radar, the PRF spectral lines become "smeared" and produce a continuous spectrum. Thus, the clutter and target signatures are no longer separated. However, if a stationary object signal of sufficient strength is present within the reflected pulse that contains the moving target signal, then these two signals will beat together in the radar's nonlinear video detector, resulting in a zero-frequency signal and a signal that appears at the Doppler frequency of the moving object. The zero-frequency signal can be filtered out, thus leaving the spectrum of the moving object. Since both the stationary and moving object signals have the same random phase relationship imparted to them by the transmitter, these random phases cancel out in the mixing process of the video detector. Similarly, no platform motion compensation is required because all false Doppler shifts due to platform motion will be imparted to both the stationary and moving objects, and will again cancel out in the video detector. The resulting video signature of the moving object can then be FFT processed in order to determine its spectrum. This is called *noncoherent pulsed Doppler processing*. A block diagram of a noncoherent pulse Doppler FFT processor is shown in Figure 6.24.

The N-point FFT processor will take the N samples of the real video signal sampled at rate R (in this case, the PRF, f_r) for a time T (where $T = N/R$), perform a discrete Fourier transform on these samples, and arrange the results at N discrete frequencies of magnitude k/T (where k is an integer), extending in frequency from $-R/2$ to $+R/2$. These results or outputs of the N-point FFT processor are N complex-valued ($2N$ real-valued) numbers (an I and Q channel for each of the N discrete frequencies). The frequency-domain magnitude and phase diagrams of the

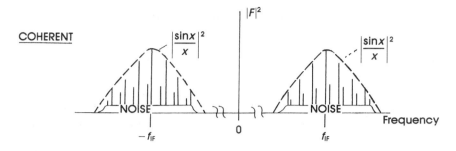

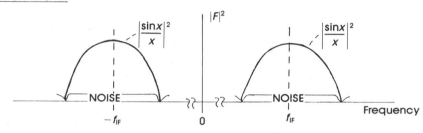

Fig. 6.23 IF power spectrum of a single channel (a) coherent and (b) noncoherent radar.

video signal can now be constructed from these N complex-valued numbers. Usually of main interest is the frequency-domain magnitude diagram (frequency magnitude spectrum), which is determined in the magnitude detector by taking the square root of $I^2 + Q^2$ for each discrete frequency. For a noncoherent radar, the magnitudes in the positive and negative frequency domains exactly duplicate, or mirror, each other. Thus, the signal from a target moving toward the radar will appear not only at the appropriate positive frequency, f_d, but also at $-f_d$. One immediate consequence of this (as shown in Section 6.2.2) is a minimum 3 dB loss in integrated SNR for the noncoherent radar as compared to the coherent radar.

The noncoherent pulse Doppler radar will not perform as well as the (motion compensated) coherent Doppler radar. For example, the video SNR for the noncoherent radar is less than the SNR for the coherent radar. However, under favorable clutter conditions, these two SNRs can be very close. To gain a better insight as to why this occurs, consider the coherent video spectrum of a signal received from a moving target and stationary clutter, shown in Figure 6.25. Here, the target power is S, the clutter power is C, and the noise power is N. The SNR is S/N. Now, consider

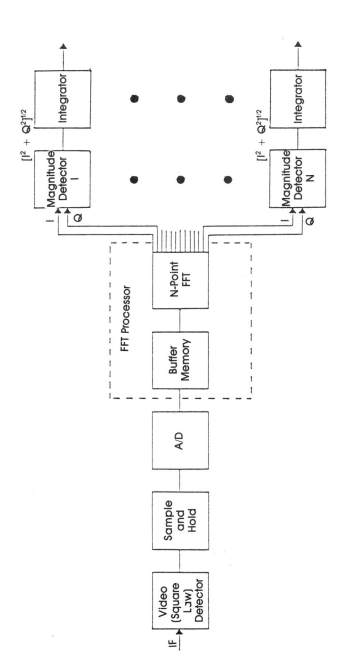

Fig. 6.24 FFT processor for a noncoherent radar.

the video spectrum of the the same signal as detected by a non-coherent radar, shown in Figure 6.26. In the nonlinear (e.g., square-law) detector that is employed by the noncoherent radar, the IF signal is convolved with itself in the frequency domain (i.e., the clutter, moving target, and thermal noise signals are mixed together). Ignoring high-frequency terms, which are easily filtered out, the power spectrum in one PRF interval out of the square-law detector consists of a zero-frequency term that is equal to the sum of the squares of the clutter and target powers ($S^2 + C^2$), two terms appearing at plus and minus the Doppler frequency, f_d, of the moving target that are equal to the product of the clutter and target powers (SC), and an increased noise power ($N^2 + 2NC + 2CN$) due to the convolution of the target and clutter signals with thermal noise, and the convolution of noise with itself. The SC terms in Figure 6.26 are now interpreted as the target signals. Thus, the resulting SNR is given by

$$\text{SNR} = \frac{2SC}{N^2 + 2SN + 2CN} = F\left(\frac{S}{N}\right) \tag{6.5}$$

where the clutter-reference loss function, F, is given by

$$F = \frac{2C/N}{1 + 2\,S/N + 2\,C/N} \tag{6.6}$$

and S/N and C/N are the IF signal-to-noise and clutter-to-noise ratios, respectively. In the presence of a strong clutter signal (i.e., $C/N \gg 1$, and $C/N \gg S/N$), $F \approx 1$. Thus, under these favorable clutter conditions, the SNR obtained with the noncoherent radar approaches that obtained with a coherent radar. Also, as the number of pulses processed becomes large, given the same favorable clutter conditions, the integration gain in a non-coherent pulsed Doppler radar approaches within 3 dB that obtained with a coherent pulsed Doppler radar.

    If a weak clutter signal is present, however, then a significant degradation in performance will result for the noncoherent radar (since now $F \ll 1$). The noncoherent radar under these conditions will suffer what is sometimes called "clutter starvation." Curves showing this loss in SNR as a function of S/N and C/N are shown in Figure 6.27.

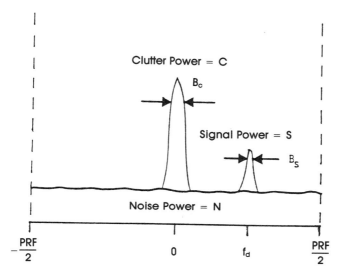

Fig. 6.25 One PRI of the coherent video spectrum received from a moving target and stationary clutter.

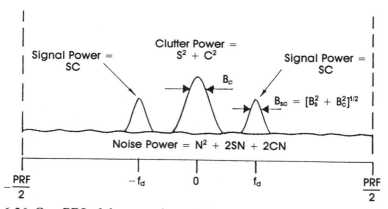

Fig. 6.26 One PRI of the noncoherent video spectrum for a signal received from a moving target and stationary clutter.

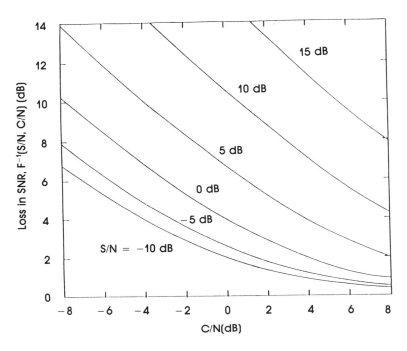

Fig. 6.27 Loss in moving target SNR for a square-law detector as a function of SNR and CNR into the detector.

Noncoherent pulsed Doppler processing is ideal for smart MMW radar sensors in their mission to identify stationary targets with moving components. Here, the stationary body of the target gives rise to an excellent stationary object reference signal for the Doppler-shifted signal from the moving component. Thus, a noncoherent MMW radar can be used, for which moving compensation techniques are not required.

There are limitations in noncoherent pulsed Doppler processing not present in coherent pulsed Doppler processing. These limitations include (1) reduction in moving object signal-to-noise ratio, (2) superposition of negative and positive Doppler shifts, (3) corruption of moving target spectrum with stationary reference object's spectrum, and (4) reduction of integration gain. However, these limitations may be more than offset by the noncoherent radar's simplicity.

6.3　MOVING TARGET IDENTIFICATION TECHNIQUES*

In the previous section on Doppler processing techniques, the discrimination of moving targets from stationary clutter was shown to be possible by applying the appropriate filtering to the MMW radar's received signal. This is possible because the moving target and clutter signatures are typically resolved in the spectrum of the received signal. This occurs even in a noncoherent MMW radar, if there is a strong clutter-reference signal in addition to the moving target signal in the backscattered return.

Additional information about the moving target can be obtained by analyzing the fine structure of its Doppler spectra. In some cases, the target's fine Doppler structure is unique enough to identify the target. In this section, two examples of how moving target identification may be accomplished by relating measured spectral features to moving components on the target are examined. The two examples are moving ground vehicles and helicopters.

6.3.1　Spectral Characteristics of Moving Ground Vehicles

A typical Doppler spectrum for a moving ground vehicle (tracked or wheeled) is shown in Figure 6.28. Most of the energy in this spectrum is in the so-called "body" line. This is the energy reflected by the parts of the vehicle moving at the "average" vehicle velocity, v_0. However, because the vehicle is vibrating on its suspension system and wheels are turning or tracks are moving, some parts of the vehicle are moving at velocities other than the average body velocity. For example, the bottom of the vehicle's wheels are not moving while the top of the wheels are moving at twice the average body velocity ($2v_0$). The same is true for tracks on a tracked vehicle. Other parts of the wheels or tracks are moving at velocities between 0 and $2v_0$. Thus, the spectra of the wheels, tracks, and other moving parts of the vehicle would be spread in the Doppler domain from 0 to $4v_0/\lambda$ Hz (where λ is the wavelength of the MMW radar's transmitted signal), as shown in Figure 6.28. This energy density, referred to as "*sidelobe*" energy in the figure, is quite small compared with that of the body line.

---

*Major portions of this section were contributed by Dr. Jim D. Echard, Principal Research Engineer, Georgia Tech Research Institute, Georgia Institute of Technology, Atlanta, GA.

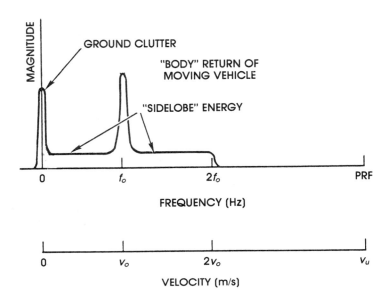

Fig. 6.28 Power density spectrum of a moving target (tracked or wheeled); single FFT spectrum. (Detailed sidelobe structure is not shown).

The level of the sidelobe energy is different for wheeled and tracked vehicles. Thus, this energy level is a potential feature for moving target identification. In addition, the fine-grained structure of the sidelobe part of the spectrum provides additional features for recognition and perhaps specific target identification. An example is shown in Figures 6.29 and 6.30, where the treads of a moving tracked vehicle are modeled as moving point scatterers (Figure 6.29) with the corresponding spectra calculated by using a FFT (Figure 6.30). A number of samples (radar pulses) were generated as they would be returned by each tread plate scatterer and an N-point FFT taken on sequential sets of N pulses. The resulting sequential FFT spectra *versus* time are plotted in Figure 6.30. The "intensity" scale used is logarithmic so that the sidelobe structure is much farther below the peak body return than it appears in the figure.

Note the systematic fine-grained structure illustrated in Figure 6.30. The slope and spacing of this structure is related to the tread velocity and tread plate spacing. Thus, a potential exists for specific target identification. The structure and level of the wheeled vehicle's Doppler spectra are quite different than those of the tracked vehicle. An illustration of wheeled

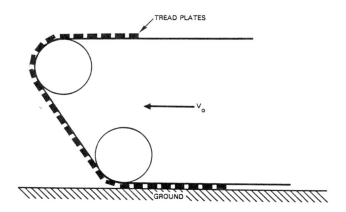

Fig. 6.29 Rotating metal plates on tracked vehicle tread.

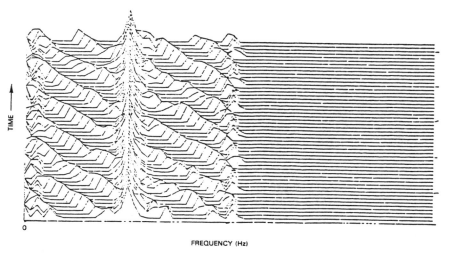

Fig. 6.30 Simulated Doppler spectra.

vehicle spectra (multiple FFTs) is shown in Figure 6.31, where the fine-grained structure in the sidelobes is not shown. In addition to the differences in the fine-grained sidelobe structure, the structure and relative amplitude of the body return is different for wheeled and tracked vehicles. All of these features can be used in moving target identification of ground vehicles.

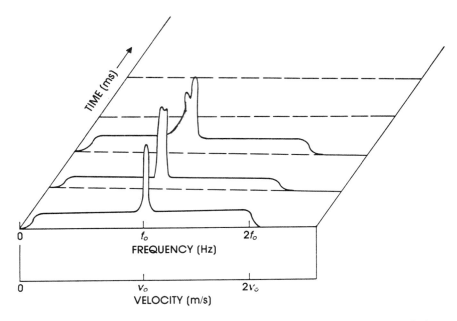

Fig. 6.31 Doppler spectra *versus* time; multiple FFT spectra; Ground clutter removed by not using low Doppler FFT output. (Detailed sidelobe structures are not shown.)

6.3.2 Spectral Characteristics of Helicopters

The dominant structures contributing to helicopter signatures are the rotor blades, especially those on the main rotor. These blades can be illuminated by the MMW radar any time the helicopter itself is illuminated, and will produce Doppler components unless the rotor disk is perpendicular to the MMW radar's line-of-sight (LOS).

When the MMW radar lies in the plane of the rotor motion, Doppler shifts (measured with respect to the body line) that represent velocities (approximately 250 m/s) of the tips of the blades are observed. A continuum of Doppler shifts from a minimum corresponding to the receding tip speed to a maximum corresponding to the approaching tip speed will be observed, and each element of the blade produces a shift proportional to its distance from the hub. Figure 6.32 shows an idealized, long-term average power spectrum showing a well defined drop in spectral energy at the point where the blades are attached to the hub.

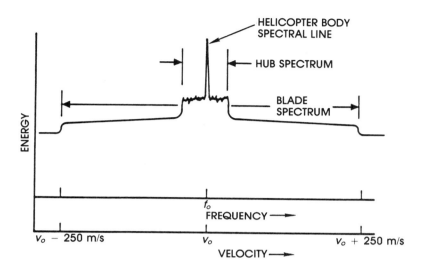

Fig. 6.32 Idealized helicopter power spectrum.

The rapid drop at the blade roots is due to the blades (being very directive scatterers) returning significant energy to the MMW radar only when they are perpendicular to the LOS. During the very short interval when the blade passes perpendicular to the LOS, it scatters a burst of energy back to the MMW radar, containing Doppler shifts all the way from the skin-line shift to plus and minus the shift due to blade tip. This phenomenon is illustrated in Figure 6.33. A simple way to explain the blade "flashes" is to consider a uniformly illuminated linear array as a simple model for a uniformly illuminated helicopter rotor blade. As shown in the figure, the blade returns reflected energy only in a relatively narrow angular beam. As the blade rotates, the energy is reflected toward the illuminating MMW radar briefly during each $1/N$ cycle, where N is the number of blades.

The hub of the main rotor generally has a more complex mechanical structure than the blade, and the hub therefore has a much broader reflected beam. In fact, the hub may simultaneously have many such beams. The result is that the hub reflections can be observed by the MMW radar during a significant portion of the radar "observation" period, while the blade reflections cannot. Thus, as indicated in Figure 6.32, the energy density reflected from the rotor hub is generally greater than that reflected from the blades, even within one observation time. The periodicity of the hub spectrum and blade spectrum will also be quite different. If the PRF is large enough, however, to provide the complete unaliased Doppler

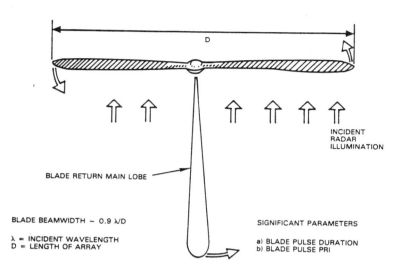

INCIDENT
RADAR
ILLUMINATION

BLADE RETURN MAIN LOBE

BLADE BEAMWIDTH ~ 0.9 λ/D

λ = INCIDENT WAVELENGTH
D = LENGTH OF ARRAY

SIGNIFICANT PARAMETERS

a) BLADE PULSE DURATION
b) BLADE PULSE PRI

Fig. 6.33 Uniformly illuminated linear array in analogy to illuminated helicopter blade.

signature, then the presence of the blades can be sensed in addition to the hub.

Figure 6.34 shows the simple mathematical relationships between the number and diameter of the blades, the blade rotation rate, blade flashes per second, and blade tip velocity. In the lower right-hand quadrant, several typical helicopters have been located in the blade-count *versus* diameter plane. In the upper right-hand quadrant, two curves of constant rotor tip velocity have been plotted in the diameter *versus* revolutions-per-second plane. The lower and upper curves are the respective, minimum and maximum tip velocities. The upper left-hand quadrant contains a number of straight lines. These are blade count multipliers to convert rotation rate to blade rate, which are plotted along the left end of the horizontal axis; the lower end of the vertical axis is the blade count scale. These relationships can be used to separate helicopters into various classes. For example, if the blade flash rate (blades per second) is measured by the MMW radar to be 17 blades per second as indicated by the dashed line in Figure 6.34, then the blade diameter must be between 14.9 m (49 ft) and 18.9 m (62 ft) for a four-bladed helicopter. Of the six helicopters shown, only one (#4) falls within this class. The other five are eliminated. If the blade count is unknown, then only three (#4, #5, #6) of the six helicopters have the specified blade flash rate.

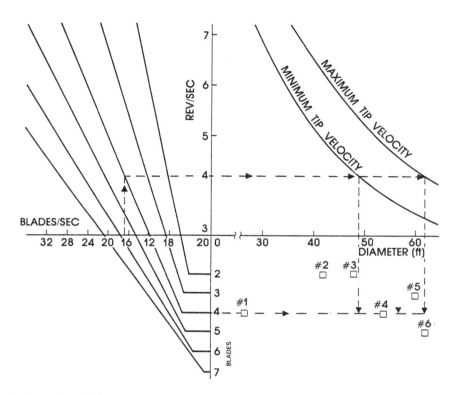

Fig. 6.34 Blade count-diameter-speed relationships.

6.3.3 Moving Target Identification with MMW Radars

MMW radars are especially well suited for the moving target identification task. As discussed above, moving target identification requires the capability of measuring the fine Doppler structure from the moving target. This means that the moving components on the target must be resolved, components which may have very small velocity differences. The velocity resolution, δv, of any radar is determined by that radar's Doppler frequency resolution, δf_d, which, in turn, depends on the target dwell time, T. Thus,

$$\delta f_d = 1/T \tag{6.7}$$

The Doppler frequency and velocity resolutions are related by

$$\delta f_d = 2\delta v/\lambda = 2\delta v f_0/c \qquad\qquad (6.8)$$

where f_0 is the frequency of the radar's transmitted signal. Thus,

$$\delta v = c/(2Tf_0) \qquad\qquad (6.9)$$

Therefore, the higher is the frequency of the transmitted signal, the smaller is δv (i.e., the greater is the target velocity resolution). Therefore, the high transmitting frequencies of MMW radars will enable high velocity resolutions, and thus allow for measurement of the fine spectral structure of the moving target.

6.4 STATIONARY TARGET IDENTIFICATION TECHNIQUES

Stationary targets are, by definition, not moving. Therein lies the problem of achieving stationary target identification with a MMW radar. None of the Doppler processing techniques used to extract Doppler features for moving target discrimination and identification are applicable. The only features that can be used for stationary targets are geometrical features. These can be obtained by high resolution or polarization techniques.

In this section, we discuss the suitability of the MMW radar for use in *stationary target identification* (STI). This is followed by a discussion of the stationary target identification problem and the various processing techniques employed for achieving STI.

6.4.1 Stationary Target Identification with MMW Radars

MMW radars are particularly well suited for the stationary target identification task due to their inherently increased cross-range resolution capability, and the increased target scatterer information that they are capable of measuring. Increased cross-range resolution is important in eliminating interference (clutter backscatter) that can occur in the return when using a wide-beamwidth, real-aperture radar. Increased scatterer information is important in maximizing the geometrical information content measured by the MMW radar. This is especially important for those MMW radars that use polarization techniques to achieve STI.

Both the increased cross-range resolution and scatterer information that can be achieved with MMW radars, as compared with conventional microwave radars, are results of the reduced wavelength, which varies from 8.6 mm (35 GHz) to 1.4 mm (220 GHz). The cross-range resolution, δR, at a range R, is approximately given by

$$\delta R = kR\lambda/L \qquad\qquad (6.10)$$

where λ is the wavelength and L is the antenna aperture dimension. The constant k depends on the details of the antenna, and is approximately given by 1.3. Thus, for a one-meter aperture, a 95 Ghz radar would have a cross-range resolution of approximately 4.1 m at a range of 1000 m. The millimeter wavelength is typically much less than the scatterer dimensions of the target being illuminated. Thus, most electromagnetic-wave and target interactions take place in the optical scattering region and yield much more target structure than can be obtained with a microwave radar.

As shown in Figure 6.35, three types of scattering, or scattering regions, can be defined for the radar-target interaction: the Rayleigh region, the Mie region, and the optical region. In the Rayleigh region, the dimensions of the targets are small compared to the wavelength, and the radar cross section (RCS) of the target nominally varies as λ^{-4}. In the Mie region, the dimensions of the target are comparable to the wavelength, and the RCS varies in an oscillatory manner as a function of wavelength. Finally, in the optical region, the dimensions of the target are much greater than the wavelength, and the target RCS is relatively independent of wavelength. Since, for most targets of interest, scattering occurs in the optical region for MMW radars, the RCS of an object is relatively insensitive to radar transmitting frequency. Thus, vehicular objects have approximately the same average RCS at, for example, 35 GHz and 94 GHz.

The RCS of a target is, however, a very sensitive function of the target viewing aspect angle, especially at millimeter wavelengths. This aspect-angle dependency is determined by the maximum cross-range extent, D_{max}, of the target. The larger D_{max} is, the more sensitive RCS is to aspect angle. For example, consider a target consisting of two scatterers separated by a distance D, as shown in Figure 6.36. Let the MMW radar system be positioned such that its viewing LOS is a perpendicular bisector of a line connecting the two scatterers of the target. From this viewing aspect angle, the returns from the two scatterers are in phase, and they constructively interfere with each other, (i.e., the target RCS is a maximum). The minimum angle, θ, the MMW radar must move such that the return from the two scatterers are out-of-phase and thus destructively interfere is given by

$$\sin(\theta) = \lambda/4D \qquad\qquad (6.11)$$

For a target with cross-range dimension of $D = 6$ m and a wavelength of 3.2 mm (94 GHz), θ is less than 0.008°! This is a problem when trying to achieve stationary target identification based on target signatures. The target signature is very sensitive to target viewing aspect angle at millimeter

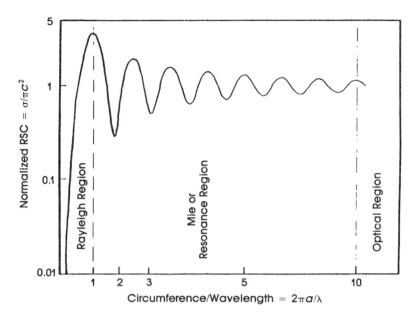

Fig. 6.35 Theoretically determined maximum RCS of a conducting sphere.

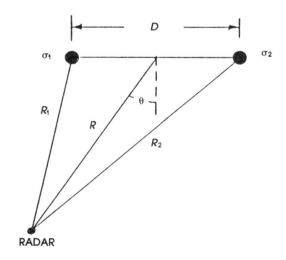

Fig. 6.36 Aspect angle fluctuations in RCS.

wavelengths. Thus, it is not unusual to find data collection efforts to support target identification programs in which target signatures are collected at aspect angle increments of 0.005° and less.

6.4.2 The Stationary Target Identification Problem

The stationary target identification problem can be divided into three phases: (1) *detection*, (2) *discrimination*, and (3) *recognition*. In the detection phase, potential targets of interest are separated from noise and benign (low cross section, distributed) clutter returns by using simple amplitude thresholding techniques (e.g., CFAR threshold). Typically, these techniques can reduce the number of clutter false alarms in an air-to-ground scenario to between five and ten (or even less) per square kilometer. In the discrimination phase, the MMW radar attempts to discriminate between targets and the remaining strong, discrete, target-like clutter returns (e.g., tree lines, utility poles, *et cetera*) that were passed in the detection phase. Finally, in the recognition phase, the MMW radar attempts to determine whether a detected target belongs to any one of the target classes of interest. The target recognition phase can be further subdivided into two phases, based on the level of performance achieved. The lower level of performance phase, sometimes referred to as target *classification*, refers to the classification of the targets based on their generic target type (e.g., tank *versus* truck). The higher level of performance phase, sometimes referred to as target *identification*, refers to the further recognition of targets within a given generic target type, or class (e.g., M-60 *versus* M-48 within the tank class). Obviously, the target identification phase is the most demanding of waveform and signal processor design, and requires the successful completion of all preceding phases. Thus, the entire process (detection, discrimination, recognition) is often referred to as simply "target identification," and is summarized in Table 6.2.

Table 6.2 Stationary Target Identification.

Phase	Purpose
Detection	Separation of targets from noise and benign clutter
Discrimination	Elimination of strong target-like clutter returns
Recognition	Recognize Targets
Classification	Determines generic target classes (e.g., trucks, tanks, *et cetera*)
Identification	Specifies targets within a generic class (e.g., M-48, T-62, *et cetera*)

6.4.3   Stationary Target Detection

In the detection phase, potential targets of interest are separated from noise and benign clutter returns by simple amplitude threshold techniques. This is usually accomplished with a CFAR processor. The CFAR processing approach is a technique employed in radar systems to provide a constant and, it is hoped, low false alarm rate. As described in Chapter 3, this processing technique determines the detection threshold setting in a resolution cell by measuring the backscatter power in adjacent range or cross-range cells. The more backscattered power there is in these threshold determination cells, the higher the threshold is set for the test cell.

The basic underlying assumption of CFAR processing is that the clutter is homogeneous in the threshold determination cells. This implies that these cells do not contain any targets, or discrete target-like clutter objects. If they do, then the CFAR processing technique results in a reduced probability of detection. This is because the threshold is set higher than normal due to the additional backscattered power from the target or discrete clutter. While this results in maintaining the required false alarm rate, it greatly reduces the probability of detecting a possible target in the test cell. The same phenomenon occurs when a discontinuity in clutter RCS is present within the threshold determination cells. The classic example of this is the transition from an open field to a wooded area. Targets positioned next to the resulting tree line will almost always escape detection by a MMW sensor employing only CFAR processing. Additional target-clutter discrimination techniques must therefore be employed to detect these targets. These MMW discrimination techniques are discussed next. A detailed discussion of CFAR detection techniques, including parameter and distribution-free CFAR processing, is contained in Chapter 3.

6.4.4   Stationary Target Discrimination

Stationary target discrimination techniques can be categorized into three classes: (1) *scalar techniques*, (2) *vector techniques*, and (3) *matrix techniques*. Scalar techniques do not use any aspect of the vector nature (i.e., polarization) of the transmitted and received waveform to effect target identification. Only the scalar quantities (amplitude, frequency, and phase) of the electromagnetic wave, as measured by a single-channel (non-polarimetric) MMW radar are used. Vector techniques use some aspect of polarization (along with the scalar quantities), but do not utilize the entire amount of polarimetric information potentially available to the MMW radar sensor. MMW radars implementing vector techniques are dual-polarimetric systems, but do not necessarily possess polarization diversity on transmit. Finally, matrix techniques use all of the polarimetric information available, and thus require measurement of the *polarization scattering matrix* (PSM). This requires a dual-channel, polarization-diverse MMW radar. Examples of techniques in all three classes are shown in Table 6.3.

Table 6.3 Target-Clutter Discrimination.

THREE CLASSES OF TARGET-CLUTTER DISCRIMINATION ALGORITHMS

Scalar Discriminants

(Do not use any aspect of polarization)

    Clutter Decorrelation Techniques (Frequency-Induced and Natural)
    Spatial Feature Algorithms
    High Resolution Techniques

Vector Discriminants

(Use some limited aspect of polarization)

    Depolarization Techniques
    Depolarization Techniques with Frequency Agility

Matrix Discriminants

(Based on measuring the polarization scattering matrix)

    Null-Polarization Techniques
    Matrix Parameter Techniques
    Mueller and Density Matrix Decomposition Techniques

6.4.4.1 Scalar Discrimination Techniques

    Scalar discrimination techniques use only the scalar information (amplitude, frequency, and phase) inherent in the radar signal to achieve target-clutter discrimination. No polarization information is utilized. Some specific scalar target identification techniques are (1) *clutter decorrelation techniques*, (2) *spatial feature algorithms*, and (3) *high resolution techniques*.

Clutter Decorrelation Techniques

    These techniques are based on the premise that the return from clutter will, or can be made to, decorrelate faster than the return from targets. As a result, the spectrum of the clutter is spread in frequency

space more than the spectrum of the target; that is, the clutter return is
more "noise-like" than the target return, and a *signal-to-clutter ratio* (SCR)
enhancement can be obtained by integrating many returns together. An
equivalent way to explain how decorrelation techniques work is by considering the *probability distribution functions* (PDFs) of the measured radar
cross sections of the clutter and target-plus-clutter. Figure 6.37 shows the
single-pulse PDF and the integrated (over several pulses) PDF for the
radar cross sections of clutter-only and target-plus-clutter. In the integrated
PDF, the clutter return is assumed to be uncorrelated from pulse to pulse.
This results in a reduction of the standard deviation, or width, in the PDF
curves. Thus, a better separation (i.e., discrimination), is obtained for the
integrated PDF curves.

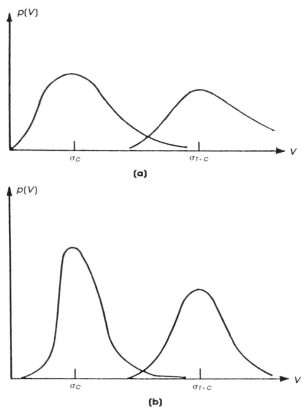

Fig. 6.37 Distributions of averaged clutter and target-clutter RCS (a) without decorrelation and (b) with decorrelation.

Clutter decorrelation either occurs naturally or is induced by changes in the transmitted frequency. Natural clutter decorrelation usually is caused by some small clutter motion (e.g., wind-blown tree canopies). Thus, the integrated SCR enhancement is obtained by dwelling on the clutter for a period longer than its decorrelation time. Frequency-induced clutter decorrelation is obtained by transmitting a wide-bandwidth (spread-spectrum) waveform, for example, a frequency-modulated (chirped) pulse.

Another spread-spectrum technique to achieve clutter decorrelation is pulse-to-pulse *frequency stepping*. The amplitude and phase of the return from a complex object (many scattering centers) will be a rapidly changing function of the transmitting frequency. Therefore, if the clutter has many more scattering centers than the target of interest, then the return signal from the clutter for a frequency-stepped waveform will fluctuate more than the return signal from the target (i.e., the clutter signal is more decorrelated). Noncoherent pulse-to-pulse frequency stepping, followed by a Fourier transform of the received signal, results in measurement of the relative (down-range) distance between scatterers. Thus, for targets with known maximum scatterer separations, a threshold can be set to eliminate all (clutter) returns with scatterer separations greater than the threshold. Coherent pulse-to-pulse frequency stepping, followed by a Fourier transform, results in a high-range-resolution profile, and is discussed in Section 6.5.

Spatial Feature Algorithms

These techniques can be grouped into two categories: (1) *spatial extent filtering* and (2) *contextual feature algorithms*. Spatial extent filtering is based on the fact that the spatial extent of many clutter types differ from the spatial extent of the targets. For example, spatially extended clutter (covering several target resolution cells) (e.g., tree lines), can be eliminated by simply filtering out returns that appear in greater than N contiguous range cells, where $N \times$ range-resolution is much greater than the target length. Contextual feature algorithms are based on the fact that the spatial distribution of targets may differ from the spatial distribution of false targets from discrete clutter. For example, the return from a truck convoy shown in Figure 6.38 can be distinguished from the return from discrete clutter based on the convoy's *deterministic* formation, compared to the *random* distribution of the clutter. Thus, these discrete clutter returns (e.g., rock out-croppings), distributed in a random fashion, can be eliminated.

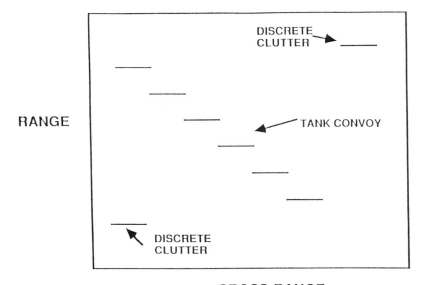

CROSS-RANGE

Fig. 6.38 Contextual features example.

High Resolution Techniques

These techniques are based on the fact that a SCR enhancement is obtained as the resolution cell is reduced (in range or cross-range), and clutter return is increasingly eliminated from the target cell. As the range resolution becomes less than the target extent (i.e., as the target becomes resolved in range), the target will appear in several contiguous range bins. Isolated range bin detections (or detections in contiguous range bins with total range extent less than that of the target) can be rejected as returns from discrete clutter (e.g., utility poles). Moderate (3–7 m) and high (0.5–3 m) range resolutions are typically used in this type of discrimination.

As the cross-range resolution also becomes less than the target extent, the target is resolved. Then, only those cells at the periphery of the target will have a finite SCR. Now, the discrimination problem becomes one of discriminating between cells containing all target return or all clutter return.

6.4.4.2 Vector Discrimination Techniques

Vector discrimination techniques are usually based on some simple polarization-dependent scattering property that differs between a target and clutter. These techniques make no attempt to separate the effects that both the radar (antenna) and the target have on the polarimetric properties of the measured backscattered signal. In fact, both effects are taken into account to achieve the desired discrimination. Most polarimetric discrimination techniques in use are vector techniques.

Vector discrimination techniques take advantage of the change in polarization of the transmitted electromagnetic wave as a consequence of scattering from the target or clutter (i.e., depolarization). For example, the most common depolarization technique is the depolarization of circularly polarized radiation by raindrops (Skolnik [14]). Since the shape of raindrops in most instances can be well approximated by spheroids, the backscatter from rain of an incident circularly polarized wave will have the opposite sense of circular polarization; that is, it will be cross-polarized. Therefore, if the MMW radar transmits a given circular polarization, but receives only the co-polarized component, then the rain clutter backscatter will have been effectively eliminated. The target return presumably will return both senses of polarization, and thus will be detected by the MMW radar.

Another vector discrimination technique involves taking the ratio of the power in the co-polarized received signal to the power in the cross-polarized received signal. For example, if a vertically (V) polarized signal is transmitted, the co-polarized receive response (VV) is the response in the vertical received channel, and the cross-polarized receive response (VH) is the response in the horizontal receive channel. The polarization ratio is then the ratio of powers in the two channels, i.e., P_{VV}/P_{VH}. The larger the cross-polarized (VH) response channel, the more the scattering object has depolarized the incident return. If clutter is more complex than the target of interest, then this depolarization should be greater for a clutter return than for a target return. Thus, the polarization ratio should be larger when the return is from a cell containing a target, and smaller when the return is from a cell containing clutter only. This technique performs better with a frequency-agile waveform, and when so used is known as the *polarization ratio discriminant* (PRD) (Hayes and Eaves [15]). The PRD has been shown to work well in sea clutter scenarios.

Another vector discrimination technique is the odd-even bounce technique. If the target consists of predominantly odd-bounce (or even-bounce) scatterers and the clutter consists of an equal number of odd- and even-bounce scatterers, and if a circularly polarized signal is transmitted, then the return signal from the target will either be predominantly cross- or co-polarized. The return signal from the clutter, however, will contain equal amounts of cross- and co-polarization. This discriminant also performs better with a frequency-agile waveform, and when so utilized is known as *pseudo coherent detection* (PCD) (Echard, et al. [16]). The PCD technique is usually implemented by transmitting a circularly polarized signal, and measuring the horizontal and vertical polarizations on receive. The relative (polarimetric) phase between the horizontal and vertical channels is then measured to determine the polarization sense (right-hand or left-hand elliptical) of the received signal. An implementation of PCD is shown in Figure 6.39.

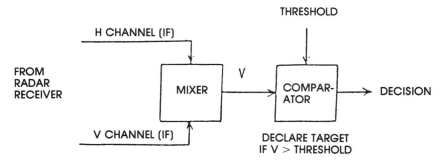

Fig. 6.39 PCD algorithm implementation.

6.4.4.3 *Matrix Discrimination Techniques*

Matrix discrimination techniques are differentiated from vector techniques in that they are based on backscattered signals in which the polarimetric effects of the radar (antenna) on the signals (other than system imperfection effects) have been eliminated, leaving only target-induced polarization effects. Matrix techniques are thus true inverse scattering techniques and require the measurement of the polarization scattering matrix.

Polarimetric matrix techniques are potentially the most powerful of all the STI discrimination and classification techniques because they require that the scattering matrix be measured. Thus, all the polarimetric information available in the backscatter can be used to achieve target identification. In a real sense, the scalar and vector techniques mentioned above can be considered as a subset of matrix techniques. Specific matrix techniques include (1) *null-polarization techniques*, (2) *matrix parameter techniques*, and (3) *Mueller*, or density, *matrix techniques*. Target-clutter discrimination through exploitation of polarimetric matrix techniques is a major signal processing research area associated with MMW radar systems. Therefore, this subject is discussed further in Section 6.6.

6.4.5 Stationary Target Recognition [17]

The *recognition* phase of STI can be considered as made up of three stages, shown in Figure 6.40. The first stage is the preprocessor stage in which the appropriate features used in the target recognition process are extracted. The second stage is the preclassifier stage in which false or alien targets (i.e., targets not of interest), are rejected. The third stage is the classifier stage in which the input feature sets are assigned to one of several classes. Although shown as a separate functional entity, the preclassifer, in practice, may be included in the classifier. For example, the classifier may assign to a feature set "class probabilities" of belonging to the various classes for which it has been trained (e.g., truck 40%, jeep 20%, *et cetera*). Any feature set not obtaining a class probability greater than some minimum threshold is then declared to be an "alien" target.

- PREPROCESSING = FEATURE EXTRACTION

- PRECLASSIFICATION = ALIEN TARGET SEPARATION

- CLASSIFICATION = ASSIGNATION OF INPUT OF ONE CLASS

Fig. 6.40 Stages in stationary target recognition.

The most important stage of the recognition phase is the preprocessor stage (i.e., the feature extractor stage). Recognition performance is determined by the quality of the target features utilized. A classifier, no matter how sophisticated, cannot compensate for poor features. Only statistical features extracted from the MMW radar's waveform, or features based on these statistical features, are considered here. Features based on information obtained from other sensors (or other information sources) could significantly improve the stationary target recognition performance of the MMW radar, but the discussion of these features, and their utilization is beyond the scope of this book.

The basic approach to stationary target recognition is to determine the appropriate N features to be extracted from the MMW radar's waveform, measure these features from the targets or target classes to be recognized at every (aspect and elevation) viewing angle anticipated, and use these features to "train" the classifier. The "training" of the classifier involves either the simple storage of the features in the memory of the classifier, or the application of some classification algorithm, like the pattern recognition algorithms shown in Table 6.4, to the measured features for optimal segregation of these features into groups that represent the various target classes. In the latter case, an unknown feature set is "recognized" when put into a particular feature group that represents a particular target class.

Table 6.4 Target Classification Techniques.

Piecewise-Linear Discriminant Function
Perception Criterion Function
Quadratic Discriminant Function
K-Nearest Neighbor Algorithm
K-NN Prototype Algorithm
Template-Matching Algorithm

Figure 6.41 shows an ensemble of high-range-resolution profiles taken from a target to be classified. The N features extracted from each profile are the amplitudes of the profile sampled at some regular interval. These N amplitudes can be regarded as the components of an N-dimensional *feature vector*. Thus, as shown in Figure 6.42 for an $N = 3$ case, each profile is now represented by a point in a three-dimensional feature space. The ensemble of target profiles is represented by an ensemble of points in feature space. Each target is thus represented by such an ensemble of points.

In the *brute-force* training method, in which all the measured feature information is retained in the memory of the classifier, a pattern recognition (PR) algorithm, like the nearest-neighbor algorithm, is used to recognize an unknown feature set. With this algorithm, an unknown target represented by a point in the N-dimensional feature space (as shown in Figure 6.43) is classified by the majority class of its nearest neighbors. For the example shown in the figure, this would be Class 1.

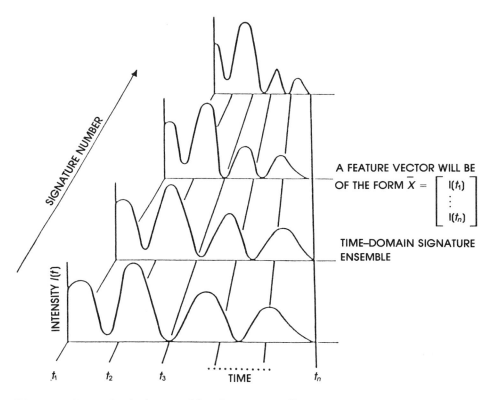

Fig. 6.41 Hypothetical ensemble of target profiles.

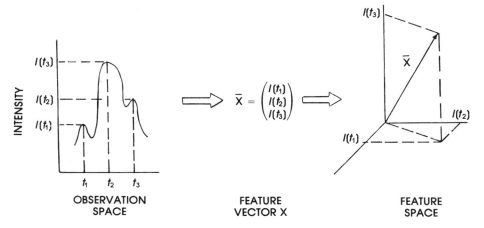

Fig. 6.42 Feature space construction.

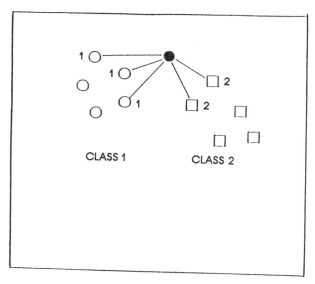

Fig. 6.43 Nearest neighbor classifier (two classes).

In the other type of training method, *decision boundaries* are determined in the N-dimensional feature space by employing the classification algorithms on the N training target features. These decision boundaries determine subspaces within the feature space which are, in turn, identified with the different target classes. This subspace determination is the training of the classifier. Target recognition is then accomplished by submitting the N features of an unknown target to the classifier, which then determines in which sub-space, or target class, they belong. For example, if the various feature-point ensembles are separated in feature space as shown in Figure 6.44 for a three-class, two-feature case, then a standard PR algorithm, such as Fisher's linear discriminant, can be employed to determine the boundaries that optimally separate the ensembles. It is the location of these boundaries, not the feature point ensembles, that are retained in the operational classifier. Such classifiers obviously require less computing capabilities than the "brute-force" trained classifier.

There are basically three MMW waveforms from which features can be extracted to achieve stationary target recognition: (1) high resolution, (2) full-polarimetric, and (3) high resolution and full-polarimetric waveforms. Polarimetric waveforms are discussed in detail in Section 6.6.

High resolution waveforms can be high-range resolution only, or high-range and cross-range resolution. In most of the applications for MMW radars, the resolution of targets-of-interest in the cross-range di-

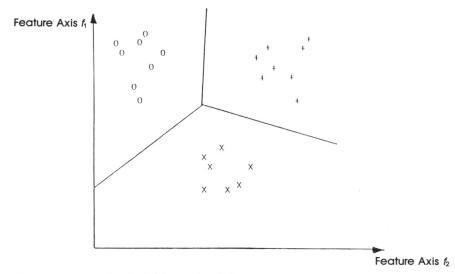

Fig. 6.44 Hypothetical (three-class) feature space.

mension requires the use of synthetic aperture radar (SAR) techniques. These techniques are now only beginning to be applied to the lower frequency regions (35 GHz) of the MMW band. Successful application of SAR techniques will allow for the application of the many radar *imaging* techniques that have been developed for stationary target identification at lower radar bands.

A high-range-resolution waveform, however, is common in MMW radars designed for stationary target identification. These waveforms tend to be of *ultra-high* resolution (i.e., on the order of 0.5 m or less). One of the more common ways of achieving ultra-high resolution in a MMW radar is by using the frequency-stepped waveform. Since this waveform has become more or less identified with MMW systems, it is discussed in detail in the next section.

An ultra-high range resolution MMW radar measures the range profile of targets as shown in Figure 6.45. Amplitude features are then extracted from the data for STI purposes. The two fundamental types of amplitude features are raw and heuristic features, as illustrated in Figure 6.46. Raw features are those which come directly from the MMW radar sensor with little or no processing or interpretation, (e.g., the amplitude values in each high-resolution-range bin). Heuristic features are derived from raw features and related to the physical characteristics of the target. Raw features are not invariant with respect to target translations, while heuristic features can be very robust with respect to these translations. Examples of heuristic features are shown in Table 6.5.

Table 6.5   Examples of Heuristic Features.

1. The distance between the highest and second highest peaks.
2. The distance between the highest and the third highest peaks.
3. The distance between the second and third highest peaks.
4. The distance between the leftmost and rightmost peaks.
5. The ratio of the distance between the highest and leftmost peak to the distance between the leftmost and rightmost peaks.
6. The ratio of the distance between the second highest and the leftmost peaks to the distance between the leftmost and rightmost peaks.
7. The ratio of the amplitude of the second highest peak to the amplitude of the highest peak.
8. The ratio of the amplitude of the third highest peak to the amplitude of the highest peak.
9. The number of peaks above 10% of the highest peak.
10. The number of peaks above 50% of the highest peak.

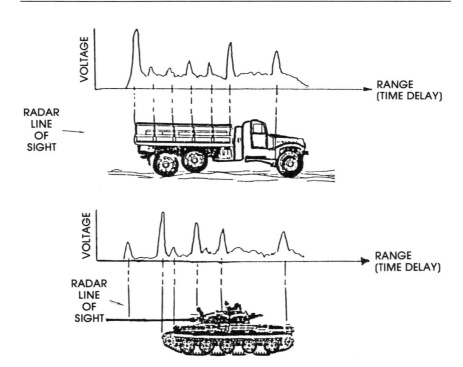

Fig. 6.45 Ultra-high-range-resolution profiles of stationary targets.

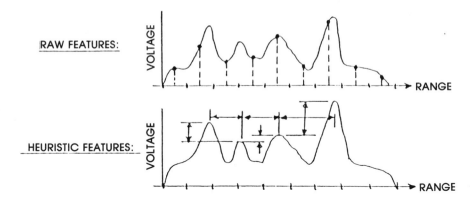

Fig. 6.46 Raw and heuristic features.

6.5 HIGH-RANGE-RESOLUTION TECHNIQUES

MMW radar systems are often operated in a high-range-resolution mode in order to obtain range images or profiles of targets for stationary target identification purposes. Often these range resolutions are ultra-high (i.e., on the order of 0.5 m or less). The large bandwidths required for such range resolutions can present a problem to the digital sampling circuitry in the MMW radar's receiver. In this section, standard techniques for obtaining high-range-resolution profiles are presented. This is followed by a discussion of the frequency-stepped high-range-resolution technique that is often employed in MMW radar systems to avoid high bandwidth sampling problems.

6.5.1 Standard High-Range-Resolution Techniques [18, 19]

The range resolution, δR, obtainable with any radar system depends on the bandwidth of the transmitted signal, and is given by

$$\delta R = \frac{c}{2B} \tag{6.12}$$

where c is the speed of light (3×10^8 m/s), and B is the bandwidth of the transmitted signal. Thus, a range resolution of 0.3 m would require a transmitted bandwidth of at least 500 MHz.

Achieving such a high range resolution (high bandwidth) can present problems for some of the conventional high-range-resolution methods (e.g., short pulse and chirped pulse compression). A 0.3 m resolution can be obtained by transmitting a very narrow pulse of approximately 2 ns. Aside from the fact that it is difficult to obtain a pulse this narrow, the resulting transmitted average power (on which target detection depends) may be too low. The average transmit power is given by

$$P_{avg} = P_t d_t \tag{6.13}$$

where P_t is the peak transmitted power, and d_t is the duty factor, given by

$$d_t = f_r \tau \tag{6.14}$$

where f_r is the PRF, and τ is the pulsewidth. To compensate for a small pulsewidth, the PRF or the peak power must be increased. The PRF is limited by the maximum unambiguous range, R_u, given by

$$R_u = \frac{c}{2f_r} \tag{6.15}$$

Thus, for example, for a R_u of 2 km, f_r can be no greater than 75 kHz. For a 2 ns pulse, this gives a duty factor of 0.00015. In this case, the peak power of the transmitter may not be able to be increased enough to give the desired average power; thus, a *peak-power-limited* condition exists.

Another problem with the narrow pulse approach involves the data acquisition sampling bandwidths. The received radar signal must be digitized for the digital processors used in today's MMW radar systems. The minimum sampling rate, B_s, for the A/D converter is given by

$$B_s = \frac{\Delta R}{\delta R} f_r \tag{6.16}$$

where ΔR is the range swath from which the backscattered signal is being received. Here, the assumption is made that return signal, which is received over some fraction of the PRI, can be stored in a buffer memory so that the digitizing (sampling) time can be extended over the entire PRI. From (6.12), (6.15), and (6.16), B_s is also given by

$$B_s = \frac{\Delta R}{R_u} B \tag{6.17}$$

Notice that if ΔR is equal to the maximum unambiguous range, then $B_s = B$, so the A/D sampling rate is equal to the radar's bandwidth (e.g., 500 MHz for 0.3 m resolution). These sampling rates are difficult to achieve for A/D converters with any appreciable dynamic range.

In the chirped pulse compression approach to high range resolution, a long pulse is transmitted and the frequency of the signal is changed (usually in a linear manner) during the transmission. The backscattered signal is processed to obtain an effectively narrow, or compressed, pulse, thus producing a high range resolution. The range resolution achieved depends on the bandwidth, B, of the frequency modulation (chirp), and is again given by (6.12). Because a long pulse is transmitted, the radar's average power can be made adequate for detection, thus avoiding a peak-power-limited situation. However, the data acquisition sampling rate problem, inherent in the real pulse technique given above, is still present.

Data sampling rate problems also can occur with the use of a frequency-modulated continuous wave (FMCW), or an interrupted FMCW approach to high range resolution, whereby a CW signal is frequency-modulated (e.g., linearly or sinusoidally), and range information is determined by taking the difference, δF, in the frequency of the transmitted wave and the received wave. Range resolution is again determined by (6.12), where now B is the bandwidth of the frequency modulation. The frequency difference as a function of range is given by

$$\delta F = \frac{\Delta R}{R_u} B \qquad (6.18)$$

Here, unambiguous range, R_u, is determined by the period of the frequency modulation. The bandwidth of this difference signal, which is equivalent to the sampling rate, B_s, will again depend on the range swath, ΔR, from which the backscattered signal is being received. From (6.18), this sampling rate is given by

$$B_s = \frac{\Delta R}{R_u} B \qquad (6.19)$$

which is identical to (6.17) for the pulsed radar. Now, the only difference is that no buffer memory storage is required. However, the same sampling rate problems that might occur for the pulsed radar also occur for the FMCW radar. In addition, range resolution in the FMCW radar depends on the uniformity of the frequency modulation from period to period. For example, for a linear frequency modulation, high range resolution requires a very linear frequency ramp, which can be difficult to obtain.

6.5.2   Frequency-Stepped High-Range-Resolution Technique

The peak-power-limited and sampling rate problems discussed above can be circumvented by implementing a frequency-stepped pulse compression technique. This technique is often employed with today's MMW radar systems. In this technique, a pulse with a constant RF is transmitted. The next transmitted pulse has an RF that is shifted in frequency by Δf from the previous pulse. The frequency of the next pulse is shifted by an additional Δf, and so on, until N pulses have been transmitted. The total bandwidth stepped through is thus $(N - 1)\Delta f$. However, the instantaneous (receiver) bandwidth is determined by the inverse of the single-pulse pulsewidth, which can be very small.

The relative phase difference, ϕ, between the transmitted pulse and the received signal from a point scatterer at range R is

$$\phi = \frac{4\pi}{c}fR \tag{6.20}$$

The relative phase difference between two consecutive pulses, ϕ', from that point scatterer is

$$\phi' = \frac{4\pi}{c}\Delta fR \tag{6.21}$$

This phase difference, ϕ', will vary, depending on the range or position of the scatterer within the range bin. The maximum it can vary is determined by the maximum variance of the scatterer position (i.e., the MMW radar's uncompressed range resolution $\delta R = c\tau/2$). Thus, the maximum phase difference is given by

$$(\delta\phi')_{max} = 2\pi\Delta f\tau \tag{6.22}$$

To obtain unambiguous measurement of scatterer range within the range bin, $(\delta\phi')_{max}$ must be less than or equal to 2π. Thus, the following relationship between the frequency step size and the pulsewidth is obtained:

$$\Delta f \leq 1/\tau \tag{6.23}$$

This relationship assumes that a perfect rectangular pulse is transmitted (i.e., no time sidelobes). When time sidelobes are present, the return from scatterers outside the range bin ($c\tau/2$) will be "folded-over" into the return from scatterers within the range bin. This problem is prevented by reducing the frequency step, Δf. Einstein [20] recommends an upper bound for Δf of $1/2\tau$ to eliminate this sidelobe fold-over problem.

To obtain high-range resolution by using the frequency-stepped pulse compression method, the received signal is simply down-converted to video, and then FFT processed. The FFT frequency bins now effectively become high-resolution-range bins inside the uncompressed ($c\tau/2$) range bin. To understand this process, consider the relative phase difference between two consecutive returns (pulses) from two scatterers, separated by a range of δR; thus,

$$\delta\phi' = \frac{4\pi}{c}\Delta f \delta R \tag{6.24}$$

The resulting frequency difference is given by

$$\delta f' = \frac{1}{2\pi}\frac{\delta\phi'}{T} = \frac{2}{cT}\Delta f \delta R \tag{6.25}$$

where T is the interpulse period. A frequency bin size is determined by the frequency resolution of the system, and is given by $(\delta f')_{min} = 1/NT$. Thus, the minimum range distance by which two scatterers can be separated and still appear in different frequency bins is given by

$$(\delta R)_{min} = \frac{c}{2B} \tag{6.26}$$

which is the range resolution of a radar with a transmitted waveform of bandwidth B. Thus, all scatterers within one high-resolution-range bin will appear in one frequency bin in the FFT output.

According to the Nyquist sampling theorem, the maximum unambiguous range of frequencies that can be measured by the radar is equal to the radar's PRF; thus,

$$(\delta f')_{max} = f_r = 1/T \tag{6.27}$$

Therefore, from (6.25), the corresponding maximum unambiguous range extent is given by

$$(\Delta R)_{max} = \frac{c}{2\Delta f} \tag{6.28}$$

If we require this maximum unambiguous range extent to be equal to or greater than the uncompressed range resolution of the radar, then the condition given by (6.23) is regained; hence,

$$\Delta f \leqslant 1/\tau \tag{6.29}$$

If we let Δf be less than $1/\tau$, this again allows for the presence of range sidelobes. If Δf is less than $1/\tau$ and no range sidelobes are present, then the backscatter return will appear only in a subset of the FFT's N frequency bins. The specific subset depends on the absolute range to the target.

6.6 POLARIZATION PROCESSING TECHNIQUES

Polarimetric matrix techniques are potentially the most powerful of all MMW radar discrimination and classification techniques. Matrix techniques require the measurement of the *polarization scattering matrix* (PSM). Therefore, an understanding of polarization matrix theory is required. A brief discussion of the pertinent theoretical facts of PSM theory is covered here, followed by a discussion specific matrix techniques. For a more complete discussion of the basics of radar polarization theory, the interested reader is referred to the literature [21, 22, 23].

6.6.1 EM Wave Polarization States

The E-field, $\vec{e}$, of a plane electromagnetic (EM) wave in a pure (i.e., completely polarized) state, propagating along the $+z$-axis, can be represented by a two-dimensional complex-valued state vector:

$$\vec{e} = \begin{bmatrix} e_x \\ e_y \end{bmatrix} \tag{6.30}$$

or a 2×2, complexed-valued coherency (or density) matrix:

$$\rho = \vec{e} \times \vec{e}\dagger = \begin{bmatrix} |e_x|^2 & e_x e_y^* \\ e_x^* e_y & |e_y|^2 \end{bmatrix} \tag{6.31}$$

where $\times$ is the direct (or Kronecker) product, $\dagger$ denotes the Hermitian adjoint, and $*$ denotes complex conjugation. The density matrix formalism is more robust than the state vector formalism, since it can also describe partially polarized, or "mixed," states. A signal with a polarization state that varies in a random fashion is said to be partially polarized. These types of signals are of most interest in polarimetric radar technology. When describing a mixed polarization state, the density matrix is written as

$$\rho = \langle \vec{e} \times \vec{e}^{\,\dagger} \rangle \tag{6.32}$$

where $\langle . . . \rangle$ is an ensemble or time average. For ρ to represent a valid polarization state, its diagonal elements and its determinant must be non-negative:

$$\rho_{ii} \geq 0 \tag{6.33}$$
$$|\rho| \geq 0$$

Any EM density matrix representing a mixed state can be uniquely decomposed into two components:

$$\rho = \rho_0 + \rho_1 \tag{6.34}$$

where ρ_0 is a pure state given by

$$\rho_0 = \left\langle \begin{bmatrix} |e_x|^2 - \epsilon & e_x e_y^* \\ e_x^* e_y & |e_y|^2 - \epsilon \end{bmatrix} \right\rangle \tag{6.35}$$

and ρ_1 is an unpolarized mixed state given by

$$\rho_i = \begin{bmatrix} \epsilon & 0 \\ 0 & \epsilon \end{bmatrix} \tag{6.36}$$

where ϵ is a real, positive number such that the diagonal elements of ρ_0 are non-negative and the determinant of ρ_0 is zero. The alert reader will notice that ϵ is determined by solving the characteristic equation for ρ:

$$|\rho - \epsilon I| = 0 \tag{6.37}$$

where I is the identity matrix. The parameter ϵ is simply the smallest root of this characteristic equation.

6.6.2 Polarization Scattering Matrix States

The relative PSM, $\mathbf{S}$ (equivalent to the PSM, to within a phase factor), can be defined in terms of the RCS, σ_{tr}, as

$$\sigma_{tr} = |\hat{e}_r \cdot S\hat{e}_t|^2 \qquad\qquad (6.38)$$

where $\hat{e}_t$ and $\hat{e}_r$ are, respectively, the normalized ($\hat{e}\cdot\hat{e}^* = 1$), two-dimensional, complex-valued transmitted and received antenna polarizations. $\mathbf{S}$ is a 2×2 complexed-valued matrix, which, for the monostatic radar case in the absence of any magnetic (e.g., Faraday rotation) effects depends on five real parameters. If a linear basis is used, for example, horizontal (H) and vertical (V) polarizations, then $\mathbf{S}$ is given by

$$\mathbf{S} = \begin{bmatrix} S_{HH} & S_{VH} \\ S_{HV} & S_{VV} \end{bmatrix} \qquad\qquad (6.39)$$

where, for example, S_{HV} is the complex target backscatter amplitude received with a vertically polarized antenna when a horizontally polarized wave was transmitted.

$\mathbf{S}$ can be thought of as representing the polarimetric response, or state, of the target. If the target is in a pure state, then the backscattered EM wave from the target is completely polarized. However, if the target is in a "mixed" state, then the backscattered EM wave is partially polarized. A mixed target state can be caused by random target motions (translational and internal), radar platform motions, and changes in the radar waveform. An ensemble of scattering matrices (instead of a single scattering matrix) must be used to represent such a mixed target state.

One method of handling mixed target states, is by considering the ensemble average of the RCS, $\langle\sigma_{tr}\rangle$, which can be written as

$$\langle\sigma_{tr}\rangle = \langle|\hat{e}_r \cdot S\hat{e}_t|^2\rangle = \hat{g}_r \cdot \langle M\rangle \, \hat{g}_t \qquad\qquad (6.40)$$

where $\hat{g}$ is the Stokes vector and $\langle M\rangle$ is the average Stokes reflection (or Mueller) matrix. The Mueller matrix is a 4×4 real, symmetric matrix that depends on nine independent parameters for a mixed target state. This reduces to five independent parameters (as it must) for a pure target state. The Mueller matrix approach in polarimetric radar theory is well documented in the literature (see the references). However, it is rather cumbersome and can be replaced with a more elegant density matrix approach.

A target density matrix formalism similar to the EM density matrix formalism can be introduced to represent mixed target states. This is accomplished by writing **S** in terms of three complex variables, *a, b, c*, thus,

$$\mathbf{S} = \begin{bmatrix} a + b & c \\ c & a - b \end{bmatrix} \tag{6.41}$$

The density matrix formalism is obtained by constructing a state vector, **S**, thus,

$$\vec{\mathbf{S}} = \begin{bmatrix} a \\ b \\ c \end{bmatrix} \tag{6.42}$$

Now, in analogy to (6.31), a density matrix representing the target state can be constructed,

$$\rho = \langle \vec{\mathbf{S}} \times \vec{\mathbf{S}}^{\dagger} \rangle \tag{6.43}$$

Now, the conditions for ρ to be a valid target density matrix are given by

$$\rho_{ii} \geqslant 0$$
$$\rho_{ii}^{(1)} \geqslant 0$$
$$|\rho| \geqslant 0 \tag{6.44}$$

where $\rho^{(1)}$ is the first-order target density matrix defined by

$$\rho_{ij}^{(1)} = \text{cofactor } \rho_{ji} \tag{6.45}$$

or

$$\rho^{(1)} = |\rho|\rho^{-1}, \quad (|\rho| \neq 0) \tag{6.46}$$

The three conditions in (6.44) can be written as a single expression

$$\rho_{ii}^{(n)} \geqslant 0, \quad n \geqslant 0 \tag{6.47}$$

where

$$\rho_{ij}^{(n)} = \text{cofactor } \rho_{ji}^{(n-1)} \tag{6.48}$$

and $\rho^{(0)} = \rho$. This is possible since $\rho^{(2)} = \rho|\rho|$.

At this point, the target density matrix can be decomposed into a component representing an average target pure state and one or more other components. However, before covering target density matrix decomposition, a discussion of pure target states is given.

6.6.3 Pure Target States

A pure target state is one in which the backscattered EM wave is fully polarized. A pure target state can be represented by the PSM, **S**, or a density matrix, ρ, where $\rho^{(1)} = [0]$. An eigenvalue analysis of the PSM yields physical insight into the scattering process. The characteristic eigenvalue problem for the scattering matrix is given by

$$\mathbf{S}\vec{\mathbf{a}} = s\vec{\mathbf{a}}^{\,*} \tag{6.49}$$

(The complex conjugate appears in this equation because the relevant eigenvalue problem here is to find the transmitted polarization that will be proportional to the optimum receive polarization of the transmitting antenna when scattered from the target.) There are two eigenvalue solutions to (6.49), $\vec{\mathbf{a}}_1$ and $\vec{\mathbf{a}}_2$, with corresponding eigenvalues, s_1 and s_2,

$$\vec{\mathbf{a}}_1 = e^{\psi J} \, e^{\tau K} \begin{pmatrix} 1 \\ 0 \end{pmatrix}$$

$$\vec{\mathbf{a}}_2 = e^{(\psi + \pi/2)J} \, e^{-\tau K} \begin{pmatrix} 1 \\ 0 \end{pmatrix} \tag{6.50}$$

and

$$s_1 = m \, e^{i2\nu}$$
$$s_2 = m \tan^2 e^{-i2\nu} \tag{6.51}$$

where

$$J = \begin{bmatrix} 0 & -1 \\ 1 & 0 \end{bmatrix} \quad K = \begin{bmatrix} 0 & j \\ j & 0 \end{bmatrix} \tag{6.52}$$

The m parameter is the amplitude of the maximum return from the target; that is, the amplitude of the optimum polarization which maximizes the radar cross section of the target. The τ parameter $[-45° < \tau < +45°]$ is a measure of target symmetry ($0°$ for fully symmetric; $+45°$ for totally nonsymmetric). The ψ parameter $[-90° < \psi < +90°]$ is measure of orientation of the target (0° for horizontal; ±90° for vertical). The ν parameter $[-45° < \nu < +45°]$ is related to the number of bounces of the reflected signal (0° for odd number of bounces; ±45° for even number of bounces). Finally, the γ parameter $[0° < \gamma < +45°]$ is related to the target's ability to polarize incident radiation (0° for fully polarized; 45° for unpolarized). For a more detailed discussion of these physically relevant parameters, the interested reader is referred to the literature (see the references).

The relative PSM can thus be expressed in terms of the five physically relevant parameters $(m, \psi, \tau, \nu, \gamma)$ as

$$\mathbf{S} = U^* \begin{bmatrix} s_1 & 0 \\ 0 & s_2 \end{bmatrix} U^\dagger \tag{6.53}$$

where U is the unitary matrix given by

$$U = [\vec{\mathbf{a}}_1 \ \vec{\mathbf{a}}_2] \tag{6.54}$$

The theory of null polarizations is now briefly discussed. The four angular parameters $(\psi, \tau, \nu, \gamma)$ also specify two polarization states, called null polarizations, of the scattering matrix, $\mathbf{S}$, such that when $\mathbf{S}$ operates on these null polarizations, the orthogonal polarizations result. Therefore, the measured backscattered RCS given by (6.38) for a target represented by $\mathbf{S}$ is zero (a *null*) when the transmitted and received antenna polarizations are one of these null polarizations. The null polarizations, and thus the *normalized* ($m = 1$) relative scattering matrix, can be represented geometrically by two points on the so-called *Poincaré sphere*. To see how the null polarizations are obtained from ψ, τ, ν, and γ, and how they are represented on the Poincaré sphere, first consider the scattering matrix as given by (6.53). The scattering matrix can also be written as

$$\mathbf{S} = U^*(\psi, \tau, \nu) \begin{bmatrix} m & 0 \\ 0 & m \tan^2\gamma \end{bmatrix} U^\dagger (\psi, \tau, \nu) \tag{6.55}$$

where we now have

$$U(\psi, \tau, \nu) = e^{\psi J} e^{\tau K} e^{\nu L} \tag{6.56}$$

and

$$L = \begin{bmatrix} -j & 0 \\ 0 & j \end{bmatrix} \tag{6.57}$$

Thus, the scattering matrix is obtained from the matrix:

$$\begin{bmatrix} m & 0 \\ 0 & m\tan^2\gamma \end{bmatrix} \tag{6.58}$$

by three successive unitary transformations (rotations): first, a v rotation; then, a τ rotation; finally, a ψ rotation.

Now, consider two line segments (as shown in Figure 6.47) drawn in the $x_1 - x_3$ plane from the center of the Poincaré sphere to the surface such that the x_1 axis bisects the angle between them. Let the angle between the two segments equal 4γ, and label the points where they intersect the sphere as N_1 and N_2. Consider a third line segment drawn from the center of the sphere to the point on the sphere that represents horizontal polarization. Let all three line segments be rigidly connected, thus producing a "fork." Now, let the three sequential rotations (v, τ, ψ) mentioned above to obtain **S** be the three rotations about the x_1, x_2, and x_3 axes by the amounts $2v$, 2τ, and 2ψ, respectively. These rotations will rotate the fork inside the Poincare sphere. After the fork has undergone these three rotations, the intersection of the handle of the fork with the sphere represents the eigenvector $\vec{a}_1$ of **S**. The polarizations $\hat{n}_1$ and $\hat{n}_2$ associated with the points N_1 and N_2 are the null polarizations of **S**. A more complete theory of null polarizations is presented in Huynen [24].

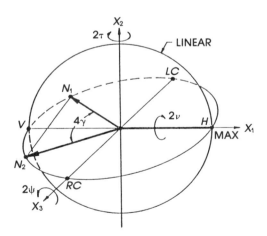

Fig. 6.47 Polarization sphere.

6.6.4   Target State Decompositions

To achieve target identification, it is often advantageous to extract an average pure target state from the measured mixed target state; that is, to decompose the target mixed density matrix into a target pure density matrix plus one or more additional (residue) components. As shown above, the pure density matrix (or, equivalently, the scattering matrix) can be expressed in terms of the five physically relevant parameters, or null polarizations.

The physical significance of performing such a decomposition of a mixed density matrix is that any time-varying (fluctuating) target can be represented by a mean stationary (nontime-varying) target plus a residue part. The time dependence of the target can be due to either the motion of the target, or to the motion of radar sensor itself. The power of such decompositions lie in their potential ability to separate motional perturbations (which are almost always present) from the target's polarimetric signature. Thus, like-targets that are observed in different states of motion can, it is hoped, be recognized as like-targets. In addition, the amount of motion variance attributed to a given target may provide some insight into the nature of that target. For example, rigid (man-made) targets may be separable from nonrigid (natural clutter) targets by examining the characteristics of their respective residue components.

Another advantage of performing these decompositions is in providing the capability to base a target identification decision on more target reflectivity information. For example, research has shown that a single MMW polarization signature of a ground vehicle at a given aspect angle is insufficient to identify that target, where, in effect, that signature would match signatures taken from many other ground vehicles. However, as more signatures are collected over some increment of aspect agile (e.g., tenths of a degree), then that ensemble of polarization signatures becomes more indigenous to that particular vehicle. This ensemble of polarimetric signatures can then be averaged together to obtain a mixed state density matrix, which, in turn, can be decomposed to obtain the average pure state density matrix that should characterize that ground vehicle.

There are an infinite number of ways to decompose a mixed target state represented by the density matrix ρ. Two decomposition methods will be discussed here. One is the *Huynen decomposition method,* and the other is the characteristic decomposition, which is similar to that of the EM density matrices presented in Section 6.6.1.

The Huynen decomposition is a physically realizable, unique decomposition, obtained by writing the mixed state density matrix as

$$\rho = \rho_T + \rho_N \tag{6.59}$$

where

$$\rho_T = \begin{bmatrix} \rho_{11} & \rho_{12} & \rho_{13} \\ \rho_{21} & \rho_{22}^T & \rho_{23}^T \\ \rho_{31} & \rho_{32}^T & \rho_{33}^T \end{bmatrix}, \quad \rho_N = \begin{bmatrix} 0 & 0 & 0 \\ 0 & \rho_{22}^N & \rho_{23}^N \\ 0 & \rho_{32}^N & \rho_{33}^N \end{bmatrix} \tag{6.60}$$

and ρ_{22}^T, ρ_{33}^T, and ρ_{23}^T satisfy the equations:

$$\begin{aligned} \rho_{11}\rho_{22}^T - |\rho_{12}|^2 &= 0 \\ \rho_{11}\rho_{33}^T - |\rho_{13}|^2 &= 0 \\ \rho_{11}\rho_{23}^T - \rho_{21}\,\rho_{13} &= 0 \end{aligned} \tag{6.61}$$

The ρ_T component is a pure target density matrix from which the five physically relevant parameters can be extracted. It represents the most "trihedral corner reflector-like" or "sphere-like" pure state target that can be extracted from the mixed target state. The ρ_N component represents a mixed state "noise" residue, and depends on four real parameters. The Huynen decomposition has physical merit and has been shown to be effective in *target discrimination* (see Section 6.4.4).

The characteristic decomposition is obtained by solving the characteristic equation:

$$|\rho - \epsilon I| = 0 \tag{6.62}$$

The three roots of this equation are the eigenvalues of ρ. There corresponding eigenvectors, $\vec{\epsilon}$, are determined by

$$\rho\,\vec{\epsilon} = \epsilon\,\vec{\epsilon} \tag{6.63}$$

Thus, ρ can be put into diagonal form:

$$\rho = U \begin{bmatrix} \epsilon_1 & 0 & 0 \\ 0 & \epsilon_2 & 0 \\ 0 & 0 & \epsilon_3 \end{bmatrix} U^{-1} \tag{6.64}$$

where

$$U = [\vec{\epsilon}_1 \, \vec{\epsilon}_2 \, \vec{\epsilon}_3] \tag{6.65}$$

Without loss of generality, the assumption can be made that $\epsilon_3 > \epsilon_2 > \epsilon_1$. The characteristic decomposition is obtained by writing ρ as

$$\rho = \rho_0 + \rho_1 + \rho_2 \tag{6.66}$$

where

$$
\rho_0 = U \begin{bmatrix} 0 & 0 & 0 \\ 0 & 0 & 0 \\ 0 & 0 & \epsilon_3 - \epsilon_2 \end{bmatrix} U^{-1}
$$

$$
\rho_1 = U \begin{bmatrix} 0 & 0 & 0 \\ 0 & \epsilon_2 - \epsilon_1 & 0 \\ 0 & 0 & \epsilon_2 - \epsilon_1 \end{bmatrix} U^{-1} \tag{6.67}
$$

$$
\rho_2 = \epsilon_1 I
$$

The density matrix ρ_0 represents a pure target state and provides the *average* target representation. The density matrix ρ_1 represents a mixed target state and provides the *variance* of the target from its average representation. Finally, the density matrix ρ_2 represents an unpolarized mixed state.

The characteristic decomposition is now applied to the example shown in Figure 6.48 which shows a rotating diplane immersed in polarization noise. The density matrix for a diplane oriented at an angle ψ is shown along with the density matrix for polarization noise. During the radar observation time the diplane rotates from an angle ψ_1 to ψ_2. During this time the averaged density matrix, ρ (of the rotating diplane and polarization noise), is measured, and the characteristic decomposition is applied. The resulting ρ_0 matrix is given by

$$
\rho_0 = \frac{a \, \sin(2\Delta\psi)}{2\Delta\psi} \begin{bmatrix} 0 & 0 & 0 \\ 0 & \cos^2 2\overline{\psi} & \sin 2\overline{\psi} \, \cos 2\overline{\psi} \\ 0 & \sin 2\overline{\psi} \, \cos 2\overline{\psi} & \sin^2 2\overline{\psi} \end{bmatrix} \tag{6.68}
$$

which is a diplane oriented at the average angle $\overline{\psi} = \frac{1}{2}(\psi_1 + \psi_2)$. The ρ_1 matrix is a mixed state of diplanes given by

$$\rho_1 = \frac{a}{2}\left[1 - \frac{\sin(2\Delta\psi)}{2\Delta\psi}\right]\begin{bmatrix} 0 & 0 & 0 \\ 0 & 1 & 0 \\ 0 & 0 & 1 \end{bmatrix}$$

and the ρ_2 matrix is the unpolarized mixed state, representing the polarization noise.

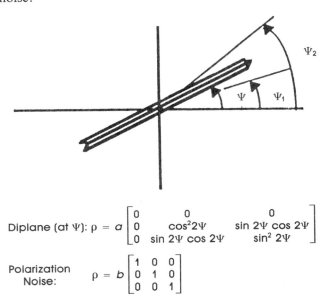

$$\text{Diplane (at } \Psi\text{): } \rho = a \begin{bmatrix} 0 & 0 & 0 \\ 0 & \cos^2 2\Psi & \sin 2\Psi \cos 2\Psi \\ 0 & \sin 2\Psi \cos 2\Psi & \sin^2 2\Psi \end{bmatrix}$$

$$\begin{array}{c}\text{Polarization}\\ \text{Noise:}\end{array} \quad \rho = b \begin{bmatrix} 1 & 0 & 0 \\ 0 & 1 & 0 \\ 0 & 0 & 1 \end{bmatrix}$$

Fig. 6.48 Characteristic decomposition example: rotating diplane in polarization noise.

6.6.5 Specific Matrix Target Identification Techniques

As was mentioned in Section 6.6.4, some of the specific matrix techniques for stationary target identification (discrimination) are (1) null polarization techniques, (2) scattering matrix parameter techniques, and (3) density matrix techniques.

6.6.5.1 Null Polarization Techniques

The null polarization techniques (Weisbrod and Morgan [25]) are often used in the discrimination phase of identification, and are based on the fact that the measured backscattered RCS for a target represented by the scattering matrix **S** is zero (a null) when the transmitting receiving antenna polarization is the null polarization of **S** (see Section 6.6.3). Therefore, if a null polarization could be found for a given clutter type, then the clutter becomes invisible to the radar by simply transmitting and receiving this polarization. Unfortunately, the radar-clutter interaction process is usually nonstationary. Therefore, the null polarizations for a given clutter type are time and space dependent. Still, if the null polarizations of the clutter remain fairly well localized on the Poincare sphere during the observation time (i.e., do not vary greatly), then an optimum transmitting and receiving polarization can be found to maximize the target-to-clutter ratio (Ioannidis [26]). Another approach suggested is adaptively changing the transmitting and receiving antenna polarizations in real time to maximize target-to-clutter ratio (Poelman [27]).

6.6.5.2 Scattering Matrix Parameter Techniques

Scattering matrix parameter techniques involve using the elements of the scattering matrix, or extracting the five physically relevant parameters $(m, \psi, \tau, \nu, \gamma)$ from the scattering matrix, and using these elements or parameters (or functions thereof) to effect target identification. One way of achieving this is to use the parameters as features in a multidimensional feature space (see Section 6.4.5) and then applying pattern recognition algorithms to these features to separate the targets (for recognition) and the targets from clutter (for discrimination) (Holm [28]). Results from the use of scattering matrix parameter techniques have been mixed. This is due, in large part, to the fact the radar return from targets and clutter is usually partially polarized. Therefore, matrix parameters tend to be "noisy," and identification performance in some cases can actually be reduced by using these techniques.

6.6.5.3 Density Matrix Techniques

Density matrix techniques involve using the elements of the density (or Mueller) matrix to effect target identification. The elements of the density matrix can be associated with geometrical attributes of the target. In addition, if several pure state density matrices representing several "looks" a stationary target are averaged together, then the resulting average density matrix can be decomposed by one of the decomposition methods mentioned above, and the components of that decomposition may be used for the target identification purposes. The discrimination performance level shown in Figure 6.49 is achieved using a discriminant extracted from density matrix components obtained after applying Huynen's decomposition to data collected from an automobile and tree line clutter. The RCS of the clutter was such that ordinary amplitude threshold techniques were ineffective for discriminating the automobile from the clutter. Choosing the threshold shown in the figure would allow 50% of the automobile signatures to pass the discrimination processor, while letting less than 2% of the clutter signatures pass (thus causing false alarms).

The physics behind the full-polarimetric (matrix) approach to target identification is that the target geometrical information inherent in the measured scattering matrix will be different for targets of differing geometry. The larger is the difference in the geometry (i.e., tank *versus* jeep), the better this approach works; conversely, the smaller is the difference (i.e., tank *versus* armored personnel carrier), the worse this approach works.

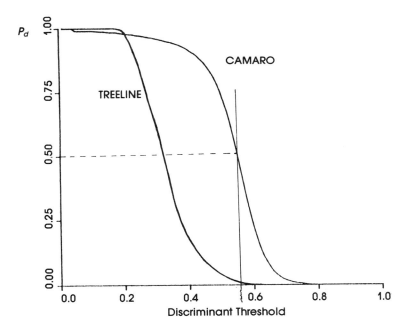

Fig. 6.49 Discrimination performance for a matrix discriminant for a Chevrolet Camaro and tree line clutter.

REFERENCES

1. M.I. Skolnik, *Introduction to Radar Systems*, 2nd Ed., McGraw-Hill, New York, 1980, p. 139.
2. W.A. Holm, "Continuous Wave Radar," in J.L. Eaves and E.K. Reedy, eds., *Principles of Modern Radar*, Van Nostrand Reinhold, New York, 1987, p. 399.
3. G.W. Stimson, *Introduction to Airborne Radar*, Hughes Aircraft Company, El Segundo, CA, 1983, p. 422.
4. M.I. Skolnik, *Introduction to Radar Systems*, 2nd Ed., McGraw-Hill, New York, 1980, p. 106.
5. C.R. Barrett, "MTI and Pulsed Doppler Radar," in J.L. Eaves and E.K. Reedy, eds., *Principles of Modern Radar*, Van Nostrand Reinhold, New York, 1987, p. 422.
6. M.I. Skolnik, *Introduction to Radar Systems*, 2nd Ed., McGraw-Hill, New York, 1980, p. 101.
7. D.K. Barton, *Radar Systems Analysis*, Artech House, Inc., Dedham, MA, 1976, p. 189.
8. G.W. Stimson, *Introduction to Airborne Radar*, Hughes Aircraft Company, El Segundo, CA, 1983, p. 424.
9. C.R. Barrett, "MTI and Pulsed Doppler Radar," in J.L. Eaves and E.K. Reedy, eds., *Principles of Modern Radar*, Van Nostrand Reinhold, New York, 1987, p. 422.
10. M.I. Skolnik, *Introduction to Radar Systems*, 2nd Ed., McGraw-Hill, New York, 1980, p. 101.
11. G.W. Stimson, *Introduction to Airborne Radar*, Hughes Aircraft Company, El Segundo, CA, Parts V, VI, and VII, 1983.
12. W.A. Holm and J.D. Echard, "FFT Signal Processing for Non-Coherent Radar Systems," *Proceedings of the 1983 International Conference on Accoustics, Speech, and Signal Processing*, Paris, 1982, p. 363.
13. P.D. Kennedy, "FFT Signal processing for Non-Coherent Airborne Radars," *Proceedings of the 1984 IEEE National Radar Conference*, Atlanta, GA, 1984, p. 79.
14. M.I. Skolnik, *Introduction to Radar Systems*, 2nd Ed., McGraw-Hill, New York, 1980, p. 504.
15. R.D. Hayes and J. L. Eaves, "Study of Polarization Techniques for Target Enhancement," AF 33(615)-2523, AD-316270, Report A-871, Georgia Institute of Technology, Atlanta, GA, August 1966.
16. J.D. Echard, *et al.*, "Discrimination Between Targets and Clutter by Radar," DAAG29-78-C-0044, Report A-2230, Georgia Institute of Technology, Atlanta, GA, December 1981.

17. N.F. Ezguerra, "Target Recognition Considerations," in J.L. Eaves and E.K. Reedy, eds., *Principles of Modern Radar*, Van Nostrand Reinhold., New York, 1987, p. 646.

18. M.N. Cohen, "Pulse Compression in Radar Systems," in J.L. Eaves and E.K. Reedy, eds., *Principles of Modern Radar*, Van Nostrand Reinhold, New York, 1987, p. 465.

19. M.I. Skolnik, *Introduction to Radar Systems*, 2nd Ed., McGraw-Hill, New York, 1980, p. 420.

20. T.H. Einstein, "Generation of High Resolution Radar Range Profiles and Range Profile Auto-Correlation Functions Using Stepped-Frequency Pulse Trains," Project Report TT-54, MIT Lincoln Laboratory, Lexington, MA, October 1984.

21. W.A. Holm, "Polarimetric Fundamentals and Techniques," J.L. Eaves and E.K. Reedy, eds., *Principles of Modern Radar*, Van Nostrand Reinhold, New York, 1987, p. 621.

22. J.R. Huynen, *Phenomenonlogical Theory of Radar Targets*, Ph.D. Dissertation, Drukkerij Bronder-Offset N.V., Rotterdam, 1970.

23. R.M. Barnes, *Detection of a Randomly Polarized Target*, Ph.D. Dissertation, Northeastern University, Boston, 1984.

24. J.R. Huynen, *Phenomenonlogical Theory of Radar Targets*, Ph.D. Dissertation, Drukkerij Bronder-Offset N.V., Rotterdam, 1970, pp. 82–86.

25. S. Weisbrod and L.A. Morgan, "RCS Matrix Studies of Sea Clutter," Teledyne-Micronetics Report No. R2-79, N00019-77-C-0494, AD B036684, January 1979.

26. G.A. Ioannidis, "Optimum Antenna Polarizations for Target Discrimination in Clutter," *IEEE Transactions on Antennas and Propagation*, Vol. AP-27, No. 3, May 1979, pp. 357–363.

27. A.J. Poelman, "The Applicability of Controllable Antenna Polarizations to Radar Systems," *Tijdschrift van het Nederlands Electronica en Radioigenootschap*, Vol. 42, No. 2, 1979, pp. 93–106.

28. W.A. Holm, "Polarization Scattering Matrix Approach to Stationary Target/Clutter Discrimination," *Colloque International sur le Radar*, May 1984, pp. 461–465.

Chapter 7
MMW Modeling Techniques

M.M. Horst and B. Perry, IV

Georgia Institute of Technology
Atlanta, Georgia

7.1 INTRODUCTION

Modeling can mean different things to different people, so this chapter will try to present a discussion of different aspects of modeling as applied to MMW radar systems. The first interpretation of "modeling" to be discussed is that of scale model measurements to characterize target backscatter. Millimeter-wave radars can be used to characterize lower frequency signatures, or lasers can be used to characterize MMW signatures. Scaling theory is presented, followed by a discussion of special problems with scale model measurements, such as appropriate scaling of electrical properties of materials. Modeling can also refer to the analytical modeling of MMW backscatter from targets and clutter. The standard high-frequency modeling techniques are described, along with the special problems and considerations of the millimeter wavelength regime. The final interpretation of modeling to be discussed is more properly called *simulation,* for it refers to computer codes which are used to characterize systems such as millimeter-wave seekers. Fundamental aspects of MMW radar software simulation are discussed. In addition, a typical hardware-in-the-loop (HWIL) simulation is described, and the limitations of the technique are discussed.

7.2 SCALE MODEL MEASUREMENTS

The first interpretation of MMW modeling refers to the measurement of radar backscatter from a physical scale model of a target. Scale model measurements are routinely made to determine various operational parameters for new vehicle designs (e.g., wind tunnel measurements on aircraft models). Principles of similitude can be developed for the reflection of electromagnetic fields, thus allowing scale model measurements for radar cross section.

7.2.1 Overview of Theory

The theory of electromagnetic modeling was developed during, and shortly after, World War II (Goudsmit and Weiss [1], Sinclair [2]), and was neatly summarized some 20 years later by Blacksmith, *et al.* [3]. A geometric scaling factor, P, is defined to describe the relationship between the full scale system and the model, as summarized in Table 7.1.

Table 7.1 Geometric Scaling Relationships.

Quantity	Full Scale System	Model System
Linear target dimension	d	$d' = d/P$
Time	t	$t' = t/P$
Frequency	f	$f' = fP$
Wavelength	λ	$\lambda' = \lambda/P$
Conductivity	σ_c	$\sigma_c' = \sigma_c P$
RCS	σ	$\sigma' = \sigma/P^2$

Note that the electromagnetic wavelength and the target dimensions are scaled by the same factor, so that the target remains the same number of wavelengths in size. The scale factor P can be any number, but it is usually an integer greater than one. The value of P is chosen for convenience, constrained primarily by the available radar parameters and target geometry parameters. The resulting scale model of the target must be large enough to be fabricated relatively easily and reliably, but small enough to fit the requirements of the measurement facility (pedestal requirements of size and weight, far field criterion, *et cetera*).

Paddison, *et al.* [4] and Burke, *et al.* [5] describe two scale model facilities that utilize MMW radars to provide scaled microwave RCS, and [6] describes a facility that uses even higher frequency sources to scale to

MMW RCS. Other similar facilities can be found worldwide, of course; these three were selected for a brief description because of their specific applicability to MMW modeling.

The US Army Mobility Equipment Research and Development Command at Fort Belvoir, Virginia, operates the MACROSCOPE facility described in [4] at a primary frequency of 98.6 GHz. Homodyne detection is used on this CW range, with the local oscillator signal provided by a sample of the transmitter signal. Signal processing results in imagery that can be superimposed on a television picture of the target, revealing the location and relative strength of the major scattering centers. The range is used primarily for diagnostic studies on ship and military vehicle targets, and has a resolution of about 0.5 inch in range and cross-range. In one example [4], a 35:1 scale model of a patrol hydrofoil was measured, yielding an equivalent full scale frequency of 2.8 GHz. An interesting experiment on the hydrofoil model involved measuring the whole model and the two halves separately, with the results summed noncoherently. Figure 7.1 shows that the noncoherent sum of the two halves provides a reasonable estimate of the RCS of the whole model.

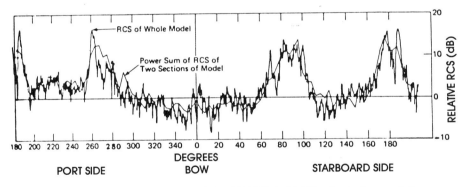

Fig. 7.1 RCS of 1/35 scale model of PGH-2 (RCS of whole model compared to two halves summed together) (from Paddison, *et al.* [4], © 1978 IEEE).

The Boeing Military Aircraft Company operates a MMW scale model measurement facility at three frequencies: 35 GHz, 60 GHz, and 95 GHz [5]. Figure 7.2 illustrates the layout of the facility. The control room houses equipment for system tuning, data recording and processing, and target control, while the radar instrumentation is located in the transceiver room at the far end of the facility. The distance between the target and the antenna is approximately 100 feet, restricting target scale models to about 12 inches. Boeing is investigating ways to increase the allowable target size, and to improve the usefulness and flexibility of the range.

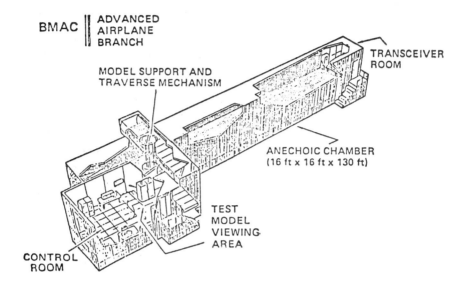

Fig. 7.2 Boeing's modeling facility (courtesy Boeing Aircraft Company).

The UK National Radio Modeling Facility [6], operated by EMI Electronics, Ltd, has much broader capabilities, with radar frequencies ranging from 30 GHz up to 2000 GHz (2 THz). Use of the higher frequencies makes it possible to scale to MMW RCS, rather than simply scaling from MMW to microwave. EMI is internationally known for the development of the instrumentation systems, including advanced RF sources and detection systems. One twist to the usual scale model measurement set-up is that the facility has the capability for either the target or the radar to move during the measurement process. The facility provides a choice of nine different radar systems and seven different model support systems for great flexibility in scale model measurements. EMI concentrates on collecting and interpreting radar scattering data, as indicated in [6].

7.2.2   Special Problems of Scale Model Measurements

Scale model measurements can be very useful, but their limitations must be recognized and understood. Most of the potential problems in scale modeling to determine radar signature fall into one of two categories: problems with the geometrical scaling of the physical dimensions of the target, or problems with the scaling of the electromagnetic properties of the materials comprising the target. These two areas are discussed below.

Scaling Physical Dimensions

The gross scaling of target dimensions by the same scale factor applied to the radar wavelength is intuitively satisfying and presents few problems to experienced model shops. However, diffuse scatter from imperfect surfaces can also affect the radar return, so the fine surface roughness of the target must be scaled appropriately to provide the model measurements with the correct balance of specular and diffuse scatter. Reasonable approximations to surface roughness can be made for most target surfaces, but the extra care and detail invested in characterizing surface imperfections inevitably results in models that are more expensive and take longer to construct. When real-world trade-offs must be made against budget and schedule constraints, the effect on scale model RCS is vital to understand.

Another more subtle phenomenon arising from purely geometric scale modeling was described by Wright [7], who compared scale model measurements with full scale RCS measurements on the FIREBEE target drone, designated MQM-34D by the Army and BQM-34A by the Air Force. The scale model measurements (Clay and Johnson [8]) were made by the Convair Aerospace Division of General Dynamics (GD/CAD) on a 0.55 scale model of the drone, and the full scale measurements [9] were made at the USAF RATSCAT facility. The comparison between the scale model and full scale measurements revealed significant differences, which Wright attributed to the physical differences between the model and the real drone.

The model drone was constructed of fiberglass and all conductive surfaces were flame-sprayed or painted with an aluminum coating. While the model was considered to be "reasonably accurate" [7], and included such small details as the riser cover and JATO thrust lug, it also possessed a smooth conductive skin that was not broken at such obvious discontinuities as hatch covers, ailerons, *et cetera*. Figure 7.3 illustrates a typical comparison between the scale model and full scale measurements on the MQM-34D. The two agree reasonably well at the large specular peaks, but in the off-specular regions, where there are no large scattering centers, the RCS of the scale model is sometimes significantly lower than that of the full scale target.

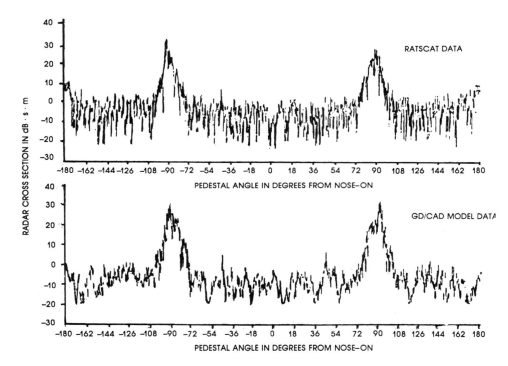

Fig. 7.3 Measured RCS of full scale (top) and a 0.55 scale (bottom) models of a BQM-34A; 0° pitch, 0° roll, vertical polarization (from Wright [7]).

The full scale target drone measured at RATSCAT was picked at random from the flight inventory at Holloman AFB. As an operational drone, it showed dents, previous repairs, and slight gaps due to imperfectly fitting hatches and joints, as opposed to the smooth, continuously conductive surface of the scale model. These differences in the target were felt to be the main source of discrepancy between the scale model and full scale measurements. Wright [7] concludes, somewhat plaintively, that

The problem of modeling techniques and accuracy from the overall electromagnetic viewpoint does not appear to have had [nor does it currently have] adequate importance or financial support. However, if model measurements are to be meaningful, the problem of model building must be solved.

In the decade since Wright made those statements, much has been accomplished in studies of the importance of gaps, joints, and surface imperfections in scale model RCS measurements. The general trend has been toward more detailed and accurate models for improved scale model RCS measurements.

Scaling Electromagnetic Properties of Materials

It is reasonably straightforward to construct a good geometric model of a target, even one that accounts for the extra expense and effort for faithful modeling of small details. The scaling of electromagnetic properties of the materials comprising the target poses another set of potential problems.

Table 7.1 lists the scaling law for conductivity as analogous to that for radar frequency:

$$\sigma_c' = \sigma_c\, P \tag{7.1}$$

For a good conductor, such as steel or aluminum, σ_c is essentially infinite over any range of reasonable scaling factors P, and common practice has been to ignore the material properties in scaling for RCS measurements. Schumacher [10, 11] has conducted a complete and rigorous derivation of the effects of scaling on the electrodynamic properties of materials. He derived three general nonlinear modeling equations in six variables. The solutions to the modeling equations indicate that complete scaling of electromagnetic properties can predict different values of scale model RCS, even for "good" conductors, than those provided by the simple scaling relations in Table 7.1.

For target materials that are not good conductors, the problem is more severe. As the technology of low observables results in more exotic materials for RCS reduction of targets, the scale model measurement problems will only increase. While it is theoretically possible to construct the scale model from materials which have the proper scaled conductivity, in practice, it may not be easy to find the proper materials while still maintaining the correct geometrical scaling, especially of surface roughness. As with the scaling of physical dimensions, trade-offs and compromises may be necessary as to scale model fidelity in electromagnetic properties. The only requirement is an understanding of the effect of compromises on model measurement results.

7.2.3 Scale Model Measurement Validation

Many of the comments about validation of analytical model predictions of RCS are appropriate for scale model measurements of RCS as well. The intended application of the RCS data drives the validation effort and sometimes leaves room for the expediency of compromise in the expectations for the scale model measurements.

As shown in Figure 7.3, a typical scale model measurement of the MQM-34D target drone matches full scale measurements well for aspects dominated by large specular returns, but does not match so well for nonspecular aspects. The full scale measurements show more rapid fluctuation of the return than that seen in the scale model measurements. However, the fine structure of the return is unimportant in a number of applications, and useful statistical averages can be obtained from a scale model measurement that may not match full scale data on the finer levels.

One interesting comment made by Wright [7] is that he expected the measured RCS of the scale model of the target drone to resemble the mathematical model of the drone more closely than either the scale model measurements or the mathematical model matched the full scale results. Of course, the scale model measurements and the mathematical model share many of the same shortcomings; their representation of the target is likely to be oversimplified to some degree and certainly more "perfect" than the capabilities of real equipment.

7.3 ANALYTICAL MODELING TECHNIQUES

Radars are intended to be detectors of reflected energy, and so the characterization of the way in which objects of interest (*targets*) tend to scatter electromagnetic waves has received attention from radar engineers since the early days of radar. Primarily concerned with monostatic radars, design engineers have sought a measure of the energy scattered by the target back to the radar (*backscatter*) that would be dependent only on characteristics of the target, and not those of the radar or the relative geometry between the radar and target. Radar designers settled on *radar cross section* (RCS), usually represented as σ with the formal definition:

$$\sigma = \lim_{R \to \infty} 4\pi R^2 \frac{|\overline{E}_s|^2}{|\overline{E}_0|^2} \qquad (7.2)$$

where R is the range between the radar and target, $\overline{E}_s$ is the electric field strength of the scattered wave, and $\overline{E}_0$ is the incident electric field strength. The limit is taken to remove the dependence on range, leaving σ for a specific frequency and polarization strictly a function of the target, but it has the added effect of implying plane-wave incidence. Thus, the problem of predicting RCS reduces to the calculation of the scattered electric field.

Exact solutions, based on vector wave equations derived from Maxwell's equations with the appropriate boundary conditions, are possible only for the few geometries which allow separation of variables in the wave equations. The results are expressed as polynomials or power series, such as Legendre polynomials and Bessel functions. In practice, numerical computation of results is complicated by problems with convergence and round-off errors, and is very time-consuming, even with modern computers, for realistic target configurations.

In source-free regions, Maxwell's equations can be expressed as a pair of surface integrals (Stratton [12]):

$$\overline{E} = \int_s \{i\omega\mu\psi\,(\hat{n} \times \overline{H}) + (\hat{n} \times \overline{E}) \times \nabla\psi + (\hat{n} \cdot \overline{E})\nabla\psi\}\,da \quad (7.3)$$

$$\overline{H} = \int_s \{-i\omega\epsilon\psi\,(\hat{n} \times \overline{E}) + (\hat{n} \times \overline{H}) \times \nabla\psi + (\hat{n} \cdot \overline{H})\nabla\psi\}\,da \quad (7.4)$$

where $\overline{E}$ and $\overline{H}$ are the total fields, $\psi = \exp(ikr)/4\pi r$ is the three-dimensional free-space Green's function, r is the distance from the surface patch of integration da to the point at which the field is to be computed, n is an outward surface normal erected on the surface patch da, and the surface of integration s is closed. The method of moments (Harrington [13]) poses a solution to the coupled integrals (7.3) or (7.4) by dividing the surfaces of integration into a collection of discrete surface patches where the form of the solution is specified as basis functions, with coefficients to be determined. This approach requires inversion of a matrix of interaction terms describing the induced current or charge on each surface patch due to the induced current or charge on every other surface patch. The inverted interaction matrix is then multiplied by a column matrix, representing the incident field at each surface element to yield the scattered field as the summation of the surface charge and current distributions in a radiation integral. Figure 7.4 illustrates the method-of-moments results for a simple shape.

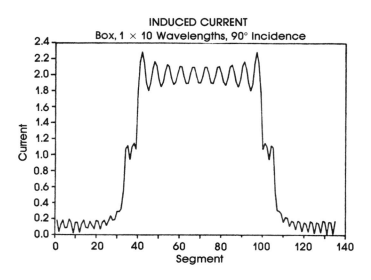

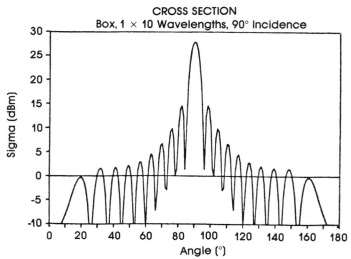

Fig. 7.4 Method-of-moments predictions for a rectangular 1 × 10 λ cylinder illuminated at θ = 90° (from Blacksmith, Hiatt, and Mack [3], © 1965 IEEE).

The method of moments is theoretically not a low-frequency technique, but it is practically unsuitable for high-frequency applications. The surface patches on the target body must have dimensions on the order of fractions of a wavelength (typically about $\lambda/5$), and the size of the matrix to be inverted is directly related to the number of surface patches employed. While computing power has greatly increased in recent years, there are still limits to the size of a matrix that may be inverted. Sparse matrix techniques have sought to extend the limits, but the method of moments remains suitable today for bodies no larger than 10 or so wavelengths in size.

Since the method-of-moments solution to the coupled integrals (7.3) and (7.4) is not very useful for practical targets in the millimeter wavelength regime, we must look for reasonable approximations to (7.3) and (7.4) that make the calculation of the scattered electromagnetic fields more tractable.

7.3.1 Overview of High-Frequency Modeling Techniques

High frequency refers not to any absolute frequency limits, but rather to the size of the target in wavelengths. As such, dimensions of the targets of interest to the MMW radar engineer are nearly always in the high-frequency region of greater than 10 wavelengths, which is also called the *optics region*.

High-frequency scattering from a target tends to be localized, with the return from each area of the target essentially independent of the return from other areas. The usual practice, then, is to represent a complex target as a collection of simple scattering shapes, to calculate the return from each individual element using techniques appropriate for that shape, and finally to sum all the contributions. If high-frequency techniques are used, this means that each element must be at least a few wavelengths in size, which is usually the case for millimeter waves.

The types of scattering shapes used in this approach are restricted to those for which exact solutions to Maxwell's equations exist, or reasonable approximations can be calculated. This means shapes such as trihedrals, dihedrals, flat plates, cylinders, cones, ellipsoids, and straight or curved edges. Fortunately, the higher the frequency, the less important the interactions among various portions of the target become. Scattering is more localized, and the assumptions of the optics techniques are reasonable.

Physical Optics

The first assumption of physical optics is that the total field on the surface is simply equal to the incident field on an infinite plane tangent to each point on the illuminated part of the surface and that there is no contribution from the shadow region. This simplifies the integral in (7.4) from a total surface integral to an integral only over the illuminated surface. Obviously, the assumption is exact when the surface is an infinite plane, so it is sometimes called the *tangent plane* approximation. If the local radii of curvature of the surface are much larger than the wavelength λ, the surface can be reasonably approximated by a plane tangent to the actual surface at the specular reflection point.

The major scattering contributions in physical optics arise from the surface regions near specular points, as shown in Figure 7.5. Physical optics theory does not include any contributions from edges, nor include any interaction between different parts of a scatterer. Physical optics effectively assumes that the dimensions of the scatterer are large compared to the wavelength λ, and that the scatterer is very far from the radar.

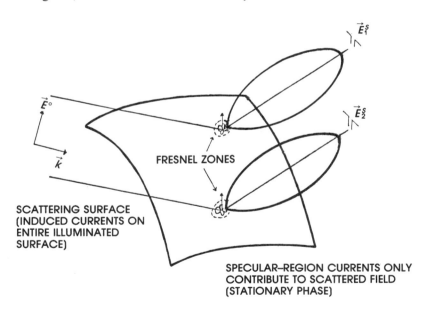

Fig. 7.5 Physical optics specular region.

With the tangent plane assumption, (7.4) reduces to

$$\overline{H}_s = \int_s (\hat{n} \times \overline{H}) \times \nabla\psi \, ds \qquad (7.5)$$

For a perfect conductor, the tangential component of the induced field, $\hat{n} \times \overline{H}$, is just $2\hat{n} \times \overline{H}_i$, where $\overline{H}_i$ is the incident magnetic field-strength. $\overline{H}_i$ can also be expressed as the incident magnetic field intensity H_0 with magnetic polarization along the unit vector direction $\hat{h}_i$.

The gradient of the Green's function, $\nabla\psi$, can be expressed rather simply as $\nabla\psi = ik\hat{s}\psi$ in the far field, where $\hat{s}$ is a unit vector in the direction of the scattered wave. If the distance to the radar is large, $\hat{s}$ will not vary over the surface of integration and the r in the Green's function can be replaced by a constant distance R_0 from the radar to some local origin, plus a surface position vector $\bar{r}$, which varies over the surface of integration.

By making these substitutions, the physical optics integral for the scattered field is

$$\overline{H}_s = -ikH_0\phi_0 \int_s (\hat{n} \times \hat{h}_i) \times \hat{s} \, e^{ik\bar{r}\cdot(\hat{i}-\hat{s})} \, da \qquad (7.6)$$

where $\phi_0 = \exp(ikR_0)/4\pi R_0$ is the far field Green's function and the surface of integration s extends only over the illuminated portion of the scatterer.

Equation (7.6) can be evaluated exactly for flat plates and cylinders, yielding results in the form of $\sin(x)/x$, and can be reasonably approximated for a few other scatterer geometries by using the method of stationary phase (Knott, Schaeffer, and Tuley [14]). Physical optics predictions for flat plates are reasonably close to measured backscatter results for incidence directions within a few sidelobes of the specular direction, but beyond that point, edge scatter becomes increasingly important to the actual backscatter and the physical optics prescription fails, as shown in Figure 7.6.

The advantage of the physical optics formalism is that it is derived from *exact* low-frequency techniques that are modified for higher frequencies, retaining the physical concept of induced surface currents reradiating the scattered electromagnetic fields. The disadvantage of physical optics theory is its inability to handle scattering from any structure other than a surface (i.e., physical optics neglect phenomena such as edge scatter or surface traveling wave scatter).

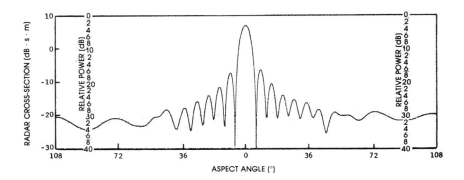

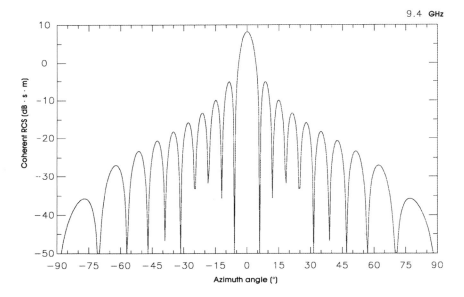

Fig. 7.6 Measured and predicted RCS of a 15 cm flat plate at 9.4 GHz; vertical polarization.

Geometric Optics

Unlike physical optics, geometric optics does not begin with exact low-frequency concepts, but rather utilizes traditional optics theory, which is simply the high-frequency limit of Maxwell's equations, to describe the scattered fields. Thus, the problem reduces to the determination of the paths followed by the electromagnetic energy (*ray tracing*). These paths may be direct, reflected, or diffracted, as shown in Figure 7.7. Geometric optics handles the direct ray paths, including spreading of rays, and the reflected rays. Diffraction is handled by the geometric theory of diffraction, to be discussed in the next section.

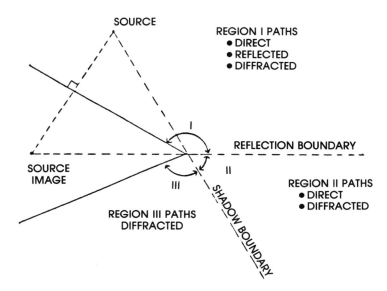

Fig. 7.7 Energy paths for a receiver in each region.

Fermat's principle determines the ray trajectory by requiring the optical path length to be an extremum, usually a minimum. This requirement does not force paths to be straight lines, in general, but, for the usual conditions of radar propagation (air as the medium of propagation, with constant index of refraction), the extremum paths are straight lines.

Snell's law (angle of reflection equals angle of incidence) applies at the specular or reflection point for any ray, and the amplitude and phase of the reflected ray may be obtained from the classic Fresnel reflection coefficients for electric fields parallel (∥) or perpendicular (⊥) to the direction of propagation:

$$R_\parallel = \frac{\mu \cos\Theta - \sqrt{\mu\epsilon - \sin^2\Theta}}{\mu \cos\Theta + \sqrt{\mu\epsilon - \sin^2\Theta}} \tag{7.7}$$

$$R_\perp = \frac{\epsilon \cos\Theta - \sqrt{\mu\epsilon - \sin^2\Theta}}{\epsilon \cos\Theta + \sqrt{\mu\epsilon - \sin^2\Theta}} \tag{7.8}$$

where Θ is the local angle of incidence. The magnitude of these coefficients is essentially unity when the surface is metallic, as is the case for many target applications.

The spreading of rays after reflection from a curved surface is handled by applying principles of conservation of energy along a ray tube as shown in Figure 7.8. If the energy density (or field) is $|\overline{E}(0)|^2$ at one end of the ray bundle with cross sectional area $A(0)$, and is $|\overline{E}(s)|^2$ at a distance s along the ray bundle, where the cross sectional area is $A(s)$, then conservation of energy requires

$$|\overline{E}(0)|^2 \, A(0) = |\overline{E}(s)|^2 \, A(s) \tag{7.9}$$

or

$$\frac{|\overline{E}(s)|^2}{|\overline{E}(0)|^2} = \frac{A(0)}{A(s)} = \frac{\rho_1 \rho_2}{(s + \rho_1)(s + \rho_2)} \tag{7.10}$$

where ρ_1 and ρ_2 are the principal radii of curvature of the wavefront at s.

RAY BUNDLE

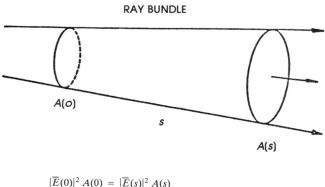

$$|\overline{E}(0)|^2 \, A(0) = |\overline{E}(s)|^2 \, A(s)$$

$$\frac{|\overline{E}(s)|^2}{|\overline{E}(0)|^2} = \frac{A(0)}{A(s)}$$

Fig. 7.8 Ray bundle energy conservation.

Inserting the results of (7.10) into the basic definition of RCS in (7.2) and letting $s = R$, we have

$$\sigma = 4\pi \, \rho_1 \, \rho_2 \tag{7.11}$$

The wavefront curvatures ρ_1 and ρ_2 can be related to the body curvature of the target at the specular point on the body (Kouyoumjian and Pathak [15]). For a sphere of radius a, for example, ρ_1 and ρ_2 are both equal to $a/2$, and (7.10) reduces to the geometric cross section:

$$\sigma = \pi a^2 \tag{7.12}$$

Thus, geometric optics provides a beautifully simple definition of RCS, one that is, in fact, independent of the radar frequency! In practical terms, the most difficult part of computing RCS by using geometric optics may be finding the specular point on an arbitrary doubly curved surface. Also if one or both of the radii of curvature are infinite, as for a cylinder or flat plate, geometric optics predicts infinite RCS. Physical optics can handle backscatter from flat plates and cylinders so long as the scattering is near specular, but it fails for large scattering angles because it ignores scattering from the edges of the plate or cylinder. Keller [16] and Ufimtsev [17] pioneered the efforts to characterize edge scatter.

Geometric Theory of Diffraction and Uniform Theory of Diffraction

The classic *geometric theory of diffraction* (GTD) was developed as a means of describing electromagnetic scattering for directions away from specular. The backscatter can be considered to have two components: one from the surface, as described by classical geometric optics, and one from the edge (or edges) of the scatterer. As in ray theory, where the scattering direction of the surface contribution is well defined by the rule that the angle of incidence and reflection must be equal, the edge diffracts the incident ray into a cone of angles for which incident and reflected angles are equal, known as the Keller cone. Its half-angle is defined by the angle between the edge and the incident ray, as illustrated in Figure 7.9.

The amplitudes of the diffracted rays on the Keller cone are derived from the exact solution to the diffraction from an infinite wedge or half plane. Sommerfeld [18, 19] solved the two-dimensional problem by defining coefficients for elementary plane waves diffracted by the wedge. Keller

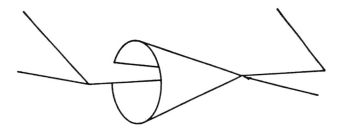

Fig. 7.9 The Keller cone of diffracted rays.

extended the results to the three-dimensional problem by modifying Sommerfeld's propagation constants to include the transverse components. The resulting diffraction coefficients, which are dependent on the polarization of the incident ray, form the basis of GTD.

Keller's GTD also includes prescriptions for the decay in intensity away from the diffracting edge and for the phase of the ray along its propagation path. The far-field expression for the diffracted field is

$$\overline{E}_d = \frac{-\Gamma\, e^{iks}}{\sin^3\beta} \,[(\hat{t} \cdot \overline{E}_i)\,(X - Y)\,\hat{s} \times (\hat{s} \times \hat{t})$$
$$+ \, Z_0(\hat{t} \cdot \overline{H}_i)\,(X + Y)\,\hat{s} + \hat{t}] \tag{7.13}$$

where Γ is a divergence factor, β is the angle between the edge and the incident ray, $\hat{s}$ is a unit vector along the direction from the edge element to the far field observation point, $\hat{t}$ is a unit vector aligned along the edge, and X and Y are the diffraction coefficients (Senior and Uslenghi [20]):

$$X = \frac{(1/n)\,\sin(\pi/n)}{\cos(\pi/n) \, - \, \cos[(\psi_s \, - \, \psi_i)/n]} \tag{7.14}$$

$$Y = \frac{(1/n)\,\sin(\pi/n)}{\cos(\pi/n) \, - \, \cos[(\psi_s \, + \, \psi_i)/n]} \tag{7.15}$$

where n is the exterior wedge angle normalized with respect to π, and ψ_i and ψ_s are defined in Figure 7.10.

Inspection of Figure 7.10 and relations (7.14) and (7.15) reveals the major drawback of GTD: the diffraction coefficients become singular as the scattering direction approaches either the shadow boundary or the reflection boundary. These singularities can be handled by modifying Keller's original GTD diffraction coefficients. Two so-called "uniform"

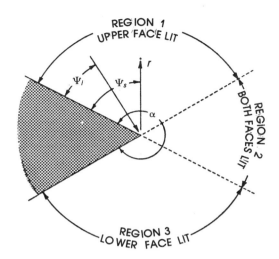

Fig. 7.10 Geometry for wedge diffraction.

theories have been developed to solve this problem: the *uniform theory of diffraction* (UTD) of Kouyoumjian and Pathak [15], and the *uniform asymptotic theory* (UAT) of Lee and Deschamps [21, 22].

UTD overcomes the singularities in the GTD diffraction coefficients by multiplying them by Fresnel integrals. At the shadow boundary or the reflection boundary, when one of the diffraction coefficients is infinite, the Fresnel integral is zero so the product is finite. The modified diffraction coefficients of Kouyoumjian and Pathak are

$$
X = \frac{1}{2} \left\{ \cot \left[\frac{\pi + (\psi_s - \psi_i)}{2n} \right] F[kLa^+ (\psi_s - \psi_i)] \right.
$$

$$
\left. + \cot \left[\frac{\pi - (\psi_s - \psi_i)}{2n} \right] F[kLa^- (\psi_s - \psi_i)] \right\} \tag{7.16}
$$

$$
Y = \frac{1}{2} \left\{ \cot \left[\frac{\pi + (\psi_s + \psi_i)}{2n} \right] F[kLa^+ (\psi_s + \psi_i)] \right.
$$

$$
\left. + \cot \left[\frac{\pi - (\psi_s + \psi_i)}{2n} \right] F[kLa^- (\psi_s + \psi_i)] \right\} \tag{7.17}
$$

where L depends on the nature of the incident wave [15], and F is the Fresnel integral:

$$F(Q) = -i2\sqrt{Q}\, e^{-iQ} \int_{\sqrt{Q}}^{\infty} e^{iz}\, dz \tag{7.18}$$

When the argument of the Fresnel integral is large, the diffraction coefficients in (7.16) and (7.17) reduce to those of (7.14) and (7.15). Figure 7.11 illustrates the effect of UTD in removing the singularities of GTD for an incident angle of 60° on the edge of a half-plane.

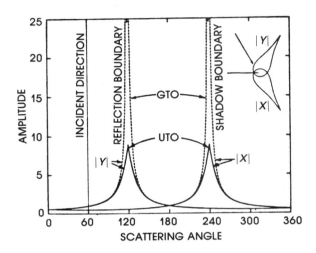

Fig. 7.11 Comparison of the diffraction coefficients of GTD and UTD for an incident angle of 60° on the edge of a half plane (from Knott, Shaeffer, and Tuley [14], © 1985 Artech House).

Another drawback of GTD, and one that is not remedied by the new diffraction coefficients of UTD, occurs for caustics such as ring discontinuities when viewed nearly on-axis. In this case, an infinity of Keller cones exist to contribute to the edge scatter, and GTD predicts an infinite result. This geometry can be quite common (e.g., re-entry bodies if viewed at nearly nose-on aspect), so it is fortunate that other theories are available to provide answers.

Physical Theory of Diffraction

Ufimtsev [17] developed the *physical theory of diffraction* (PTD) as a way to improve the accuracy of the physical optics prescription for scattered fields. He postulated the total surface current to be the sum of the physical optics (surface) current and a fringe (diffracted) current. The unique twist to PTD is that Ufimtsev used the exact solution of the half-plane problem to determine the edge contribution by subtracting the physical optics contribution from the exact (total) solution to the semi-infinite problem. Then, to obtain the complete solution for finite structures, he added the finite physical optics contribution to the edge contribution for the final answer. While the approach seems somewhat circuitous, it results in diffraction coefficients that are generally well behaved and solutions that avoid most of the difficulties of GTD. The major limitation of PTD is its restriction to scattering directions on the Keller cone, a restriction that Mitzner [23] sought to overcome with his incremental length diffraction coefficients.

Method of Equivalent Currents

Ryan and Peters [24] first applied the notion of fictitious equivalent currents to approach the problem of the axial caustics of GTD. Filamentary electric and magnetic currents are postulated to exist around the edge contour, and the current distribution is summed in a far-field radiation integral to yield a finite result for the diffracted field. The equivalent currents are fictitious because they depend on the scattering direction, but they work when GTD does not, allowing computation of scattering in directions not on the Keller cone.

The far-field radiation integral is

$$\overline{E}_d = -ik \, \psi_0 \int [I_e \, \hat{s} \times (\hat{s} \times \hat{t}) + I_m \, (\hat{s} \times \hat{t})] \, e^{ik\hat{r}\cdot\hat{s}} \, dt \qquad (7.19)$$

where ψ_0 is the far field Green's function (as in (7.6)), $\hat{t}$ is a unit vector aligned along the contour, and the equivalent currents I_e and I_m are given by

$$I_e = i2 \, (X - Y) \, (\hat{t} \cdot \overline{E}_i)/kZ_0 \qquad (7.20)$$

$$I_m = i2 \, (X + Y) \, (\hat{t} \cdot \overline{H}_i)/kY_0 \qquad (7.21)$$

Knott and Senior [25] carried the method of equivalent currents one step further by applying a stationary phase evaluation of the radiation integral in (7.19) and comparing the result with the exact GTD solution for directions on the Keller cone. The difference was a factor $\sin^2\beta$, where β is the angle between the incident ray and the local edge. For scattering

directions not on the Keller cone, they also postulated that the $\sin^2\beta$ factor
was in fact $\sin\beta_i$ $\sin\beta_s$, where the subscripts i and s refer to the angle
between the edge and the incident and scattered ray directions.

With these modifications, the radiation integral of (7.19) is now the
contour integral:

$$\overline{E}_d = -2E_0\,\psi_0\int_c \frac{e^{ik\hat{r}\cdot(\hat{i}\,-\,\hat{s})}}{\sin\beta_i\,\sin\beta_s} \times [(\hat{t}\cdot\hat{e}_i)\,(X\,-\,Y)\,\hat{s}\times(\hat{s}\times\hat{t})$$

$$+\,(\hat{t}\cdot\hat{h}_i)\,(X\,+\,Y)\,\hat{s}\,\times\hat{t}]\,dt \qquad\qquad (7.22)$$

where $\hat{e}_i$ and $\hat{h}_i$ are unit vectors aligned along the incident electric and
magnetic fields.

The most recent equivalent currents approach is that of Michaeli
[26]. Rather than assuming filamentary equivalent currents, he considered
an endpoint asymptotic integration of the exact solution at scattering by
an infinite half-plane. Michaeli's equivalent currents are then based on
generalized diffraction coefficients, including a dependence of the equivalent electric current on the incident magnetic field as well as the incident
electric field. Michaeli's prescription is much more complicated than that
of Ryan and Peters, or Knott and Senior, and it was more rigorously
derived. Any of the equivalent current theories represents a significant
improvement over Keller's GTD, both in allowing calculation for axial
caustics and in allowing scattering directions not on the Keller cone. Still,
the complexity introduced by the method of equivalent currents makes it
less desirable for targets comprised of large numbers of facets, where
simple solutions are more practical in terms of computer run times.

Incremental Length Diffraction Coefficients (ILDC)

Mitzner's extension of PTD is similar to Michaeli's extension of GTD.
Mitzner assumes that the incremental form of his results can be integrated
over the illuminated portion of any edge to provide answers for the general
case. As expected, Mitzner's ILDC reduce to Ufimtsev's PTD coefficients
for scattering directions on the Keller cone. As with Michaeli's prescription, Mitzner's is complicated, but accurate and reasonably well-behaved.
The ILDC remain finite in the transition regions by judicious cancellation
of singularities in the physical optics coefficients and Michaeli's coefficients.
As with Michaeli's approach, however, surface contributions and edge
contributions must be calculated separately in a procedure that remains
time-consuming for very complicated targets.

7.3.2 Special Problems of MMW Target Modeling

While the previous sections have given an overview of general analytical modeling techniques, there are two other important considerations in modeling the backscatter from targets at millimeter wavelengths. The first is that realistic targets in realistic geometries are likely to be in the near field of a millimeter wave radar. While some theories can automatically handle this situation (*cf.* the divergence factor Γ in equation (7.13) for GTD), other theories are based on plane-wave incidence and special care must be taken for near-field scenarios. The other problem specific to millimeter wave systems is the effect of local surface imperfections on backscatter. The analytical theories for surface scatter assume that the local surface is *perfectly* smooth, which is usually interpreted to mean smooth within some fraction of a wavelength over a given area. Even fairly relaxed standards for smoothness are hard to meet on real targets for MMW radars, and so some provision must be made in an analytical prediction for the diffuse scatter attributed to surface roughness.

These two topics, near-field considerations and diffuse scatter from surface imperfections, are discussed next.

Near-Field Models

The definition of radar cross section in (7.2) contains the requirement to calculate σ in the limit as the range R approaches infinity. This is the mathematical equivalent of requiring plane wave incidence at the target and has the effect of removing any dependence of σ on R. In any practical situation, however, the distance between the radar and the target is finite, and some measure must be made of whether the electromagnetic wave at the target is "acceptably" planar.

Assuming the radar to be a point source, we can use the geometry of Figure 7.12 to calculate the distance h over target dimension d at a range r:

$$h = r\sqrt{1 - [1 - (d/2r)]^2} \tag{7.23}$$

Assuming $d \ll 2r$, we can approximate

$$h \approx d^2/8r \tag{7.24}$$

The phase deviation between the center and ends of the target is given by $kh = 2\pi h\lambda$. The most common requirement is that this phase deviation be less than $\pi/8$ rad, or 22.5°, leading to the well known far-field condition:

$$r \geq 2d^2/\lambda \tag{7.25}$$

Despite its popularity, (7.25) is an arbitrary criterion, since it is based on the arbitrary choice of $\pi/8$ rad for the maximum allowed phase deviation over the target. Still, it is widely acknowledged and used to define a minimum range for *far-field* target RCS measurements.

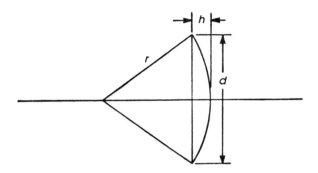

Fig. 7.12 Phase deviation over a transverse aperture due to spherical incident wavefronts (from Knott, Shaeffer, and Tuley [14], © 1985 Artech House).

The dimension d in (7.23) to (7.25) is generally accepted as the largest target dimension. A cursory application of the far-field criterion in (7.25) to a typical ground vehicle target with $d = 3$ m for a 95 GHz MMW radar yields a minimum far-field range of more than 6 km. Many MMW applications would require target RCS at ranges much closer than 6 km, so let us consider what happens to target RCS in the near field.

For very simple targets, such as a single flat plate, the far-field RCS pattern shows a main lobe specular return with sidelobes following a $\sin^2(x)/x^2$ pattern, as shown in Figure 7.13. As the range to the target decreases, the nulls are filled in and the sidelobe amplitudes are increased. At extremely short ranges, the main lobe amplitude is significantly decreased, the nulls disappear entirely, and the sidelobes are seen as shoulders in the pattern.

For complicated targets, however, the picture is somewhat different. A complex target such as a tank or ground vehicle is composed of literally hundreds (if not thousands) of high-frequency scattering centers, each of which scatters more or less independently of the others. As the dimensions of each scattering center approach the near field criterion, the effects for simple targets come into play (nulls filled in, sidelobes reduced). Because

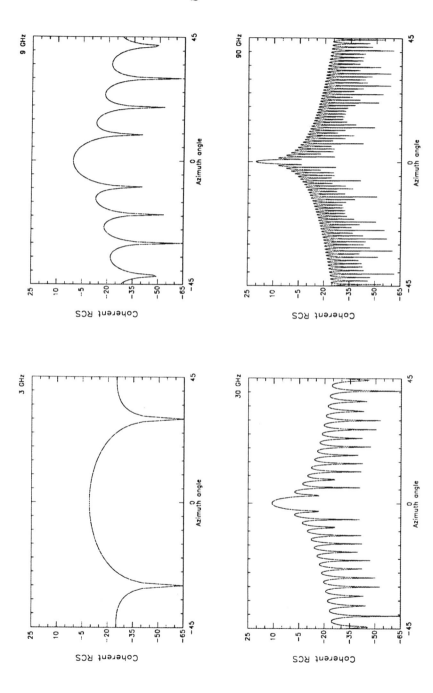

Fig. 7.13 RCS of a 10 cm square flat plate at several frequencies.

typical scatterer dimensions are much less than overall target dimensions, however, these near field effects occur at ranges much closer than the far-field minimum calculated from (7.25) using overall target size. Thus, the major effect on total target RCS may be a slight shifting around of the lobes and nulls of the composite pattern as compared with the true far-field pattern. Target statistics, such as mean or median RCS determined over some number of target aspects, are unlikely to be affected by moderate near range effects.

Analytical codes to estimate RCS (glint, range profiles, images, *et cetera*) by representing the target as a collection of simple scatterers can include provisions for calculating the correct (spherical wave) distance from the radar to each individual scatterer and can deal with any individual scatterer that is too large for the far-field criterion in a given geometry. One way to handle large scatterers in the near field is to compare the linear dimensions of the scatterer with the far field criterion. If a scatterer is too large, it can be automatically subdivided into two or more pieces. The process continues until all the pieces individually meet the far-field criterion, and computation continues. Alternatively, the user could be informed whenever a scatterer's linear dimensions do not meet the far-field criterion. The user would then decide whether the application required subdivision of offending scatterers or could tolerate near-field effects on some scatterers.

Diffuse Scatter

The analytical techniques described in the previous section shared one important assumption: that the simple target shape be perfectly smooth. Just as "far field" need not mean "infinite range," so, too, is there some latitude in "perfectly smooth" for real-world targets. Senior [27] conducted measurements on rough spheres at four radar frequencies, and concluded that good data could be obtained on targets with roughness on the order of a thousandth of a wavelength or less. By applying this criterion, we do not find very many smooth targets for MMW applications!

Surface texture can be described by a number of parameters: waviness, lay, flaws, and roughness, to name a few. The manufacturing processes used to produce the target can result in different kinds of surface roughnesses, as summarized in Table 7.2. Manufacturing specifications may also influence the final surface roughness seen in a particular target. For typical operational ground vehicle targets, for example, average surface roughnesses on the order of 1000 microinches or larger are not uncommon, and these values are well above Senior's thousandth of a wavelength at frequencies of 95 GHz and above.

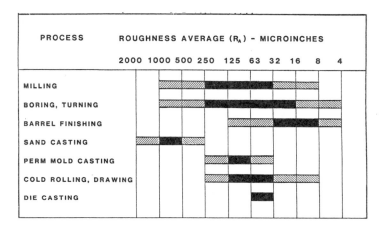

PROCESS	ROUGHNESS AVERAGE (R_A) - MICROINCHES									
	2000	1000	500	250	125	63	32	16	8	4
MILLING										
BORING, TURNING										
BARREL FINISHING										
SAND CASTING										
PERM MOLD CASTING										
COLD ROLLING, DRAWING										
DIE CASTING										

Table 7.2 Surface Roughness Produced by Common Production Methods (from ANSI B 46.1-1977).

Rough surface scattering can be considered in three separate regimes: roughness much smaller than a wavelength, roughness on the order of a wavelength, and roughness much greater than a wavelength. For MMW scatter, roughness much smaller than wavelength (by three orders of magnitude or more) can usually be ignored. Structures that exhibit roughness much greater than a wavelength can be handled by classical techniques like Bragg scatter. The in-between region, roughness on the order of a wavelength, is not amenable to classical techniques, occurs frequently for millimeter wave radars and real targets, and cannot legitimately be ignored.

Figure 7.13, given above, shows the changes in the physical optics RCS predicted for a 10 cm square smooth flat plate as the radar frequency increases. The main lobe becomes narrower and the specular return becomes larger as the frequency increases. At 95 GHz, the main lobe is less than a degree in width, which makes it difficult to orient the plate exactly enough to measure the specular peak. Thus, the peak specular return predicted by physical optics for a smooth plate will be much larger than measured on a typical "rough" plate. Surprisingly enough, the sidelobe return predicted by physical optics for the smooth plate will probably be lower than measured. The difference is the component of the return that is scattered diffusely by a real plate.

One approach to the problem of modeling the diffuse component of backscatter for MMW radar systems which has been used at Georgia Tech (Tuley [28]) was developed by Beckmann [29] and extended by Barton [30]. The coherent scatter is calculated for smooth metallic shapes using physical optics, geometric optics, or method of equivalent currents techniques as described above. That calculated RCS is then modified by a Fresnel reflection coefficient and roughness factor [29] to yield the specular part of the total backscatter:

$$\sigma = \Gamma^2 \, \rho_s^2 \, \sigma_c \qquad\qquad (7.26)$$

where

$$\rho_s^2 = \exp\left[-\tfrac{1}{2} \left(\frac{4\pi\sigma_h \, \cos\Theta}{\lambda} \right)^2 \right] \qquad\qquad (7.27)$$

The diffuse component of the backscatter is assumed to be distributed in a Gaussian beam about the specular direction. The power available for the diffuse component is that which is not accounted for in the specular component above, and the diffuse scatter is formulated as a diffuse reflection coefficient ρ_d^2 times the specular RCS [30], where

$$\rho_d^2 = \sqrt{(1 - \rho_{s1}^2)\,(1 - \rho_{s2}^2)} \, f\,(\lambda, \, \Theta, \, \beta_0, \, \text{size})$$

$$\exp\left[-\frac{\tan^2 \beta}{\tan^2 \beta_0} \right] \qquad\qquad (7.28)$$

where ρ_{s1} and ρ_{s2} are the specular reflection coefficients for the incident and scattered angles.

The form of the function $f(\lambda, \, \Theta, \, \beta_0, \, \text{size})$ depends on the type of scatterer being considered. The phase of the diffuse component of the RCS is assumed random with a uniform distribution.

Figure 7.14 shows the result of this approach to diffuse scatter calculations. The methodology has the desired result of reducing the specular peak and raising the sidelobe levels as the surface roughness increases, until the scatter is entirely diffuse at normal incidence for an rms surface roughness of $\lambda/2$ in Figure 7.14(d).

7.3.3 Comparison of Models with Measured Data (Validation)

The overview of high-frequency modeling techniques in the previous section has provided some general guidelines for appropriate methods to model various types of targets. Smooth, doubly curved surfaces are easily and accurately modeled with the simple theories of geometric optics, while flat plates and singly curved surfaces are better handled with physical optics. However, as the scattering direction moves further away from the

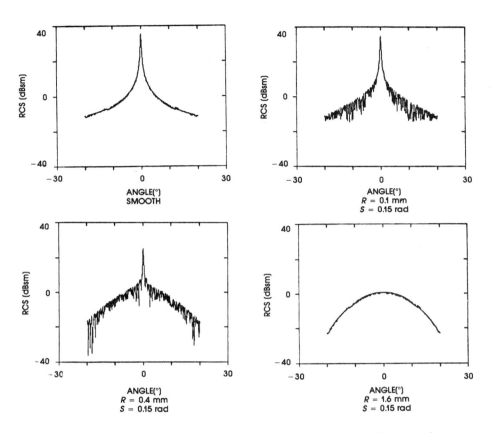

Fig. 7.14 Predicted flat plate patterns at 94 GHz, for rms surface roughness
R and rms surface slope S.

specular, edge effects become more important and theories such as Keller's
geometric theory of diffraction and Ufimtsev's physical theory of diffraction
must be seriously considered. Further refinements in the theory are available through the uniform theory of diffraction and Michaeli's and Mitzner's
extensions to GTD and PTD, but a price must be paid in complexity of
the theory for improvement in the prediction.

None of the foregoing high-frequency prediction techniques addresses the problem of the surface traveling wave. All of the various
diffraction techniques consider the edge in isolation, while the surface
traveling wave involves the interaction between the entire surface and a
discontinuity. There is as yet no routine way to predict this component of
target backscatter, although repeated application of diffraction theories

can yield reasonable approximate results, at least to the extent that the major contributors to the surface traveling wave are the edges. However, for most targets of practical interest at MMW frequencies, traveling waves should not normally be a major contributor to the RCS.

Model validation for the different analytical techniques can be considered on several levels. First, and most basically, the user is interested in how well a given technique performs on simple shapes in isolation. Figure 7.15 illustrates the performance of physical optics theory and geometric theory of diffraction (with multiple diffraction) in predicting the radar cross section of a flat plate 5λ by 5λ in size (Ross [31]). The predictions are compared with measured data for vertical and horizontal polarization. Within about 30° of normal incidence, both theories perform reasonably well. The GTD predictions are better than the physical optics for incidence angles greater than 40° off specular, but both theories fail near edge-on incidence.

While correct RCS prediction of simple shapes is of interest in developing analytical theories, the user is usually interested in a practical application to more complex bodies. In the high frequency regime, where interaction between different scattering centers on a complex body is minimal, the individual scattering patterns are less important than the composite total scattering from the body. Looking again at Figure 7.15, the theoretical prediction for flat plate RCS is acceptable for levels down to some 30 dB below the specular return. If that flat plate were one of dozens, hundreds, or even thousands in a representation of a complex target, its RCS prediction would be correct for near specular incidence, where it might be expected to provide a major contribution to total target RCS. For incidence angles farther from specular, it would not be expected to be a major contributor to total target cross section. Even if the theoretical contribution from this particular plate is calculated 10 or 15 dB below its "correct" (measured) value, the level of that contribution is probably still many decibels below the total target RCS and should not affect the total.

Figure 7.16 shows the measured RCS of an Oldsmobile Cutlass automobile at 69 GHz (Martin [32]). The RCS of a complex target typically exhibits the kind of fluctuations seen in Figure 7.16, with pulse-to-pulse variations in the data of greater than 20 dB at times. No analytical model can be expected to replicate these precise fluctuations, since they are caused by factors that are beyond the scope of analytical modeling. What then constitutes adequate model validation for complex targets?

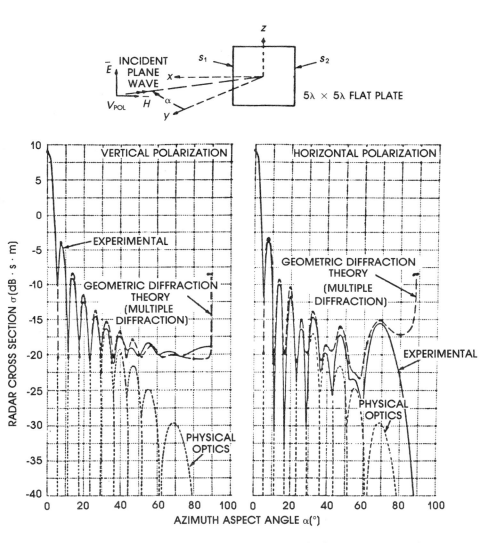

Fig. 7.15 RCS of a square flat plate (from Ross [31], © 1966 IEEE).

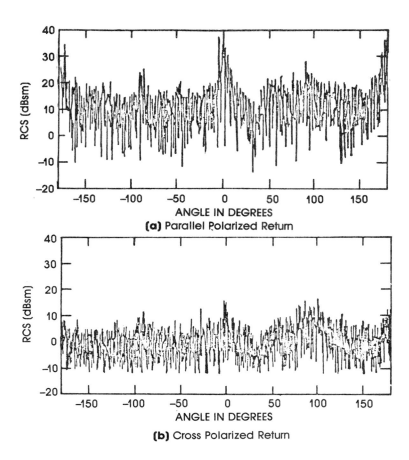

Fig. 7.16 RCS of an Oldsmobile Cutlass automobile at 69.8 GHz for horizontally polarized transmission (from Martin [32]).

The answer depends, of course, on the application for which the RCS predictions are being generated. If the application is a study of RCS reduction possible in various target configurations, then absolute RCS accuracy may not even be a requirement if the intent is simply to rank the configurations from highest RCS to lowest. Most commonly, however, validation consists of comparing measured versus predicted statistical quantities (usually mean and standard deviation) of the RCS over the same parameter intervals (e.g., aspect angle). Unless the theory being tested is very simple indeed, the measured data statistics will usually be formed from many more independent samples than the predicted data can generate in reasonable computer run times. Care must be taken to ensure that enough model data points are run for valid comparison. This can be accomplished by computing statistics for progressively smaller measured data sets to determine the optimum size for the predicted data set.

It is always advisable to use a measured data set recorded specifically for model validation purposes, since that data set will be designed to comfortably bracket the most important parameters of the intended application. In the real world, however, with fiscal as well as physical constraints, this may not always be possible. Evaluation of the measured data set is critical, especially when the measurements were not made specifically for a particular model validation (Currie, *et al.* [33]).

7.4 MMW CLUTTER MODELING

MMW clutter modeling can be approached in a variety of ways depending on the purpose of the simulation. Clutter models can generally be separated into two major categories: phenomenological and statistical. Phenomenological (or deterministic) clutter models are based upon basic scattering theory. The approach is to develop closed form solutions for the scattering from various surfaces based on one or more theoretical approaches. When validated, these deterministic models are inserted into a sophisticated signal-based radar simulation to determine the effect of the clutter backscatter on radar performance. Key aspects of phenomenological clutter modeling are discussed in this section.

An alternative approach to phenomenological clutter modeling often used at MMW frequencies is to represent the clutter statistically. In this case, the amplitude and frequency characteristics of the clutter are specified in terms of statistical distributions. While considerably simpler than phenomenological modeling, statistical modeling is still quite complex, since the clutter statistics can vary with clutter type, radar parameters, illumination geometry, propagation conditions, *et cetera*. Statistical clutter models can be analytical, empirical, or a combination of the two. Examples of all three types of statistical clutter models are presented in this section.

7.4.1 MMW Surface Clutter Modeling

While similar from a theoretical point of view, sea and land clutter are sufficiently different to warrant separate discussions. In the following, the basic theory of phenomenological clutter modeling, which applies to both terrain types, in discussed first. Sea clutter modeling is considered next, and an example of a combined analytical and empirical model is then presented. Finally, an entirely empirical model for MMW land clutter is discussed.

Phenomenological MMW Clutter Modeling

The application of electromagnetic scattering theory to clutter surfaces differs considerably from its application to metallic target objects. One basic difference is that edge effects are extremely important, and in some cases dominant, in target modeling. While edge effects are also of concern in clutter modeling, there are additional considerations that apply to clutter, which are absent in the analysis of targets. Targets can generally be assumed to be specular, while returns from clutter, which is dielectric rather than metallic, can range from entirely specular to entirely diffuse. In addition, clutter surfaces may depolarize reflected energy. Finally, at shallow grazing angles, portions of the clutter surface can mask or shadow other portions. Nonetheless, the starting point for clutter reflectivity modeling is a consideration of rough surface scattering. The following discussion is derived from "A Survey of Scattering from Rough Surfaces" by Michael T. Tuley [34].

Reflection and transmission of EM waves at a smooth dielectric boundary are well documented. Unfortunately, few clutter surfaces can be considered smooth, particularly at MMW frequencies. Lord Rayleigh developed a simple and widely used criterion for determining if a surface is smooth or rough. A surface may be considered rough if

$$h > \lambda/(4\sin\Theta) \tag{7.29}$$

where

h = rms surface roughness,
λ = EM wavelength,
Θ = radiation grazing angle.

At 94 GHz and a 10° grazing angle, for example, a surface may be considered rough if the rms roughness is greater than approximately 5 mm. Clearly, in the MMW regime most surfaces must be treated as rough.

The Rayleigh criterion provides a crude but useful measure of the effective roughness of a surface. If, for a problem of interest, approximating the solution as though reflection occurred at a smooth surface provides insufficient accuracy, then some formal mathematical method must be chosen to evaluate the scattering. Historically, three methods have generally been applied to the solution of rough surface scattering problems. These are geometrical optics, physical optics and perturbation theory. Since, in the limit, rough surface scattering degenerates to the smooth scatterer case, a perturbation method holds promise as an analytical technique for slightly rough surfaces. The first use of the method to analyze the scattering problem is credited to Rice [35], and Peake [36] used Rice's formulation to derive scattering cross sections from terrain. Wright [37] and Valenzuela [38] later used the perturbation method to predict the depolarization effects due to a slightly rough surface, and this method has been widely used to predict scatter from the sea at VHF and below (Barrick [39] and Bass, *et al.* [40]).

Unfortunately, perturbation theory is only applicable if the surface can be considered slightly rough. While this assumption may be valid for the sea surface at VHF, MMW frequencies present another matter entirely. If the surface can be categorized as very rough, electromagnetic scattering can be modeled using the techniques of physical or geometric optics.

The Kirchhoff or physical optics method approximates the field at any point on the surface by the field which would be present on an infinite plane of the same material tangent to that point (the tangent plane approximation). For that assumption about the surface fields to be valid, the rough surface must be restricted as follows:

(a) The radii of curvature at each point on the surface must be considerably larger than a wavelength.

(b) The surface slopes must be relatively small, so that shadowing and multiple scattering can be neglected.

(c) For a roughness correlation length T and an illuminated area A, $T^2 \ll A$, implying that the surface must be truly rough, rather than having only a few irregularities.

Physical optics is described in great detail in Section 7.3 of this chapter and by Knott, Schaeffer, and Tuley [14]. For situations that match the required criteria, physical optics will provide the comforting feeling of a solution based on extension of exact EM scattering theory. However, the formal solution of the physical optics integral is extremely complicated, and can only be obtained using approximations. Another method for the solution of rough surface scattering can be derived by extending the physical optics theory to the optical limit of $\lambda \to 0$. In this limit, all the returns occur from specular points, and traditional optics theory can be utilized. This method, called geometric optics, has also been discussed in previous sections.

Geometric optics, the limiting case of physical optics, greatly simplifies the solution to the scattering problem. For a given rough surface, it is only necessary to find the specular points for the desired directions of incidence and scattering, the appropriate tangent plane reflection coefficients, and the radii of curvature. The total return is then the sum of the returns, with appropriate consideration of the phase difference between specular points.

Geometric and physical optics are related through their use of the local tangent plane. However, the important differences between the two is that geometric optics approximates the scattered field based upon the tangent plane (i.e., all energy is scattered specularly), while physical optics approximates only the boundary conditions. If there are no specular points occurring for a given direction, geometric optics predicts no field scattered in that direction, while the fields on the surface approximated in physical optics generally radiate in all directions (diffuse scattering).

Thus, if it is believed that a majority of the scattering in a specified direction is from specular points on the surface, analysis is greatly simplified the geometric optics approximation.

The theories thus considered are restricted in their use to very specific types of surfaces. Physical optics can only be applied to what could be described as gently undulating surfaces, while perturbation theory will handle small but abrupt irregularities. Theories of sea clutter (Wright [41], Long [42]) and terrain return (Ruck, *et al.* [43]) have been postulated which assume that the surface is composed of large scale undulations upon which is impressed a small-scale roughness. Several authors have used a blend of physical optics and perturbation theory to treat scattering from such composite surfaces.

Assume that a rough surface may be described by $f(x, y) = Z(x, y) + s(x, y)$, where $Z(x, y)$ represents a large-scale roughness to which the tangent plane approximation of physical optics applies, and $s(x, y)$ is a small scale roughness to which perturbation theory applies. Further assume that Z and s are generated by independent random processes. A first-order approximation to the scattered power is given by a simple addition of the average incoherent scattered power from each process. Such a model has been proposed and analyzed (Semenov [44]), and provides reasonable results under restricted conditions.

All of the theories we have discussed are well developed, but they are not easily applied to actual modeling of surface clutter effects. It is possible to apply a composite model to a reasonably homogeneous surface, such as the sea, but at MMW even this would be difficult. For dry land, which contains many terrain types in an infinite variety of arrangements, theoretical surface clutter modeling is almost untenable. For this reason, the majority of surface clutter models are statistical. It is these types of models which will be considered next.

Sea Clutter Modeling

As mentioned in the previous section, the surface of the sea lends itself to theoretical modeling under some conditions, and at some wavelengths (Ewell, *et al.* [45]). At millimeter wavelengths, however, even very tiny structures become important, and it is no longer possible to concentrate on gravity waves and ignore surface ripples and similar fine structure effects. The following discussion will present an example of an empirical-analytical model developed by researchers at Georgia Tech (Horst, *et al.* [46]). This is by no means the only such model, and others can be found in the references.

The problem of developing a model for sea backscatter is a complicated one at best, since it is dependent upon radar frequency, sea state, incidence angle, wind-sea direction, polarization, and other factors. Researchers at Georgia Tech designed an analytical model to describe sea clutter backscatter using empirical constants derived from measured data.

For ease of reference, Tables 7.3 and 7.4 present a list of the variables in the Georgia Tech sea clutter model and a summary of the basic equations of the model for 10 to 100 GHz. The equations in the first section of Table 7.4 serve to define the incidence angle, taking into account both the curvature of the earth and the possibility of ducting or enhanced propagation conditions. The propagation of an electromagnetic wave in a surface duct of height h_d can be effectively modeled by clamping the incidence angle in this manner. The equations in the second and third sections will be explained in detail below, while the equations in the last section provide a means of expressing the total clutter cross section seen by the radar as the product of the clutter cross section per unit area times the radar clutter cell size.

The clutter model calculates $\sigma°$ as the product of three variables: multipath, sea direction, and wind speed (Horst, Dyer, and Tuley [47]). Each of these factors is, in turn, a function of the appropriate independent variables. The factor describing the multipath effect, or interference, between the direct and scattered fields from the surface for horizontal polarization is derived from forward scatter theory assuming a Gaussian distribution of surface height with standard deviation c_h. A surface reflection coefficient of -1 is assumed for all angles of incidence. A roughness parameter σ_ϕ, may be defined as

$$\pi_\phi = \frac{4\pi \sin\alpha \ \sigma_h}{(\lambda + 0.015)} + \frac{2\sqrt{2\pi} \ \alpha \ h_{av}}{(\lambda + 0.015)} \tag{7.30}$$

Table 7.3 Variables of the Sea Clutter Model.

Symbol	Definition	Restrictions
h_a	Radar Antenna Height	—
R	Range	—
A_e	Effective Earth Radius	—
λ	Radar Wavelength	0.003 m–0.3 m (0.1 ft to 1 ft)
h_d	Duct Height Constant	—
α	Incidence Angle	0.1°–10°
h_{av}	Average Wave Height	0 m–4 m (0 to 13 ft)
Φ	Angle between Boresight and Upwind	0°–180°
τ	Pulsewidth	50 ns–2 μs
Θ_a	3 dB Azimuth Beamwidth (One-Way)	—
c	Speed of Light	—
$\sigma°$	Average Clutter Cross Section per Unit Area	—
A_c	Area of Radar Resolution Cell	—
σ_c	Average Clutter Cross Section	—
A_i	Interference Factor	—
A_u	Upwind-Downwind Factor	—
A_w	Wind Speed Factor	—

where λ is radar wavelength, h_{av} is average wave height, and the angle of incidence α is small enough that $\sin\alpha \approx \alpha$.

The interference term A_1, is then given by

$$A_i = \sigma_\phi^4/(1 + \sigma_\phi^4) \tag{7.31}$$

At microwave frequencies, the behavior of $\sigma°$ *versus* incidence angle falls into two distinct regions: a low grazing angle region, where $\sigma°$ is a strong function of angle, and a plateau region, where $\sigma°$ is approximately independent of angle. These two regions are separated by a critical angle α_c, which is generally considered to be proportional to λ and $1/h_{av}$:

$$\alpha = \frac{\lambda}{Kh_{av}} \tag{7.32}$$

where K is a proportionality constant.

Table 7.4 Sea Clutter Model Equations—10 to 100 GHz.

Metric Units
h_a, h_d, λ, h_{av} (m) R, A_e (km) V_w(m/s)
$\alpha' = h_a/1000R - R/2A_e$ $\alpha = \left[\alpha'^2 + \left(\dfrac{\lambda}{4h_d} \right)^2 \right]^{1/2}$
$\sigma_\phi = 2\sqrt{2\pi}\ \alpha h_{av}/(\lambda + 0.015)$ $A_i = \sigma_\phi^4/(1 + \sigma_\phi^4)$ $A_u = \exp[0.25 \cos\Phi\ (1 - 2.8\alpha)\ \lambda^{-0.33}]$ $qw = 1.93\lambda^{-0.04}$ $V_w = 8.67h_{av}^{0.4}$ $A_w = [1.94V_w/(1 + V_w/15.4)]^{qw}$
$\sigma_{HH}^\circ = 10 \log(5.78 \times 10^{-6}\ \alpha^{0.547}\ A_iA_uA_w)$ $\sigma_{VV}^\circ = \sigma_{HH}^\circ - 1.38 \ln(h_{av}) + 3.43 \ln(\lambda) + 1.31 \ln(\alpha) + 18.55$
$A_c = 10 \log\left(\dfrac{1000R\Theta_a c\tau}{2\sqrt{2}} \right)$
$\sigma_c = \sigma^\circ + A_c$

No data could be found to define the critical angle at millimeter wavelengths, since α_c would probably be well below the angles practically possible in a measurements program. Therefore, only σ° data clearly from the plateau region are used in determining parametric dependencies for the model.

The sea direction term is based on upwind-downwind ratio data, although the reference for the aspect angle, ϕ, should be the sea wave propagation vector rather than the wind vector. The up-down ratio is dependent on incidence angle, α, increasing as α decreases, but approaching a finite value at $\alpha = 0$. Since there are insufficient data to determine clearly the functional form of the dependence, a simple circular function, $\cos\phi$, is chosen where ϕ is the angle between antenna boresight and the upwind direction. The upwind-downwind term, A_u, is then given by

$$A_u = \exp[0.156(1 - 2.8\alpha)\ (0.5 \cos\Phi)c_1\lambda^{c_2}] \tag{7.33}$$

To determine the wavelength coefficient and exponent, the relation between the upwind and downwind backscatter is modeled by the equation:

$$\frac{\sigma^\circ_{up}/\sigma^\circ_{down}}{1 - 2.8\alpha} = c_1\lambda^{c_2} \tag{7.34}$$

Due to a lack of upwind, downwind, and crosswind data for MMW frequencies, the following assumption is also made:

$$\sigma^\circ_{up}/\sigma^\circ_{down} = 2(\sigma^\circ_{up}/\sigma^\circ_{down}) \tag{7.35}$$

With the additional upwind-crosswind data, the wind direction term becomes

$$A_u = \exp[0.25 \cos\phi(1 - 2.8\alpha)\lambda^{-0.33}] \tag{7.36}$$

The dependence of σ° on sea state has been shown to be more strongly a function of wind speed than of wave height. The apparent saturation of σ° with increasing wind speed, V_w, is accounted for using a wind speed term, A_w, which is computed in two steps under the assumption that

$$\sigma^\circ \propto (V_w)^\beta = (V_w)^{c_1\lambda^{c_2}} \tag{7.37}$$

First, wind speed exponents β are found using σ°_{HH} grouped by incidence angle, wavelength, and source. This provides nine β values. Then, the least-squares fit to

$$\log_{10}\beta = \log_{10} c_1 + c_2 \log_{10}\lambda \tag{7.38}$$

is computed by using these β values and their associated wavelengths to produce the result:

$$\sigma^\circ = V_w^{1.93\lambda^{-0.04}} \tag{7.39}$$

The small exponent on λ indicates little wavelength dependence in the wind speed term.

By using the three model components thus computed, the backscatter σ° is found by making a fit to

$$10^{\sigma^\circ_{HH}/10}/\alpha^{0.4} A_i A_u A_w = c_1\lambda^{c_2} \tag{7.40}$$

and by using the least squares approach on

$$\log_{10}(10^{\sigma^\circ_{HH}/10}/\alpha^{0.4} A_i A_u A_w) = \log_{10}c_1 + c_2 \log_{10}\lambda \tag{7.41}$$

which results in an expression for σ_{HH}°:

$$\sigma_{HH}^\circ = 10 \log_{10}(3.5 \times 10^{-6}\lambda^{0.01}\alpha^{0.4}A_iA_uA_w) \tag{7.42}$$

The wavelength exponent is approximately zero, and it is removed for simplification purposes. Without λ in the equation, a new angle dependence can be determined by making a least-squares fit to

$$\log_{10}(10^{\sigma_{HH}^\circ/10}/A_iA_uA_w) = \log_{10}c_1 + c_2 \log_{10}\alpha \tag{7.43}$$

This produces the final result for σ_{HH}°:

$$\sigma_{HH}^\circ = 10 \log_{10}(5.78 \times 10^{-6}\alpha^{0.54}A_iA_uA_w) \tag{7.44}$$

The correction factor for vertical polarization is assumed to follow the form of a linear fit to sea state (average wave height), wavelength, and incidence angle. Only data for which σ_{HH}° and σ_{VV}° exist at approximately the same sea state, wavelength, and angle are used to obtain the coefficients:

$$\sigma_{VV}^\circ - \sigma_{HH}^\circ = -1.38 \ln(h_{av}) + 3.43 \ln(\lambda) \\ + 1.31 \ln(\alpha) + 18.55 \tag{7.45}$$

As an example of the predictions of the Georgia Tech sea clutter model, Figures 7.17 and 7.18 present σ_{HH}° and σ_{VV}° predictions as a function of frequency from 10 to 100 GHz for average wave heights of 0.15, 0.6, 1.5, and 3.5 m, an incidence angle of 1.2°, and crosswind direction. In an effort to show more points for comparison, data from five sources, including σ° data at incidence angles of 1.0° and 1.4°, and from an upwind direction, were plotted with the predictions. The data should be just above the knee of the curve into the plateau region, so the variation in incidence angle should have negligible effect on the comparison. The difference between upwind and crosswind could account for some of the differences between measured and predicted σ° values, but the wind direction contribution should be less than the spread of the data within one experiment.

Figures 7.17 and 7.18 illustrate that the σ° model equations from 10 to 100 GHz do provide a reasonable fit to the data considering the scarcity of data points and the spread of the data. The rms error between measured and predicted σ° values was 3.7 dB for horizontal polarization and 5.9 dB for vertical polarizations. Since the model equations were designed using all of the available data, these rms errors reflect the quality and consistency of the data base more than the accuracy of the model.

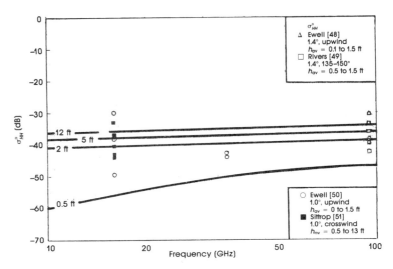

Fig. 7.17 Comparison of MMW radar sea clutter data with the Georgia Tech sea clutter model for 1.2° incidence angle, crosswind direction, horizontal polarization, and several average wave heights (from Horst, *et al.* [47], © 1978 IEEE).

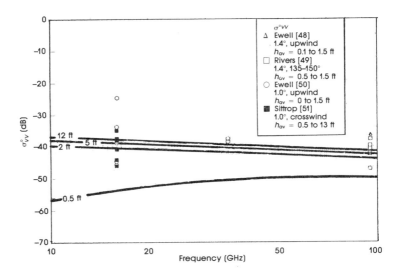

Fig. 7.18 Comparison of MMW radar sea clutter data with the Georgia Tech sea clutter model for 1.2° incidence angle, crosswind direction, vertical polarization, and several average wave heights (from Horst, *et al.* [47], © 1978 IEEE).

Land Clutter Modeling

Modeling land surfaces is even more complex than modeling the sea, and it is not possible in this limited space to examine all the implications of MMW land clutter modeling. The basics of theoretical modeling of surfaces was presented earlier, and a facet terrain simulation could be generated using these techniques. However, for most practical applications, a more limited model is desirable. In this section we will present an example of an entirely empirical model for land clutter backscatter which is reasonably accurate and can be applied to a variety of terrain types (Zehner and Tuley [52], Currie [53]).

Measured backscatter from land surfaces is affected by many factors. The type and shape of the terrain is a major influence, as is the incidence angle at which the ground is illuminated, but characteristics of the measuring radar system itself affect the nature of the data collected. Some aspects of radar design which can affect the clutter backscatter include the radar frequency, the antenna beamwidth, the pulse length, and the polarization.

Frequency effects on $\sigma°$ have been reported by many sources [54–57]. These are mostly associated with the relationship between terrain roughness and radar wavelength. In general, radar backscatter increases with increasing frequency. Polarization of the transmitted signal can also influence the measured return, but this dependence is not clear for general terrain at low angles.

The behavior of $\sigma°$ is strongly affected by characteristics of the illuminated surface area. Surface roughness, moisture content, and dielectric properties, as well as shape, orientation, and distribution of clutter targets all contribute to the terrain impact on $\sigma°$. Unlike radar parameters, terrain factors cannot be controlled, and surface conditions are difficult to define and measure.

Although there are many factors affecting ground clutter backscatter, the Georgia Tech model assumes that the major tendencies can be accounted for using a model based on terrain roughness and depression angle. The fundamental equation of the model is

$$\sigma° = A\,(\Theta + C)^B \exp\left[-D\Big/\left(1.0 + \frac{0.1\sigma_h}{\lambda}\right)\right] \qquad (7.46)$$

where

Θ = depression angle, in radians,
σ_h = rms surface roughness;
A, B, C, D = empirically determined constants;
λ = wavelength.

For very rough surfaces ($\sigma_h/\lambda \gg 1$), (7.46) becomes

$$\sigma^\circ = A(\Theta + C)^B \tag{7.47}$$

For very smooth surfaces ($\sigma_h \ll 1$), (7.46) becomes

$$\sigma^\circ = A(\Theta + C)^B \exp[-D] \tag{7.48}$$

When using this general form, we must account for different terrain types and different radar frequencies by selecting different constants: A, B, C, and D. At the MMW radar bands the terrains examined were sufficiently rough that the constant D, which modeled relatively smooth surfaces, was essentially zero. Consequently, the form of the backscatter model was reduced to (7.47).

Values for the three remaining constants were generated for a variety of terrain types using least square fitting of the data. A list of coefficients for these terrain types is presented in Table 7.5. These values were based on data collected by Georgia Tech, the Rome Air Development Center, the University of Kansas Center for Research, and Goodyear Aircraft Corporation [57–63].

Table 7.5 Georgia Tech MMW Land Clutter Model Parameters.

Clutter Type	Frequency (GHz)	A	B	C .
Wet Snow	35	0.195	1.7	0.002
	95	1.138	0.83	0.002
Dry Snow	35	2.45	1.7	0.002
	95	3.6	0.83	0.002
Trees	35	0.036	0.59	0.001
	95	0.36	1.7	0.012
Grass, Crops	35	0.301	1.5	0.012
Short Grass	35	0.125	1.5	0.012

Figures 7.19 and 7.20 depict a comparison of the land clutter model and tree clutter data collected by Georgia Tech. As we can see in the figures, there is a considerable variance in the reflectivity data used for the model. Some of this variance is due to intrinsic variability of clutter backscatter, but some is due to the fact that wet and dry tree clutter data were combined to produce one curve.

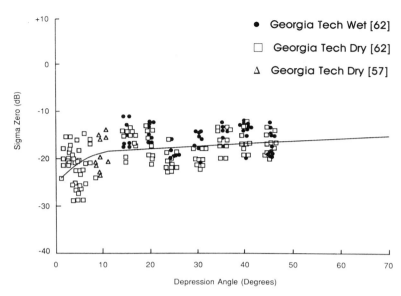

Fig. 7.19 Comparison of tree backscatter data with Georgia Tech land clutter model at 35 GHz (from Currie [53]).

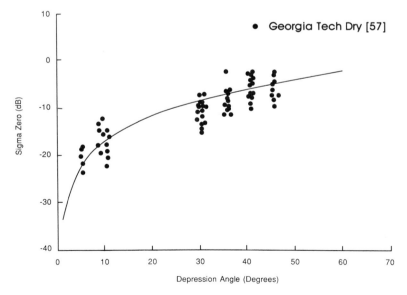

Fig. 7.20 Comparison of tree backscatter data with Georgia Tech land clutter model at 94 GHz (from Currie [53]).

7.4.2 MMW Volume Clutter Modeling

While volume clutter can refer to chaff or dust, the primary sources of volume clutter are hydrometeors, and at millimeter wavelengths the most important of these is rain. As with surface clutter, rain clutter can be modeled phenomenologically or statistically. Because rain drops can be modeled with reasonable accuracy as spheres, phenomenological rain models are somewhat better developed than surface clutter models.*

Reflectivity of an individual rain drop can be computed by modeling the drop as a sphere. Theoretical calculations on reflections from a sphere indicate that there are three regions: one, when the wavelength is much longer that the circumference of the sphere; two, when they are comparable; three, when the wavelength is much smaller. Vakser, Malyshenko, and Kopilovich [64] computed the reflectivity of three different drop sizes as a function of transmitting frequency (Figure 7.21). In the Rayleigh region, when the wavelength is long, the reflectivity of the drop is linear with frequency. When the wavelength and drop size are on the same order, the reflectivity does not continue to increase, but begins to oscillate around the optical limit: this is called the Mie region. For many rain drop sizes, MMW reflectivity values fall into this region.

While Mie scattering theory can be used to accurately predict the reflectivity of rain drops, this reflectivity is a function of size, shape and the index of refraction of the water. The shape of the rain drops does not have a great impact on the reflectivity, but the size is of critical importance, as is apparent in Figure 7.21. Unfortunately, most rain precipitation information is recorded in rain rates, and drop size distributions cannot be uniquely related to rain rate since they are a function of the type of rain storm, the wind speed and other factors. Consequently, theoretical modeling of rain reflectivity requires selection of a drop size distribution that is considered most reflective of the geography being simulated. A drop size distribution that is often used by meteorologists was derived by Laws and Parsons in 1943 [65].

Once a drop size distribution has been selected, computation of the rain reflectivity can be performed by summing the contributions of the various drop sizes weighted by their prevalence in the distribution. Such rain reflectivity modeling at MMW frequencies has been performed by many authors, such as Mitchell [66] (discussed in Chapter 5) and Wilcox [67]. Unlike some investigators, Wilcox computed rain backscatter for a variety of drop size distribution functions, and two reflection coefficients, thus providing bounds on the reflectivity predictions.

---

*A more extensive discussion of this subject can be found in Richard and Kammerer [77].

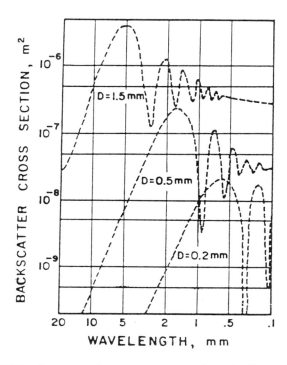

Fig. 7.21 RCS of water spheres *versus* wavelength (from Vakser, Maly-shenko, and Koplivich [64]).

Even more useful rain models have been reported in the open Soviet literature. In particular, Krasyuk, Rozenberg, and Chistyakov have computed rain reflectivity versus transmit frequency for different rain sources. They compared rain produced by melting granular snow, nongranular snow, and soft hail. The resulting reflectivity values are depicted in Figure 7.22 for a variety of rain rates. It is interesting to note that while the rain produced by the different types of snow have predominantly similar reflectivities, rain produced by melting hail demonstrates much higher reflectivities, particularly with increasing frequency. The presumed reason for this behavior is that melting hail generally contains larger drop sizes than melting snow due to the size and structure of the hail itself.

Statistical modeling of rain reflectivity usually depends primarily on empirically based models of reflectivity *versus* rain rate. Examples of such rain curves were presented in Figure 5.37. These curves can be used to

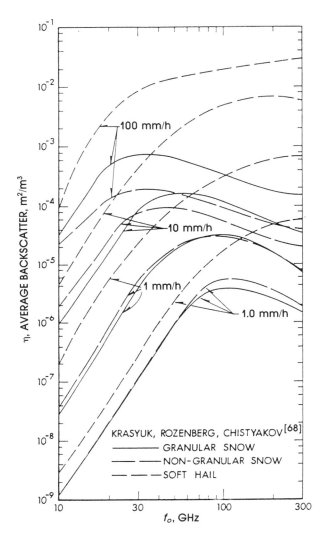

Fig. 7.22 Calculated rain backscatter coefficient by Krasyuk, Rozenberg, and Chistyakov from rain, snow, and hail (from Richard and Kammerer [77]).

model the mean rain reflectivity, but they say nothing about the amplitude distribution of the rain. This subject is discussed briefly in Chapter 5, where the point is made that while Rayleigh distributions have been assumed for rain PDF's in the past, MMW measurements suggest that lognormal distributions are a better fit to the data. In general, the distribution that is most appropriate depends on the nature of the rain, the parameters of the radar, and the geometry under which the measurements were taken.

Models of rain spectral characteristics are considered in some detail by Nathanson, who also gives references for additional information (Nathanson [69]). Basically, his calculations are based on assumptions concerning the phenomenological nature of wind and rain. As such, the models are based on rain velocities, and are thus independent of transmit frequency. While these techniques are probably a reasonable approximation for long range radars, they are limited in their applicability to the kind of short range scenarios in which most MMW radars operate. Under these conditions, phenomenological simulations may not be feasible, and the radar analyst may be forced to use empirical models. Examples of such models, based on actual data, were presented in Chapter 5, and will not be discussed here.

7.5 MMW RADAR SIMULATIONS

One very important application of MMW modeling is in the assessment of MMW radar or seeker performance. Radar system simulation can provide a powerful tool for the design of MMW radars. Simulation of radar performance can be used to compare tradeoffs between the many aspects of radar design without the need for any actual MMW hardware. In addition, once the MMW radar design has been completed, more sophisticated models can be used to test the performance of the MMW radar under a wide variety of conditions. This allows location of design weaknesses prior to construction of the actual hardware, and can save considerable time and expense. Finally, hardware-in-the-loop (HWIL) simulations allow realistic testing of the actual hardware, without the expense and complexity of field or flight tests. HWIL simulations are particularly useful in testing the overall performance of the hardware and software configurations of the system while the opportunity to modify these designs is still available.

The following section discusses aspects of MMW radar modeling. Simulations contained entirely in computer software will be examined first, followed by a discussion of HWIL simulations.

7.5.1 Software Simulations

7.5.1.1 Aspects of Radar Performance Modeling

Computer simulation of MMW radar systems can be performed at almost any level of complexity and sophistication, from 100 line implementations of the radar range equation to 10,000 line simulations of actual missile performance. The level of sophistication required depends upon the type of performance being simulated, and the level of accuracy required in the results. A simple model might be useful for determining the peak power required, but would be inadequate for testing various tracking algorithms. This section discusses alternate methods of radar modeling, the components of the radar model, and the various radar functions which might be the objects of performance analysis.

A typical MMW radar engagement scenario is depicted in Figure 7.23. In this case, the MMW radar is mounted in the nose of an air-to-ground missile, and the missile's target is an armored vehicle, such as a tank or an APC. A simulation of this engagement scenario must model all aspects of the situation. These simulation components can be separated into the four basic categories listed in Table 7.6. The simulation must model the components of the radar system itself, such as the antenna, detector, and signal processor. Target reflection characteristics must also be considered, whether as a simple RCS or as a complex faceted object. The environment must also be considered, as this can have a significant affect on MMW seeker performance. The environmental characteristics can be further segmented into surface effects and volume effects. The surface of the earth affects the radar performance in two ways: clutter interference and multipath interference. Clutter interference is caused by energy reflected back toward the radar from the surface, while multipath interference is caused by energy being reflected forward toward the target. Which effect is predominant depends on many aspects of the scenario, and modeling of both may be required.

Atmospheric phenomena also affect radar performance. Refraction, while significant at long ranges, does not greatly affect MMW radar performance because of the short ranges over which such systems usually operate. More important to MMW systems is the effect of attenuation of the atmosphere, and attenuation and backscatter caused by hydrometeors. This latter phenomenon can be particularly severe at short millimeter wavelengths. Some account of all of these environmental effects must be taken into account if the simulation is to be used to predict radar performance in a real scenario.

Table 7.6 Radar Simulation Components.

1. *Radar System Characteristics*
   a. Antenna Design
   b. Transmitter and Receiver Hardware
   c. Signal Processing Techniques
2. *Target Reflectivity Characteristics*
3. *Environmental Characteristics*
   a. Surface (Clutter, Multipath Interference, *et cetera*)
   b. Volume (Clutter, Attenuation, Refraction, *et cetera*)
4. *Geometric Relationships*
   a. Radar platform location and motion
   b. Target location and motion

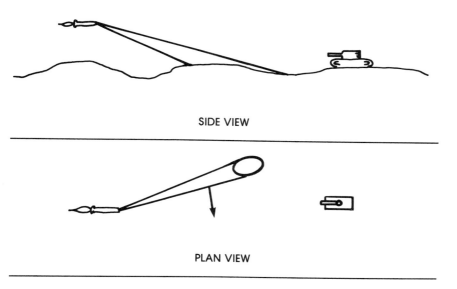

SIDE VIEW

PLAN VIEW

Fig. 7.23 Typical MMW missile seeker engagement.

Finally, since the radar and target positions may both vary with time, some consideration must be given to the geometrical relationship between them. The complexity required depends on the nature of the scenario being simulated. For a ground based radar and a slowly moving target, the geometrical relationship may be extremely simple. However, many problems in MMW radar analysis involve situations in which the radar platform or the target are airborne, and moving at high speeds. In this case, the effect of flight dynamics on the radar performance can be considerable and cannot be ignored. For example, if the complete missile engagement of Figure 7.23 had to be accurately simulated from launch to impact, the nature of the missile aerodynamics would be just as important as the MMW seeker performance. Before a simulation can be constructed, it is important to consider what aspect of the radar performance is to be evaluated, and what type of simulation approach is required.

There are two primary methods of modeling radar system performance: *radar function modeling* and *radar signal modeling*. Radar function modeling is a power-based representation of radar operation. In this case the received power from the target is computed and compared with either the ambient noise in the receiver or the interfering signal returned from the clutter. The radar system itself is reduced to parameters which are used as inputs to the radar range equation. Additional aspects of the radar such as MTI signal processing are modeled separately. In the case of functional radar modeling, targets and clutter are usually represented by statistical distributions, and the probabilities of detection and false alarm are computed statistically.

An alternate approach is modeling of the actual received signal in the radar. This approach is primarily a voltage-based representation of the radar problem. As such, it is a complex valued function, and requires modeling of all aspects of the situation as complex valued. In this case radar components are not simply parameterized, but are modeled as transfer functions which affect the received signal. Target and clutter scattering functions must also be complex, and consequently, these objects are usually modeled deterministically rather that statistically as is done in a functional approach. Radar signal modeling is a much more sophisticated approach to radar performance evaluation than radar function modeling. A detailed examination of the techniques used to model radar signals and radar components can be found in *Radar Signal Simulation* by Richard L. Mitchell (Artech House, 1976). Additional information concerning deterministic modeling of targets and clutter can be found elsewhere in this chapter and in the references.

Radar systems can be used to perform a wide variety of functions, and specific radar designs will reflect the signal processing techniques associated with these functions. Table 7.7 lists six of the functions most often associated with radar systems. In addition, this table relates the radar functions to specific signal processing techniques which would be used to accomplish these functions, and therefore must be simulated in any model of such a radar system.

Table 7.7   Radar Functions and Signal Processing Techniques.

Radar Function	Signal Processing Techniques
Detection and False Alarm Reduction	Amplitude Thresholding Automatic Gain Control (AGC) Constant False Alarm Rate (CFAR) Moving Target Indication (MTI) Frequency Diversity
Target Discrimination and Classification	Polarimetric Processing Spatial Filtering Doppler Spectral Processing
Target Location and Tracking	Angle Processing Range-Gate Tracking Doppler-Gate Tracking

A primary purpose of any radar system is the detection of target signals in the presence of interference. This is accomplished by detection and thresholding of the received signal. Provided that the target and clutter signals can be adequately represented by statistical models, detection performance prediction can usually be performed adequately using radar function modeling. Even sophisticated detection oriented techniques such as moving-target-indication (MTI) processing can be simulated using this method.

Some radar systems may be required to perform more complex types of signal processing in order to discriminate the target from surrounding clutter. In addition, a radar might conceivably need to identify a specific type of target automatically. These processes of target discrimination and classification are based on the derivation of additional information from the received signal. Polarimetric processing, for example, uses the complex nature of the received signal to determine its polarimetric properties. Computer modeling of such sophisticated techniques clearly requires equally complex modeling of the received signals, and could only be performed using deterministic target and clutter models and complex signal representations in the radar. Other target classification techniques utilize other target characteristics, such as the size and shape of the radar return; once again, a more complex radar simulation would be required.

Finally, many radar systems, particularly in missile applications, must acquire and track targets once they are detected. Simple functional models can be used to determine the average tracking error of the system as a function of target and radar parameters. However, such results are only statistical averages, and do not represent the actual performance of the tracker in a dynamic environment. Performance simulation of a true tracking system, particularly in a missile application, would require deterministic signal modeling of the situation. Clearly, there are many factors which impact the type of simulation to be used.

Table 7.8 is a partial list of factors which might affect the approach to the MMW radar simulation design. The primary driver is the objective of the simulation. If it is only desirable to determine the feasibility of a particular design, a simple model might be adequate. Such a model might also be used for trade studies between radar parameters, such as radar frequency or antenna size. More complex models may be necessary for development of improved radar signal processing algorithms or detailed evaluation of existing ones. The flexibility and fidelity required of the simulation also affect the type of approach selected and the depth to which sophistication is pursued. In addition, the available budget and time must be considered. The nature of software development is such that 95% of the work is performed before any result is obtained; time and money must be considered carefully before any particular approach is taken. Finally, the size and speed of the simulation can affect the nature of the model, particularly if the host computer has limited computing capability. This is especially true of deterministic models which may require tens of thousands of operations to be performed for each simulation update.

Table 7.8   Factors Affecting the Modeling Approach.

1. *Objective*
    a. Feasibility Analysis, Trade Study, *et cetera*
    b. Algorithm Development Tool
    c. Software Integration Testing
2. *Budget*
3. *Flexibility*
4. *Fidelity or Accuracy*
5. *Time Schedule*
6. *Size and Speed Limitations*

The fundamentals of computer simulation of MMW radar systems using a functional approach is discussed in the following section. The complexity of the task is such that one or more volumes could be written on the subject. Consequently, the following discussion presents an overview of the components of the radar simulation and the methods of modeling these components. More detailed discussions and specific examples of MMW models can be found in the references [46, 52, 70].

7.5.1.2 *MMW Radar Function Modeling*

MMW radar function modeling is primarily used to predict the detection performance of the system. While this is the most basic function of a radar system, it is also pivotal in determining the feasibility of more sophisticated radar operations. A detection performance model revolves around the radar range equation, although the form in which the equation is used may depend upon the nature of the radar system being modeled. A block diagram of a generalized detection performance model is depicted in Figure 7.24.

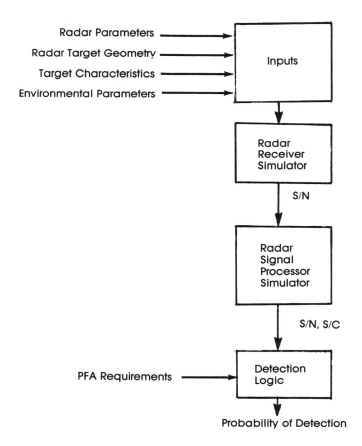

Radar Parameters ⟶

Radar Target Geometry ⟶

Target Characteristics ⟶

Environmental Parameters ⟶

Inputs

↓

Radar Receiver Simulator

S/N

↓

Radar Signal Processor Simulator

S/N, S/C

↓

PFA Requirements ⟶ Detection Logic

↓

Probability of Detection

Fig. 7.24 Simplified block diagram of a radar detection performance simulation.

Input parameters to the simulation include radar parameters, geometric parameters, target characteristics, and environmental characteristics. These data are then processed in two different operations. The radar range equation is used to simulate the radar receiver and detector. The output of this computation is the signal-to-noise ratio (S/N) of the target as a function of target range. A simple form of this calculation can be used to determine performance in free space. However, a more realistic computation would include the affects of atmospheric attenuation and multipath interference.

A second model component simulates the video signal processing used by the radar to remove clutter interference. Such techniques include MTI, Doppler filtering, and CFAR processing. The output of this section is the signal-to-clutter ratio (S/C) as a function of target location.

The signal-to-noise and signal-to-clutter ratios are measures of radar system performance. However, it is usually desirable to translate these signal ratios into probabilities of detection. The process in which the detection logic is implemented depends on the target, clutter, and noise statistical characteristics, but the output of this computation is the *probability of detection* (P_d) as a function of *probability of false alarm* (PFA).

The individual components of this detection performance simulation are considered in more detail in the following discussion.

Radar System Modeling

Millimeter-wave radar systems can be configured in a wide variety of ways, depending upon the desired application, and the MMW radar simulation design must reflect the operation of the system it is emulating. In this brief overview it would be impossible to discuss all of the possible MMW radar configurations, or how they should be modeled. The following discussion will concentrate on basic aspects of MMW radar design which affect the nature of the radar simulation. Specific consideration of MMW radar system design and MMW components will be discussed in other chapters.

Simulation of a radar system in a functional model requires reduction of the radar components to parameters which can be used in the performance calculations. A list of radar parameters for a typical pulsed radar system is presented in Table 7.9. For a simple pulse system, the pulse-length, τ, is the reciprocal of the noise bandwidth (B_n), and the bandwidth correction factor (CB) is 0 dB. However, many MMW systems use pulse compression to increase the effective peak power. In this case, the compressed peak power and compressed pulselength can still be used, but a bandwidth correction factor (1 to 2 dB) must be included to account for the mismatch loss through the pulse compression filter.

Table 7.9   Typical Pulsed Radar Parameters.

Symbol	Definition
P_T	Peak Power (W)
λ	Transmitting Wavelength (m)
G	Monostatic Antenna Gain
B_n	Receiver Noise Bandwidth (MHz)
$\overline{NF}$	Receiver Noise Figure
L_T	Transmission Loss
L_R	Receive Loss
L_s	Scanning Loss
CB	Mismatch Filter Correction Factor
L_m	Miscellaneous Loss
—	Polarization
PRF	Pulse Repetition Frequency (Hz)
τ	Pulse length (μs)
Θ_{AZ}	Azimuth Beamwidth (rad)
Θ_{EL}	Elevation Beamwidth (rad)
F_s	Scan Frequency (Hz)
Ω	Coverage Solid Angle (steradians)

Using the radar parameters of Table 7.9 and the mean target radar cross section (σ), the single pulse signal-to-noise can be computed as a function of range using the radar range equation,

$$S/N = \frac{P_T G^2 \lambda^2 \sigma}{(4\pi)^3 R^4 k T_0 \overline{NF}\, B_n L_T L_R L_s\, \text{CB}\, L_m} \qquad (7.49)$$

For an FMCW radar system, the noise bandwidth B_n is the portion of the sweep which is coherently processed. If a total sweep of F_s MHz was divided into M frequency agile subbands, the bandwidth would be F_s/M.

The result of this calculation is the single pulse (or sweep), free-space signal-to-noise ratio of the radar system. The effects of attenuation and multipath interference will be considered in a later section.

As discussed in Chapter 3, very few detection decisions are based on a single pulse. For a scanning radar, the total number of pulses returned from a target on each scan is a function of the PRF, the scan coverage, the scan frequency, and the antenna beamwidths. Thus,

$$n = \text{PRF} \cdot \frac{\Theta_{AZ}\Theta_{EL}}{\Omega F_s} \qquad (7.50)$$

It is, of course, possible that not all of these pulses are used in each detection decision, and, in fact, a sophisticated radar system may use a complex form of pulse management. In these cases, the appropriate number of pulses should be used in the computation of probability of detection. The process of combining the S/N and the number of pulses to determine the detection performance will be considered later.

Target Modeling

A significant portion of this chapter has already considered the process of modeling the returns from complex targets. Such simulations are extremely useful in deterministic signal modeling of radar system performance. However, for radar function modeling, it is usually considered adequate to represent the target reflectivity characteristics statistically in terms of RCS.

Historically, target reflectivity fluctuations have been represented by the four Swerling models discussed in Chapter 3. In these models, the target amplitude probability density function (PDF) is represented either as a Rayleigh distribution (Swerling's cases 1 and 2) or as a chi-squared distribution (Swerling's cases 3 and 4). The rapidity of the fluctuation is modeled either as pulse-to-pulse (2 and 4), or scan-to-scan (1 and 3) decorrelation of the return. A major advantage to these models is that closed form solutions for the detection performance have been generated. However, as mentioned in Chapter 3, they do not adequately represent complex targets at millimeter wavelengths.

Extensive experimental measurements of ground vehicles at millimeter wavelengths suggest that MMW target PDFs exhibit a greater percentage of high amplitude returns than is represented by any of the Swerling models. This high amplitude "tail" is apparent in the PDF of many ground vehicles. Such an amplitude distribution is more accurately represented by a lognormal distribution. The log-normal function (3.46) is characterized by a mean (or median) RCS and a standard deviation.

In addition to exhibiting different amplitude distributions from the Swerling models, MMW target returns possess extremely complex temporal fluctuation characteristics. Depending on the radar engagement scenario, the period of decorrelation of target returns can vary from tens to hundreds of milliseconds. As discussed in Chapter 3, rapid decorrelation of the target return results in a potential increase in the number of independent samples collected during each scan. This can affect the probability of detection. The effect of these target reflectivity characteristics on detection performance modeling is considered later in this section.

Environment Modeling

Characteristics of MMW surface and volume clutter have been considered in great detail in Chapter 5, and aspects of clutter modeling were discussed in the previous section of the present chapter. The effect of atmospheric attenuation has also been discussed elsewhere. All that remains is to explain how these clutter and attenuation models are used in a radar performance simulation.

Surface Effects

The amount of power received from a clutter cell is a function of the reflectivity and the area illuminated. In a simulation, the illuminated area is computed from the geometrical parameters and the radar resolution parameters using (5.16) or (5.18). The clutter RCS competing with the target return is then given by

$$\sigma_c(\text{sm}) \ = \ \text{area} \cdot \sigma° \tag{7.51}$$

The reflectivity of the surface ($\sigma°$) generally varies with incidence angle, as discussed in Section 7.4. In a radar performance simulation, the angle of incidence can vary from 0° to 90°, and it will frequently vary continuously throughout the simulation. Consequently, the mean reflectivity *versus* incidence angle must also be modeled in the computer program. Depending on the surface over which the simulation is being exercised, any one of the reflectivity models presented in 7.4 could be used.

As with target returns, surface clutter reflectivity is characterized by a probability amplitude density function. At long wavelengths and large coverage areas surface clutter has often been adequately represented by a Rayleigh distribution. However, as shown earlier, MMW clutter data also exhibit the long tails apparent in target returns. Consequently, MMW surface clutter backscatter is often modeled by using either log-normal or Weibull amplitude distributions. Once again, the type of distribution model selected affects the approach to detection probability estimation.

The final aspect of surface clutter which must be modeled in the simulation is the spectral nature of the clutter return. Many radar systems utilized some form of Doppler signal processing to improve detection of moving targets in a background of higher amplitude cutter returns. In order to model the effect of this signal processing upon the clutter interference, the spectral nature of the clutter must be described. Usually, surface clutter spectra are assumed to be centered around zero hertz. While this is not true if the radar platform is moving, it is possible to account for this motion in the signal processor. Consequently, it is only necessary to determine the shape and width of the zero-mean spectrum.

There are three potential factors which can contribute to the spectral width: intrinsic motion, antenna scanning, and platform motion. The intrinsic component (σ_i) is caused by internal motion of the clutter scattering centers and has been discussed in Chapter 5. For a scanning radar system, there is an additional component caused by the motion of the antenna beam past the clutter cell. This scanning component is given by

$$\sigma_s(\text{Hz}) = 0.265 \, \dot{\Theta}/\Theta_{AZ} \tag{7.52}$$

where

   Θ = antenna scan rate,
   Θ_{AZ} = azimuthal 3 dB beamdwidth.

If the radar platform is itself in motion, there is an additional component caused by the spread in radial velocities across the antenna beam. This component can be computed from the equation

$$\sigma_p(\text{Hz}) = 0.6 \, (V/\lambda)\Theta_{AZ} \sin\phi \tag{7.53}$$

where

   V = radar platform velocity (m/s),
   ϕ = azimuth aspect angle (deg.)

The antenna scanning and platform motion effects on the clutter spectrum can usually be represented by a Gaussian distribution in frequency. If the intrinsic clutter component can also be modeled this way, the total clutter spectral standard deviation can be shown to be

$$\sigma_T(\text{Hz}) = (\sigma_i^2 + \sigma_s^2 + \sigma_p^2)^{1/2} \tag{7.54}$$

If the intrinsic clutter spectrum is not Gaussian, and neither component is negligible, the net surface clutter spectrum can only be derived by convolving the two spectral functions together.

In addition to backscatter interference, forward scatter off the earth's surface can adversely affect radar operation. This phenomenon is known as multipath interference. The nature of the effect of multipath interference is discussed in great detail in a number of sources [Long [55], Barton, *et al.* [71], and will not be considered here. Although multipath can be a significant problem in some situations, the severity of the effect is inversely proportional to the transmitting frequency, and multipath is rarely a problem at millimeter wavelengths.

Volume Effects

There are two atmospheric factors which affect radar performance: attenuation and backscatter. Models of both of these effects have been discussed in Section 7.4. The effect of attenuation is to reduce the target signal strength received by the radar, and it consequently affects the detection performance against noise. This can be modeled as a reduction in the free space signal-to-noise ratio computed in (7.49):

$$(S/N)_1 = (S/N)_1 \exp(-2\alpha R) \qquad (7.55)$$

where α = attenuation coefficient (m^{-1})

While attenuation through clear air may be negligible over short distances, the effect of hydrometeors on MMW propagation can be severe, and must be included.

As with surface clutter, the radar cross section of volume clutter is a function of the volume illuminated and the intrinsic reflectivity of the airborne scatterers. While volume clutter could be produced by dust or chaff, the primary source of interference is rain. The volume illuminated by the radar is given by (5.21), and the total rain RCS is given by

$$\sigma_V(sm) = volume \cdot \eta \qquad (7.56)$$

In most situations, the reflectivity of rain (η) is independent of the antenna direction, but it is directly related to the rain rate within the illuminated volume. The curves depicted in Figure 5.27 can be used to model rain reflectivity as a function of rain rate and transmitting frequency.

In the past, a Rayleigh distribution has generally been used to describe rain clutter probability amplitude density functions. While this may be true when the illuminated volume is large, data collected at millimeter wavelengths suggest that a lognormal distribution would be a better fit. The model that is selected for simulation should be based upon consideration of the illumination geometry of the radar scenario.

The spectrum of the rain returns must be modeled to determine the effect of Doppler filtering upon volume clutter interference. This is all the more important in the case of rain because the rain clutter spectrum is not necessarily centered around zero hertz, and thus may not be removed by a simple MTI processor. Rain spectral characteristics are much more difficult to model than surface clutter characteristics because the sources of rain spectral components are complex and can vary with altitude and time.

Figures 5.11 and 5.12 depict rain spectral data at 35 and 95 GHz for two rain rates. The curves shown in the figures are representative of the data, and could be used to model the nature of the rain spectra. However, these data are somewhat misleading, and are representative of the danger of clutter models. The data in these figures were collected in the summer in Florida using a narrow-beam radar over a relatively short distance (1 to 2 km). The antenna was ground-based and pointed at shallow elevation angles. If any one of these conditions were different, the nature of the rain spectra could have changed significantly. For example, rain velocity is strongly affected by wind speed, and wind speed is often a function of altitude, being weakest at the surface and increasing with height. Had the data of Figures 5.11 and 5.12 been collected with an airborne platform, the measured spectra might have been considerably different. While these data are perfectly adequate for many circumstances, it is very important to consider the nature of the simulation scenario when selecting a rain clutter model.

Signal Processor Simulation

Signal processing refers to operations performed on the received IF or video signal to enhance the detection or discrimination of the desired targets. The topic of signal processing covers a wide range of operations, from moving target indication (MTI) to target discrimination and classification. The more sophisticated techniques involve discrimination between the desired target characteristics and those of clutter or other target-like objects. Accurate modeling of these methods would require extremely accurate and complex target and clutter reflectivity models. Such depth is beyond the scope of the present discussion, which will be limited to consideration of the two most basic signal processing techniques: Doppler processing and CFAR processing.

Doppler processing or moving target indication (MTI) involves filtering the received signal in such a manner that interfering clutter signals are suppressed while moving target signals are unaffected. While some MTI techniques can be performed noncoherently, the great majority require the use of synchronous detection in the receiver (Chapter 3) to provide both inphase and quadrature signal components for processing. For the purpose of radar function modeling, the numerous filter configurations that can be designed can be reduced to two categories: bandpass filters and narrowband filter banks.

Bandpass filtering can be performed at IF, but is more common at the video level. The most common type of bandpass MTI design uses delay-line cancelers (DLC) with either feedback or feedforward loops which can be used to adjust the shape of the DLC frequency response. The design of DLC circuits has been described in great detail in several references, and will not be further considered here (Skolnik [56]). Modeling of DLC operation can be performed either in the time domain or the frequency domain, but for functional modeling the latter allows much simpler computations.

The frequency response of many DLC designs can be represented by a roll-off function:

$$H(f) = \frac{1}{1 + \left|\frac{f_c}{f}\right|^n} \text{ for } |f| < \frac{\text{PRF}}{2} \tag{7.57}$$

where

f_c = filter cut-off frequency (Hz),
n = number of filter poles.

One PRF interval of the frequency response of a typical DLC filter is depicted in Figure 7.25. Superimposed on the DLC response are sample target, surface clutter and volume clutter spectra. For many applications, the target spectrum can be represented by a delta function located at the Doppler frequency corresponding to the target's radial velocity. The surface clutter spectrum is assumed to be centered around zero, but the rain spectrum may be shifted to reflect the Doppler associated with a prevailing wind direction. The areas under the clutter spectra correspond to the total surface and volume clutter RCS before signal processing.

The affect of the DLC filter can be modeled simply by multiplying the filter response times the signal spectral components. The result of this filtering process is depicted in Figure 7.26. Because the target was within the bandpass of the filter, it remains unchanged. The surface clutter was almost entirely removed by the filter notch centered around zero. However, the rain clutter spectrum extended beyond the filter notch, so a significant portion of the rain clutter signal was unaffected. This effect exposes one of the weaknesses of bandpass MTI processing.

The second type of Doppler processing involves the use of a bank of narrow-band filters covering the unambiguous PRF interval. Such a set of filters can be created using fast Fourier transforms (FFT), finite impulse response filters, or a bank of actual filter circuits. A typical set of such filters is depicted in Figure 7.27, along with the target and clutter spectra.

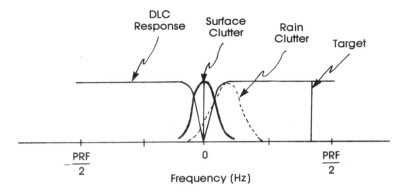

Fig. 7.25 One PRF interval of a typical DLC response and target and clutter spectra before filtering.

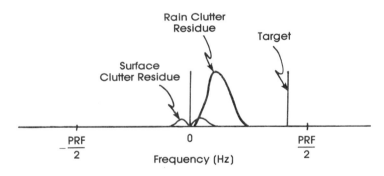

Fig. 7.26 Target and clutter spectra after filtering by a DLC.

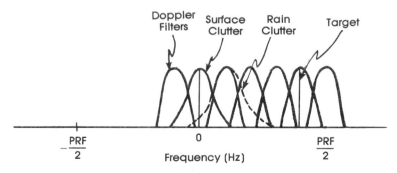

Fig. 7.27 One PRF interval of a typical Doppler filter bank and target and clutter spectra before filtering.

In an actual system, the output of each filter would be individually thresholded, so modeling of such a filter bank requires examination of the effect of each filter separately. Since there is no way of knowing whether the signal out of a filter is caused by a target or by clutter, there must be some a priori process by which some filter outputs are rejected. Surface clutter, for example would presumably be contained in filters close to zero, so these filters would be ignored. Rain clutter covers a larger frequency region, and cannot be as easily discarded. However, if the filters are sufficiently narrow, there will be very little rain clutter passed by the filter containing the target. If CFAR is used to suppress the detection in the filters containing only rain, the target should be easily detected. In this case, the output of such a filter bank would appear as in Figure 7.28.

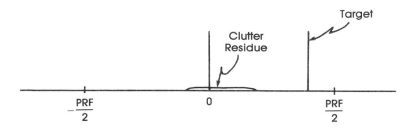

Fig. 7.28 Target and clutter spectra after filtering by a Doppler filter bank.

Doppler filter banks are more complex to implement than simple bandpass filters, but they provide a significant improvement in clutter filtering, and they have two added advantages. If the width of the filters is sufficiently narrow, the velocity of the target can be accurately determined from knowing which filter it occupied. This information could be useful in tracking or identifying the target. In addition, because of their narrow filter passbands, Doppler filter banks provide a coherent S/N integration gain. Bandpass filters only remove a small portion of the noise power, and do not provide this improvement.

Using the frequency representations of target and clutter signals, and the filter response, the signal processor simulator can determine the effect of MTI filtering on the signal-to-clutter ratio (S/C). Once the filter response and the signal spectra have been combined (Figure 7.26 or 7.28), the S/C out of the signal processor can be computed by taking a ratio of the power in the filtered target spectrum divided by the total power in the filtered clutter spectra. This output S/C is then passed on to the detection model of the radar simulation.

The other aspect of signal processing which will be considered here is CFAR processing. As described in detail in Chapter 3, CFAR processing involves averaging the signal from resolution cells in the vicinity of the test cell in order to set the threshold for the test cell. While this is a very important operation in an actual radar system, it is very difficult to model in a simulation. CFAR processing is required because the nature of surface clutter signals varies from point to point. However, for the sake of convenience, clutter models frequently assume a homogeneous clutter background. In such a situation it is not possible to model actual CFAR operation, and a CFAR can only be represented by the additional loss it induces. If it were desirable to model a CFAR processor, it would be necessary to provide a detailed and complex facet model of the clutter surface. While more accurate, such a clutter map can be much more complicated to create, and will increase the computational time required. As will be discussed later, this situation can be rectified by the use of empirical data in the simulation.

Detection Modeling

So far, we have simulated two aspects of the MMW radar system: the receiver model, which produced the single pulse signal-to-noise and the number of pulses integrated, and the signal processor model, which computed the signal-to-clutter ratio. A key issue remaining is how these values are used to determine the probability of detection. The process of statistical analysis of radar detection has been discussed in great detail in Chapter 3. This section will consider the approaches to statistical modeling of the detection process.

There are three contributing sources for the signal detected in the receiver, the target signal S, the clutter signal C, and the residual noise N. Each of the types of signals present are characterized statistically by an amplitude and an autocorrelation function. The PDF determines the nature of the amplitude returns, and the autocorrelation determines the rate at which they decorrelate. Both functions affect the detection performance. The problem is further complicated by the fact that in a functional radar, the detection problem is not to detect the signal in the presence of noise or clutter, the problem is to detect $S + N + C$ in the presence of $C + N$.

Computer modeling of this complicated problem requires numerous assumptions and simplifications. The four Swerling models, for example provide a convenient closed form computation of detection probability under the assumptions that the target is Rayleigh or chi-squared, and the interference is Rayleigh. An additional attraction to these models is that Blake has modeled them in closed form in a FORTRAN computer simulation that is readily available (Blake [72]). However, MMW targets and clutter are not well represented by these models, and use of them could produce misleading results.

MMW targets and clutter frequently exhibit amplitude distributions that can be closely represented by a log-normal distribution (3.46). D.C. Schleher has derived closed form solutions for probabilities of detection and false alarm for a constant target in log-normal clutter and for a log-normal target in log-normal clutter (Schleher [73]). The accuracy of the latter model is quoted to be within 1 dB over a wide range of signal-to-clutter ratios, and less than 0.3 dB for high S/C. However, the model implicitly assumes that the signal-plus-clutter distribution can be adequately represented by the signal distribution alone; this assumption implies that the signal-to-clutter ratio is fairly high. The model accounts for independent samples of the clutter signal but not for the target signal. In addition to considering detection probability, Schlerer has designed a CFAR processor for lognormal clutter, and determined the CFAR loss as a function of the number of independent sample cells. These equations are only valid for situations when the ratio $(S + C + N)/(C + N)$ can be represented accurately by S/C. In this relatively high S/C, clutter-limited scenario, the model could be extremely useful, but it should not be extended to other situations.

For situations when the target, noise or clutter signals must all be included in the detection calculation, the probabilities of detection and false alarm can be determined using a Monte Carlo simulation. Such a simulation would proceed as follows. Each signal type would be characterized by a PDF, a mean RCS and a standard deviation (SD). The clutter-plus-noise distribution would be constructed by generating Rayleigh noise and lognormal clutter samples, adding a uniform phase shift, and collecting the resulting set of samples to form a histogram which approximates the $C + N$ PDF. Once the $C + N$ PDF was formed, the PFA could be determined by integrating the area under the PDF above a set threshold. The threshold for a specific PFA could be determined in a reverse fashion. A similar technique could be used to generate a $S + C + N$ distribution, from which the probability of detection could be computed.

The Monte Carlo technique thus described possesses the great advantage that it is independent of the statistical characteristics; the method is equally accurate for any target and clutter PDFs. Computational time is increased significantly over closed form solutions since several thousand samples of each distribution must be collected to make the combined PDF sufficiently accurate. However, with modern computing capabilities, such a detection model is quite practical, as long as real time simulation is not required.

One aspect of the detection problem is missing from the Monte Carlo model: target, clutter and noise autocorrelation functions. Using the target, clutter, and noise statistics as defined, the output of the Monte Carlo simulation would correspond to the single-pulse $(S + C + N)/(C + N)$ detection performance. No account has been taken in the detection process of the independent samples produced by decorrelation of these signals.

System noise bandwidth is assumed to be determined by the bandwidth of the IF receiver. Consequently, the noise signal decorrelates completely between samples. If a total of n noise samples is collected for one detection, the normalized standard deviation and mean of the noise are reduced as follows:

$$SD_n = \frac{SD_1}{n} \tag{7.58}$$

and

$$\langle N \rangle = \frac{SD_n}{(4/\pi - 1)^{1/2}} \tag{7.59}$$

Integration of target and clutter signals poses a somewhat more difficult problem. The rate at which target and clutter signals decorrelate depends on many factors: transmit frequency, frequency agility, observation time, radar-target geometry, *et cetera*. In addition to the natural decorrelation of signals through temporal fluctuations, independent samples can be generated using frequency agility. The number of independent samples of target and clutter signals which are available for each detection must be determined for the specific radar scenario being simulated. The effect of these independent samples on detection computation is also uncertain and dependent upon the signal PDFs.

One model has been developed to account for integration of independent samples (Einstein [74]). By using this model, the net target or clutter variance after integration is given by

$$SD^2 = SD_s^2 + \frac{SD_i^2}{N} \tag{7.60}$$

where

SD_i = intrinsic standard deviation for each clutter cell (dB),
SD_s = spatial standard deviation between cells (dB),
N = number of independent samples.

This model is based on the following logic. A radar scans across a group of clutter cells. The returns from any one cell can be decorrelated by transmitting a wideband frequency agile signal. Integrating the decorrelated returns from this cell is equivalent to estimating the mean RCS of that cell: if the bandwidth were infinite, the estimate would be exact. However, as the radar scans to a different cell, the mean RCS changes. This is a result of the fact that the clutter signal is nonstationary as well as being random. The spatial distribution of means is represented by SD_s, which is unaffected by integration. SD_s is a function of the environment and must be measured; typical values for land clutter range from 1 to 2 dB. The intrinsic SD_i for each cell can be reduced by integration, and the net variance is computed by using (7.12). SD_i can be derived analytically by assuming a type of distribution and a receiver response. For a Rayleigh distributed signal and a logarithmic receiver, T.H. Einstein derives SD_i to be 5.57 dB [74]. A different signal distribution would produce different values. This same integration model can be applied to target returns; in this case, SD_s is defined relative to different aspect angles on the target.

Radar Engagement Geometry Modeling

The method used to model the geometry of the radar-target engagement depends on both the nature of the engagement and the purpose of the simulation. However, whether the geometry is simple or complex it must be considered carefully, because it can have a great impact on the radar performance, and even on the radar system design. In order to demonstrate the effect of engagement geometry on the radar system performance, the following discussion examines the scenario depicted in Figure 7.23.

Figure 7.23 displays an air-to-ground missile in level flight using an active MMW seeker to scan the area for armored targets. Such a scenario is typical of many modern small missile seeker concepts. While the actual missile engagement would be considerably more complex, for the purpose of detection performance prediction, the simple geometry of Figure 7.23 may be adequate. In this scenario the missile is flying at a constant velocity along a level linear flight path. The target is assumed to be stationary on the ground or moving slowly relative to the missile. Even in this simple scenario, the specific geometric parameters can have a serious effect on the radar performance, as can be seen in the following.

Given a missile altitude, H_a, the range to the illuminated area is given by

$$R = H_a/\sin\Theta \tag{7.61}$$

where Θ = the antenna depression angle.

The smaller is the depression angle Θ, the larger is the range to the target. An increased range allows the missile a longer time to acquire and track the target, but also decreases the S/N and the S/C, and, consequently, the detection performance. The depression angle can be increased to reduce R, but this reduces the range extent of the illuminated region, which is given by

$$R_i = R\,\Theta_{EL}\,\csc\Theta \tag{7.62}$$

The illuminated region is separated in the radar processor into resolution cells. Each of the cells must be processed for detection, and the more cells there are, the more computing capability the radar must have. So, it might seem desirable for two reasons to reduce the illuminated area. However, there is an additional aspect that has not yet been considered.

Since the radar scans across an angular region, $\Delta\Theta$, in search of targets, it takes a certain amount of time to return to the same point. This scan time is given by the equation

$$t_s = \Delta\Theta/\dot{\Theta} \tag{7.63}$$

where

$\Delta\Theta$ = the scan coverage angle,

$\dot{\Theta}$ = the antenna scan rate.

It is certainly necessary for the radar to scan over the entire search region, and in many cases it is desirable for each target to be detected on more than one scan. However, if the illuminated range R_i is very short, the radar must scan extremely quickly to avoid missing coverage on a target. Since the PRF is determined by other factors, a higher scan rate will result in fewer pulses on target which in turn results in a lower probability of detection.

This brief discussion demonstrates that even in a very simple engagement scenario the geometric parameters selected can have a serious effect on seeker performance. In fact, many of the effects work at cross purposes, and final specification of the geometry may require a trade-off study to determine the optimum combination. In addition, the motion of the target itself, which has not been considered in this discussion, can affect the design and must be included in the analysis.

The simple geometry of Figure 7.23 may be adequate for analysis of detection performance, but it would not support a more detailed examination of acquisition or tracking capability. For detection performance prediction, the seeker can usually be considered as a separate entity, independent of the missile design. For analysis of the functions of acquisition and track, this is no longer true. Both processes are very sensitive to the radar-target geometry, and are, in fact, dependent upon the aerodynamics of missile flight. Any detailed tracking analysis would require specification of the missile guidance and control characteristics, missile flight instabilities, actual missile velocity and position profiles, and much more. In addition, missile seeker tracking effects involve time constants on the order of fractions of seconds, so the frequency of simulation computations must be extremely high. In general, simulation of missile acquisition and tracking capabilities should be performed using a radar signal modeling approach rather than the radar function model approach being considered here.

Use of Empirical Data in MMW Radar Simulations

When available, the use of empirical data can greatly enhance the effectiveness of radar performance modeling. In general, such data consist of detailed target or clutter reflectivity values that can be used to replace analytical models. Such detailed empirical data can be used in signal-based radar modeling to replace complex and sometimes unreliable deterministic models with more trustworthy representations. More important for the present discussion, the use of empirical data can greatly improve the accuracy of function-based radar models without a significant increase in computation requirements or in software complexity.

There are an increasing number of MMW radar data bases available in the United States. One of the most comprehensive is the Target and Background Information Library System (TABILS), administered by the Air Force Armament Laboratory at Eglin AFB, Florida (Lane [75]). This data base contains multiple directories of target and clutter data collected with MMW and IR sensors. Some of the MMW data directories contain detailed target signatures as a function of aspect angle for a wide variety of radar parameters. Such data are useful in testing target detection, discrimination, and classification techniques, and could also be inserted in a functional simulation as a target "model."

Other directories, such as TABILS 5 and 11, contain high resolution maps of targets and clutter collected with airborne radar systems. The data of TABILS 5 and 11 were collected at 35 GHz and 95 GHz, at 12° and 30° depression angles. A functional radar simulation can use such a MMW map to test various features of the seeker design. By degrading the resolution to match the parameters of the radar being tested, and scaling the received signal values to match the signal-to-noise that would be present in the test seeker, the detection and false alarm performance of the seeker can be examined.

Many more signal processing techniques can be modeled using an actual MMW scenario map than would be possible with an analytical model. Because the clutter background consists of a real map, an actual CFAR algorithm can be tested. Recall that previously CFAR could only be accounted for as an additional loss. With a clutter map, various CFAR sizes and types can be tested easily. In addition, such a two dimensional map allows testing of more sophisticated spatial filters which are a part of many modern airborne radar designs. Such spatial extent discriminants can be used to reduce the effect of clutter false alarms, particularly in regions of large clutter reflectivity variations, such as at tree lines. These techniques cannot be simulated using the types of clutter models previously discussed. Finally, the map can be used to test target array discriminants which are designed to identify targets deployed in typical battlefield arrangements.

Empirical data bases can clearly be used to enhance the effectiveness of MMW radar modeling, and to test aspects of the radar system design that cannot be easily simulated in any other fashion. It is important to remember, however, that empirical data is a snapshot of the real world, and as such can only be used in a simulation to test radars against one specific scenario. In a different deployment environment, the radar performance could be quite different. Radar simulations of all kinds provide only an estimate of radar performance, not a measure of it.

7.5.2   Hardware-in-the-Loop (HWIL) Simulations

HWIL Theory

Figure 7.29 illustrates a typical HWIL simulation facility, the Terminal Guidance Laboratory (TGL) at Boeing Aerospace Company in Seattle, Washington (Bland, *et al.* [76]). The MMW seeker hardware is placed on a three-axis flight table with appropriate simulators to handle the missile motion and rotational dynamics. The target (or target and clutter) signal is radiated from a large array at the proper radar frequency and with the desired target dynamics and signature characteristics. The array at Boeing's TGL has 1280 gas discharge tubes to generate target radiation for passive radiometric MMW seekers. The array is programmed to provide apparent target motion and growth with range closure.

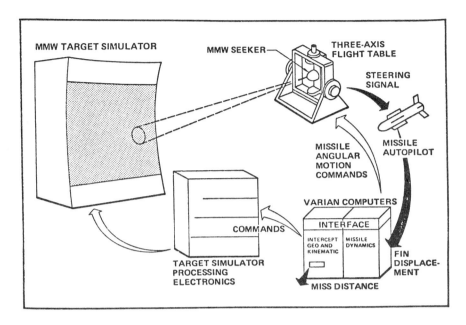

Fig. 7.29 Hardware-in-the-loop (HWIL) simulation for testing MMW seekers (from Bland, *et al.* [76]).

HWIL simulation can be used at almost any time in MMW seeker development. Tests can be run on subassemblies in open loop mode to develop software versions of the hardware transfer function for each subassembly. Closed loop data can be generated with available combinations of hardware subassemblies and the software models, resulting in gradually improving models. Final optimization occurs when all the missile hardware is flown in the loop for many runs covering the range of conditions expected in flight testing. Following initial flight testing, further optimization might be indicated for either the hardware or the software in the simulation.

Special Problems of HWIL Simulations

The concept of HWIL simulations promises to provide extensive seeker performance data under a much wider variety of conditions than can be expected to be measured in flight testing. The precise control of test conditions available to the test engineer in an HWIL simulation makes it a powerful diagnostic tool.

Possibly the most serious limitation of HWIL simulation at the present time is the lack of sophisticated, detailed target and clutter signal simulations to feed to the seeker. A major constraint on the complexity of the target and clutter signal simulations is the need to radiate the signals in real time, which requires computer control and speeds sometimes beyond current capabilities. Parallel processing offers hope for improvement in this area, but better statistical representations of the appropriate signals are also needed.

The greatest utility of HWIL simulations lies in their ability to optimize subassemblies such as the seeker control loop, the seeker output filter, the guidance filter, and the autopilot. Hardware-in-the-loop simulations can assist in pinpointing hardware-software inadequancies in the seeker and can be used to evaluate possible remedies.

Validation of HWIL Simulation

Validation of a HWIL simulation should consist of comparison of HWIL simulation data with well-controlled flight test data. The best approach is probably to define a test plan, conduct the flight tests, following the plan as closely as possible, and then replicate the flight test conditions in the HWIL simulation. The only problem with that approach is that the greatest utility of the HWIL simulation lies in the contribution that it can make in the early design stages of a millimeter wave seeker, well before any flight tests are possible for validation.

Still, the HWIL simulation concept can be validated during the seeker design by comparing HWIL simulation results with software model results, as shown in Figure 7.30. The quantities compared are the computer *pitch* line-of-sight (LOS) rate (QLOS) with the measured pitch LOS rate from the seeker, and the computed *yaw* LOS rate (RLOS) with the measured yaw LOS rate from the seeker. The general agreement between computed and measured LOS rates shown in Figure 7.30 indicates proper seeker operation.

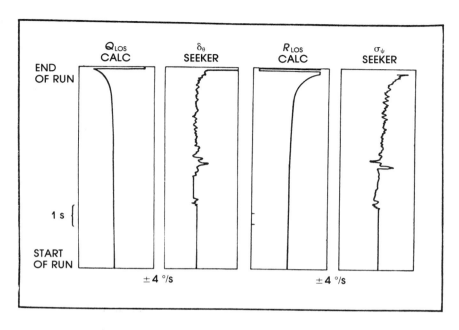

Fig. 7.30 Hardware-in-the-loop test data (from Bland, *et al.* [76]).

REFERENCES

1. S.A. Goudsmit and R.R. Weiss, "Possible Measurement of Radar Echoes by Use of Model Targets," Massachusetts Institute of Technology, Radiation Laboratory, Report RL-196, March 4, 1943.
2. G. Sinclair, "Theory of Models of Electromagnetic Systems," *Proceedings of the IRE*, Vol. 36, No. 11, November 1948, pp. 1364-1370.
3. P. Blacksmith, Jr., R.E. Hiatt, and R.B. Mack, "Introduction to Radar Cross-Section Measurements," *Proceedings of the IEEE*, Vol. 53, No. 8, August 1965, pp. 901-920.
4. F.C. Paddison, *et al.*, "Radar Cross Section of Ships," *IEEE Transactions on Aerospace and Electronic Systems*, Vol. AES-14, No. 1, January 1978, pp. 27-34.

5. H.S. Burke, *et al.*, "A Millimeter Wave Scattering Facility," paper presented at the 1980 Radar Camouflage Symposium, Orlando, FL, 18-20 November 1980; Technical Report No. AFWAL-TR-81-1015, Avionics Laboratory, Wright-Patterson AFB, OH, pp. 327-336.

6. L.A. Cram and S.C. Woolcock, "Review of Two Decades of Experience Between 30 GHz and 900 GHz in the Development of Model Radar Systems," *Millimeter and Submillimeter Wave Propagation and Circuits*, AGARD (Advisory Group of Aerospace Research and Development) Conference Proceedings No. 245, September 1978.

7. J. Wright, "Analysis of Radar Cross Section Measurements of a Scale Model of the MQM-34D," Technical Report RE-75-30, US Army Missile Command, Redstone Arsenal, Alabama, 12 May 1975.

8. R.W. Clay and J.K. Johnston, "BQM-34A Polarization Characteristics—Radar Cross Section and Glint," General Dynamics Report No. FZE-1312, General Dynamics Convair Aerospace Division, Fort Worth, TX, Final Report on Contract No. DAAH01-73-C-11D2, February 1974.

9. "Radar Signature Measurements of BQM-34A and BQM-34F Target Drones," AFSWC-TR-74-01, Air Force Special Weapons Center, 6585th Test Group (RX), Holloman AFB, NM, January 1974.

10. C.R. Schumacher, "Electrodynamic Similitude and Physical Scale Modeling, Part I: Nondispersive Targets," Technical Report DTNSRDC-86/023, David W. Taylor Naval Ship Research and Development Center, Bethesda, MD, April 1986.

11. C.R. Schumacher, "Electrodynamic Similitude and Physical Scale Modeling, Part II: Dispersive Targets," David W. Taylor Naval Ship Research and Development Center, Bethesda, MD (work in preparation).

12. J.A. Stratton, *Electromagnetic Theory*, McGraw-Hill, New York, 1941, pp. 464-470.

13. R.F. Harrington, *Field Computation by Moment Methods*, Macmillan, New York, 1968.

14. E.F. Knott, J.F. Shaeffer, and M.T. Tuley, *Radar Cross Section: Its Prediction, Measurement and Reduction*, Artech House, Dedham, MA, 1985, pp. 119–130.

15. R.G. Kouyoumjian and P.H. Pathak, "A Uniform Theory of Diffraction for an Edge in a Perfectly Conducting Surface," *Proceedings of the IEEE*, Vol. 62, No. 11, November 1974, pp. 1448-1461.

16. J.B. Keller, "Diffraction by an Aperture," *J. App. Phys.*, Vol. 28, No. 4, April 1957, pp. 426-444.

17. P. Ia. Ufimtsev, "Approximate Computation of the Diffraction of Plane Electromagnetic Waves at Certain Metal Bodies: Pt. I. Diffraction Patterns at a Wedge and a Ribbon," *Zh. Tekhn. Fiz.* (USSR), Vol. 27, No. 8, 1957, pp. 1708-1718.
18. A. Sommerfeld, "*Mathematische Theorie der Diffracton,*" *Math. Ann.*, Vol. 47, 1896, pp. 317-374.
19. A. Sommerfeld, "Lectures on Theoretical Physics," *Optics*, Vol. 4, New York, Academic Press, 1964.
20. T.B.A. Senior and P.L.E. Uslenghi, "High-Frequency Backscattering from a Finite Cone," *Radio Science*, Vol. 6, No. 3, March 1971, pp. 393-406.
21. S.W. Lee and G.A. Deschamps, "A Uniform Asymptotic Theory of Electromagnetic Diffraction by a Curved Wedge," *IEEE Transactions on Antennas and Propagation*, Vol. AP-24, No. 1, January 1976, pp. 25-34.
22. G.A. Deschamps, J. Boersma, and S.W. Lee, "Three-Dimensional Half Plane Diffraction: Exact Solution and Testing of Uniform Theories," *IEEE Transactions on Antennas and Propagation*, Vol. AP-32, No. 3, March 1984, pp. 264-271.
23. K.M. Mitzner, "Incremental Length Diffraction Coefficients," Technical Report No. AFAL-TR-73-296, Northrop Corporation, Aircraft Division, April 1974.
24. C.E. Ryan, Jr. and L. Peters, Jr., "Evaluation of Edge-Diffracted Fields Including Equivalent Currents for Caustic Regions," *IEEE Transactions on Antennas and Propagation*, Vol. AP-17, No. 3, May, 1969, pp. 292-299. See also correction in Vol. AP-18, No. 2, March 1970, p. 275.
25. E.F.Knott and T.B.A. Senior, "Equivalent Currents for a Ring Discontinuity," *IEEE Transactions on Antennas and Propagation*, Vol. AP-21, No. 5, September 1973, pp. 693-695.
26. A. Michaeli, "Equivalent Edge Currents for Arbitrary Aspects of Observation," *IEEE Transactions on Antennas and Propagation*, Vol. AP-32, No. 3, March 1984, pp. 252-258. See also correction in Vol. AP-33, No. 2, February 1985, p. 227.
27. T.B.A. Senior, "Surface Roughness and Tolerances in Model Scattering Experiments," *IEEE Transactions on Antennas and Propagation*, Vol. AP-13, No. 4, July 1965, pp. 629-636.
28. M.T. Tuley, Georgia Tech Research Institute, Atlanta, GA, private communication.

29. P. Beckmann and A. Spizzichino, *The Scattering of Electromagnetic Waves from Rough Surfaces*, Artech House, Norwood, MA, 1987.

30. D.K. Barton, "Low Angle Radar Tracking," *Proceedings of the IEEE*, Vol. 62, No. 6, June 1974, pp. 687-704.

31. R.A. Ross, "Radar Cross Section of Rectangular Plates as a Function of Aspect Angle," *IEEE Transactions on Antennas and Propagation*, Vol. AP-14, No. 3, May 1966, pp. 329-335.

32. E.E. Martin, "Dual Polarized Radar Cross Section Measurements of Vehicles and Roadway Obstacles at 16 GHz and 69 GHz," Final Technical Report on General Motors Industrial Contract, Georgia Institute of Technology, Atlanta, GA, 1973.

33. N.C. Currie, ed., *Techniques of Radar Reflectivity Measurement*, Artech House, Dedham, MA, 1984, pp. 218-219.

34. M.T. Tuley, "A Survey of Scattering from Rough Surfaces," Qualifying Exam Paper, Georgia Institute of Technology, Atlanta, GA, 1975.

35. S.O. Rice, "Reflection of Electromagnetic Waves from Slightly Rough Surfaces," *Comm. Pure and Appl. Math*, Vol. 4, 1951, pp. 351-378.

36. W.H. Peake, "Thoery of Radar Return from Terrain," *IRE International Convention Record*, No. 7, Part 1, 1959, pp. 27-41.

37. J.W. Wright, "Backscatter from Capillary Waves with Application to Sea Clutter," *IEEE Transactions on Antennas and Propagation*, Vol. AP-14, November 1966, pp. 749-754.

38. G.R. Valenzuela, "Depolarization of EM Waves by Slightly Rough Surfaces," *IEEE Transactions on Antennas and Propagation*, Vol. AP-15, July 1967, pp. 552-557.

39. D.E. Barrick, "First Order Theory and Analysis of MF/HF/VHF Scatter from the Sea," *IEEE Transactions on Antennas and Propagation*, Vol. AP-16, September 1968, pp. 554-559.

40. F.G. Bass, *et al.*, "Very High Frequency Radiowave Scattering by a Disturbed Sea Surface," *IEEE Transactions on Antennas and Propagation*, Vol. AP-16, September 1968, pp. 554-559.

41. J.W. Wright, "A New Model for Sea Clutter," *IEEE Transactions Antennas and Propagation*, Vol. AP-16, pp. 217-223, March 1968.

42. M.W. Long, "On a Two-Scatterer Theory of Sea Echo," *IEEE Transactions on Antennas and Propagation*, Vol. AP-22, pp. 662-666, September 1974.

43. G.T. Ruck, *et al.*, *Radar Cross Section Handbook* Vol. 2, Plenum Press, New York, 1970.

44. B.I. Semenov, "Computation of Scattering of Electromagnetic Waves by a Rough Surface for Arbitrary Angles of Observation," *Radio Eng. and Elect. Physics*, Vol. 15, 1970, pp. 505-508.
45. G.W. Ewell, M.M. Horst, and M.T. Tuley, "Predicting the Performance of Low-Angle Microwave Search Radars—Targets, Sea Clutter, and the Detection Process," Proceedings of OCEANS, September 1978, pp. 373-378.
46. M.M. Horst, *et al.*, "Radar Sea Clutter Model," International Conference on Antennas and Propagation, Part 2, 1978.
47. M.M. Horst, F.B. Dyer and M.T. Tuley, "Radar Sea Clutter Model," *URSI Digest*, 1978 International IEEE AP/S URSI Symposium, College Park, MD.
48. G.W. Ewell, *et al.*, "Radar Sea Clutter at Millimeter Wavelengths," *Proceedings of the Sixth DARPA/Tri-Service Millimeter Wave Conference*, Harry Diamond Laboratories, Bethesda, MD, 1977.
49. W.K. Rivers, "Low Angle Radar Sea Return at mm Wavelength," Georgia Institute of Technology, Atlanta, GA, Final Report on Contract N62267-70-C-0489, 1970.
50. G.W. Ewell, *et al.*, "OWEX II Radar Tests," Georgia Institute of Technology, Atlanta, GA, Final Technical Report on Contract N00123-74-C-5415, 1975.
51. H. Sittrop, "X- and K_u-Band Radar Backscatter Characteristics of Sea Clutter," Parts I and II, Physics Laboratory of the National Defense Research Organization, The Hague, The Netherlands, 1975.
52. S.P. Zehner, and M.T. Tuley, "Development and Validation of Multipath and Clutter Models for Tac Zinger in Low Altitude Scenarios," Georgia Institute of Technology, Atlanta, GA, Final Technical Report, Contract No. F49620-78-C-0121, 1979.
53. N.C. Currie, private communication, 1987.
54. D.E. Kerr, *Propagation of Short Radio Waves*, Boston, Technical Publishers, 1964, pp. 396-434.
55. M.W. Long, *Radar Reflectivity of Land and Sea*, 2nd Ed., Artech House, Dedham, MA, 1983.
56. M.I. Skolnik, *Introduction to Radar Systems*, McGraw-Hill, New York, 1962.
57. N.C. Currie, F.B. Dyer, and R.D. Hayes, "Radar Land Clutter Measurements at Frequencies of 9.5, 16, 35, and 95 GHz," Technical Report No. 3 on Contract DAA25-73-C-0256, Engineering Experiment Station, Georgia Institute of Technology, Atlanta, GA, 2 April 1975.

58. N.C. Currie, F.B. Dyer, and G.W. Ewell, "Radar Millimeter Back-scatter Measurements on Snow," *1976 IEEE-AP/S International Symposium*, Amherst, MA, October 1976.

59. W.H. Stiles, and F.T. Ulaby, "Microwave Remote Sensing of Snow-packs," Final Technical Report on Contract NAS5-23777, The University of Kansas Center for Research, Lawrence, KS, June 1980.

60. T. Bush, *et al.*, "Seasonal Variations of the Microwave Scattering Properties of the Deciduous Trees as Measured in the 1-18 GHz Spectral Range," Final Technical Report on Contract NAS9-10261. The University of Kansas Center for Research, Lawrence, KS, June 1976.

61. "Radar Terrain Return Study," Goodyear Aircraft Corporation, Final Report on Contract NOas-59-6186-C, GERA-463, September 1959.

62. N.C. Currie, "Characteristics of Millimeter Radar Backscatter from Wet/Dry Foliage," *1979 IEEE-AP/S International Symposium*, Seattle, WA, June 1979.

63. W.H. Stiles, F.T. Ulaby, and E. Wilson, "Backscatter Response of Roads and Roadside Surfaces," Final Technical Report on Contract DE-AC04-76POD789, The University of Kansas Center for Research, Lawrence, KS, January 1979.

64. I.Kh Vakser, Yu.I. Malyshenko, and L.E. Kopilovich, "The Effect of Rain on the Millimeter and Submillimeter Radiowave Distribution," *Atmospheric and Oceanic Physics*, Vol. 6, No. 9, September 1970, pp. 568-570.

65. J.O. Laws, and D.A. Parsons, "The Relationship of Raindrop Size to Intensity," *Transactions of the Geophysics Union*, Vol. 24, Part II, 1943, pp. 452-460.

66. R.L. Mitchell, "Radar Meteorology at Millimeter Wavelengths," Report No. SSD-TR-66-117 on Contract AF 04(695)-669, Aerospace Corporation, Los Angeles, CA, June 1966.

67. F.P. Wilcox, and R.S. Grazino, "Millimeter Wave Weather Peformance Projections," GERA-1989, Goodyear Aerospace Corp., Litchfield Park, Arizona, 1 March 1974, Also Vol. 1, *Proceedings 1974 Millimeter Waves Techniques Conference*, NELC/TD 308, March 1974, pp. 26-28.

68. N.P. Krasyuk, V.I. Rozenberg, and D.A. Chistyakov, "Attenuation and Scattering of Radio Waves by Raindrops of Various Origins," *Atmospheric and Oceanic Physics*, Vol. 4, No. 11, November 1968, 1209-1213.

69. F.E. Nathanson, *Radar Design Principles*, McGraw-Hill, New York, 1969.

70. B. Perry, *et al.*, "MERGE: An Analytical Radar Performance Model," Georgia Institute of Technology, Atlanta, GA, Final Technical Report, Contract No. DAAK10-81-R-0006, 1986.
71. D.K. Barton, *et al.*, *Radars*, Vol. 4, *Radar Resolution and Multipath Effects*, Artech House, Dedham, MA, 1975.
72. L.V. Blake, "A FORTRAN Computer Program to Calculate the Range of a Pulse Radar," NRL Report 7448, 1972.
73. D.C. Schleher, "Harbor Surveillance Radar Detection Performance," *IEEE Journal of Oceanic Engineering*, Vol. OE-2, No. 4, October 1977.
74. T.H. Einstein, "Effect of Frequency-Averaging on Estimation of Clutter Statistics Used in Setting CFAR Detection Thresholds," Technical Report TT-60, Contract No. F19628-80-C-0002, MIT Lincoln Laboratory, Lexington, MA, 1982.
75. T.L. Lane, "95 GHz Clutter Backscatter TABILS V Data Base Summary," Final Technical Report, Boeing Aerospace Company Letter Contract No. BB3041, Georgia Institute of Technology, Atlanta, GA, April 1982.
76. J.G. Bland, *et al.*. "Development Testing and Flight Certification Testing of Terminally Guided Submissiles," AGARD Conference Proceedings No. 292, Guidance and Control of Tactical Air-Launched Missiles, p. 25-1, Eglin AFB, FL, 6-9 May 1980.
77. V.W. Richard and J.E. Kammerer, "Rain Backscatter Measurements and Theory at Millimeter Wavelengths," Report No. 1838, US Army Ballistic Research Laboratories, Aberdeen Proving Ground, MD, October 1975.

Part III
COMPONENTS AND SUBSYSTEMS

Chapter 8
MMW Solid-State Sources

R.W. McMillan

Georgia Institute of Technology
Atlanta, Georgia

8.1 INTRODUCTION

As in all areas of solid-state technology, there is rapid advance in the development of solid-state MMW sources. Improvements in the areas of theory, materials, and fabrication techniques have resulted in the availability of a wide variety of devices which meet every application need formerly satisfied by electron tubes with the possible exception of those requiring extremely high powers such as radar transmitter applications, which will probably always be met by tube sources. Even in this case, solid-state transmitters have been built to meet short-range, ground-based radar requirements where high power is not required. The advantages of using solid-state sources as compared with tube sources are similar to the advantages of using solid-state devices in any application; namely, low cost, reliability, and simplicity of operation. The rapidly evolving technology of MMW solid-state sources will no doubt lead to further improvements in each of these areas as new concepts are explored.

The types of sources used in most MMW applications at the present time are IMPATT (*impact ionization avalanche transit time*) diodes, Gunn-effect devices, and field-effect transistors (FETs). Many interesting and exciting developments are taking place in other areas, however, including advances in the technology of TUNNETT (*tunneling transit time*) and MITATT (*mixed tunneling and transit time*) devices. Perhaps the most exciting developments are in the technology of FETs, which promise to provide means of building low-noise amplifiers throughout most of the

397

useful MMW spectrum in addition to being the basis of the very interesting, exciting, and rapidly evolving field of MMW integrated circuits. Each of these device types is discussed in subsequent sections of this chapter.

Until very recently, the use of solid-state sources to fulfill radar requirements has been confined to applications as local oscillators, which were met solely by Gunn devices. As recently as 1980, suitable means did not exist to employ solid-state sources as local oscillators at frequencies above about 100 GHz, which was the upper frequency limit of available Gunn oscillators. Recent advances in phase-locking technology have also made IMPATTs available as local oscillators, and improvements in multiplier circuits and Gunn device power outputs have extended the range of multiplied Gunn sources to 220 GHz and higher. The advances made in indium phosphide Gunn devices hold the promise of devices with the frequency capability of IMPATTs together with the low noise performance of Gunn sources. Improvements in materials and fabrication techniques have led to high power IMPATT diode sources which serve as useful radar transmitters, especially when several units are combined in appropriate power-combining circuits. As the technology advances, integrated circuits will play a larger role in MMW technology, and will indeed provide the basis for MMW systems that would not otherwise be built. It is perhaps safe to say that MMW systems will never be deployed to any great extent, especially in military applications, until suitable integrated circuits become available; fabrication of large numbers of systems using waveguide technology is too expensive because of the high precision requirements of this method of implementation. Each of these types of sources and technologies will be discussed in subsequent sections of this chapter.

8.2 SOLID-STATE OSCILLATORS

8.2.1 IMPATT Devices

As the name implies, IMPATT devices employ impact-ionization and transit-time properties of semiconductor structures to give negative resistance, thus leading to oscillation at MMW frequencies. Typically, a $p^+ - n - n^+$ device structure is biased slightly into avalanche breakdown by an external voltage (Chaffin [1], Sze [2]). Avalanche electrons, generated by impact ionization near the p^+ region, drift across the n region with a transit time determined by the device parameters. This transit time causes the current to lag the voltage, and oscillation is possible if this delay exceeds one-quarter cycle. It is assumed that the high-frequency components of the voltage associated with thermal noise start these oscillations, which are then sustained by the negative resistance of the device. Excellent discussions of IMPATT theory have been given by Kuno [3, 4] and Kramer [5], who also give an extensive list of references to this subject.

Figure 8.1 [3] is a schematic diagram of two manifestations of IM-PATT device fabrication. Figure 8.1(a) shows the *single-drift* structure discussed above, and Figure 8.1(b) shows the so-called *double-drift* structure. In the single-drift device, which is characterized by a $p^+ - n$ junction, only the *n* region contributes to IMPATT operation, but in the double-drift structure, which has a $p - n$ junction, both *p* and *n* regions contribute to operation. As might be expected, the single-drift device is better for higher frequencies, since the mobility of electrons, which are the primary current carriers in the drift region, is greater than that of holes. However, the efficiency of the double-drift device is greater at the lower frequencies because of the greater current density of this device for a given physical cross section.

IMPATTs are packaged as shown in Figure 8.2 [3], with the indicated structure being sealed for devices used up to 110 GHz and open for reduced parasitics [4] at higher frequencies. Figure 8.3 [3, 4] shows how the device of Figure 8.2 is typically mounted in its waveguide circuit. Figure 8.4 (Kramer [6]) is a photograph of a typical sealed IMPATT package shown schematically in Figure 8.2.

The current technology in IMPATT power output capability as of October 1982 is shown in Figures 8.5 and 8.6 (Ying [7, 8]) for pulsed and CW devices, respectively. These figures are a compilation of published data from the IMPATT producers, Hughes, Plessey, and Raytheon. The regions of f^{-1} and f^{-2} slope are caused by thermal and circuit limitations, respectively [3].

Efficiencies of IMPATT devices approaching 8 percent at 140 GHz have been reported. Figure 8.7 shows the efficiencies of both double- and single-drift IMPATTs at frequencies up to 140 GHz, as measured by Gokgor, *et al.* [9]. This figure shows the roll-off in double-drift efficiency caused by hole mobility limitations at about 100 GHz, an effect that does not occur for single-drift devices until frequencies have exceeded approximately 140 GHz. This figure also shows the greater efficiency of double-drift devices at lower frequencies resulting from the participation of holes in the power generation process at these frequencies. The theoretical upper limit of these single-drift devices is about 300 GHz, and operation has been demonstrated at frequencies up to 255 GHz. Limitations of IMPATTs are based on the ability of the semiconductor to dissipate heat from the small volumes required for high-frequency operation. An efficiency of 0.5% at 230 GHz has been achieved for IMPATT oscillators [5].

Since IMPATTs depend for their operation on avalanche breakdown in a semiconductor junction, they are inherently broadband, noisy devices.

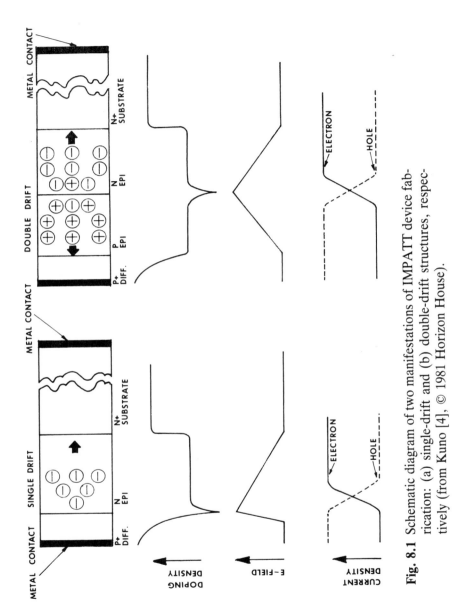

Fig. 8.1 Schematic diagram of two manifestations of IMPATT device fabrication: (a) single-drift and (b) double-drift structures, respectively (from Kuno [4], © 1981 Horizon House).

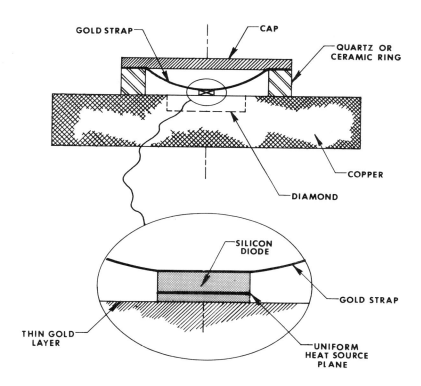

Fig. 8.2 IMPATT diode packaging techniques (from Kuno [4], © 1981 Horizon House).

For many applications, such as receiver local oscillators, this noise output may be prohibitive, but it is possible to phase-lock these diodes so that they are useful in many applications. Currently available IMPATTs may even be used as local oscillators if carefully phase-locked, because their FM and AM noise outputs are comparable to those of phase-locked Gunn oscillators at frequencies sufficiently removed from the carrier to be useful as practical MMW intermediate frequencies. Methods of phase-locking IMPATT oscillators will be discussed in Section 8.6.

Single-port CW IMPATT amplifiers capable of power outputs equal to those of equivalent oscillators have also been devised [5]. An IMPATT device is configured as an amplifier by using a three-port circulator as an

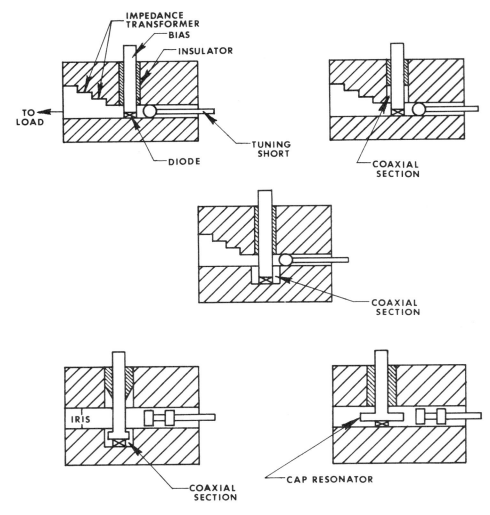

Fig. 8.3 Five methods of mounting IMPATT devices in waveguide circuits showing methods of matching to the higher waveguide impedance (from Kuno [4], © 1981 Horizon House); similar mounting methods are used for Gunn oscillators.

input device as shown in Figure 8.8. Pulsed amplifiers are more difficult to stabilize, and injection-locked oscillators generally replace amplifiers for pulsed applications. Solid-state MMW amplifiers are discussed in Section 8.4.

IMPATT DIODE

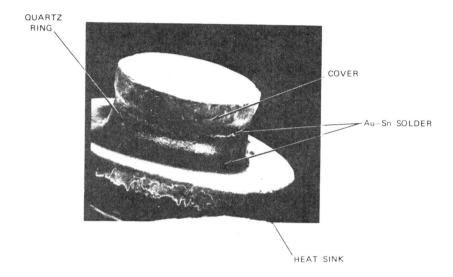

QUARTZ RING

COVER

Au—Sn SOLDER

HEAT SINK

Fig. 8.4 Photograph of MMW IMPATT package (courtesy Hughes Aircraft Company).

The TUNNETT (Pan and Lee [10]) is a variant of the IMPATT, which uses quantum-mechanical tunneling to generate carriers, which, in turn, generate millimeter waves through negative-resistance, transit-time effects, in a manner similar to that in which millimeter waves are generated in the IMPATT. These devices are still in the experimental stage, but show promise of combining the power-bandwidth capability of the IMPATT with the quiet operation of the Gunn devices. The MITATT oscillator uses a mixture of tunneling and impact-ionization effects to generate carriers, and so it may be considered as a cross between the IMPATT and the TUNNETT.

TUNNETTs have been fabricated by Nishizawa, *et al.* [11], which operate at 338 GHz with an efficiency of 0.12%. Elta, *et al.* [12] have reported the fabrication of a MITATT which had a CW output of 3 mW at 150 GHz. Pan and Lee [10] emphasize that TUNNETTs and MITATTs should operate most efficiently in the range 100 to 800 GHz, and their calculations predict an efficiency of 5% at 500 GHz. TUNNETTs are difficult to fabricate because of the required abrupt junction doping profile (Pan and Lee [13], Elta and Haddad [14]), but these devices have the potential for good efficiency at submillimeter wavelengths.

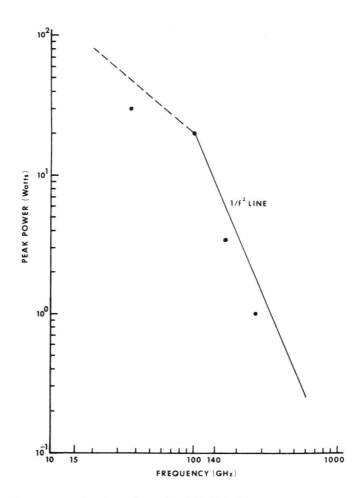

Fig. 8.5 Current technology in pulsed IMPATT power capability based on data from Hughes, Plessey, and Raytheon (from Ying, [8], © 1983 Horizon House).

8.2.2 Gunn Oscillators

The Gunn oscillator is a solid-state device that depends on the bulk properties of the semiconductor for its operation, unlike the IMPATT, which is a junction device [1, 2]. Its operation is based on electric-field-induced differential negative resistance, caused by a transfer of conduction band electrons from a low-energy, high-mobility valley to a higher energy,

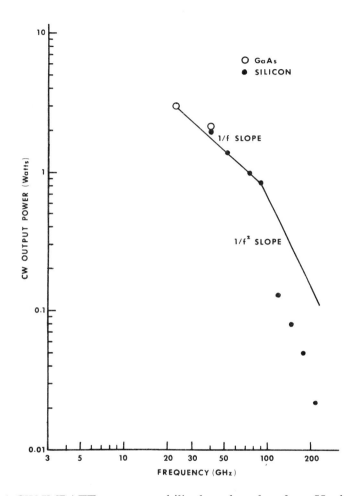

Fig. 8.6 CW IMPATT power capability based on data from Hughes, Plessey, and Raytheon (from Ying [8], © 1983 Horizon House).

low-mobility valley. This reduction in carrier mobility with increasing electric field is a negative-resistance effect. A bulk semiconductor exhibiting differential negative resistance is inherently unstable, since a random fluctuation of carrier density within the semiconductor causes a momentary space charge which grows exponentially in space and time. These negative-resistance-induced space-charge fluctuations move through the bulk device, and thus give rise to microwave oscillations. Figure 8.9 is a simplified schematic diagram of the energy levels in a GaAs Gunn oscillator.

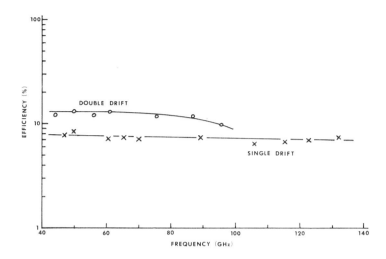

Fig. 8.7 Efficiencies of single-drift and double-drift IMPATTs at frequencies up to 140 GHz (from Gokgor, *et al.* [9], © 1981 Electronics Letters).

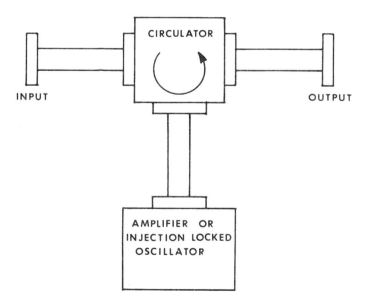

Fig. 8.8 An IMPATT configured as an amplifier or injection-locked oscillator; a similar configuration is used for Gunn amplifiers (from Ying [8], © 1983 Horizon House).

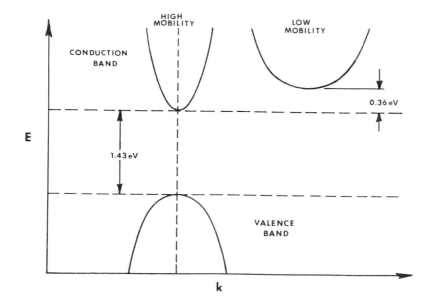

Fig. 8.9 Simplified schematic diagram of the energy levels in a Gunn oscillator; negative resistance results when carriers are excited into
the higher energy, low mobility level.

The upper frequency limit of GaAs Gunn oscillators is about 110
GHz, a limitation imposed by carrier mobility. However, Gunns made
from InP have exhibited power outputs of more than 100 mW (CW),
average, and 250 mW (pulsed), peak, at 100 GHz, and appear to have an
upper frequency limit of about 200 GHz [15–19] because of the higher
mobility of carriers in this InP material.

Since Gunn sources depend for their operation on bulk rather than
junction semiconductor effects and do not operate in the avalanche mode,
these devices are much quieter than IMPATTs. For this reason, Gunn
devices are generally useful as receiver local oscillators. It is also possible
to phase-lock these devices in the same manner as IMPATTs, as will be
discussed in Section 8.6. The useful frequency range of these devices can
also be extended by frequency multiplication, so that the advantage of low
noise resulting from the use of a Gunn oscillator can be extended to
frequencies as high as 230 GHz. Frequency multipliers will be discussed
in Section 8.3.

Figure 8.10 [5] shows the CW power output and efficiency achieved
as a function of frequency for GaAs Gunn oscillators. These devices are
not generally operated in pulsed mode because the pulsed power output

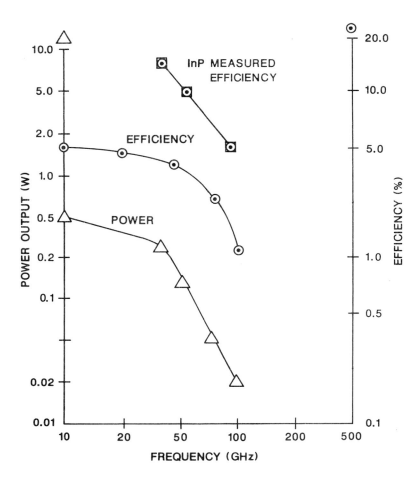

Fig. 8.10 Power output and efficiency of GaAs Gunn oscillators (from
Kramer [5], © 1981 Academic Press). Recent results for InP
Gunns are shown for comparison. (Courtesy of B. Fank, Varian
Associates, 1986.)

is not much greater than that of the CW mode. Recent measurements of
InP Gunn efficiency are plotted on the same graph for comparison. Figure
8.11 shows the projected power output as a function of frequency for CW
InP Gunn devices (solid line). Power outputs achieved by GaAs and InP
sources are shown on the same graph. Table 8.1 (Fank [19]) shows the
CW performance of several InP Gunn oscillators in the 89–100 GHz range,
and Table 8.2 [19] shows InP pulsed performance in the range 89–94 GHz.

Note that the points of Table 8.1 agree well with the performance projections of Figure 8.11. Comparison of Tables 8.1 and 8.2 shows that the outputs of pulsed InP Gunn sources is about twice that of comparable CW sources.

Table 8.1 Performance of CW InP Gunn Oscillators
(from Fank [19], Microwave Systems News, February 1982.
Reprinted with permission.)
(© 1982 EW Communications, Inc. All rights reserved.)

Frequency (GHz)	*Power Output* (mW)	*Efficiency* (%)
85.5	125	3.3
89.6	107	3.5
90.1	100	2.8
93.1	91	3.0
93.2	79	2.8
94.5	71	2.5
94.8	68	2.5
94.9	63	2.4
100.5	44	1.5
87.7	35	4.7

Table 8.2 Performance of Pulsed InP Gunn Oscillators
(from Fank [19], Microwave System News, February 1982.
Reprinted with permission.)
(© 1982 EW Communications, Inc. All rights reserved.)

Frequency (GHz)	*Power Output* (mW)	*Efficiency* (%)
89.8	240	2.7
92.0	215	2.7
90.0	248	2.7
89.8	255	1.4
90.3	236	4.3
89.8	195	3.6
93.7	155	3.3

Indium phosphide Gunn oscillators have shown great promise as amplifiers, and have exhibited full waveguide band performance in the 40–60 GHz and 50–75 GHz bands. Although higher frequency amplifiers are not yet available, there is no physical limitation on their being built, but the lack of availability of wideband circulators at the higher frequencies would limit the bandwidth of such devices if the amplifier configuration of Figure 8.8 is used.

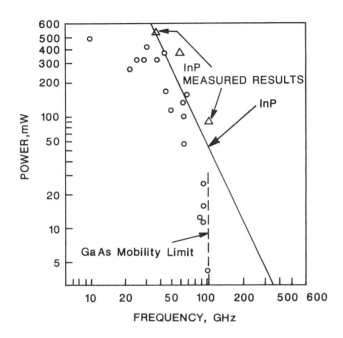

Fig. 8.11 Predicted power output as a function of frequency for CW InP Gunn devices (solid line). The circles indicate performance achieved by CW GaAs Gunn oscillators for comparison. Recent experimental results for InP Gunns are also shown. (Courtesy of B. Fank, Varian Associates, 1986.)

8.2.3 Field-Effect Transistor Sources

Perhaps some of the most exciting current developments in the area of MMW source technology are those for field-effect transistor (FET) sources. It is interesting to observe the evolution of FET technology, from the early devices which were useful only at audio frequencies to the recent development of a FET oscillator capable of output at 110 GHz (Niehenke [20], Tserng and Kim [21]). The availability of useful amplifiers based on FETs has had a significant effect on defense electronics, especially in the area of low-noise, broadband amplifiers useful in electronic warfare and other applications, and is likely to have even more of an effect in the future. Although power FETs are not yet available at MMW frequencies, the availability of one watt amplifiers [22] at 20 GHz is an indication that MMW power devices are in advanced stages of development.

FET devices designed for microwave and MMW applications generally have the MESFET (*Metal Semiconductor* FET) structure shown in Figure 8.12 (Fischer and Morkoc [23]). Source and drain contacts are applied to a highly doped channel, which lies on a substrate, as shown. A metal gate is applied between these electrodes, forming a Schottky barrier, which depletes the carriers in the channel beneath the gate. Application of a potential to this gate modulates the depth of this depletion layer thus modulating the conduction of carriers of the channel region. A signal applied to the gate will then be amplified in much the same way as are signals in a vacuum-tube pentode.

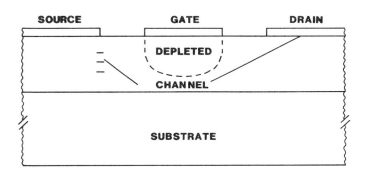

Fig. 8.12 Schematic diagram of MESFET structure (from Fischer and Morkoc [23], © 1985 IEEE).

One of the keys to good high-frequency performance of a FET amplifier or oscillator is in the length of the gate electrode. Millimeter-wave FET devices have gate lengths as short as 0.25 μm, a dimension corresponding to the wavelength of ultraviolet light, which requires electron-beam lithography to fabricate, and electron microscopy to visualize.

The MODFET (*Modulation-Doped* FET) [23] is a fairly recent development in FET technology, and promises to improve the noise figure performance level of 3 dB at 35 GHz achieved by the MESFET. Figure 8.13 is a schematic diagram of the MODFET structure. The undoped GaAs material and the AlGaAs layer form a heterojunction which traps a *two-dimensional electron gas* (2DEG) between the conduction-band discontinuity and the bending of the conduction band due to charge structure. Application of a potential to the gate modulates the density of electrons in the 2DEG, and thus modulates current flow through the device and achieves amplification [23]. The MODFET performs better than the MESFET because the electrons in the 2DEG are confined in pure GaAs, and

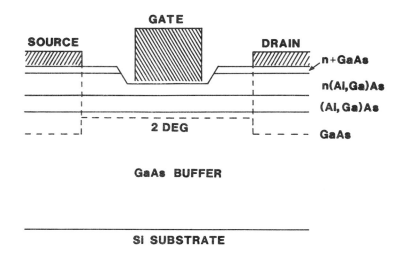

Fig. 8.13 Schematic diagram of MODFET structure; carrier mobility is higher because conduction occurs in the 2DEG, which is in intrinsic GaAs (from Fischer and Morkoc [23], © 1985 IEEE).

thus have higher mobilities; the electrons travel unhampered by donor impurities in this region, whereas in the MESFET the velocity-reducing donors are present. As of this writing (*circa* 1987), MMW results have not yet been obtained with MODFETs, although a noise figure of 1.3 dB has been measured at 18 GHz (Mishra, *et al.* [24]), and switching times of 12.2 ps have been achieved. It is possible that MODFET oscillators and amplifiers operating at frequencies greater than 100 GHz will be available within a few years.

At the lower MMW frequencies, FET oscillators compete favorably with Gunn and IMPATT devices in terms of efficiency, although power outputs achieved are limited to a few tens of milliwatts. A FET oscillator has been developed [21] that has an efficiency of 12 percent at 35 GHz and will oscillate at frequencies up to 110 GHz. This efficiency at 35 GHz compares favorably with the efficiency of the double-drift IMPATT at 40 GHz, shown in Figure 8.6.

MMW FET oscillators are generally varactor tuned, and have the potential of covering very wide bandwidths. However, these devices have phase noise that has a $1/f$ dependence due to up-conversion of baseband noise. Apparently, the broadband units are worse in this regard, while dielectric resonator-tuned oscillators are quietest, although this tuning method has apparently not been used at MMW frequencies.

Perhaps the most important potential use of MMW FET devices is in the area of integrated circuits. Because of the capability of fabricating various devices based on GaAs FET technology, and the compatibility of GaAs and silicon, the best approach to the implementation of MMW ICs appears to be that of using the combination of GaAs and FET technology. Few MMW ICs have been fabricated at any frequency with any material as of this writing, but microwave ICs are becoming increasingly prevalent, and it is virtually certain that MMW devices will follow within a short time. It does not appear likely that MMW techniques in great measure will ever be applied to the practical problems of radar, communication, remote sensing, and other commercial and military uses, unless the low cost, reliability, and small size inherent in integrated circuits can be applied to practical multipurpose devices.

8.3   HARMONIC GENERATORS

Solid-state diode multipliers are also useful MMW sources, whether driven by solid-state or tube oscillators. Multipliers in a crossed-waveguide (cross-guide) configuration have long been used for spectroscopic applications, where power requirements may be only on the order of a few microwatts. This arrangement is inherently broadband, and may typically cover the entire bands of the waveguides used in the device. Useful power outputs at frequencies greater than 600 GHz have been obtained by using this approach.

The basic cross-guide multiplier has been greatly improved by Archer, *et al.* [25–27], who used a suspended-substrate quartz stripline filter to couple the fundamental power to a Schottky-barrier varactor diode situated in the output waveguide. By optimizing the tuning and bias for each operating frequency, Archer *et al.* were able to achieve a conversion loss of about 20% over the output range of 90–124 GHz, as shown in Figure 8.14. If the tuning and bias are held fixed, a minimum conversion loss of 10% is achieved over the narrower output range of 80–120 GHz. In more recent developments, Faber, *et al.* [27] have constructed a frequency doubler with an efficiency of 32% between 97 and 102 GHz, with a peak of 35% at 98 GHz. Fixed-tuned varactor multipliers have recently been used with good results. These multipliers are two-port devices in which fundamental power enters one port and multiplied power exits from the other port. The multipliers are carefully tuned to suppress higher harmonics, and are therefore very efficient, but are also not able to be tuned over a range greater than a few tenths of one percent. Efficiencies of 35% and 15%, respectively, have been obtained in doubling from 70–140 GHz and from 90–180 GHz for input powers of 43 and 100 mW. Because of the nature of these devices, they are highly nonlinear, with output power increasing rapidly with input near the optimum operating point. The

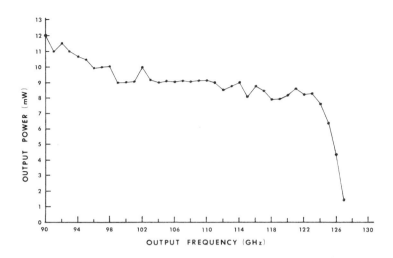

Fig. 8.14 Power output of a cross-guide multiplier over the range 90–124 GHz. This curve represents an efficiency of 20% over this range (from Archer [26], © 1982 IEEE).

70–140 GHz result was obtained using a solid-state oscillator. Figure 8.15 is a photograph of the 90–180 GHz device.

Fig. 8.15 Photograph of fixed-tuned, narrow-band doubler capable of 15% efficiency in doubling from 91.5 to 183 GHz with 100 mW input.

8.4 SOLID-STATE AMPLIFIERS

All of the solid-state MMW oscillators discussed in this chapter may also be configured as amplifiers, and these amplifiers can be combined by using the same techniques as used for oscillators to give significant and useful power outputs of up to several tens of watts, depending on the frequency of operation. These amplifiers are of two different types: two-port devices, which are FETs, and single-port amplifiers, which are Gunns and IMPATTs. These single-port amplifiers require a circulator for coupling in and out, as shown in Figure 8.8, and their performances are sometimes limited by these circulators. FET amplifiers are just beginning to be used in the lower MMW bands as low-noise receiver amplifiers, because their maximum power outputs are about 10 mW, but these devices have the potential of being incorporated into MMW integrated circuits, a step which will greatly enhance their usefulness and lead to rapid improvement of their capabilities; and indeed will revolutionize the entire MMW field of endeavor.

Millimeter-wave Gunn and IMPATT solid-state amplifiers operate in either a stable amplifier mode or an injection-locked oscillator (ILO) mode [28]. The differences between operation in these two modes is in the circuit tuning, and minor circuit changes may enable the same device to operate in either mode. The mode of operation chosen will depend on the application, since both modes have unique advantages and disadvantages. For operation as a stable amplifier, the negative resistance of the device must be less than the circuit impedance, otherwise the device becomes unstable and oscillates. Fortunately, this condition prevails over a broad bandwidth, typically an octave or more, so that solid-state amplifiers may have very broad bandwidths, and their performances are typically limited by circulator performance. For example, InP Gunns having full waveguide bandwidths in the 40–60 GHz range have been fabricated, and similar performance could be achieved at higher frequencies if adequate circulators were available. A 44 GHz InP Gunn amplifier has been recently reported (Fank [29]) to have a small-signal gain of 35 dB, a noise figure of 12 dB, and a bandwidth of 3 GHz.

The stable amplifier mode is perhaps the most common for Gunn or IMPATT amplifiers, because it has the advantage that the output goes to zero when the input is removed, which is, of course, a desirable amplifier property. Further advantages include the absence of the locking time, which is a function of power input, inherent in injection-locked oscillators, the ability to reproduce amplitude, phase, or frequency modulation; capability of reproducing complex waveforms due to multiple signals; and low amplitude modulation to phase modulation (AM-PM) conversion for small signals.

The injection-locked oscillator has the advantage of having a greater output power and efficiency capability. These devices also have higher gain over a wider bandwidth, are better adaptable to power combining, and are generally more reliable because of the higher efficiency and power output. Perhaps the main disadvantage of the ILO, that of output without input, can be overcome by switching on the bias to the device when output is desired, since the turn-on time is almost instantaneous, especially when the bias is reduced to just below threshold instead of being removed completely. Since the ILO is phase-locked to its driver, it will reproduce chirped signals very well, but will not, of course, reproduce AM signals. If the locking input to the ILO is near the minimum required to maintain lock, AM-PM conversion will occur in the device, because the device tends to free-run when its input is reduced. This problem may be avoided by providing adequate drive power. Because of its higher output power and efficiency, the ILO is more widely used than the stable amplifier. IMPATT ILOs having power outputs of up to 1 W at 95 GHz are available [28]; power combiners using these devices have achieved power outputs of more than 10 W. Figure 8.16 [28] shows the bandwidth as a function of locking gain for a W-band IMPATT ILO.

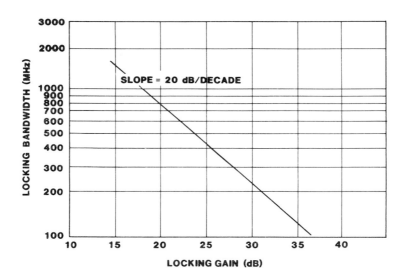

Fig. 8.16 Locking bandwidth as a function of locking gain. This graph shows that an ILO will have a wider capture range for higher input power (courtesy Hughes Aircraft Company [28]).

Although FET amplifiers are limited to low power at MMW frequencies, they are treated here because of their potential as low-noise broadband amplifiers for both transmitter and receiver applications, and also because of their potential as active components in MMW integrated circuits. Low-noise receiver amplifiers with useful gains to 44 GHz are presently available as catalog items, and a recently developed FET has 3.7 dB noise figure and 6.5 dB gain at 40 GHz (Sando [30]), while further developments have led to the same gain at 60 GHz [20]. A three-stage amplifier with 18 dB gain over the 58–60 GHz range has also been built [20]. These high-frequency devices are made possible by the capability of fabricating extremely short gate lengths by using electron-beam lithography. This technology applies also to FET mixers, which promise to be a useful complement to the amplifiers.

8.5 POWER COMBINING METHODS FOR SOLID-STATE SOURCES

It is in the area of power output that solid-state sources are most deficient as compared with tube sources. At this writing, and in light of the significant advances being made in high power tube technology, it appears that this deficiency will never be made up, barring an unprecedented breakthrough in solid-state technology. In the frequency range up to 140 GHz (30–140 GHz for the purpose of this discussion), some of the power output deficiencies of solid-state sources relative to tube-type sources have been overcome by the use of power combining of several sources. When considered on the bases of reliability and circuit simplicity, a power-combined solid-state source might be a better choice for a MMW systems application than a tube source. This section briefly discusses the methods of power combining which have proven useful in MMW applications; namely, resonant cavity, hybrid, spatial, and resonant cap devices.

Resonant-cavity combiners are based on a design by Kurokawa and Magalhaes [31] originally used for X-band IMPATT oscillators. In this device, the individual oscillators are placed in an oversized rectangular waveguide a distance of $\lambda/2$ apart with the end devices placed $\lambda/4$ from the end wall or from an iris, as shown in Figure 8.17. This combiner has high efficiency (≈ 90 percent), is amenable to use at frequencies up to 300 GHz, and has built-in isolation between diodes (Chang and Sun [32]). However, bandwidth is limited to a few percent because of the diode spacing requirement, and the number of diodes that can be used at the higher frequencies is limited because the number of cavity modes increases with waveguide dimensions. Variations of the Kurokawa combiner have been used with IMPATTs to generate 20.5 W pulsed from two diodes at

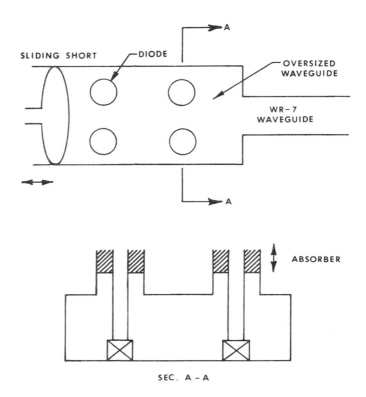

Fig. 8.17 Resonant cavity power combiner used for solid-state oscillators (from Chang and Sun [32], © 1983 IEEE).

92.4 GHz with 82% combining efficiency, 40 W from four diodes at 80% efficiency at the same frequency, and 9.2 W pulsed from four diodes at 140 GHz with 80% efficiency [32]. This design has also been scaled up to 217 GHz to give 1.05 W from two diodes with a combining efficiency of 87% (Chang, *et al.* [33]). A modified Kurokawa combiner has been used by Thoren and Virostko [34] to generate 1.3 W peak over a 10-percent mechanical tuning range at W band (75–110 GHz).

Hybrid-coupled combiners are generally used as amplifiers or injection-locked oscillators, and these circuits offer 5% bandwidth and inherent isolation between sources. Figure 8.18 is a schematic diagram of this type power combiner. Power injected at port 1 is split between ports 2 and 3 where it is amplified. The amplified power combines in-phase at port 4 and out-of-phase at port 1 if the sources are properly matched. If port 1 is terminated, it is possible to combine oscillators using this technique. It is also possible to use sources combined by another method, such as the

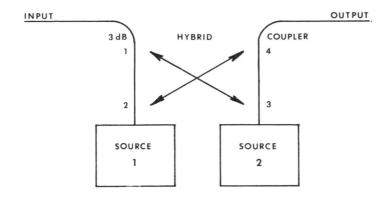

Fig. 8.18 Schematic diagram of a hybrid coupled power combiner configured as an amplifier. If port 1 is terminated, this circuit would
be used to combine oscillators (from Chang and Sun [32],
© 1983 IEEE).

resonant cavity method, as the sources on ports 2 and 3 of the hybrid. At
W band, a hybrid combining scheme was used to combine four two-diode
combiners to give 63 W peak power (Yen and Chang [35]).

The combining efficiency for a two-source hybrid circuit configured
as an amplifier is given by [36]:

$$\eta = \frac{1 + 10^{(D/10)} + (2 \cos\Theta) \, 10^{(D/20)}}{2[1 + 10^{(D/10)}]} \tag{8.1}$$

where D is the power difference between sources in decibels, and Θ is the
phase angle deviation from the proper phase relationship required for
optimum combining. The phase error is more critical than the power difference in attaining good efficiency, since greater than 50% combining
efficiency can be achieved for a wide range of power differences if the
phase deviation can be kept below 30° [32].

Spatial combiners use radiating elements having the proper phase
relationship to combine power from many such elements in space. This
approach was first demonstrated at UHF frequencies, but has not been
extensively used at MMW frequencies except recently by Wandinger and
Nalbandian [37], who combined the outputs of two Gunn oscillators in a
Fabry-Perot resonator at 60 GHz to achieve 54% combining efficiency.
This work is mentioned because of the possible application of this technique
to higher frequency sources. In some very recent work, Mink [38] describes
a quasioptical spatial combiner which combines multiple sources in a planar

array by means of a semiconfocal Fabry-Perot resonator. Such quasioptical techniques have proven useful for solving MMW circuit problems on many occasions. In this regard, it should be noted here that phased-array antenna systems are actually spatial power combiners, and were perhaps the first such devices devised at either MMW or microwave frequencies.

Cap resonators sometimes used for mounting Gunn oscillators may also be used to combine these devices. Such a combiner is shown in Figure 8.19 [32], in which two resonant cap structures are mounted in a common waveguide with a common movable short. It is also possible to place both oscillators under the same cap. This technique has been used to combine four 90 GHz InP Gunn oscillators to give 260 mW power output and 93% combining efficiency. Table 8.3 [19] shows results obtained at 90 GHz for two-diode and four-diode resonant cap combiners in several different cases.

Table 8.3 Performance of CW 2-Device and 4-Device Combining Circuits Using InP Gunn Oscillators (from Fank, [19], © 1982 Microwave Systems News).

Number of Devices	Frequency (GHz)	Power Output (mW)	Efficiency (%)	Combining Efficiency (%)
2	85.6	170	2.9	93
2	90.3	150	2.7	82
2	91.8	97	1.6	106
4	90.6	260	1.6	94
4	90.8	230	1.4	107

Chip-level combining, in which the individual sources are combined on a common substrate or heat sink, is a potentially useful method of power combining. This approach has been used up to 40 GHz and shows some promise of being useful at higher frequencies if the thermal and parasitic problems can be solved. Rucker, *et al.* have achieved a chip-level combining efficiency of 82% in combining two IMPATTs at 40 GHz [39, 40]. Figure 8.20 is a diagram of the technique used to achieve these results, showing the diode chips mounted on diamond heat sinks and the stabilization capacitors. This approach to power combining has the advantage of being more compact and not requiring the detailed machining and assembly required by the other combining methods. For these reasons, this method will probably receive more attention in the future.

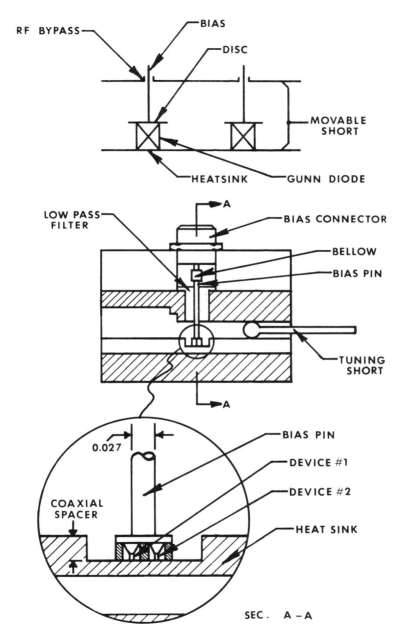

Fig. 8.19 A cap resonator arrangement used to combine solid-state oscillators (from Chang and Sun [32], © 1983 IEEE).

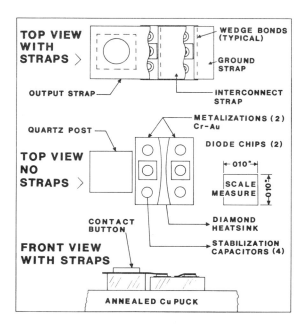

Fig. 8.20 Method used to combine MMW IMPATT oscillators (from Rucker, Amoss, and Hill [40], © 1981 IEEE).

8.6 SYSTEM ASPECTS

8.6.1 Phase and Frequency Control of MMW Solid-State Sources

General Concepts

The increasing use of MMW frequency bands for communication, radar, and measurement has created the need for more sophisticated and higher frequency methods of precisely controlling the frequency and phase of sources of radiation in these bands. Part of this need arises because the MMW bands are becoming more crowded, but the major requirement for better frequency control results from the needs imposed by more sophisticated signal processing schemes for the detection of targets. The availability of higher frequency sources and more sensitive receivers has pushed functional MMW technology to the limit of its usefulness as determined by the atmospheric transmission bands, and the need for precise phase and frequency control of these sources has kept pace with this development. Coherent radar systems have for years relied on phase-locked or injection-locked transmitters as well as phase-locked receiver local oscillators to give the degree of frequency control required for determination of target velocities by the Doppler effect. Furthermore, the very sophisticated methods of data processing, which are necessary for target detection in the presence of clutter, rely heavily on precise methods of frequency and phase control of the transmitter and local oscillator.

Several methods of frequency and phase control are used for solid-state MMW sources. The frequency discriminator may be used for frequency control to approximately a few parts in 10^6. The phase lock will control frequency to the accuracy of the phase-locked reference, which may be as good as one part in 10^{11} if a suitable reference is used, and phase locking will also control the relative phase of the source and reference to an accuracy of a few degrees, after a suitable warm-up period. The most powerful phase-locking techniques combine the wide capture range of the discriminator with the precise phase control of the phase lock to result in a frequency control system which has excellent frequency stability and good immunity to perturbations that cause loss of phase lock. Injection-locking of solid-state sources is also a useful technique, and this approach is usually employed in conjunction with phase or frequency locking. All of these methods of phase and frequency control are discussed further in the following paragraphs.

Phase and Frequency Control Fundamentals

A block diagram of a basic phase-locking system is shown in Figure 8.21 (Wetenkamp and Wong [41], Gardner [42]). The output of the *voltage-controlled oscillator* (VCO) mixes with that of the reference oscillator in a phase detector mixer to generate a phase error signal, which is, in turn, fed back through the loop filter to control the phase of the VCO. The VCO is a phase integrator, and its transfer function is K_0/s, where K_0 is the VCO constant in rad/s per V, and s is the Laplace transform differential operator. K_d is the phase detector constant in V/rad, and $F(s)$ is the loop filter transfer function, chosen to give the best combination of phase-locked-loop frequency stability and phase noise performance. Note that the feedback fraction in Figure 8.21 is unity, because the phase output of the mixer is always the same as that fed back to the VCO, although the frequencies may differ due to down-conversion of the VCO output.

Frequency control of a VCO is effected by a discriminator, which is a device that has an output voltage proportional to the difference between the VCO frequency and the reference frequency. Since discriminator error voltage is proportional to frequency rather than phase, frequency-control

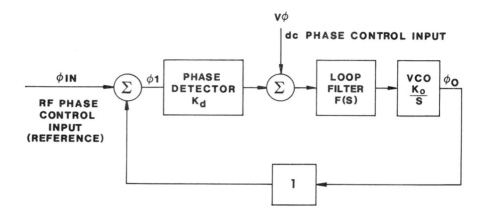

Fig. 8.21 Block diagram of a basic phase-locked voltage-controlled oscillator.

loops are not as "tight" as phase-control loops, and the frequency errors are therefore greater.

We mentioned that the combination of a frequency-control and a phase-control loop provides a very powerful phase-locking method. Such a circuit has been devised by Henry [43], who designed a circuit to lock klystrons for use in radio astronomical applications, although the same techniques apply to MMW sources in general. Figure 8.22 is a block diagram of the Henry phase lock showing the phase-control and frequency-control loops. Assuming that the source is not locked, it will generally be oscillating at a frequency such that the difference between the IF and the reference oscillator is outside the capture range of the phase-locked loop. In this case, the discriminator captures the source and pulls it within range of the phase lock, where the discriminator is disabled. In this way the phase-control and frequency-control loops do not interfere with each other, but act in a complementary way to combine the wide capture range of the discriminator with the precise frequency and phase control of the phase-locked loop.

With the availability of high-speed *emitter-coupled logic* (ECL) counters and high-frequency phase and frequency detectors, digital phase-locked loops are being used in many applications. Pickett [44] has modified Henry's circuit to operate with a digital phase-frequency detector, and has found that the lock is more reliable. Other workers (Guillory and McMillan [45]) have used digital circuitry and obtained similar results in the area of reliability, but they have also found that the phase noise obtained with the

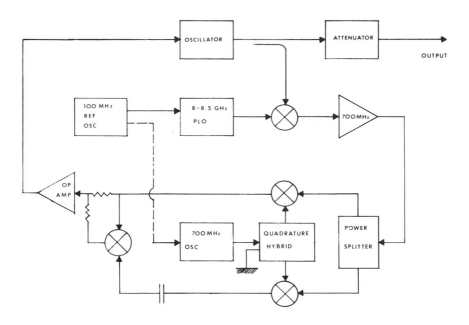

Fig. 8.22 Block diagram of the Henry phase and frequency control circuit. The devices represented by crossed circles are mixers (from Henry [43], © 1976 Review of Scientific Instruments).

digital circuitry is not as good as the linear circuit results. Figure 8.23 is a block diagram of a typical digital phase lock, which we can see is similar to the linear loop, with the exception of the IF and detection circuits. The IF is counted down by the ECL prescaler to a frequency at which the phase-frequency detector will operate, usually less than 100 MHz. The counter output is compared to a phase reference in the detector, the output of which is a pair of out-of-phase rectangular waveforms with a period that is proportional to the phase error. These signals are fed into a differential integrator, which usually serves as the loop filter in addition to driving the frequency-control electrode on the source.

Another approach to extending the capture range of the phase-locked loop involves activating a sweep-search mode if the source loses lock. Since most phase-locked loops have a much narrower capture range than sources have electronic tuning range, a source will likely remain unlocked if it has been unlocked because of a perturbation or upon initially applying power. If this problem occurs, the sweep-search mode is initiated to sweep the source through the proper frequency repetitively until lock is regained.

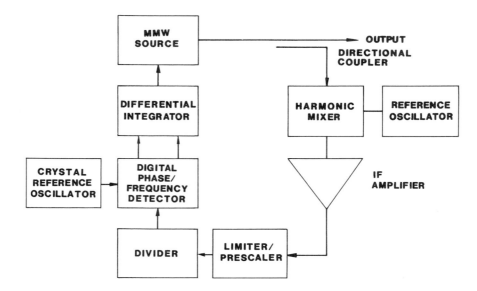

Fig. 8.23 Block diagram of a digital phase-locked loop (from Guillory and McMillan [45]).

The sweep is usually a 1 Hz ramp that is applied to the frequency control input of the source when diagnostic circuitry senses that lock is broken. Figure 8.24 is a block diagram of a circuit, used extensively to lock solid-state oscillators (Crandall and Bernues [46]), which incorporates a sweep-search mode for wide frequency capture range. The sweep-search circuitry is incorporated in the loop electronics and driver block.

In locking solid-state sources, the phase-correction signal is applied to either a varactor diode in the cavity of the source, or the bias input of the device. Varactor tuning is usually employed at frequencies below about 55 GHz, where waveguides are larger and resulting losses are acceptable. Above this frequency, bias tuning is generally used. Bias-tuned oscillators generally have narrower electronic tuning ranges than varactor-tuned sources.

Injection-locking is a method of phase control in which power from the reference oscillator is injected directly into the output of the oscillator to be controlled by means of a circulator (Cadwallader, *et al.* [47]). It is generally used for pulsed sources, because the methods of phase control discussed earlier are more easily used for CW sources. The rationale for injection-locking is that radiation from all oscillators builds up initially from broadband noise in the device. The injected signal provides a coherent

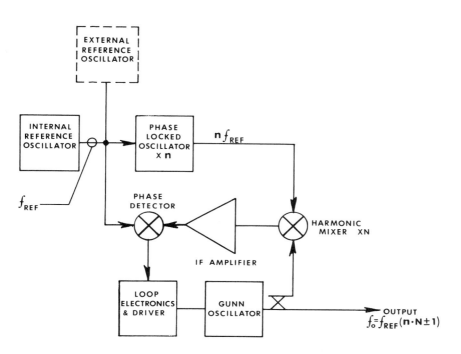

Fig. 8.24 Schematic diagram of a circuit used extensively to lock solid-state oscillators (from [49], Courtesy of Hughes Aircraft Company).

basis for this buildup, causing the source to oscillate in phase with this reference signal. The reference signal comes from a phase-locked oscillator generally controlled by the methods described above.

The frequency of a phase-locked source is characterized by both long-term and short-term stability. Long-term stability is almost totally determined by the quality of the reference source, and is usually specified as parts per million of frequency deviation per unit time. Times for this specification are usually long—one hour and one day are common specification times. Short-term stability is determined by the quality of the reference source, the inherent FM noise of the phase-locked source, and the response of the phase-locked loop. Short-term stability is usually characterized as phase noise, which is a measure of the FM noises in a specified bandwidth at a given displacement from the carrier. The displacement is normally measured in decibels below the carrier peak, and, for convenience, the specified measurement bandwidth is sometimes chosen to be 1 Hz.

Another property of a phase-locked source is that it can never have a better phase-noise spectrum than its reference oscillator; in particular, it is not difficult to show [41] that the phase noise of the MMW source can never be better than 20 log$N$ plus the phase noise of the reference, where N is the frequency multiplication ratio. Figure 8.25 shows the relationships between reference phase noise and MMW source phase noise. At frequencies at which the phase-locked loop has control, the 20 log$N$ relationship is maintained, as shown. When the loop begins to lose control, the phase margin of the loop decreases because of phase shift in the loop filter and transport lag, and the phase noise of the locked MMW source may actually be worse than that of the unlocked source, as shown in the figure. However, it is possible to minimize the phase noise peak by careful design. When the loop finally loses control, the phase noise reverts to that of the unlocked source, again as shown in the figure.

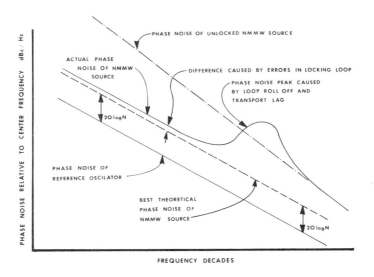

Fig. 8.25 Relationship between phase noise outputs of reference oscillator and MMW source.

Phase-Locking Results

Figure 8.26 shows the spectrum of a phase-locked Gunn oscillator, which is locked by using the circuit of Figure 8.24. The corresponding phase noise spectrum of this source is shown in Figure 8.27 [46], which

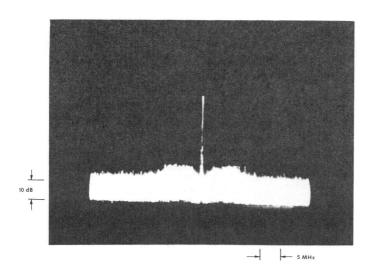

Fig. 8.26 Spectrum of phase-locked 94 GHz Gunn oscillator. Scales for 5 MHz (horizontal) and 10 dB (vertical) are indicated on the figure.

illustrates the relationships discussed in Section 8.2. It has also been possible to lock IMPATT oscillators at frequencies up to about 220 GHz by using this circuit [47]. By virtue of these techniques, it has been possible to build broadband, sophisticated, IMPATT-based frequency synthesizers with 10 GHz bandwidths at frequencies up to W band (70–110 GHz) (Fortunato and Ishikawa [48]).

Injection-locked IMPATT oscillators have been used in an all-solid-state, pulsed or CW, coherent, W-band instrumentation radar [49]. Figure 8.28 is a block diagram of this radar, which may have a power output of up to about 10 W in pulsed mode, and Figure 8.29 shows the measured spectrum of the pulsed transmitter. Note that this spectrum departs little from the theoretically ideal $\sin^2(x)/x^2$ shape for a rectangular pulse.

8.6.2 Modulators for MMW Solid-State Sources

There are two general types of modulators of potential use in systems based on MMW solid-state sources: (1) high-power pulse modulators used for radar transmitters; (2) amplitude, phase, or frequency modulators used for MMW communication, or for such purposes as FMCW radar systems.

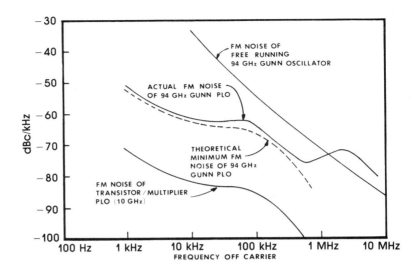

Fig. 8.27 Measured phase noise of 94 GHz Gunn oscillator showing relationships between the reference source, the phase-locked oscillator, and the free-running oscillator (from [49], Courtesy of Hughes Aircraft Company).

There has recently been a great deal of interest in communications applications. In particular, there is interest in secure communications over short ranges at frequencies for which the band of oxygen absorption lines near 60 GHz provides a high level of absorption of the transmitted power. Although the use of the MMW bands for communication is appealing from the point of view of high achievable information-carrying capacity based on potentially wide bandwidth, the use of frequencies much higher than 60 GHz will probably be limited by atmospheric turbulence, which also renders the visible and near-infrared bands essentially useless for communication over atmospheric paths. The amplitude modulation effects of turbulence increase as frequency raised to the 7/6 power, so that the higher frequencies are especially vulnerable to turbulence problems. For this reason, and for the reason that the subject of this book is MMW radar, only modulators used for radar transmitters will be discussed in this section.

Modulators for FMCW radar systems, or for other applications not requiring high peak power, are implemented by using linear solid-state circuitry. Gunn oscillators are generally driven by voltage source power supplies, and IMPATTs are powered by current sources. Because of the linear nature of these supplies, modulation is usually superimposed on the current or voltage operating point of the device at a low level. Since both

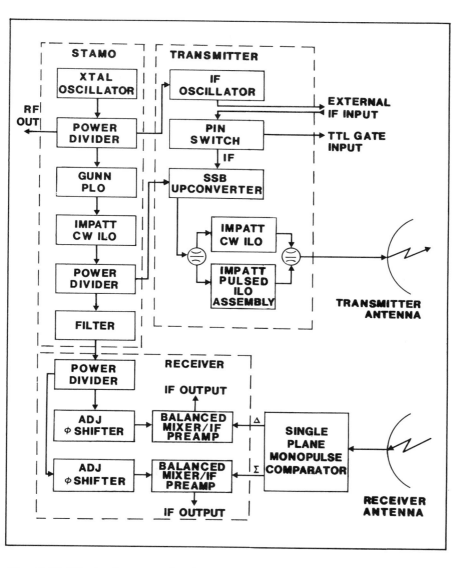

Fig. 8.28 Block diagram of coherent CW-pulsed MMW instrumentation radar (from [49], Courtesy of Hughes Aircraft Company).

Gunn and IMPATT oscillators are sensitive to bias voltage changes, frequency modulation can be accomplished by simply modulating the bias. There are some problems associated with this technique, however, because

Fig. 8.29 Spectrum of pulsed 94 GHz, 10 W, injection-locked solid-state transmitter. This photograph was made with a pulsewidth of 60 ns and a PRF of 100 kHz. The spectrum analyzer was set for 300 kHz bandwidth, 20 MHz/div horizontal scale, and 10 dB/div vertical scale (from [49], Courtesy of Hughes Aircraft Company).

bias changes generally cause both frequency and amplitude variations, so that a modulator must be well designed, or be able to operate over a narrow range of frequencies, so as to avoid excessive power changes associated with frequency modulation.

External ferrite modulators are employed in AM applications, but their use is usually limited to level control applications. Useful ferrite modulators are available at frequencies up to about 100 GHz, although ferrite devices with more loss (≈ 2 dB) can be obtained at frequencies up to about 140 GHz. Above this frequency, these devices are likely to be lossy and expensive. A possible application for AM radar systems is in short-range applications, such as proximity detection, in which the phase of the received AM signal is compared with that of the modulator to measure distances less than a certain ambiguity range, but most applications of this type use optical devices instead of MMW radars.

It is also possible to implement frequency modulation of solid-state sources by using varactor modulators, although this approach is usually confined to Gunn oscillators. Varactor-tuned oscillators are available at frequencies up to 100 GHz with 3 GHz tuning bandwidth. Varactor tuning might be used, for example, in an FM CW system for which a low-level Gunn oscillator stage is varactor-tuned and used to drive a higher level output device such as an IMPATT power amplifier. In practice, varactor tuning is used in such broadband applications as synthesizers, in which the varactor may be used for both tuning and phase-locking. Such a configuration might be used in a frequency-agile radar front end, for example.

Modulators for pulsed, high-power, solid-state sources are implemented by simply providing a means for applying or removing the bias voltage for the device to be modulated. To minimize power dissipation and maximize duty factor, the voltage is generally removed completely between pulses. It is also possible to implement frequency modulation of solid-state sources by varying the voltage during the pulse, as discussed elsewhere in this chapter. Frequency modulation by varying the intrapulse voltage also provides the means for frequency chirping of solid-state sources, or compensating for temperature-induced frequency chirp.

General requirements for modulators of pulsed, high-power solid-state sources are that these modulators must furnish high currents with fast rise and fall times and do so at a stable output voltage. This voltage may be deliberately varied during the pulse to achieve some degree of frequency chirp, or to compensate for chirp introduced by heating of the source during the pulse, but the voltage must be stable and predictable. Figure 8.30 [50] shows an example of an approach that is widely used for modulation of pulsed Gunn and IMPATT oscillators. The lumped-element transmission line is charged between pulses by a suitable efficient charging circuit. At the proper time, the transmission line is discharged into the source by using a silicon controlled rectifier. Circuit requirements for most high-power IMPATTs dictate that this modulator be capable of furnishing currents up to 20 A. Deliberate frequency chirp or compensation for chirp due to temperature must be accomplished by properly varying the transmission line parameters to achieve the desired modulation.

Recent investigations into methods for improving MMW solid-state source modulators have emphasized the use of power FETs as circuit elements. Several of these devices are operated in parallel to achieve the required current levels. The advantages of this approach include better current or voltage control through linear operation and easier implementation of safety circuits for device protection.

It is extremely important to provide proper safeguards against overvoltage or double pulsing in modulators for high-power solid-state sources (Ying [50]). To achieve good efficiency, these sources operate with very narrow margins of stability, and exceeding these limits, even by a small amount, will cause the device to be destroyed. The demonstrated

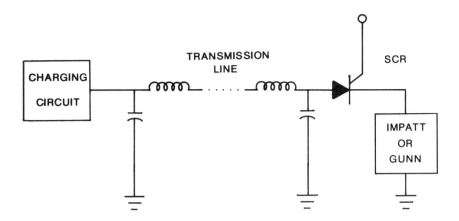

Fig. 8.30 Schematic diagram of one embodiment of a modulator used for high-power, pulsed, solid-state sources. It is important that proper safeguards be employed in circuits of this type to prevent damage to the source by overvoltage or double pulsing.

reliability of solid-state sources indicates that modulator circuit designers have done a good job of engineering design for these protective circuits.

Consider a pulsed, high-power, solid-state amplifier or injection-locked oscillator driven by a stable phase-locked oscillator. It is necessary to ensure that the phase noise of the oscillator is not degraded by the output power stage; such a combination might be used, for instance, in a pulsed Doppler radar. If the modulator voltage or current is maintained stable within a few percent during the pulse, the degradation of phase noise will be negligible during the main body of the pulse, but there will almost always be some degradation of phase noise performance at the beginning and end of the pulse because of the finite rise and fall times of the modulator voltage or current. It is at the beginning and end of the transmitter output signal that the most severe degradation of phase noise occurs. This degradation may be minimized by minimizing the transition times at the beginning and end of the pulse, so that the time of phase noise degradation is minimized. Alternatively, the beginning and end of the pulse might be switched out by a suitable high-speed *pin* switch. A method of reducing the transition time, which works well for injection-locked oscillators, is to bias the output stage to a level just below threshold, then superimpose a small voltage or current on this pedestal to drive the source above threshold and into oscillation. In this way, a very short transition time can be achieved because the superimposed signal may have a very small amplitude and resulting rapid transition time. This method dissipates more power, but dissipation can be reduced by applying the pedestal voltage or current a short time before output is desired, and removing the pedestal shortly after the end of the pulse. Since the modulator pedestal is not continuously applied, power dissipated in the device may be considerably reduced.

The design of the modulator is a very important part of the design of any radar system. In tube transmitters, the problem is one of reliably delivering a stable high voltage to the transmitter, while the delivery of a stable high current is the usual requirement in solid-state transmitters. Not surprisingly, tube sources use tube modulators and solid-state sources use solid-state modulators. We hear a great deal about solid-state reliability, but it is probably safe to say that, at this stage of development, MMW vacuum-tube transmitters are more reliable than their solid-state counterparts because of the great wealth of experience that transmitter designers have accrued over the last few years. Because of the inherently higher reliability of solid-state sources and modulators, this advantage is likely to disappear as we develop better solid-state devices and learn how to use them to design better transmitter and modulator circuits.

8.7 MMW INTEGRATED CIRCUITS

We mentioned earlier in this chapter that the development of MMW integrated circuits (ICs) is considered by many researchers in the MMW and microwave area to be the key to large-scale deployment of MMW systems in both commercial and military applications. This assertion is based on the fact that the fabrication of waveguide circuits becomes increasingly difficult and expensive at higher frequencies because of the close tolerances required for good, low-loss performance. MMW ICs would also be expensive at first for the same reasons, but given the history of semiconductor device development, we could expect prices to drop dramatically as methods of improving yields were developed. Another disadvantage of MMW waveguide circuits is their size. Although MMW components are approximately smaller, by the ratio of wavelengths, than microwave devices, MMW devices are still extremely bulky according to the standards by which they must be judged, namely, that of considering them for applications requiring small size, such as in missile seekers.

At the present time it appears that most reported MMW ICs have been implemented to perform only one function. For example, monolithic mixers have been built, which work well, but they must be pumped by external, discrete-circuit, waveguide local oscillators. In other instances, monolithic local oscillators or transmitter master oscillators have been fabricated, which might pump or excite waveguide components, or possibly other integrated circuits. Few instances have been reported of MMW ICs combining more than one function. Obviously, this capability must exist if MMW ICs are ever to be a viable concept, but it is also true that single-function circuits must come first, and recent progress has been gratifying in this area.

An exception to the single-function IC mentioned above is a monolithic receiver at 35 GHz (Seashore [51]). This receiver has a double-sideband (DSB) noise figure of 5.2 dB, and was designed for missile seeker applications. In other recent developments in the multifunction MMW IC area, Knust-Graichen and Bui [52] built a suspended stripline integrated circuit receiver comprising Gunn local oscillator, balanced mixer, phase-locked harmonic mixer, diplexer, coupler, and various filters. A separate IC included the IF amplifier and its internal voltage regulator. This receiver was designed to operate at 60 GHz on the strong atmospheric oxygen molecular absorption for spaceborne applications. The receiver had a single-sideband (SSB) noise figure of 6.5 dB or less over the 59–64 GHz frequency range, a result which compares favorably with the best obtained with discrete-circuit receivers. Figure 8.31 is a block diagram of this receiver. Another example of a multifunctional MMW IC is a 30 GHz monolithic mixer with an integrated dipole antenna, reported by Nightingale, *et al.* [53]. This device gave a conversion loss of 6 dB at 30 GHz with a 1 GHz IF.

W-band IC mixers have also been designed by Tahim, *et al.* [54], and by Yuan and Asher [55]. Both of these devices used stripline circuitry, although Tahim's was fabricated on dielectric stripline by using GaAs beam-lead diodes, and the latter was a monolithic circuit fabricated on GaAs with *in-situ* diodes. Tahim, *et al.* also used the same techniques at Q band, and they fabricated some of their circuits with finline techniques. Tahim's stripline mixer achieved a conversion loss of less than 7.2 dB over a 20 GHz instantaneous bandwidth at W-band, while the fin-line mixer operated over 28 GHz instantaneous bandwidth with 7.6 dB conversion loss in W band. Yuan and Asher's GaAs mixer achieved a conversion loss of less than 8 dB over the range 73.6 to 83.6 GHz with an IF from 8 to 18 GHz. Figure 8.32 is a schematic diagram of this mixer. GaAs beam-lead diodes have also been used by Forsythe, *et al.* in a stripline subharmonically pumped down-converter. Point-contact versions of these mixers have yielded a DSB noise figure of 5.5 dB over a 1 GHz bandwidth at 183 GHz, while pumped by a 91 GHz local oscillator [56]. An $f/4$ version of this mixer achieved a DSB noise figure of 8.5 dB at 225 GHz with a local oscillator operating near 55 GHz (McMillan, *et al.* [57]), and a 340 GHz version of this device is under construction as of this writing.

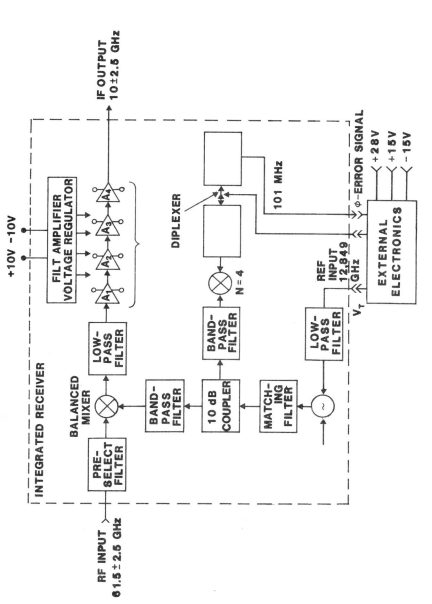

Fig. 8.31 Block diagram of an integrated circuit 60 GHz receiver (from Knust-Graichen and Bui [52], © 1985 Horizon House).

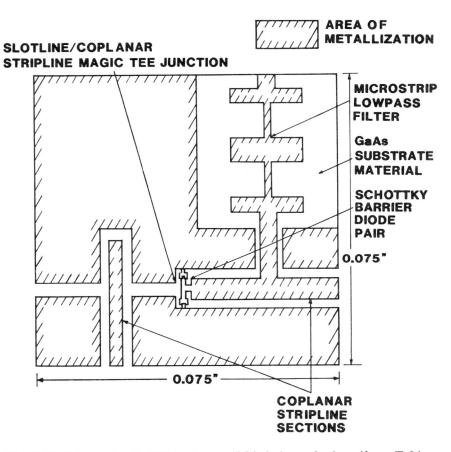

Fig. 8.32 Schematic of a W-band monolithic balanced mixer (from Tahim, Pham, and Chang [54], © 1985 Horizon House).

Examples of IC implementation of transmitter sources are rarer than those of receiver implementations. Bayraktaroglu and Shih [58] have constructed monolithic IMPATT oscillators capable of up to 1.25 W of CW power output at 32.5 GHz with 27% efficiency. Figure 8.33 is a schematic of an implementation of one of these devices with distributed-element impedance-matching circuits.

FET devices are expected to play a large role in the development of MMW ICs in the future as a result of their demonstrated low-noise capability in discrete circuits, and because oscillators, mixers, and IF amplifiers can be implemented with FET devices. There is also some precedent

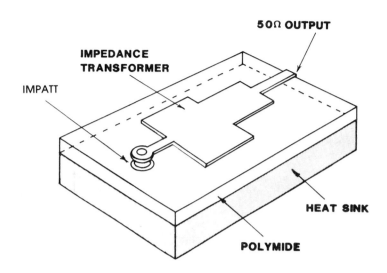

Fig. 8.33 Diagram of a 1.25 W, 32.5 GHz monolithic IMPATT oscillator (from Bayraktaroglu and Shih [58], © 1985 IEEE).

for making this assertion as is evident by the increasing number of FET microwave integrated circuits becoming available.

Note that in all of the examples of IC technology referenced above, the performance of the ICs was at least as good as that of the best discrete circuits, a circumstance that bodes well for future applications of these potentially very useful devices. It appears that the next few years will bring interesting and exciting improvements in MMW ICs, a development that must indeed come to pass if MMW technology is to provide viable solutions to many of the problems which it has been projected to solve.

8.8 SUMMARY

The history of employing solid-state techniques for MMW applications closely parallels the use of these techniques for electronics applications in general. Low-frequency Gunn and IMPATT sources were the first devices available, in analogy with the first low-frequency germanium point-contact and alloy-junction transistors. The development of solid-state devices now transcends all electronic fields of endeavor, and it is probably safe to say that similar progress will be made in the area of MMW solid-state devices, as great progress has indeed been made in this area already.

We do not intend to claim here that solid-state sources will totally replace tube-type sources in all applications. Indeed, there will always be requirements which solid-state devices simply cannot meet. The power output capabilities of tubes are growing faster in terms of percentage than those of solid-state sources, although significantly less money is being spent on tube development. Barring fundamental and unforeseen developments (which have certainly happened in the past), it is unlikely that the power output capabilities of solid-state sources will ever approach the very high levels achieved by gyrotrons and free-electron lasers (FELs), for example (McMillan [59]).

The advantages of solid-state devices for such applications as receiver local oscillators and low-level transmitter stages include small size and the significant advantage of requiring only a few volts for operation. As we mentioned earlier, all-solid-state radars have been built at frequencies up to W band for ground-based applications, which do not require much power because the maximum line of sight may be only 1 or 2 km. For applications requiring higher power, spatial power-combining techniques, such as those used in implementing phased array antenna systems, can be used to achieve high powers, although there appear to be no examples of this technique being used at MMW frequencies to date.

We also mentioned that the greatest promise of viable applications of solid-state MMW technology lies in the area of ICs, and developments in this area will lead to deployment of MMW systems for many applications, especially those requiring short ranges and higher resolution than obtainable with microwave techniques. Combinations of IC techniques with the phased array system mentioned above might lead, for example, to an all solid-state array element that would be a miniature solid-state radar front end on a single chip, which could comprise master oscillator, local oscillator, mixer, and transmit-receive (T-R) switch. Useful power outputs will be obtained by combining large numbers of these devices, which will be possible only when they can be made cheaply by using IC techniques. It is likely that IC techniques will play a large part in the development of MMW technology during the next few years, just as these techniques have revolutionized the electronics industry in general since the 1960s.

REFERENCES

1. R.J. Chaffin, *Microwave Semiconductor Devices: Fundamentals and Radiation Effects,* Wiley-Interscience, New York, 1973, Ch. 8, 9.
2. S.M. Sze, *Physics of Semiconductor Devices*, Wiley-Interscience, New York, 1969, Ch. 4, 5.

3.   H.J. Kuno, "IMPATT Devices for Generation of Millimeter Waves," Chapter 2 in *Infrared and Millimeter Waves*, Vol. 1, K.J. Button, ed., Academic Press, New York, 1979.

4.   H.J. Kuno, "Solid State Millimeter-Wave Power Sources and Combiners," *Microwave Journal*, vol. 24, pp. 21-34, June 1981.

5.   N.B. Kramer, "Sources of Millimeter-Wave Radiation: Traveling-Wave Tube and Solid-State Sources," Chapter 4, *Infrared and Millimeter Waves*, Vol. 4, K.J. Button and J.C. Wiltse, eds., New York: Academic Press, 1981.

6.   N.B. Kramer, Hughes Aircraft Company, private communication, 1982.

7.   R.S. Ying, "Millimeter-Wave Solid State Transmitter Sources," in *Proceedings of the Military Microwaves '82 Conf.* (London, October 1982), pp. 468-471.

8.   R.S. Ying, "Recent Advances in Millimeter-Wave Solid-State Transmitters," *Microwave Journal*, Vol. 26, No. 6, June 1983, p. 69.

9.   H.S. Gokgor, I. Davies, H.M. Howard, and D.M. Brookbanks, "High-Efficiency Millimeter-Wave Silicon IMPATT Oscillators," *Electron. Lett.*, Vol. 17, October 1981, pp. 744-745.

10. D.S. Pan and N. Lee, "GaAs Abrupt Junction MITTATT and TUN-NETT," paper presented at the *6th International Conference on Infrared and Millimeter Waves*, Miami Beach, FL, 7-12 December, 1981, paper M-5-3.

11. J. Nishizawa, K. Motoya, and Y. Okuno, "Submillimeter Wave Oscillation From GaAs TUNNETT Diode," paper presented at the *9th European Microwave Conference*, Brighton, UK, 17-20 September, 1979.

12. M.E. Elta, H.R. Fetterman, W.V. Macropoulous, and J.J. Lambert, "150 GHz GaAs MITATT Source," *IEEE Electron Device Letters*, Vol. EDL-1, June 1980, pp. 115-116.

13. D.S. Pan and N. Lee, "Investigation of Narrow Band-Gap Semiconductors for TUNNETTS," paper presented at the *6th International Conference on Infrared and Millimeter Waves*, Miami Beach, FL, 7-12 December, 1981, paper M-5-8.

14. M.E. Elta and G.I. Haddad, "High-Frequency Limitations of IMPATT, MITATT, and TUNNETT Mode Devices," *IEEE Transactions on Microwave Theory and Techniques*, Vol. MTT-27, No. 5, May 1979, pp. 442-449.

15. J.J. Gallagher, "InP: A Promising Material for EHF Semiconductors," *Microwaves*, Vol. 21, February 1982, pp. 77-84.

16. F.B. Fank and J.D. Crowley, "Gunn Effect Devices Move Up in Frequency and Become More Versatile," *Microwave Journal*, Vol.

25, September 1982, pp. 143-147.

17. J.D. Crowley, J.J. Sowers, B.A. Janis, and F.B. Fank, "High Efficiency 90 GHz InP Gunn Oscillators," *Electronics Letters*, Vol. 16, 1980, pp. 705-706.

18. I.G. Eddison, I. Davies, and D.M. Brookbanks, "Indium Phosphide Proves Itself For Millimeter Applications," *Microwave Systems News*, Vol. 12, February 1982, pp. 91-96.

19. F.B. Fank, "InP Emerges as Near-Ideal Material for Prototype Millimeter-Wave Devices," *Microwave Systems News*, Vol. 12, February 1982, pp. 59-72.

20. E.C. Niehenke, "GaAs: Key to Defense Electronics," *Microwave Journal*, Vol. 28, No. 9, September 1985, pp. 24-44.

21. H.Q. Tserng and B. Kim, "110 GHz GaAs FET Oscillator," *Electronics Letters*, Vol. 21, No. 5, February 1985, pp. 178-179.

22. Raytheon Company, advertisement, *Microwaves and RF*, Vol. 24, No. 6, June 1985.

23. R. Fischer and H. Morkoc, "New High Speed (AlGaAs) Modulation Doped Field-Effect Transmitter," *IEEE Circuits and Devices Magazine*, Vol. 1, No. 4, July 1985, pp. 35-38.

24. U.K. Mishra, S.C. Palmateer, P.C. Chao, P.M. Smith, and J.C. Hwang, "Microwave Performance of 0.25 μm Gate Length High Electron Mobility Transistors," *IEEE Electron Devices Letters*, Vol. EDL-6, 1985, pp. 142-145.

25. J.W. Archer, B.B. Cregger, R.J. Mattauch, and J.D. Oliver, "Harmonic Generators Have High Efficiency," *Microwaves*, Vol. 21, March 1982, pp. 84-88.

26. J.W. Archer, "A High Performance Frequency Doubler for 80–120 GHz," *IEEE Transactions on Microwave Theory and Techniques*, Vol. MTT-30, No.5, May 1982, pp. 824-825.

27. M.T. Faber, J.W. Archer, and R.J. Mattauch, "A Frequency Doubler with 35% Efficiency at W Band," *Microwave Journal*, Vol. 28, No. 7, July 1985, pp. 145-152.

28. Hughes Aircraft Company, *Millimeter-Wave Products 1985 Catalog*, Microwave Products Division, Torrance, CA, pp. 39–40. This catalog contains an excellent treatment of MMW solid-state sources.

29. F.B. Fank, "First InP Gunn Sources Put the Heat on GaAs," *Microwaves and RF*, July 1985, pp. 129–131.

30. S. Sando, "GaAs FET's Small Gates Yield Big Performance," *Microwaves and RF*, June 1985, pp. 141-144.

31. K. Kurokawa and F.M. Magalhaes, "An X-band 10 Watt Multiple-IMPATT Oscillator," *Proceedings of the IEEE*, Vol. 59, No. 1, January 1971, pp. 102–103.

32. K. Chang and S. Sun, "Millimeter-Wave Power Combining Techniques," *IEEE Transactions on Microwave Theory and Techniques*, Vol. MTT-31, No. 2, February 1983, pp. 91-107.

33. K. Chang, W.F. Thrower, and G.M. Hayashibara, "Millimeter-Wave Silicon IMPATT Sources and Combiners for the 110-260 GHz Range," *IEEE Transactions on Microwave Theory and Techniques*, Vol. MTT-29, No. 12, December 1981, pp. 1278-1284.

34. G.R. Thoren and M.J. Virostko, "A High-Power W-Band (90–99 GHz) Solid-State Transmitter for High Duty Cycles and Wide Bandwidth," *IEEE Transactions on Microwave Theory and Techniques*, Vol. MTT-31, No. 2, February 1983, pp. 183-188.

35. H.C. Yen and K. Chang, "A 63 Watt W-band Injection-Locked Pulsed Solid-State Transmitter," *IEEE Transactions on Microwave Theory and Techniques*, Vol. MTT-29, No. 12, December 1981, pp. 1292-1297.

36. J.R. Nevarex and G.J. Herokowitz, "Output Power and Loss Analysis of 2^n Injection-Locked Oscillators Combined Through an Ideal and Symmetric Hybrid Combiner," *IEEE Transactions on Microwave Theory and Techniques*, Vol. MTT-17, No. 1, January 1969, pp. 2-10.

37. L. Wandinger and V. Nalbandian, "Millimeter-Wave Power Combiner Using Quasi-Optical Techniques," *IEEE Transactions on Microwave Theory and Techniques*, Vol. MTT-31, No. 2, February 1983, pp. 189-193.

38. J.W. Mink, "Quasi-Optical Power Combining of Solid-State Millimeter-Wave Sources," *IEEE Transactions on Microwave Theory and Techniques*, Vol. MTT-34, No. 2, February 1986, pp. 273-279.

39. C.T. Rucker, J.W. Amoss, and G.N. Hill, "Chip Level IMPATT Combining at 40 GHz," *1981 IEEE MTT-S International Microwave Symposium Digest*, June 1981, pp. 347-348.

40. C.T. Rucker, J.W. Amoss, and G.N. Hill, "Chip Level IMPATT Combining at 40 GHz," *IEEE Transactions on Microwave Theory and Techniques*, Vol. MTT-29, No. 12, December 1981, pp. 1266-1270.

41. S.F. Wetenkamp and K.J. Wong, "Transportation Lag in Phase-Locked Loops," Tech-Notes, The Watkins-Johnson Co., Palo Alto, CA, 1978.

42. F.M. Gardner, *Phaselock Techniques*, 2nd Ed., John Wiley and Sons, New York, 1979.

43. P.S. Henry, "Frequency Agile Millimeter-Wave Phase Lock System," *Review of Scientific Instruments*, Vol. 47, September 1976, pp. 1020-1025.

44. H.M. Pickett, "Locking Millimeter Wavelength Klystrons With a Digital Phase-frequency Detector," *Review of Scientific Instruments*, Vol. 48, June 1977, pp. 706-707.

45. D.M. Guillory and R.W. McMillan, "Frequency Stabilization of Millimeter Wave Sources," *10th International Conference on Infrared and Millimeter Waves*, Orlando, FL, December 1985.

46. M. Crandell and F.J. Bernues, "Oscillators Lock and Tune at W Band," *Microwave Systems News*, Vol. 10, December 1980, pp. 54-60.

47. J.M. Cadwallader, M.M. Morishita, and H.C. Bell, "217 GHz Phase-Locked IMPATT Oscillator," *Microwave Journal*, Vol. 25, No. 8, August 1982, pp. 106-109.

48. M.P. Fortunato and K.Y. Ishikawa, "A Broadband Solid State Millimeter-Wave Synthesizer," *IEEE MTT-S International Microwave Symposium Digest*, Dallas, TX, June 1982, pp. 494-496.

49. *Millimeter-Wave Products Catalog*, Hughes Aircraft Company, Torrance, CA, 1985.

50. R.S. Ying, Hughes Aircraft Company, private communication, May 1987.

51. C.R. Seashore, "Missile Guidance," Chapter 3 in *Infrared and Millimeter Waves*, Vol. 4, K.J. Button and J.C. Wiltse, eds., Academic Press, New York, 1981.

52. R.A. Knust-Graichen and L. Bui, "60 GHz Low Noise Wideband Receiver," *Microwave Journal*, Vol. 28, No. 7, July 1985, pp. 179-181.

53. S.J. Nightingale, M.A.G. Upton, B.K. Mitchell, U.K. Mishra, S.C. Palmateer, and P.M. Smith, "A 30 GHz Monolithic Single Balanced Mixer with Integrated Dipole Receiving Elements," *IEEE Transactions on Microwave Theory and Techniques*, Vol. MTT-33, No. 12, December 1985, pp. 1603-1610.

54. R.S. Tahim, J. Pham, and K. Chang, "MM-Wave Integrated Circuit Wideband Downconverter," *Microwave Journal*, Vol. 28, No. 7, July 1985, pp. 131-141.

55. L.T. Yuan and P.G. Asher, "A W-Band Monolithic Balanced Mixer," *IEEE MTT-S International Microwave Symposium Digest*, St. Louis, MO, June 1985, pp. 113-115.

56. R.E. Forsythe, V.T. Brady, and G.T. Wrixon, "Development of a 183 GHz Subharmonic Mixer," *IEEE MTT-S International Microwave Symposium Digest*, Orlando, FL, May 1978.

57. R.W. McMillan, R.A. Bohlander, D.S. Ladd, R.E. Forsythe, A. McSweeney, J.M. Newton, O.A. Simpson, M.J. Sinclair, J.C. Butterworth, "Near Millimeter Wave Radar Technology," Final Report

on Contract DAAK70-79-C-0108, US Army Night Vision and Electro-Optics Laboratory, Ft. Belvoir, VA, October 1981.
58. B. Bayraktaroglu and H.D. Shih, "High Efficiency Millimeter Wave
Monolithic IMPATT Oscillators," *IEEE MTT-S International Microwave Symposium Digest*, St. Louis, MO, June 1985, pp. 124-127.
59. R.W. McMillan, "Near-Millimeter-Wave Sources of Radiation,"
Proceedings of the IEEE, Vol. 73, No. 1, January 1985, pp. 86-108.

Chapter 9
High-Power MMW Transmitters

J.C. Butterworth and T.V. Wallace

Georgia Institute of Technology
Atlanta, Georgia

9.1 INTRODUCTION

The design of a high-power transmitter at MMW frequencies is based on the same parameters that govern transmitter design at microwave frequencies: for example, peak and average power, PRF, pulsewidth, bandwidth, and tuning ability. However, due to the small physical size of RF structures and high losses in waveguides and their components, the design of a MMW transmitter is much more complex than systems designed to operate at lower frequencies.

Special care must be taken, for example, in the mechanical layout of the transmitter to minimize losses in waveguide runs, which at 95 GHz can be as high as 6 dB/m. Insertion losses in typical waveguide components, such as isolators and circulators, are on the order of 1 to 2 dB at 95 GHz. Obviously, the expected benefit from a high power source in signal-to-noise performance of the radar can be seriously degraded unless losses are minimized.

The selection of a high-power source and modulator power supply suitable for a particular application is of prime importance in transmitter design. Tube selection depends on peak and average power requirements, bandwidth, and availability, to name a few. A modulator power supply topology must be chosen to meet the PRF, pulsewidth, and stability requirements. This chapter will discuss the major high-power sources of energy on the market or in development, and also includes a discussion of modulators and MMW tube protection.

9.2 MILLIMETER-WAVE TRANSMITTING TUBES

The design of a MMW radar system is heavily dependent on the availability of the transmitter power source. The power levels offered by solid-state sources are naturally much lower than levels available from vacuum-tube sources, although the MMW spectrum is well covered by both types. Most of the power sources and other hardware components have been developed in the regions of the *atmospheric windows,* and commercial devices are more readily available in these bands. However, by using *scaling* techniques, a source can be developed at any frequency within the MMW band.

Millimeter-wave tubes are usually classified into two main categories, *fast-wave* and *slow-wave,* with the slow-wave devices being subdivided into *crossed-field* and *linear-beam* classifications. The crossed-field devices include *crossed-field amplifiers* (CFAs) and *magnetron* devices, while the linear-beam devices include tubes like the *klystron, traveling wave tube* (TWT), and *backward-wave oscillator* (BWO). Primarily, these have been scaled down in size from existing microwave sources and exhibit similar operating characteristics. However, frequency scaling of a device results in a shrinking of the frequency-determining elements, and imposes constraints on fabrication because of size and tolerance as well as on power because of the smaller dimensions and associated lower voltage breakdown. The fabrication of devices in the higher MMW region is virtually impossible with conventional manufacturing techniques. Fast-wave devices, such as *gyrotrons, laddertrons,* and *peniotrons,* represent new technology with advancements occurring as a result of ongoing research. Fast-wave interactions occur in large, smooth, multimode waveguide pipes, and thus involve higher order, overmoded resonators with large magnetic field requirements. This configuration allows these devices to achieve much higher efficiency and greater output power levels as compared with conventional slow-wave type devices.

With the rapidly advancing technology in MMW tubes, as many references as possible should be consulted when seeking more detail on the most recent technological developments.

9.2.1 Magnetrons

The magnetron is a crossed-field device (i.e., magnetic field at right angles to the electric field) which was developed during the pre-World War II era. Millimeter-wave magnetrons are usually scaled down versions of their microwave counterparts. Magnetrons are especially useful as pulsed oscillators in simple, lightweight systems, and in airborne radar systems. In the MMW band, most of the commercially available devices are pulsed, rather than CW, and the peak power output is on the order of a few kilowatts (see Figure 9.1). Magnetrons are available in the power range from 5 to 100 kW peak at 8 mm wavelength, to 5 kW peak at 4 mm wavelength, and to 2 kW peak at 3.2 mm wavelength. The typical duty factor of a magnetron is from 0.0001 to 0.001. For CW operation, sources of other designs, such as solid-state (Gunn or IMPATT) or linear-beam (TWT), are more practical. Developmental work is underway to extend the frequency range of magnetrons as well as the reliability, frequency stability, and tuning ability in both the US and the USSR. Figure 9.1 illustrates the success that the Soviets have had with magnetron designs at high millimeter-wave frequencies.

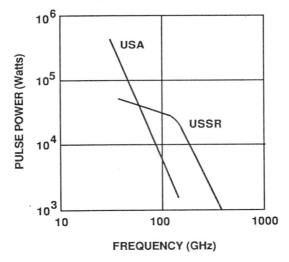

Fig. 9.1 Power output of millimeter-wave magnetrons (adapted from Bhartia and Bahl, [1], ©1984 John Wiley & Sons, Inc.).

At MMW frequencies there are three types of multicavity traveling wave magnetrons. They are the *rising sun magnetron,* the *coaxial magnetron,* and the *inverted* (or inside-out pulsed anode) *magnetron.* The rising sun seems to be the most popular type, although no clear performance superiority has been shown. References to Soviet literature (Bhartia and Bahl [1]) indicate that a MMW "surface-wave," or low-field, magnetron device provides approximately 7 dB more peak power than the available US counterpart.

Magnetron Operating Characteristics

One of the major problems associated with the operation of a magnetron is *moding*. Moding occurs when the tube oscillates at more than one frequency or mode. The rate of rise of the modulator voltage pulse is one of the most critical factors, in addition to source and load impedance, in controlling the moding problem.

Other problems associated with magnetrons are frequency pushing and pulling, arcing, and missing pulses. Also, the physical construction of conventional MMW magnetrons dictates small cathodes, and, therefore, high cathode current densities, which shorten tube life. The inverted magnetron might be a solution to the problem of high current density, but only a few designs have been explored.

Rising Sun Magnetrons

The conventional magnetron consists of a cylindrical cathode surrounded by an anode formed by a series of resonant cavities, which are regularly spaced around the circumference and concentric with the cathode. Because MMW magnetrons are scaled versions of microwave devices, the most practical form of a resonant anode structure is the rising sun, which consists of alternating deep and shallow slots that form the anode. This structure, with a simple cathode and impedance-matching waveguide section, form the entire vacuum device. A permanent magnet provides a magnetic field parallel to the tube axis. However, this type of structure has two sets of resonators, with the operating frequency falling somewhere in between these frequencies (see Figure 9.2). This results in some of the modes resonating at wavelengths above the π-mode, (where the phase difference between any two adjacent anode cavities is π radians) and the remainder at wavelengths below the π-mode. By properly adjusting the number of resonators and the anode and cathode dimensions, a practical magnetron can be designed in the MMW range, although moding difficulties can be experienced with this type of device. With a negative high-voltage pulse applied to the cathode, electrons emitted from the cathode are accelerated in the crossed-field region between the anode and the cathode, and follow a spiral or cycloidal path on their way to the anode. As these electrons spiral toward the resonant anode structure, energy is coupled into the slots and forms a rotating standing wave, or *bunching* of the electrons. Electrons that emerge at the right time to fall in-phase with the bunched beam will reinforce the RF field that is being generated, while electrons that are emitted at a low density time of the beam will simply fall back to the cathode.

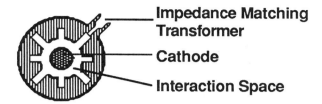

Fig. 9.2 Rising sun magnetron.

The bunching process is called *phase focusing,* and it is responsible for the high efficiency of magnetrons [1]. Scaling of an existing design is the simplest way to achieve a new frequency magnetron for a particular requirement, although this approach may not always be practical. For a rising sun pulsed magnetron, the vane thickness is approximatly 0.01 inches at 30 GHz, and approximately 0.003 inches at 100 GHz [1], which results in construction difficulties as well as heat dissipation problems. Therefore, special considerations must be given to the design of MMW magnetrons to achieve optimum power output and life expectancy. Useful lifetimes for MMW magnetrons operating in the 80 to 100 GHz frequency range can be expected to be between a few tens of hours and a few hundred hours.

Coaxial Magnetrons

Some of the limitations of vane magnetrons can be overcome by the coaxial magnetron using a built-in stabilizing cavity to improve mode or frequency stability, spectral quality, and pulse-to-pulse jitter. The individual cavity resonators are no longer the frequency-determining element, and conventional coaxial-cavity modes become the frequency-determining factor. However, the complicated inner structure of a coaxial magnetron (Figure 9.3), when scaled to MMW frequencies, becomes difficult to manufacture, and so this type of magnetron has not been widely explored.

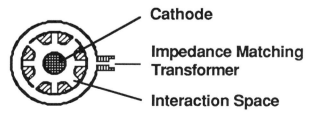

Fig. 9.3 Coaxial magnetron.

Two limitations of the rising sun and the coaxial magnetron are voltage breakdown in the window and waveguide areas that prevent high-power operation. The inverted magnetron (Figure 9.4) uses a cylindrical waveguide operating in the TE_{01} mode, which has the characteristic that attenuation decreases with increasing frequency, and the breakdown limitation is also greatly reduced. However, the pulsed anode heat dissipation becomes a limitation to its average power capability. Another limitation of the inverted magnetron is its reduced overall efficiency because of the large heater dimensions and overall power requirements. Typically, the inverted magnetron is the heaviest of the the three magnetrons for a given power output.

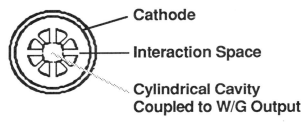

Fig. 9.4 Inverted magnetron.

9.2.2 Klystrons

Klystrons are linear-beam devices in which the interaction between an RF field and an electron beam causes velocity modulation and bunching of the electrons. The bunched beam causes induced currents to flow in the output circuit, and energy is thus extracted from the device. There are several types of tubes that fall under the general category of klystrons: *reflex oscillators, multiple-cavity oscillators,* and *amplifiers.*

Reflex Klystron

Reflex klystron oscillators consist of a single resonator cavity. An electron beam is injected into the gap and passes through it twice — first, after emission from the electron gun; second, after being reflected or repelled by the reflector. Electron velocity modulation occurs within the cavity, and produces RF energy. The electrons are collected by the cavity, which is a small and delicate structure at MMW frequencies. The power output of the reflex klystron is limited to about 1 W at 50 GHz down to 10 mW at 220 GHz. Voltage potentials of the cathode, cavity, and reflector determine the beam velocity, and therefore have specific values for a particular cavity design and thus are several of the determining factors of the operating frequency. The dimensions of the cavity, which usually can be varied by mechanical means, and the spacing of the elements along with the applied voltages cause the reflex klystron to oscillate at a specific frequency. Figure 9.5 is a schematic representation of a reflex klystron. Due to the low-power capability of reflex klystrons, they are generally used as local oscillators in radar systems or short-range alarm systems.

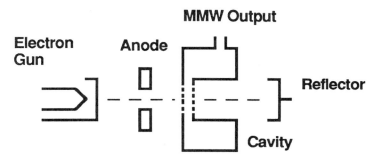

Fig. 9.5 Reflex klystron.

Multicavity Klystrons

Conventional multicavity klystrons consist of two or more cavities separated by drift tubes through which the electron beam passes. Figure 9.6 illustrates a schematic of a three-cavity klystron amplifier. With the proper feedback from the last cavity to the first cavity, this device can serve as an oscillator. As an amplifier, each cavity represents a resonant circuit at a particular frequency. Hence, different device characteristics result when the cavities are tuned in different ways. For instance, detuning the second cavity to a frequency lower than the other two results in lower gain, but an increase in bandwidth. Higher powers are achievable in multicavity klystrons as compared with reflex klystrons, since a separate collector element is used to dissipate the electron beam energy. Powers on the order of 1 kW CW have been produced in the 30 GHz region by multicavity klystrons.

9.2.3 Extended Interaction Oscillators

Extended interaction oscillators (EIOs) consist of a single cavity with multiple gaps. These devices are physically shorter than other linear-beam devices, which simplifies the magnetic focusing and permits smaller magnets to be used. Early extended, or distributed, interaction oscillators used

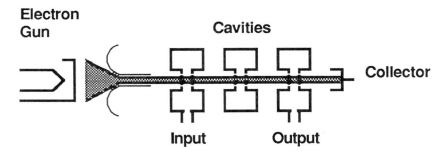

Fig. 9.6 Multicavity klystron amplifier.

little or no magnetic field to focus the beam, but depended on space-charge-focused electron beams.[1] These early EIOs produced lower output power than the modern kilowatt oscillators. A few examples of modern EIOs are given in Table 9.1, and a schematic drawing of an EIO is given in Figure 9.7. Since only one cavity is used, tuning is relatively simple as compared with multicavity klystrons, and is commonly achieved by mechanically changing the cavity dimensions. The EIO may also be electronically tuned by controlling the beam voltage. Tubes with mechanical tuning ability have a typical tuning range of 1% to 3%. The electronic tuning range may be on the order of 0.5% to 1% at 94 GHz. The beam voltage sensitivity which allows electronic tuning also imposes tight stability requirements on the power supply. However, this beam voltage sensitivity can be used to stabilize the frequency with the use of an *automatic frequency control* (AFC). Frequency modulation sensitivities typically range from 0.04 MHz/V at 35 GHz to 0.3 MHz/V at 220 GHz. Careful modulator design and construction must be maintained to obtain a flat modulating pulse shape for minimum frequency modulation.

Table 9.1 Typical EIO Characteristics
(Manufactured by Varian Canada).

Frequency (GHz)	Pulsed or CW	Peak Power	Duty Factor	Average Power	Type
30–50	P	2 kW	0.1	200W	VKQ2422G
30–40	CW	—	—	1 kW	VKQ2435E
50–80	CW	—	—	70 W	VKE2436E
93–96	P	1 kW	0.05	5 W	VKF2443T
110–140	CW	—	—	10 W	VKT2438G
140–170	P	100 W	0.005	0.5 W	VKT2444T
220	P	60 W	0.005	0.3 W	VKY2429R

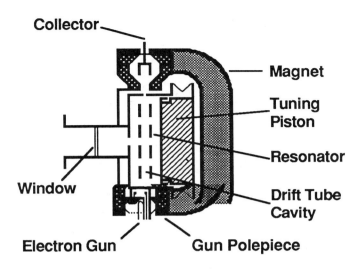

Fig. 9.7 EIO schematic.

As with all pulsed oscillators, the EIO has a natural delay in onset of oscillation after all proper operating voltage parameters have been met. This delay can vary from a few nanoseconds to as many as 20 ns, depending on pulse shape. Short-pulse transmitters can be designed to produce 3–4 ns RF outputs, despite the variable random start-up time of the oscillator, by injecting a low-power (1–10 mW) priming signal into the output waveguide prior to the application of each voltage pulse.

This priming signal should be near the operating frequency (F_0) of the EIO, with the power requirements being minimum at F_0. Satisfactory priming can be accomplished up to 100 MHz from the carrier frequency. A simplified schematic for a method of injection priming is given in Figure 9.8. The priming oscillator is a pulsed, solid-state Gunn device, and it needs at least 60 dB isolation from the typical 1 kW MMW transmitter.

9.2.4 Extended Interaction Klystron Amplifiers

Extended interaction klystron amplifiers (EIKA, or EIA) have been produced at 35 and 95 GHz with peak power outputs of up to 2.8 kW. At 95 GHz, an EIA was reported to have 38 dB gain when synchronously tuned, and 400 MHz bandwidth when broadband tuned (Acker [2]). The EIA has been packaged with a samarium cobalt magnet, and has a useful mechanical tuning range of 2 GHz, with an overall efficiency of 18% to 20%. However, the tuning of an EIA, as with any multicavity klystron, is extremely critical. Figure 9.9 is a photograph of a 95 GHz EIA with a

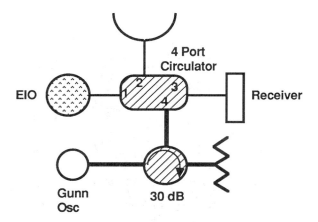

Fig. 9.8 Schematic of primed EIO transmitter.

one-inch scale for size comparison. Varian Canada has developed a line of compact MMW transmitter subsystems that use pulsed EIOs and EIAs, which operate in discrete frequency bands from 30 to 220 GHz [3]. Table 9.2 gives some characteristics of a few EIAs that are available as of this writing (*circa* 1987).

Table 9.2   Typical EIA Characteristics (Manufactured by Varian Canada).

Frequency (GHz)	Power (W)	Duty (%)	Gain (dB)	Bandwidth (MHz)	Model
50–80	100	10	—	—	VKE2406
94–96	1000	10	30	200	VKB2400T
95	30	100	30	150	—
220	60	1.7	—	—	—

9.2.5   Traveling Wave Tube

The helix traveling wave tube (TWT) was introduced by R. Kompfner in 1944. The basic configuration of the helix TWT has changed little since its development. With improvements in manufacturing technology, new techniques have been developed to suppress backward-wave oscillations, increase power output, and increase bandwidth. The helix TWT is ideally suited for broad-bandwidth, high-gain amplification of MMW signals. Due to its thermally fragile construction, however, output powers are generally limited to hundreds of watts at 35 GHz and tens of watts at 95 GHz. The

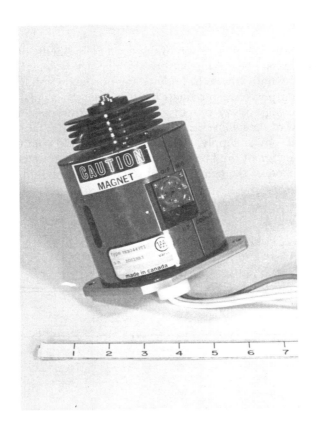

Fig. 9.9 Photograph of a 95 GHz EIA.

coupled-cavity TWT, on the other hand, is generally capable of an order
of magnitude greater average power output than the helix version due to
its greater mechanical and thermal ruggedness. The coupled-cavity interaction circuit, however, does not afford the wide bandwidth of the helix
circuit.

The TWT is a linear-beam device in which the direction of the electric
and magnetic fields are parallel. The typical advantages of a linear-beam
device as compared with a crossed-field tube like the magnetron or crossed-
field amplifier are higher gain, higher bandwidth, and lower noise figure.
Some disadvantages are lower efficiency (around 10%) and a higher voltage
requirement for a comparable output power in a crossed-field device.
Traveling wave tube efficiency is quite often improved by depressed collector operation.

TWT Operational Characteristics

Figure 9.10 is an illustration of the main components of a TWT. An electron gun emits a beam of electrons through a slow-wave RF circuit at a velocity set by the cathode to body voltage or beam voltage. Input and output coupling ports are provided for the input RF signal and the extraction of the high-power output signal. A separate collector electrode intercepts most of the beam current, and hence dissipates most of the input power from the power supply. An axial magnetic field confines the beam within the diameter of the helix (or drift tubes of the coupled cavities) to prevent dispersion of the beam and subsequent tube meltdown.

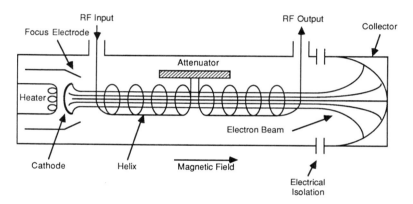

Fig. 9.10 Configuration of a typical TWT.

The electron beam is velocity modulated at the input by the injected RF signal. This velocity modulation or electron bunching thus induces higher amplitude RF currents on the slow-wave structure. The resulting RF field hence provides tighter electron bunching, and so on, in a regenerative amplification process in which energy is transferred from the beam to the RF field. The attenuator shown in Figure 9.10 is present to suppress oscillations caused by reflections within the slow-wave circuit. The attenuator suppresses any forward or backward traveling RF signal, but does not affect the forward traveling electron bunches, which begin the amplification process anew as they emerge from the attenuator.

Beam modulation is typically accomplished by the pulsing of a modulating anode or grid. Gridded tubes allow the lowest possible voltage control of the electron beam, and thereby provide fast rise and fall times for short-pulse operation. Intercepting control grids can only be used in low-power tubes, due to thermal limitations. Nonintercepting *shadow-grids* have been developed that allow grid modulation of high-power tubes (Pit-tack [4]). Such a grid is constructed by placing one grid structure very close to the cathode and at cathode potential. Another geometrically similar structure is then used as the control grid, which is physically "shadowed" from beam interception by the first grid structure.

The method of focusing employed affects the maximum average power capability of the tube. For very high average power tubes, solenoidal focusing is required to confine the beam within the narrow helix tunnel diameter or cavity drift tubes to minimize beam interception. Beam trans-mission requirements can be as high as 99%, and magnetic fields of 8 kG may be required. *Periodic-permanent-magnet* (PPM) *focusing* is attractive for lower frequency or lower power tubes because the samarium cobalt magnets used are light in weight. Such focusing structures are generally limited to producing fields on the order of 4 kG (Amboss [5]).

Figure 9.11 is an idealized transfer curve for a TWT that shows gain and output power as a function of input drive. In the small-signal or linear region, the output power is proportional to input drive. As the input drive is increased, however, the gain begins to drop, and, at the point of saturated output power, the gain is typically down about 6 dB [6]. If the input drive level is increased further, output power will decrease. Outside of the linear amplification region, the TWT acts as a nonlinear device, and intermod-ulation products can become a problem in broadband applications. Am-plitude modulation to phase modulation (AM/PM) conversion, which will be discussed later in this section, is also more pronounced in the nonlinear region of the drive characteristic.

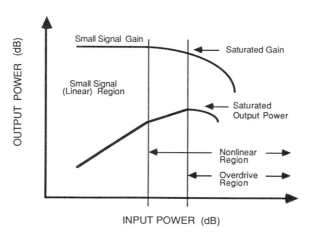

Fig. 9.11 Transfer characteristics for a typical TWT.

Coupled Cavity Versus Helix Slow-Wave Structures in Traveling Wave Tubes

The extremely wide bandwidth of the helix structure is by far its most prominent claim to fame, and is unparalleled by the other available circuit types. The circuit provides a constant phase velocity, which greatly reduces distortion and interfering modes. Due to the lack of large opposed metal surfaces, there is lower stored energy in the propagating wave than for a coupled-cavity circuit. Also, the RF electric field within the electron beam is approximately uniform, which provides strong interaction between the beam and RF field. The wide bandwidth of the helix structure makes possible the multioctave bandwidth TWTs required by the electronic warfare and satellite communication industries.

Helix TWTs suffer from low-to-moderate average power-handling capability. Heat is transferred to the helix primarily by interception of beam electrons. Since the helix is electrically isolated from the walls of the tube, the only thermal path through which this heat may flow is through the isolating support rods. Typical materials used for such support rods are boron nitride and beryllium oxide, with experiments also performed using diamond support rods for maximum thermal conductivity (Hamilton, *et al.* [7]). Shrink fitting of the tube barrel around the helix and support rods has also been used for tight, low thermal impedance interfaces (Cascone [8] and Wada, *et al.* [9]).

Coupled-cavity circuits exhibit less bandwidth than helix circuits, but, due to better mechanical and thermal ruggedness, coupled cavities are generally capable of one to two orders of magnitude greater average power with some dependence on whether PPM or solenoidal focusing is used. The two most popular approaches to the coupled-cavity circuit use circular and rectangular cavities with staggered or alternating aligned coupling holes or slots. The circular-cavity approach easily lends itself to PPM focusing, but there are problems in machining the individual circuit components (100 to 200 per tube) to specified tolerances and in stacking the pieces together. For example, tolerances required for the Hughes 982H 95 GHz cavity parts were ±35 microinches (Davis and Vaszeri [10]). The rectangular, staggered slot, coupled-cavity circuit reduces these critical construction tolerances, and allows cold test adjustment of the cavities prior to final assembly. This circuit was employed in the Varian VTA-5700 35 GHz TWT, which produces 30 kW peak power at 10% duty cycle. It is reported, however, that this rectangular cavity circuit may not be considered for frequencies above 50 GHz [2].

A number of other slow-wave structures have been derived from the basic helix and coupled-cavity approaches, and these structures are being investigated by various tube manufacturers to overcome some of the power-handling, bandwidth, or construction limitations of the two basic configurations. The *comb-quad* structure is an attempt to reduce the number of individual parts for a cavity tube from several hundred to less than 10 (Karp [11]). A *folded-waveguide* circuit is being investigated as having good thermal ruggedness and a wider bandwidth than the standard coupled cavity circuit. *Electric discharge machining* (EDM) is used to manufacture the circuit to acceptable tolerances at reasonable cost. Another advantage to this circuit is that the input and output sections are direct waveguide transitions (Phillips [12]). The *Hightron* (acronym for *helix in guide hybrid*) circuit uses an oversized electron-beam hole diameter to solve some of the problems of beam interception, and allows higher average power due to the larger and more robust RF circuit. The circuit can be constructed from a coiled double-ridged waveguide [12]. Required operating voltages are high, and the bandwidth is on the order of 5%, but relatively low magnetic fields permit PPM focusing. Another problem with this circuit is the control of unwanted modes due to the oversized RF structure [2,12].

TWT Efficiency and Depressed Collector Operation

One method of improving the efficiency of a TWT is by reducing the collector voltage to a value below that of the slow-wave or body structure (depressed collector). The kinetic energy of the electrons striking the collector is then reduced by the lower potential difference seen by the electrons between the cathode and collector. Therefore, less heat is dissipated by the collector structure, which will then require less cooling. This can be seen schematically in Figure 9.12. Typical collector-voltage depression is on the order of one-half of the beam voltage for a single-stage collector. Some applications, such as satellite communication and extremely high power radars, require multiple-stage collectors with different levels of depression for even higher efficiency.

As an example of the increase in efficiency that can be realized by depressing the collector, consider a TWT with a beam voltage of 20 kV and a beam current of 100 mA operating at 50% duty, with a peak output power of 100 W. The average input power from the power supply is 1000 W, and the average output power is 50 W, for an efficiency of 5%. If, however, the collector is depressed to 10 kV and the tube has 90% transmission (90 mA collector current, 10 mA intercepted body current), the input power is then calculated as (20 kV × 10 mA + 10 kV × 90 mA) × 50%, or 550 W, which results in an efficiency of 9.1%. The power dissipated by the collector is reduced from 1 kW to 450 W, which results in a substantial reduction in cooling requirements. If, however, this tube were operated at 1% duty, the reduction in average input power from 20

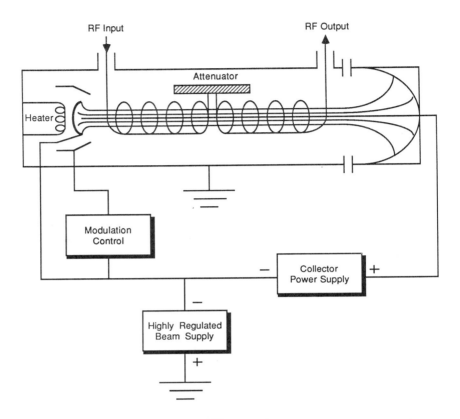

Fig. 9.12 Depressed collector TWT operation.

to 11 W hardly warrants the cost and complexity of an extra high voltage power supply. Therefore, each individual application must be examined to determine the need for depressing the collector.

There is another important advantage to be gained from depressing the collector. The phase sensitivity of the cathode-to-collector voltage is much lower than that of the cathode-to-body voltage. Therefore, the collector power supply need not be as highly regulated as the cathode-to-body supply. Since only the intercepted body current need be highly regulated, this reduces the complexity and cost of the beam power supply. The majority of the beam current is then regulated through the higher power, but lower performance, collector supply.

Care should be taken when depressing the collector of a TWT so as not to exceed a maximum value of collector voltage. If the electrons see too low a potential as they approach the collector, some electrons may actually be turned around and strike the output section of the slow-wave structure, resulting in an increase of body current and decrease of efficiency.

Another technique for improving the efficiency of a TWT is known as *velocity resynchronization* [6]. During the amplification process, the electron beam loses energy to the RF wave, and thus slows the velocity of the beam. Therefore, the beam and RF wave begin to lose synchronism, and the phasing between the RF wave and the electron bunches is no longer optimum for energy transfer. It is no surprise that a loss of efficiency results. This phenomena can be accounted for by designing a tapered-velocity slow-wave circuit. This can be implemented, for example, by altering the pitch of the helix along the length of the tube, thus slowing down the axial velocity of the RF wave to keep it in synchronism with the beam as it slows down. Another method of velocity resynchronization involves the acceleration of the electron beam rather than the deceleration of the RF wave by the introduction of a *voltage jump* near the output of the circuit. This technique, however, is rather difficult to implement [6].

TWT Phase Sensitivity

Coherent radar amplifiers require minimum pulse-to-pulse phase variations due to the transmitter chain. From a design standpoint, this translates into careful specification and design of all power-supply voltages and the grid or anode modulator. Variations in dc voltages or pulse voltage result in phase perturbations determined by the phase sensitivity of the particular electrode in question. Table 9.3 gives typical sensitivities for a high-power TWT. Sensitivities for individual devices may be different from those in the table, where typical values are given to illustrate the relative sensitivities of the different electrodes of a TWT. As one can see, the cathode-to-body or beam voltage is the most critical, and it must be highly regulated and have low ripple content. For example, to keep the phase shift through a TWT due to a 20 kV beam voltage power supply down to 1°, ripple must be kept below 0.03% or 6 V, which can be a substantial design burden for a very high power supply. Cathode-to-collector voltage phase sensitivity is at least an order of magnitude less than that for the beam voltage, and this is used to advantage in depressed-collector operation.

For a cathode-to-grid voltage sensitivity of 6° per 1% voltage change, a 1 kV grid voltage requires 0.16% or 1.6 V maximum pulse voltage slope or jitter to keep the phase shift to 1°. This generally requires a hard-tube modulator for tight control of the pulse shape. For lower pulse voltages

Table 9.3   Typical TWT Phase Sensitivities.

Electrode Voltage	Phase Sensitivity
Cathode to Body (Beam Voltage)	32°/1%
Cathode to Grid	6°/1%
Cathode to Anode	2°/1%
Cathode to Collector	1°/1%
Heater	0.005°/1%
Drive Power	3°/1 dB

afforded by higher μ gridded electron guns, a solid-state modulator may be used.

Amplitude modulation to phase modulation (AM-PM) conversion refers to the phase shift through the tube due to a change in RF drive level, and is expressed in degrees per decibel. This phenomena is more sharply pronounced in the saturation, or nonlinear, region of operation. The phase shift is caused by the reduction of beam velocity due to the increased energy exchange between the beam and higher level input RF signal. The practical implication of this effect is that, especially for wide band coherent systems, the drive level should be relatively constant over the operating bandwidth.

Traveling Wave Tube Examples

Table 9.4 gives some examples of recently developed MMW TWTs. Note that for high output powers, large magnetic fields and, hence, heavy solenoids are required to focus the electron beam to minimize body current interception. The heavy solenoids preclude the use of these tubes for all but ground-based radar applications. The EEV N10043 and Stantel W09MW1K are lightweight, wideband, helix tubes that have been developed primarily for electronic warfare, radar, and communication applications. The Varian VTA-5700 is one of the highest powered 35 GHz slow-wave devices developed to date, and it has extensive potential for use in high-power radars. The VTA-5700 is illustrated in Figure 9.13 without the focusing solenoid magnet. Maximum dimensions (with the solenoid) are 63.5 cm in height and 38 cm in diameter. The Hughes 982H TWT, illustrated in Figure 9.14, uses PPM focusing, and this device has dimensions of 41 cm in length and 10 cm in diameter.

Table 9.4 Recent Millimeter-Wave TWT Developments.

Vendor	Model	Frequency (GHz)	Pulsed or CW	Peak Power (kW)	Average Power (W)	Gain (dB)	Structure	Focusing	Weight (lbs)
EEV	N10043	18–40	CW	—	20	40	Helix	PPM	2.4
Stantel	W09MW1K	26.5–40	CW	—	10	30	Helix	PPM	5.5
Hughes	8900H	32–35	P	0.08	16	43	Helix	PPM	5.5
Hughes	921H	33.5–36.5	P	4	220	46	Cavity	PPM	17.5
Raytheon	QKW1995	Ka-Band	CW	—	1000	43	Cavity	PPM	6
Varian	VTA-5700	34.5–35.5	P	30	3000	53	Cavity	Solenoid	350
Hughes	982H	93.75–95.75	P	0.1	50	60	Cavity	PPM	14
Varian	VTW-5795	94.5–95.5	P	5	500	46	Ladder	Solenoid	220

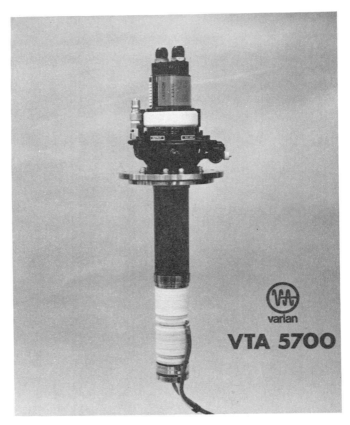

Fig. 9.13 Photograph of the VTA-5700 35 GHz traveling wave tube (photo courtesy of Varian Associates, Inc.).

9.2.6 The Gyrotron

Of all the types of thermionic sources of MMW energy, the gyrotron has perhaps the greatest potential as the premier generator of high power. The gyrotron operates on the principle of RF field and electron-beam interaction, with coupling provided by the phenomenon of electron cyclotron resonance. Thus, the gyrotron is also known as the *cyclotron resonance maser* (CRM). The gyrotron can be constructed with interaction cavities or circuits that are large compared to the wavelength of operation, thus increasing the average and peak power capability of the device. Gyrotrons typically require high operating voltages (50 to 100 kV) and superconducting magnets to provide the extremely high magnetic fields necessary

Fig. 9.14 Photograph of the 982H 95 GHz traveling wave tube (photo courtesy of Hughes Aircraft Company, Electron Dynamics Division, Torrance, CA).

for fundamental-mode operation. Gyrotrons that have been built and tested have provided efficiencies on the order of 40% and CW power levels of 200 kW at 60 GHz.

Gyrotron Operational Characteristics

Figure 9.15 illustrates the major components of a gyrotron oscillator. The electron gun emits a hollow electron beam with helical electron trajectories controlled by the gun anode and the field generated by the gun magnet coil. The beam is tightly compressed by the main magnetic field, passes through the interaction cavity, and is finally deposited on the walls of the collector. The RF created in the interaction cavity follows the tapered waveguide structure, and is extracted from the tube through the vacuum window.

Energy is transferred from the electron beam to the RF field by angular electron phase bunching, as opposed to longitudinal bunching that occurs in slow-wave, linear-beam tubes like the TWT. This energy transfer requires, however, that the majority of the electron energy be in the

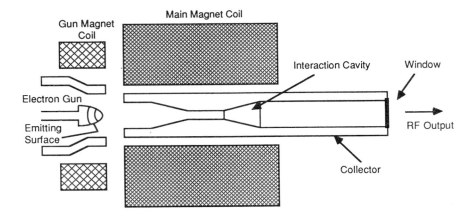

Fig. 9.15 Schematic representation of a gyrotron oscillator.

transverse direction, which is accomplished by the use of *magnetron injection guns* (MIGs), or Pierce guns used with so-called "beam wigglers" (Berry [13]).

The magnetic field required by a gyrotron is proportional to the cyclotron resonance frequency. Operation at the nth cyclotron harmonic, however, reduces the required magnetic field by a factor of $1/n$. Fundamental resonance provides the strongest interaction between the beam and RF field and thus higher efficiency. Efficiency at the second harmonic remains relatively high, and such operation has been accomplished with one-half of the magnetic field required for the fundamental mode. Fundamental resonance at 28 GHz requires a 10 kG magnetic field, and a 34 kG field is required at 95 GHz. Thus, for high-frequency operation, either harmonic operation or superconducting magnets are required. Due to the direct relationship between the two, high-frequency stability requires tight control of the magnetic field.

Experimental Results

There are several versions of the gyrotron that have been designed, built, and tested. The basic gyrotron oscillator has been studied for several decades, and impressive output powers and efficiencies have been achieved. Interest in the generation of high power for coherent applications has spawned the development of amplifier versions of the gyrotron, which are similar in nature to the conventional klystron and traveling wave tube.

The primary impetus for the development of the gyrotron oscillator has been the need for high power at high frequencies for plasma heating in fusion experiments, where CW power levels of 200 kW at 60 GHz and 38% efficiency and 500 kW peak power for 30 ms pulses at 100 GHz have been achieved (Berry [13] and Ferguson, *et al.* [14]). Operation at 375 GHz has achieved 120 kW pulsed power at 100 ms and 15% efficiency (Granatstein and Park [15]). Figure 9.16 is a photograph of a Varian 140 GHz, 100 kW, CW gyrotron oscillator with its superconducting solenoid.

Fig. 9.16 Photograph of a 140 GHz gyrotron oscillator in its superconducting magnet (photo courtesy of Varian Associates, Inc.).

The *gyroklystron* amplifier consists of several resonant cavities in which the cyclotron resonance interaction occurs. The electrons in the beam are velocity modulated at the input cavity by the input RF signal, and phase bunching occurs. Further interaction, and hence amplification, occurs in the following cavities. Such a device has been built at 28 GHz, and has achieved 65 kW peak power with 30 dB gain at 5% duty, 10%

efficiency, and 0.2% bandwidth [13]. The gyroklystron is expected to exhibit less bandwidth but higher gain, higher power, and higher efficiency than the gyro-TWT. Efficiencies as high as 40% may be achieved by tapering the magnetic field along the length of the interaction region (Arfin, et al. [16]).

The gyro-TWT is based on the interaction of the beam and RF field in a nonresonant structure, such as a circular waveguide. A device built at 95 GHz has achieved 30 dB gain, peak output power of 20 kW, efficiency of 8%, and 2% bandwidth [15]. Experiments have achieved 13% bandwidth at 35 GHz, and the objective is to approach the full waveguide bandwidth (Barnett, et al. [17]). Such devices with high output powers and wide bandwidth may have important applications in MMW radar systems. However, for gains on the order of 30 dB and output powers of 10 to 100 kW, high-power conventional TWTs will have to be used as drivers in the amplifier chain. Such a transmitter will be large, complex, and costly.

9.3 SYSTEMS CONSIDERATIONS

Radar systems that require extremely high range and azimuth resolution will benefit from the use of MMW frequencies. This allows the design engineer to utilize relatively small antenna apertures to achieve narrow beamwidths.

The primary application for magnetrons is currently noncoherent radar equipment. Tuned magnetrons are available with both mechanical and electronic (piezoelectric) tuning control. By combining the high resolution qualities and frequency agility capabilities of MMW magnetrons, several applications become apparent: high resolution mapping radar, airport surveillance radar, radar scaling measurements, antiglint and clutter reduction radar, and radar ECCM applications.

In comparison with the magnetron, the cost of an EIO is similar for a given power level. However, magnetrons are somewhat smaller and lighter in weight, and they use simpler, more compact modulator-power supplies. EIOs can operate at higher duty cycles, longer pulsewidths (up to 25 μs as well as CW for some models), higher average power, and have longer lifetimes.

TWTs and EIAs are the principal sources for high-power, coherent, MMW radar systems. Distinct differences in capability exist between the two types of tubes, however, that drive the choice for a particular application. Pulsed MMW EIAs currently have bandwidths on the order of

0.2% to 0.3%, whereas high-power pulsed MMW TWTs can have band-widths on the order of 2% to 5%. Therefore, for a system requiring wide bandwidth, such as that required for pulse-compression and frequency-agile systems, the use of a TWT is desirable. Typical gains for an EIA are on the order of 30 dB, as compared with 50 dB to 60 dB for a TWT. Therefore, if RF input drive power is limited, use of a TWT may be necessary.

EIAs have produced several kilowatts of peak power at 95 GHz, which is greater than that produced by comparably sized TWTs. Extremely high power TWTs require heavy solenoidal magnets for focusing. There-fore, if high peak power is required, use of an EIA may be desirable when size and weight are important factors.

Figure 9.17 is a block diagram of a high-power, coherent, 95 GHz transceiver (Lane, et al. [18]). The radar can operate in one of two modes: a 2 GHz bandwidth, 100 W output power mode, using a TWT; a 350 MHz bandwidth, 1 kW output power mode, using an EIA as the output stage. The system is pulse-to-pulse polarization agile, using a stepped frequency waveform (SFWF), and frequency agility. This system will provide polar-ization scattering matrix data, and will have range resolution of less than 10 cm.

The primary use for gyrotrons to date has been in plasma heating for fusion experiments. There are several problems that will have to be addressed if the gyrotron is to be considered as a candidate radar trans-mitter. Among these are output modal purity (Kreischer, et al. [19]) and the development of MMW components to handle the high-power levels. The high magnetic field requirements dictate the use of heavy conventional or superconducting solenoids. Therefore, size and weight limits the use of gyrotrons to stationary and semimobile ground-based radar applications.

Tube efficiency will determine the output power requirements of the high-voltage power supply (HVPS), and hence its size, complexity, and cost. Magnetrons are available with efficiencies of 20% at 35 GHz but only about 5% at 95 GHz. TWT efficiency can be increased by depressed collector operation, as described earlier, with the added advantage of the lower ripple and regulation required for the majority of the beam current. EIOs and EIAs are generally limited to efficiencies between 5% and 10%, but a 95 GHz EIA has achieved 2.3 kW peak output power at 21 kV beam voltage and 650 mA beam current for an efficiency of 16%. Gyrotrons may be the most efficient thermionic tube source to date within the MMW band, and efficiencies on the order of 40% have been demonstrated.

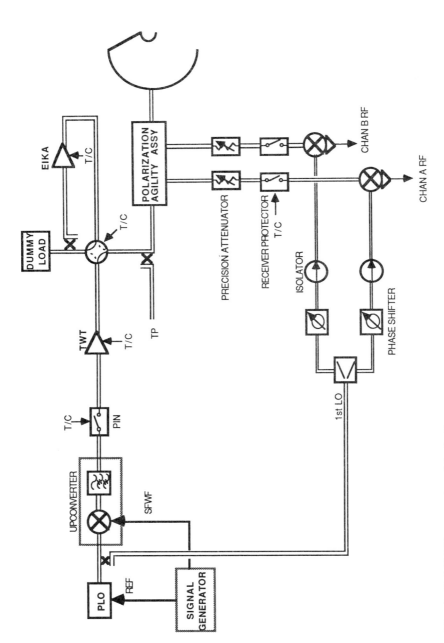

Fig. 9.17 MMW transceiver block diagram.

9.4 MMW TUBE PROTECTION

The high-power transmitting tube is typically the most expensive and least available single component in a MMW radar system. Protection from damage due to overvoltages or other misuse is important to prevent premature failure of a MMW tube, and there is a spectrum of protective measures to be considered in its use.

There are several physical protective steps that should be taken to ensure the safety of personnel and the tube itself. The mechanical packaging should include interlocks to prevent personnel from reaching high voltage during operation. For a floating deck modulator, this means totally enclosing large amounts of circuitry. Care must be taken not to "degauss" the focusing magnet of the tube by placing it near other high magnetic fields, or working on the tube with magnetic tools. For very high voltages, around 30 kV and above, x-ray shielding should be provided for personnel safety.

In addition to physical safety, the design of the power supply and modulator should include a number of items to protect the tube from electrical damage. The design of the HVPS should limit the amount of stored energy in the output filter capacitors and magnetic components to a safe value. Current-limiting resistors should be placed in series with the HVPS to limit arcing current in the event of an electron gun-to-body arc. Although these resistors will normally have little voltage dropped across them, they should be high-voltage resistors so that they do not arc over or break down if an arc does occur.

A number of currents, voltages, and physical quantities should be monitored, and appropriate actions ought to be taken if they exceed specified values. Beam voltage and collector voltage (if appropriate) should be cut off quickly if they exceed maximum values. Collector and body current should be monitored separately, since excessive body current is an early indication of beam defocusing, which can lead to catastrophic tube meltdown. Tube heater voltage and any other critical modulator voltages should be monitored with low and high set-points. If the voltage levels exceed those bounds, then all voltages should be removed from the tube. If appropriate, vacuum ion-pump power supply voltage and current should also be monitored. Tube body or collector temperature should be measured, and beam current ought to be cut off if the temperature is too high. Coolant flow, whether it be air or water, should be monitored and the power supply must be de-energized if the flow is broken.

For very high power systems, the output waveguide window of the tube is another point of concern. Reflected power can be measured by a directional coupler, and the transmitter is to be shut off if the load VSWR becomes too high.

The sequencing of turning on power to the various electrode power supplies must also be addressed. Heater and negative grid bias must be applied first, and beam voltage is to be applied only after sufficient heater warm-up time. Logic circuitry should also be included to limit the maximum pulsewidth and duty cycle of the pulse train to the modulator.

9.5 MODULATORS FOR MMW TRANSMITTING TUBES

The modulator and high-voltage power supply for any radar transmitter are critical elements, and they must be properly specified and designed. Overall frequency stability for noncoherent systems and phase stability for coherent systems are dependent on the performance of the modulator and power supply. Individual tube electrode sensitivities determine the regulation required of dc voltage supplies and modulator pulse shape (flatness, jitter, *et cetera*). Other factors that affect modulator design include pulsewidth, PRF, operating voltage and current, and size and weight restrictions.

There are two major types of modulator topologies: *line-type modulators* and *hard-tube modulators*. Line-type modulators are primarily used for pulsing magnetrons, since these tubes exhibit lower frequency pushing or pulling figures than other MMW tubes, like TWTs, EIOs, and EIAs. Hard-tube modulators are more easily designed to provide the extremely flat pulses required by tubes with stringent electrode voltage sensitivities. With the advent of high-voltage MOSFET technology and high-μ gridded transmitter tubes, some hard-tube modulator configurations can be converted to solid-state pulsers using FETs as the high-voltage switches. At the present time, the use of solid-state modulators is limited to the 1 kV pulse voltage region.

Due to the high cost and generally low availability of high-power MMW tube sources, tube protection is a critical factor in the design of the modulator-power supply. Design of the power supply should ensure that stored energy is kept to a value below that which would damage the tube in the event of an arc. Protective circuits should be incorporated to prevent the application of high voltage to the tube in the event that any critical parameters, such as heater voltage, beam voltage, body or collector current, and body or collector temperature, are out of safe operating range.

9.5.1 Line Modulators

The line modulator is so named because the generation of the output pulse is caused by the discharge of an energy storage element through a switch to the load. The energy storage element may be considered as an open-circuited transmission line of impedance Z_0 and one-way propagation length T, as shown in Figure 9.18. If the transmission line is initially charged to a voltage V, at the instant switch SW is closed, a pulse of length $2T$ and magnitude $V/2$ will be generated across a matched load of impedance Z_0. In practice, the transmission line is replaced by a network of lumped inductive and capacitive elements, called a *pulse-forming network* (PFN). The design of PFNs has been well treated in the literature (Ewell [20]), and so will not be addressed here.

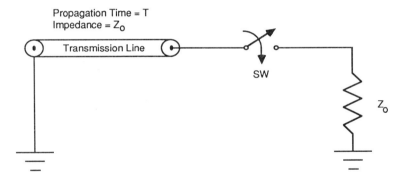

Fig. 9.18 Generic representation of the line-type modulator's pulse generation.

Figure 9.19 is a simplified schematic of a line-type modulator. The dc power supply charges the PFN through the charging choke during the interpulse period. The series diode prevents the PFN from discharging once it is charged. A pulse transformer is frequently used to step up the pulse voltage to the load, thereby reducing high-voltage requirements on the power supply, diodes, and switch (in this case, a hydrogen thyratron). When the switch is closed, the PFN is discharged into the load through the pulse transformer.

Figure 9.20 is a representation of the output pulse from a line-type modulator and delineates distinct parts of the pulse that must be considered

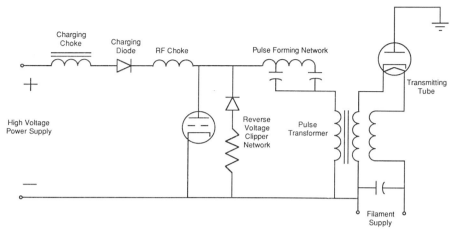

Fig. 9.19 Simplified schematic of a line-type modulator.

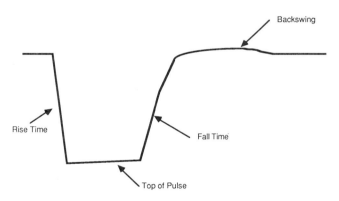

Fig. 9.20 Idealized output pulse of a line-type modulator.

in the design process. The rise time should be short enough to produce a good pulse shape, but magnetrons can suffer from moding and starting problems if the rate of rise of applied voltage is too fast. The top of the pulse should be sufficiently flat so as not to produce undesirable frequency shifts or modulation. The fall time should not produce excessive backswing, which could damage the tube. All of these requirements must be taken into account in the design of the PFN and pulse transformer.

Line-type modulators can typically be designed for small size, light weight, and high efficiency. Line-type pulsers, however, suffer from a lack of wide range of adjustability of the interpulse period or PRF, which is limited by the recharging circuit (power supply voltage V and charging inductance L). In addition, adjustable pulsewidths are available only if special techniques are used to add extra components and complexity to the circuit [20]. Long pulsewidths require large PFNs and pulse transformers. This disadvantage is rarely encountered, since most MMW magnetrons are limited to pulsewidths of less than 1 µs.

9.5.2 Hard-Tube Modulators

Hard-tube modulators are used when better control of the pulse shape (flatness, rise and fall times, *et cetera*) is needed. This need arises in the linear-beam tubes (TWTs, EIOs and EIAs), which have higher electrode voltage sensitivity factors than magnetrons. Hard-tube modulators also benefit from high flexibility in pulsewidth and PRF, items which are limited in line-type modulators.

Figure 9.21 is a simplified schematic of a floating deck hard-tube modulator. Such a configuration is used for pulsing of sources with modulating anodes or grids (EIOs, EIAs, TWTs), or for direct cathode pulsing of tubes, without the use of coupling capacitors or inductors. High-voltage isolation must be provided for the control pulses to the switch tubes. This can be provided by transformer or capacitive coupling, or by way of a fiber optic link from a ground-referenced trigger circuit. To turn on the MMW tube, switch tube A is pulsed on, while switch tube B is off. At the end of the pulse, switch tube A is turned off and switch tube B is turned on, and it acts as a "tail biter" at the trailing edge of the pulse to provide a fast fall time. Without the use of the tail biter tube, the fall time would be set by the exponential discharge of the modulating anode (or grid) to cathode capacitance plus stray circuit capacitance (which should be minimized through careful mechanical layout) through the pull-down resistor R. Resistor R keeps the transmitting tube in the cut-off region when switch tubes A and B are both off. Through this arrangement, extremely fast rise and fall times as well as flat pulses can be obtained.

9.5.3 Solid-State Modulators

Some high-µ gridded source tubes require only hundreds of volts of pulse swing, instead of thousands of volts, to bring the tube from cut-off to conduction. Such voltages can be handled by solid-state switches, and thus preclude the necessity for bulkier vacuum tubes and separate heater power supplies, with a subsequent reduction in size, weight, cost, and

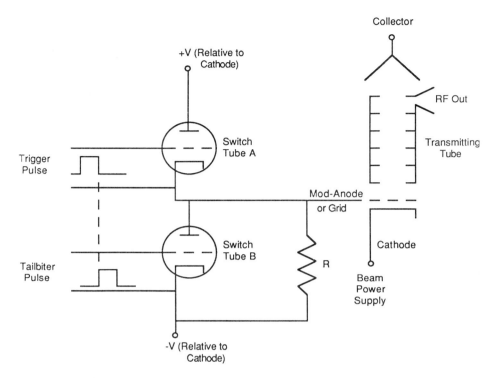

Fig. 9.21 Floating deck hard tube modulator.

complexity. The reduction in size and weight can be extremely important for MMW systems, where the choice of millimeter waves is indeed made for compact and lightweight hardware.

Of the available solid-state switches (MOSFETS, BJTs, and SCRs), MOSFETS seem to have an advantage in the application of high-voltage, high-speed switching. Although BJTs are available with breakdown voltages to approximately 1.5 kV, these devices suffer from storage time problems, which decrease maximum switching speed. SCRs can be permanently damaged by excessive dv/dt or di/dt, which make them difficult to use for high-speed switching. MOSFETs are available with breakdown voltages to 1 kV. The switching speed of MOSFETS is set by the current available to charge the total input capacitance, which can be several thousand picofarads. This input capacitance consists of the gate-to-source capacitance plus the reverse transfer or Miller capacitance associated with any high-gain amplifier.

MOSFETs, however, suffer a few disadvantages as compared with vacuum tubes for high-voltage switching. Typical output capacitance for a high-voltage MOSFET is on the order of thousands of picofarads, while interelectrode capacitances for planar triodes are on the order of only a few to 10 picofarads. Switching losses P can be approximated by $P = (C)(V^2)(\text{PRF})$, where C is the output capacitance and V is the pulse voltage swing. Therefore, for the same PRF and pulse voltage, the switching losses in a MOSFET will be about two orders of magnitude higher than those for a vacuum tube.

9.5.4 Modulator Efficiency

Modulator, HVPS, and tube efficiencies directly affect the size and weight of the transmitter assembly. Line-type modulators are generally very efficient due to the absence of lossy components. The vacuum tubes for hard-tube modulators and MOSFETS for solid-state modulators require heat sinks with appropriate thermal impedances. For a modulated-anode or gridded pulsed tube, most of the losses in the modulator are switching losses, since the load consists primarily of only a capacitor. For high-PRF systems, losses in a solid-state modulator may be excessive. For dc power supplies, lossy linear regulators should be avoided if possible. High-efficiency, high-frequency switch mode converters should be used instead. Many design examples of transmitter power supplies are available in the literature [21–26].

REFERENCES

1. P. Bhartia and I.J. Bahl, "Millimeter Wave Sources, Part 1—Tubes," Chapter 2, *Millimeter Wave Engineering and Applications,* John Wiley and Sons, New York, 1984.
2. A.E. Acker, "Interest in Millimeter Waves Spurs Tube Growth," *Microwaves,* Vol. 21, No. 7, July 1982.
3. "Millimeter Wave Transmitter Subsystem," Technical Data Sheet, Varian Canada Microwave Division, Georgetown, Ontario, Canada, 1984.
4. U.J. Pittack, "Advances in TWTs Promise Higher Power, Smaller Packages," *Microwave Systems News,* June 1983.
5. K. Amboss, "The Current Art of Millimeter Wave Solid State and Tube Type Power Sources," *Proceedings of Military Microwaves,* London, October 1980.
6. Hughes Aircraft Company, *Hughes TWT and TWTA Handbook,* Hughes Electron Dynamics Division, Torrance, CA.
7. J.J. Hamilton, R. Harper, and U.J. Pittack, "Recent Advances in

High Power Millimeter Wave Traveling Wave Tubes," *6th International Conference on Infrared and Millimeter Waves,* Miami Beach, FL, December 1981.

8.   M.J. Cascone, "Unbraze Your Helix for High Power," *1984 Microwave Power Tube Conference,* Monterey, CA, May 1984.

9.   G. Wada, R.H. Ohtomo, J.T. Benton, M.H. Pohlneg, "Recent Developments in Millimeter-Wave Helix TWTs," *1984 Microwave Power Tube Conference,* Monterey, CA, May 1984.

10. J.A. Davis and J.P. Vaszari, "New Developments in Gridded Coupled Cavity Millimeter Wave TWTs," *International Electron Devices Meeting,* San Francisco, CA, December 1982.

11. A. Karp, " 'Comb-Quad' Millimeter Wave Coupled-Cavity TWT Interaction Structure," *International Electron Devices Meeting,* Washington, DC, December 1981.

12. R.M. Phillips, "Some Surprising Helical Interaction Circuits May Hasten Millimeter Waves," *International Electron Devices Meeting,* San Francisco, CA, December 1982.

13. B. Berry, "Gyro-Devices Fulfill Important Tasks in Microwave Systems," in *The Microwave Systems Designers Handbook, Microwave Systems News,* Vol. 14, No. 7, July 1984.

14. P. Ferguson, *et al.,* "A High Efficiency High Power 100 GHz Gyrotron," *1984 Microwave Power Tube Conference,* Monterey, CA, May 1984.

15. V.L. Granatstein and S.Y. Park, "Survey of Recent Gyrotron Developments," *International Electron Devices Meeting,* Washington DC, December 1983.

16. B. Arfin, *et al.,* "Gyro-Klystron Amplifier Design and Performance," *1984 Microwave Power Tube Conference,* Monterey, CA, May 1984.

17. L.R. Barnett, *et al.,* "A Wideband Fundamental Mode Millimeter Gyrotron TWA Experiment," *International Electron Devices Meeting,* San Francisco, CA, December 1982.

18. T.L. Lane, *et al.,* "Coherent High Power 95 GHz Radar Development," Interim Technical Report No. 1, prepared by the Georgia Tech Research Institute for US Army MICOM, under Contract No. DAAH01-34-C-0853, October 1985.

19. K.E. Kreischer, *et al.,* "Multimode Oscillation and Mode Competition in High-Frequency Gyrotrons," *IEEE Transactions on Microwave Theory and Techniques,* Vol. MTT-32, No. 5, May 1984.

20. George W. Ewell, *Radar Transmitters,* McGraw-Hill, New York, 1981.

21. J.W. Williams, "High Reliability High Voltage Power Processors," *IEEE Power Electronics Specialists Conference,* Culver City, CA, 9–11 June 1975.

22. S.R. Peck and R.L. Rauck, "A Pulsed TWT Power Supply and Test Console," *IEEE Power Electronics Specialists Conference*, Palo Alto, CA, 14–16 June 1977.
23. W. Mueller and W. Denzinger, "Square Wave AC Power Generation and Distribution of High Power Spacecraft," *IEEE Power Electronics Specialists Conference*, Palo Alto, CA, 14–16 June 1977.
24. B.P. Loraelson, et al., "A 2.3 kv High-Reliability TWT Power Supply: Design Techniques for High Efficiency and Low Ripple," *IEEE Power Electronics Specialists Conference*, Palo Alto, CA, 14–16 June 1977.
25. B.F. Farber, *et al.*, "A High Power TWT Power Processing System," *IEEE Power Electronics Specialists Conference*, Murray Hill, NJ, 10–12 June 1974.
26. S.R. Peck, "A 100 Watt TWT Power Conditioning System," *IEEE Power Electronics Specialists Conference*, Culver City, CA, 9–11 June 1975.

Chapter 10
MMW Radar Receivers

J.A. Scheer

Georgia Institute of Technology
Atlanta, Georgia

10.1 INTRODUCTION

The function of a MMW radar receiver is the same as that of a microwave radar: to sense the signals returning to the system from various targets and provide an output to a display device or processor. This is accomplished through a series of processes including amplification, frequency conversion, filtering, and detection. The basics of radar receiver technology and a description of the componentry available for millimeter wave radar receivers are presented in this chapter. Typical parameters, indicative of the performance achievable at both 35 and 95 GHz are given, but, since the current MMW technology is rapidly improving, it is suggested that the system designer consult with the various vendors to determine the optimum configuration of components for his or her particular need. For example, the designer of an instrumentation radar may demand a superior performance in noise factor, whereas the environmental performance could be compromised to achieve the design goals.

Before discussing the details of the available technology for MMW receivers, it is appropriate to discuss the general design considerations of a radar receiver, and more particularly to address the features of a MMW system which make these considerations unique to that class of radars.

10.2   BASIC RECEIVER DESIGN FUNDAMENTALS

10.2.1   Block Diagram

Figure 10.1 is a simplified block diagram of a radar receiver, showing its relationship to the other major radar subsystems. The received signal arrives at the antenna and is applied to the receiver through the transmit-receive (TR) switch. The function of this switch is to direct the transmitted energy to the antenna and the received energy to the receiver. It protects the receiver from the effects of the relatively high-power transmitted signal. In the case of a high-power transmitter, more typical at microwave frequencies than at MMW frequencies, the switch may be implemented using a hybrid and a gas-fired TR tube, as shown in Figure 10.2. The high-power signal from the transmitter, depicted as a solid line, causes the gas in the tube to ionize, creating a short circuit in the waveguide. This short circuit causes the transmit signal to reflect from the tube and travel back through the hybrid to the antenna port. The low-power received signal, depicted by a dotted line, will not have sufficient energy to ionize the gas, so the signal penetrates the tube to be applied to the receiver.

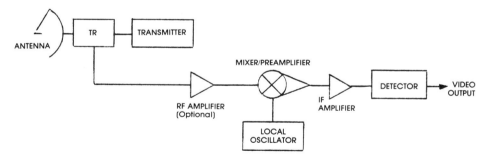

Fig. 10.1 Simplified radar system block diagram.

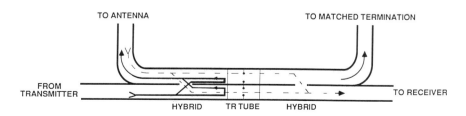

Fig. 10.2 TR switch using hybrids and TR tubes.

In the case of low-power transmitters, typical at MMW frequencies, the TR device may be a circulator, which is a three-port ferromagnetic device that directs energy entering input port 1 to port 2, and energy entering port 2 to port 3. Some circulators have a fourth port, which is normally terminated.

There is always some amount of unintentional leakage signal from the transmitter entering the receiver through the TR device. In some cases, this will be enough energy to damage or destroy the sensitive receiver mixer diodes in the first down-converter stage. Even if the leakage energy is not sufficient to destroy the mixer diodes, it is undesirable from the point of view that it nearly always has sufficient energy to saturate the receiver. When this occurs, the receiver requires some time (several microseconds) to recover from the saturation state, during which time it has reduced sensitivity. To minimize the effects of the unintentional leakage, a receiver protection switch is often employed between the circulator and the mixer as shown in the block diagram of Figure 10.3.

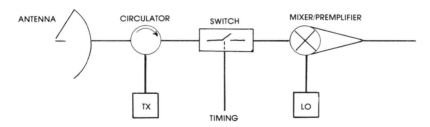

Fig. 10.3 Use of a receiver protection switch.

If the expected leakage power level is high, a ferrite switch having a relatively high power-handling capability may be employed. For a low-power leakage signal, a solid state switch using a positive-intrinsic-negative (*pin*) diode may be used. The switch is controlled in synchronism with the transmitted pulse such that it is in the high attenuation mode during the transmitted pulse and in the low attenuation mode, otherwise.

The switch normally takes some time to change from the high to the low attenuation mode. This will limit the minimum range, but not as severely as would the saturation recovery time. The *pin* diodes are used when fast switching is desired. These diodes can switch in less than 100 ns, giving a 15 m minimum range. The *pin* diodes often introduce more loss, however, than a ferrite switch, which may require a microsecond to switch. The trade-off required in choosing between the two types of switches depends on the desired minimum range and the required sensitivity.

For most radar systems, the received RF signal is converted to an equivalent signal at an IF, where the subsequent functions (i.e., amplification, gain control, filtering, and detection) are performed better than they would be at RF. In some cases, a stage of RF amplification may be employed before the down-conversion process, depending on whether the noise performance of the system can be improved by doing so. In general, the noise figure of MMW amplifiers is considerably worse than that of the mixer-preamplifiers available, so all amplification and subsequent processing is generally done at IF.

The down-conversion to IF is performed by the frequency subtracting action of the mixer. The two inputs to the mixer are the received signal, which is nominally at the transmitted frequency, and a local oscillator frequency, which is near the RF, but at some designed offset value. The output of such a mixer has components resulting from the sum, the difference, and each of the input freqencies. The sum and the two original frequencies are suppressed by the low-pass filtering action of the IF amplifier circuits. The difference frequency is the intentional signal, and therefore it is the one which is processed. Common IF frequencies for microwave radar systems are 30 and 60 MHz, and other typical frequencies are 45 and 70 MHz. Any convenient IF can be used, but there is a greater availability of processing components at the more popular frequencies.

Millimeter-wave radar systems often use a higher IF than lower frequency radars because higher IF bandwidths are usually employed, requiring a higher IF center frequency. Typical intermediate frequencies for MMW radars include 100, 120, 300, and 3000 MHz. In fact, there is no particular standard IF frequency for wideband millimeter wave radars as is the case for microwave systems.

The signal received from the target is usually of very low amplitude. Amplification is required before any meaningful processing can be applied to the signal. Signal power gain is performed at IF, using either wideband amplifiers or tuned amplifiers. Tuned amplifiers can usually provide more gain per stage of amplification for a given gain-bandwidth product capability because they have a narrower bandwidth than untuned wideband amplifiers. The use of wideband amplifiers, however, provides the designer with more flexibility in the subsequent processing and filtering.

The output of the preamplifier is amplified and processed by whatever means that the system requires. In the least complex systems, the IF signal is simply detected for an oscilloscope, or *plan-position-indicator* (PPI) display. In more sophisticated systems, a signal processor may be used to provide target detection and recognition. Typical processes include MTI, FFT Doppler processing, CFAR thresholding, and pattern recognition algorithms. Most of these are implemented in the digital domain in which case the detected video is converted from analog before being applied to the processor.

10.2.2 Down-Conversion

Single Conversion—Spurious Signal-Image Considerations

The intentional output from the mixer in a down-converter stage of a radar is the difference between the local oscillator (LO) frequency and the RF applied to the mixer. In fact, other frequencies are present at the output. The sum frequency and each of the input frequencies are the most obvious of these undesired frequencies, but there are others as well. Since the mixer contains nonlinear devices, harmonics of the various inputs are generated, and these harmonics and the differences among them are also present. Careful selection of the frequencies must be made to ensure that the undesired signals are not within the passband of the subsequent receiver stages. The lower order harmonics are of the most interest because the higher order harmonics are reduced in amplitude, as we can see from Table 10.1, which shows the relative levels for the various harmonics from a typical family of mixers. A complete discussion of the use of Table 10.1 is given in Watkins-Johnson Company's *RF Signal Processing and Components Catalog* (1985–1986).

Dual Conversion

Dual conversion is characterized by the use of two up-conversion, or down-conversion mixing processes, in series, to develop the transmitted RF or the final received IF for processing. Figure 10.4 is a block diagram demonstrating this technique.

In some systems, for reasons often associated with the transmitting function, dual up-conversion is employed, requiring dual down-conversion. This is typically true for wideband, frequency-stepped systems in which dual up-conversion is employed to simplify the filtering of the carrier and unwanted sidebands. Typically, the first up-conversion is performed using a fixed-frequency coherent oscillator (COHO) and a frequency-stepped frequency synthesizer signal. The second up-conversion mixes this frequency-agile waveform with an RF stable local oscillator (STALO) to the desired RF.

In the first receiver down-conversion stage, the LO is typically the STALO, which is at a fixed frequency. In the case where the RF is stepped in frequency, the resulting IF is varying in frequency over the same bandwidth as the RF. In the second mixer function, the second LO is split in much the same way as the first, and then applied to the mixer LO ports. Since the second is typically a tracking LO (that is, it changes frequency on a pulse-to-pulse basis with the transmitter), the IF at this point is a fixed frequency, typically a relatively low IF such as 60, 160, or 300 MHz.

Table 10.1 Example of an Intermodulation Chart Showing the Power Level of Various Intermodulation Products Relative to IF Output Power.*

Harmonics of f_R \ Harmonics of f_L	0	1	2	3	4	5
7	79	>99	>99	69	>99	78
6	90	>99	>99	86	>99	66
5	72	93	73	70	96	82
4	80	96	80	79	91	55
3	51	63	58	49	73	66
2	69	68	67	72	76	74
1	25	25	0	0	29	16
0	24	23	27	26	31	14

Harmonics of f_R (rows) — Harmonics of f_L (columns)

*Watkins-Johnson Company, *RF Signal Processing and Components Catalog*, 1985–1986, p. 663.

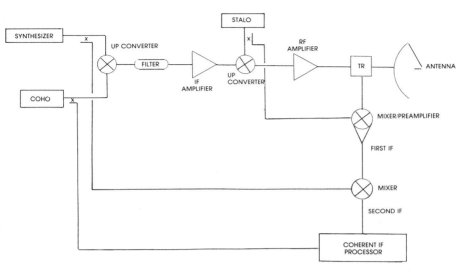

Fig. 10.4 Dual conversion radar block diagram.

This makes the design of the subsequent IF processor simpler than if it had to handle wideband signals.

10.2.3 Gain Control Considerations

Often, the dynamic range of the received signals exceeds the instantaneous dynamic range of the radar receiver and display device. Thus, the ratio between the smallest signal expected, usually at the noise level, and the largest signal expected, from very close targets of large cross section, is greater than the range of signal levels which can be displayed or processed, without the strongest signal saturating the system, or the weakest signal being below the detection threshold, or both. The dynamic range of a system can be extended by employing a gain control function, which allows the dynamic range of the input signals to be matched to the dynamic range of the system processor or display.

IF amplifiers often provide a gain control function, allowing the gain of the receiver to be adapted for various signal levels. In many cases, the increasing or decreasing signal strength associated with a target is predictable, and the receiver gain can be manually adjusted such that the instantaneous dynamic range of the received signal does not exceed the dynamic range of the processor or display. In these cases, the operator has a *manual gain control* (MGC) function, which he or she adjusts for the condition in which the dynamic range of the received signal matches the dynamic range of his display.

Automatic gain control (AGC) is a mode in which the gain of a receiver is automatically controlled, usually such that the target of interest (identified as the target in the range gate) is kept at a constant output level. This is usually done to optimize the performance of some processing functions to follow, such as range tracking and angle tracking, which are usually designed to perform best for a constant signal level.

If the scenario is dynamic, as is the case with airborne radar systems, the gain control function is controlled automatically, such that the target strength is kept relatively constant. In the AGC mode, the target isolated in the range gate is the one the radar system keeps at constant amplitude. The block diagram of a typical AGC function is shown in Figure 10.5. The video amplitude out of the video amplifier is sampled at the range-gate time to establish a dc estimate of its value, and compared to a reference voltage in a subtracting operational amplifier. The difference represents the *gain error*. This error is integrated, and applied to a gain control device, which adjusts the gain to keep the video of interest equal to the reference voltage.

Sensitivity time control (STC) is another gain control mode in which the gain of the receiver is varied as a function of time, within an interpulse period, beginning with reduced gain at transmit time and increasing at some rate to maximum gain at some time after transmit time.

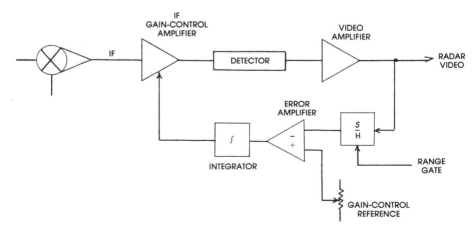

Fig. 10.5 Automatic gain control (AGC) block diagram.

The STC gain control mode quickly changes the gain according to a prescribed function during each pulse interval. This mode corrects for the expected strength of the close-in targets relative to that of the farther targets, due to the R^4 term in the denominator of the radar range equation. Correction is done in an open-loop fashion; that is, the video output is not sensed to control the gain. The gain is reduced by some amount, usually determined by the operator, at transmit time, and allowed to increase at some slope, again usually controlled by the operator, until the maximum gain is achieved.

Figure 10.6 shows the gain as a function of time for a typical STC implementation. A system pretrigger usually sets the close-in gain to some level (amplitude), and immediately after transmit time the gain increases toward the maximum gain at some rate (slope). The amplitude and slope are normally controllable by the operator who adjusts these parameters to optimize the presentation. The influence is most important in a system that uses a PPI display, wherein the strong targets will produce a very bright spot at the center of the display if not suppressed.

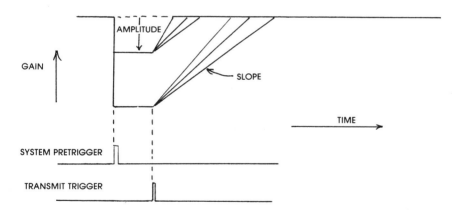

Fig. 10.6 Sensitivity time control (STC) operation.

10.2.4 Noise Figure—Minimum Detectable Signal

Of primary concern to the radar system designer is the ability to detect targets in the presence of competing interfering signals. Interfering signals arise from thermal noise, electromagnetic interference (EMI) and nontargets received by the radar, such as weather and land clutter returns, assuming that these are not targets of interest.

The system noise figure (sometimes called noise factor) is normally established at the RF amplifier or mixer-preamplifier stage. It is equal (in decibels) to the conversion loss of the mixer added to the noise figure of the preamplifier, assuming that there is sufficient gain in the preamplifier to suppress the effects of noise added by subsequent stages. It is important to note that losses prior to this point in the system affect the signal-to-noise ratio, while effects of resistive losses and attenuation subsequent to this point generally are reduced. For this reason, it is prudent to minimize the incorporation of extraneous RF components ahead of the mixer-preamplifier stage.

At this point, the concept of *minimum detectable signal* (MDS) will be introduced. MDS is defined as the minimum signal level which gives reliable detection in the presence of Gaussian noise. The detection performance is related to the received SNR. The statistical nature of the detection process necessitates the need to express detection in terms of probabilities of detection and false alarm. This topic is discussed in Chapters 2, 3, and 13. It is appropriate here to present the method for determining the received SNR.

The noise power with which the target of interest must compete can be determined from the available noise power, P_n, as predicted by (Blake [1]):

$$P_n = kT_sB_n \tag{10.1}$$

where

k = Boltzmann's constant (1.38×10^{-23}J/K),
T_s = the system noise temperature,
B_n = the noise bandwidth.

The 3 dB bandwidth of the receiver is often used as an approximation for the noise bandwidth because it is considerably easier to determine the 3 dB bandwidth and the approximation is good for most typical radar receiver designs.

The radar system designer is not normally armed with the information regarding the system noise temperature. The designer is more likely to know the noise figure of the receiver system, which is a term relating the actual noise at some point in the receiver to the expected noise power as predicted by (10.1). To expedite the determination of system noise power, a substitution of T_0NF_0 for T_s is often made, where T_0 is the standard temperature (290 K) and NF_0 is the system operating noise figure. The most "seat-of-the-pants" approach to predicting the system noise power is to use the receiver noise figure (NF), as determined from vendor data or laboratory tests on the receiver components, and the expression:

$$P_n = kT_0B_n\text{NF} \tag{10.2}$$

Usually, the noise figures associated with individual devices, such as the RF amplifier (if one is employed), the mixer-preamplifier, and subsequent IF amplifiers, are known. The equivalent noise figure, F, of a set of cascaded components can be determined from (Skolnik [2]):

$$F = F(1) + [F(2) - 1]/G(1) + [F(3) \\ - 1]/G(1)G(2) + \ldots \tag{10.3}$$

where

$F(n)$ = the noise figure of the nth stage,
$G(n)$ = the power gain of the nth stage.

It is evident from (10.3) that substantial power gain in a given stage reduces the effect of high noise figures in the subsequent stages. For this reason, the noise figure of an RF amplifier or the first mixer-preamplifier element (in the absence of an RF amplifier) establishes the noise performance of the radar receiver, assuming that reasonable gains are exhibited.

As an example, for the parameters shown in the block diagram of Figure 10.7, the total noise figure, considering all three stages is 5.07 (7.05 dB), which is only an increase of a factor of 1.01 (0.05 dB), relative to the noise figure of the mixer-preamplifier. Notice that the relatively high gain of the first stage (100) has diminished the effect of the relatively noisy subsequent stages. If the gain of the preamplifier had been low, the total system noise figure would have been higher.

It is important to note that most component vendors measure and specify the double-sideband noise figure of their equipment. When used in a radar system, the single-sideband noise figure, determined by doubling the double-sideband noise figure, is the proper parameter to use when predicting SNR.

10.2.5 Bandwidth

The instantaneous receiver bandwidth must match that of the received signal, which is about the same as the transmitted signal. A rectangular RF pulse of width τ has a spectrum with a shape of $\sin(x)/x$, where the null-to-null bandwidth is $2/\tau$. The RF signal is mixed down to an IF signal having the same spectral characteristics as the RF signal. A first-order approximation for the 3 dB bandwidth to receive a pulse of width τ with reasonable fidelity, with an optimum SNR, is found from

$$B = 1/\tau \tag{10.4}$$

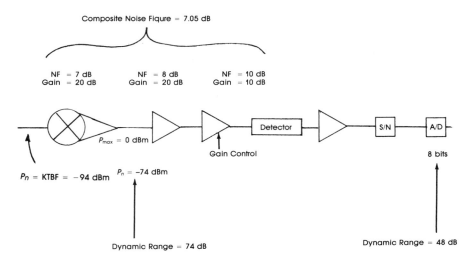

Fig. 10.7 Block diagram showing noise figure and dynamic range considerations.

Depending on the detailed shape of the pulse and the receiver bandpass response, the optimum SNR may not occur with a bandwidth of $1/\tau$, but this approximation is commonly used. Some designers use a bandwidth of $1.2/\tau$, which produces a more optimal SNR for some cases (Barton [3]). Multiple stages produce a composite bandwidth less than either of the individual stages. In this case, it is appropriate to employ bandwidths of the individual stages that are somewhat larger than $1/\tau$ at IF and $1/2\tau$ at video.

After the IF processor, the signal is usually converted down to baseband (video) by a conventional amplitude detector or a coherent I/Q detection process. In either case, the bandwidth of the signal at video is about one-half of the bandwidth at RF or IF because the signal is folded around dc. That is, the signal having a given bandwidth is centered at zero hertz. Half of the signal bandwidth is from zero to $1/2\tau$, and the other half is from zero to $-1/2\tau$. The negative half appears in the positive frequency region. Therefore, the video bandwidth needs to be only half as wide as the predetection IF bandwidth to maintain pulse fidelity.

10.2.6 Dynamic Range and Gain Control Considerations

Dynamic range is defined as the ratio (usually expressed in decibels) of the highest signal that can be handled (received and processed) to the noise level of the stage being considered, without extending beyond the linear response of the system. There are typically two ways to specify the dynamic range of a system, instantaneously and time variably. Both methods are widely used, and often employed for the same radar system. *Instantaneous dynamic range* refers to the range between the noise level and the saturation level without regard to varying the gain of any of the receiver stages. Total dynamic range is typically greater than the instantaneous dynamic range because the gain of one or more stages is allowed to be varied in establishing this parameter. For example, in the case where a system is configured for detection of the lowest level of expected signals, the gain of some intermediate stage may be quite high. At the time when the system is to detect a higher signal level without introducing distortion (i.e., when the target is at a closer range), some gain reduction may be introduced. Figure 10.7 is the block diagram of a typical receiver, showing the components that contribute to the limitation of the dynamic range. The noise level, which represents the receiver "floor" is established at the mixer-preamplifier. The noise power, as discussed above is determined by the factor $kTBF$.

Referring to Figure 10.7, the gains of the stages are the RF-to-IF gain of the mixer-preamplifier, the gain of the IF amplifier, the detector loss, and the video amplifier gain. Gain control is normally employed at the IF amplifier, the minimum normally being that required to match the system saturation point with the maximum signal out of the mixer-preamplifier, and the maximum being that required for noise to become detectable.

The maximum signal level which can be tolerated is determined by the points at which any of the various amplifier stages become saturated, or nonlinear to the extent that the outputs are not usable for the system application. The maximum signal level must be calculated at each stage. When this level is compared with the noise level at each stage, the dynamic range at each stage can be determined. The instantaneous dynamic range is equivalent to the smallest of these values. Often, the output display device or an A/D converter is the limiting item in determining the instantaneous dynamic range.

In the case of the example depicted in Figure 10.7, the input experiences a dynamic range of 74 dB (a noise level of −94 dBm and an input signal of −20 dBm are required to saturate the preamplifier output). The dynamic range of the A/D converter is 48 dB. Therefore, gain control of 26 dB is required to allow proper detection of the full range of input signal levels.

10.2.7 Detection

In most radar systems, the signal processing or display functions are applied to video (baseband) signals. In very few systems is processing of the IF signal employed. For this reason, some form of detection of the IF signal is required. The two basic detection approaches are *noncoherent* and *coherent detection*. Noncoherent detection involves detecting the envelope of the waveform, providing only amplitude information, while coherent detection involves maintaining the phase information in the received signal. Coherent detection requires that a phase reference be maintained in the radar system, while noncoherent detection does not. Each approach will be described below.

Noncoherent Linear and Log Detection

Noncoherent detection is performed in one of a number of ways. Two basic classes of noncoherent detection to be discussed here are *linear* and *logarithmic detection* (Blake, Taylor, and Mattern [4]). Simple diode detection is sometimes employed for linear detection, as shown in Figure 10.8(a), where the diode is biased with a dc trickle current to improve its sensitivity. Another approach is to use a synchronous detector, as shown on Figure 10.8(b).

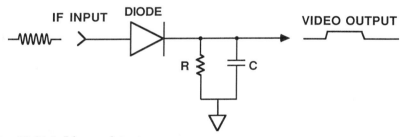

Fig. 10.8(a) Linear detector.

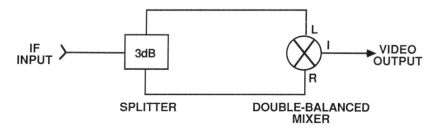

Fig. 10.8(b) Square law detector.

Logarithmic detection is often employed to increase the dynamic range of the detection system. In this mode, the output is proportional to the log of the input, where volts out represent dBm in. A wide dynamic range of input signal levels is compressed into a smaller dynamic range of output signal levels, which is appropriate for matching the dynamic range of a typical display or processor. A transfer curve of a typical log detector is shown in Figure 10.9 [5]. The dynamic range of such a system is usually in excess of 60 dB, with some systems being 80 dB.

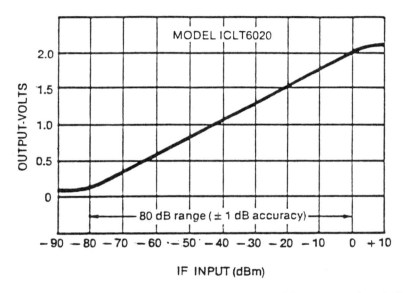

Fig. 10.9 Typical transfer curve for an IC log amplifier. Input signals from −80 to +10 dBm can be accommodated with video outputs compressed to 0.2 to 2.0 V (20 dB). Note that the ±1 dB accuracy obtained over an 80 dB portion of the dynamic range (from [5], courtesy of RHG Electronics, Inc.).

Figure 10.10 is a block diagram of a typical implementation of a log detector. The output from each of the IF amplifier stages is detected and summed with the outputs of the other stages. The IF stages are designed to all have the same gain and to all saturate at some output signal level. As the signal input increases from some small value, the stages progressively saturate, starting with the last stage. As stages progressively saturate, they contribute no more to the summed output video, having the effect of reducing the gain for progressively increasing input signal level. The stages do not reach saturation in an abrupt manner, so the gain gradually

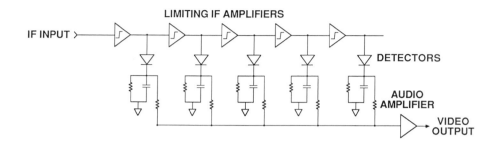

Fig. 10.10 Logarithmic amplifier-detector block diagram.

reduces, creating the smooth response shown in Figure 10.9. At signal levels below which any of the stages saturate, the response is linear. For this reason, such a system is sometimes referred to as a *lin-log detector*. Log detection is not normally employed in a coherent mode because of the difficulties in maintaining the precise delays that would be required to sum the IF signals at the appropriate phase, or to keep a predictable phase shift through the various stages with a wide dynamic range of input signal levels.

Coherent I/Q Detection

The coherent detection treated here produces *in-phase* (I) and *quadrature* (Q) video. These signals are typically used for coherent processes, such as integration, MTI filtering, and Fourier transform processing. The I and Q components are derived by mixing the received signal with in-phase and quadrature components of the COHO signal. Since the outputs of the balanced mixers used for this function typically produce a large undesired IF component and a larger component at twice the IF frequency (Kurtz [6]), low-pass filtering is required at this point. Care must be taken not to use a filter having too steep a slope, since such a filter time response can create a ringing condition.

Detection of both the I and Q components of the received vector has at least two advantages over detection of just one of the two. The most obvious is associated with the fact that ambiguities in the phase and Doppler frequency measuring capability are resolved. This is seen when considering the expressions for the two components of the vector as found from:

$$I = A \cos(\Theta) \tag{10.5}$$

and

$$Q = A \sin(\Theta) \tag{10.6}$$

where A is the vector amplitude and Θ is the phase. When measuring only one of these, the phase cannot be determined without ambiguity, and when performing a spectral analysis of a sequence of the returned signal samples, the Doppler frequency will have sign ambiguity. That is, the direction of target velocity cannot be determined.

Besides solving the Doppler ambiguity, there is another advantage to processing both I and Q. The SNR of the system is improved by 3 dB when both are processed (integrated). Although the system complexity is increased when two detector circuits are used, and the recording or processing load increases, these are often warranted by the improved performance achieved.

The I/Q detector network provides a phase-sensitive detector mode, which exploits the coherent nature of the radar. Figure 10.11 is a block-diagram of an I/Q detector network. There is one of these networks for each channel in the receiver subsystem. The received IF from the linear IF amplifier is divided to provide the input to two doubly balanced mixers, which are used as synchronous detectors. That is, the outputs are functions of the input signal levels and relative phase between the two. Since the COHO input is at a fixed high level, which is optimum for the operation of the mixer, the output level is dependent on the amplitude of only the signal input. Due to the function of the COHO quadrature splitter, the COHO inputs to the two mixers are 90° out of phase with each other. The integrity of the 90° is improved by the use of line stretchers or phase trimmers. The absolute phase of these signals is coherent (stable) with the transmitted waveform, which allows coherent detection of the targets. In particular, any Doppler frequency shift of the targets, due to target motion or frequency stepping, (which manifests itself as a pulse-to-pulse phase shift) can be detected and measured. The Doppler shift will show up on both the I and Q signals. The two components are used to resolve the direction of the velocity vector.

The outputs of the doubly balanced mixers provide the video signal representative of the two orthogonal components of the received signal vector. Unfortunately, also present at this output is some amount of the input IF signal, and a very large signal at twice the IF frequency. Both of these signals are unintentional and undesired. The low-pass filters following the mixers attenuate this undesirable IF signal. To maintain the 48 dB integrity associated with an 8-bit A/D conversion process, the unwanted signals must be suppressed at least 48 dB. A design of 60 dB ensures ample suppression. The IF signal will be suppressed by at least 20 dB by the mixer itself, so 40 dB additional suppression at IF is required. There is no mixer suppression of the second harmonic in the IF, so the low-pass filter must suppress this component by 60 dB. The three-pole low-pass filter, which has a break point at about $1/\tau$, usually satisfies these attenuation characteristics without introducing undue ringing on the video signal.

The video outputs at this point are applied to a data acquisition or display system, depending on the ultimate application of the radar.

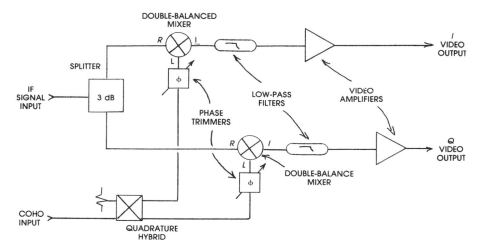

Fig. 10.11 I/Q detector network.

10.2.8 Phase Noise Considerations

The objective of having coherence in a radar system is usually associated with the desire to detect the Doppler frequency of moving targets, or to detect the synthetic Doppler frequency associated with a synthetic aperture radar or a stepped frequency waveform system. An important consideration in determining the performance of such a coherent radar system is the phase noise of the system, introduced by the various stable oscillators in the transmitter exciter and receiver subsystems. The object of coherence in a radar is to measure the absolute phase of the target return on each transmitted pulse. The ability to do so is limited by the phase stability of the reference oscillators. A perfectly stable target will appear to exhibit a phase scintillation if the oscillator which is used as a reference for measuring that phase is not perfectly stable. Single-sideband (SSB) phase noise is a measure of the stability of such oscillators. This parameter is normally specified in terms of power in a given bandwidth, i.e. dB relative to the carrier power (dBc) per hertz. When the noise power density expressed in these terms is multiplied by the processing bandwidth of the system, the total phase noise power can be determined. It is important to note that the phase noise is typically a function of the frequency separation between the oscillator frequency and the center of the processing band. The amount of phase noise exhibited in the system is found by integrating the phase noise of the oscillator (usually expressed in dBc/Hz) over the processing bandwidth.

Figure 10.12 is a plot of a typical phase noise specification for a phase-locked MMW oscillator [7]. In this example, if the phase noise of a STALO is − 82 dBc/Hz at 1 kHz offset, and the processing bandwidth is 1000 Hz, as would be the case for a 128 point FFT processor for a radar with a 128 kHz PRF, then the total phase noise due to the oscillator would be − 52 dBc. Generally, the phase noise contributions of the LO and transmitter add, increasing the phase noise by 3 dB. This would produce a 49 dB maximum on the clutter attenuation of a coherent Doppler radar. The dependence of the phase noise on the offset frequency means that phase noise varies depending on the Doppler bin of interest.

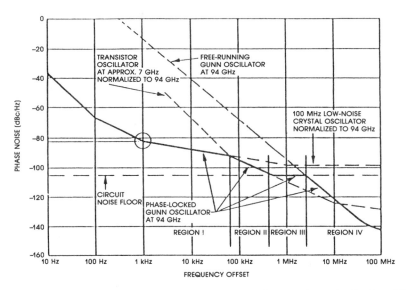

Fig. 10.12 Typical phase noise characteristics for a phase-locked milli-meter-wave oscillator (from [7], courtesy of Hughes Aircraft Company).

In many cases, the COHO, LO, or RF results from phase-locking a source to a harmonic of a stable crystal source. In these cases the basic phase noise is a multiple of the phase noise of the oscillator. For example if a $\times 10$ multiplier is used, the phase noise power of the oscillator is multiplied by 10^2, and degrades the power spectral density by 20 dB.

10.2.9 Multichannel Considerations

Some consideration must be given to isolation between channels in a radar system that has multiple channels. Depending on the application, it is usually important to not have the signal received by one channel contaminating the other channel. For example, in a monopulse tracking radar, where there are usually three channels, an error would be developed in the elevation track if some of the energy from the azimuth channel or sum channel were to corrupt elevation channel. In some cases this could be severe enough to cause a loss of target track. As another example, in a polarimetric radar in which two orthogonal polarizations are simultaneously received, erroneous data on target backscatter properties could be determined if there is insufficient isolation between receiving channels. Figure 10.13 shows the extent to which various levels of isolation will create errors in adjacent receiver channels due to crosstalk. The vertical axis shows the amount of error a channel will experience as a function of the interchannel isolation shown on the horizontal axis. This curve assumes that signals are of equal strength in the two channels. If one channel is different in signal level that the other, as is usually the case for the sum and error channels of a monopulse system, the error may be more severe. The amount of error can still be determined from the curve, but the isolation must be assumed to be less than that for equal signals, if the amount of the unintentional signal exceeds the intentional signal. For example, if the sum signal is 6 dB above the error signal and the channel isolation is 30 dB, we use a figure of 24 dB (30 dB − 6 dB) isolation to determine the amount of contamination, in this case about ±0.5 dB.

Several elements combine to limit the isolation which a radar receiver will exhibit. The antenna and radome related effects arise from radome depolarization, antenna depolarization, and isolation of the orthomode transducer and monopulse comparator. Receiver electronics related elements include local oscillator distribution isolation, power supply decoupling, and IF and video circuit separation and shielding. Care must be taken in design of the system to provide channel isolation necessary to avoid unacceptable channel-to-channel contamination. LO and COHO distribution networks should have isolators in the output paths and use an isolated splitter, such as a hybrid or magic *tee*. Good shielding and power supply decoupling is required in all stages of the system.

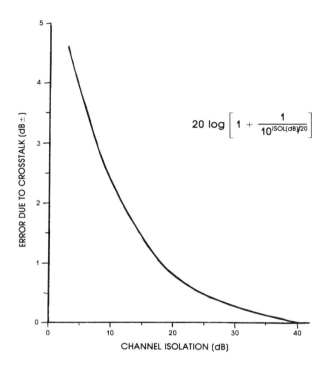

The equation shown in the figure:

$$20 \log \left[1 + \frac{1}{10^{ISOL(dB)/20}} \right]$$

Fig. 10.13 Crosstalk error as a function of channel isolation.

10.3 GENERAL MILLIMETER-WAVE RADAR RECEIVER CONSIDERATIONS

Figure 10.14 shows a block diagram of a receiver system employing a wide variety of features associated with the topics discussed in the preceding section. This figure is meant as an example and not to imply that it is typical. A typical system might employ some of the features, but probably not all of those shown in the example. The receiver system shown is dual-polarized. Another configuration would be a monopulse system, using three channels instead of two, but otherwise having many of the same features. Also, the system shown is coherent, that is, there is a phase reference from which the receiver can determine if the phase of the signal from the target is constant or changing on a pulse-to-pulse basis. This type of system provides valuable target motion information at the expense of incorporating expensive and complex coherent oscillators in the transmitter and receiver subsystems. For applications in which coherence is not required, considerable simplification of the receiver (as well as the transmitter) can be realized.

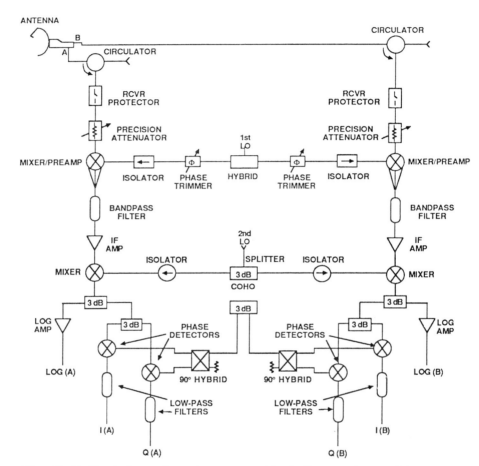

Fig. 10.14 Dual-channel receiver system block diagram (example).

Several classes of system applications which are particularly pertinent to MMW radars include instrumentation, tracking, and air-to-ground terminal homing. Each of these applications creates a particular set of design conditions which may make the design a challenging experience.

For example, instrumentation systems need to have a wide variety of functions and usually need to be versatile in their operation. These systems are used to collect data for operational systems designers. For these data to be of general value, all modes of operation need to be considered in the instrumentation sensor design. The radars often must be coherent, wideband, polarization-agile, dual-polarized, and have wide dynamic range. Since the data are used to understand target and clutter reflectivity characteristics, it is usually important to have a high single-pulse SNR. This means that a relatively high peak power requirement exists for these systems, sometimes creating a receiver protection problem. Calibration functions, in the form of calibrated attenuators and phase shifters, have to be implemented in the systems to provide the most useful data.

Air defense tracking applications of MMW radars arise because of the low angle capability of a MMW system compared to that for a lower frequency system for a given antenna size. The MMW system's narrow beamwidth reduces the multipath effects, and provides a more accurate track. Generally, a MMW radar is used synergistically with a more conventional microwave radar. The wider beam, higher power microwave system provides the acquisition and long range tracking, while the MMW system provides the more accurate lower angle track. Modern radar systems almost universally employ a monopulse tracking technique, requiring the use of a monopulse comparator network and attendent phase-matched electronics section; implications of this requirement on receiver design are addressed elsewhere in this chapter.

Air-to-ground applications of radar systems sometimes create another set of design challenges associated with the need for radar systems that occupy very little space and consume small amounts of power. Use of a single antenna is almost a universal requirement, necessitating the use of a circulator and sometimes a receiver protection device. The minimum range requirement for some of these systems requires a short receiver recovery time, which is a concern with the receiver protection device as well as with the recovery of the receiver amplifiers.

One of the advantages of MMW systems is the improved cross-range resolution achievable for a given aperture. The system designer often wants an associated improvement in range resolution. It is natural to expect such an improvement because for an increase in RF frequency, a given percentage bandwidth will yield an improved resolution. This increased bandwidth at RF must be matched at IF to realize the increased resolution. This implies that the IF center frequency is often higher than that for most microwave systems. In fact, an IF on the order of 3 GHz is not uncommon for MMW systems.

Many MMW systems employ some sort of pulse-to-pulse frequency-hopping technique to realize the wide bandwidth. This eliminates the data acquisition problems associated with very narrow pulses (either real or generated by using pulse compression techniques). Such a frequency-stepped system would probably employ a dual down-conversion approach to simplify the filtering requirements associated with eliminating images.

10.4 COMPONENTS FOR MILLIMETER-WAVE RECEIVERS

For the most part, the technology used at MMW frequencies is the same as that used at microwave frequencies. There are some cases, however, in which some special techniques are employed, due to the space limitations imposed by the small waveguide dimensions. In this section, a review of current technology in MMW radar receiver componentry is presented. Typical performance parameters available at the time of publication are given, but the reader is advised to consult with the various vendors to determine whether his or her particular requirements can be met.

10.4.1 Passive Components

10.4.1.1 Waveguide Components

There are a variety of waveguide components which are used in radar receivers to tailor and condition the signal for given purposes. Couplers are used for purposes such as test and sample ports. Attenuators and phase shifters are used for local oscillator amplitude and phase conditioning. These functions can be implemented in a number of ways. Variable attenuators and phase shifters are used where it is required to easily adjust these parameters. Some component vendors supply a commercial device with a manual adjustment for controlling the attenuation or phase. However, these devices can be made much smaller and cheaper in an integrated waveguide assembly for a production system where the requirement for adjustment is limited to the initial adjustment required for system set-up at the factory. In this case, a simple screw adjustment would be typical.

Figure 10.15 shows an ensemble of standard MMW waveguide components. So, going counterclockwise, starting from the top right, they are a direct reading precision attenuator, hybrid ring, broadband directional coupler, and E-plane *tee*. Starting from the bottom right, toward the left, the components are E/H-plane tuners, H-plane bend, Y-junction isolator, and broadband coupler.

Attenuators

Waveguide attenuators are available in three general types, level-setting, precision variable, and fixed. The level-setting attenuators are adjustable over a range of about 25 dB, and they have a maximum insertion loss of about 0.2 to 0.3 dB from 35 to 95 GHz. They consist of a waveguide

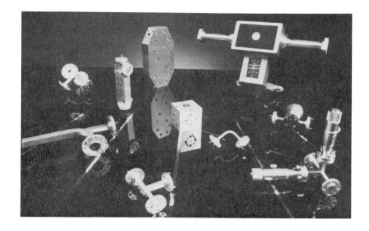

Fig. 10.15 Ensemble of MMW waveguide components (courtesy of Alpha Industries).

section with a flange at each end. A slot in the guide allows the controlled insertion of a resistively coated dielectric card. The insertion of the card is accomplished by rotation of a multiturn control mechanism. A locking device is usually employed to keep the card from inadvertently being moved after the proper setting has been made. Figure 10.16 is a photograph of several types of MMW attenuators. The larger ones are the precision calibrated types and the smaller ones are the uncalibrated style, which is used for level setting.

Fig. 10.16 Examples of MMW attenuators.

Precision attenuators are implemented by the incorporation of a re-sistive vane, which is installed within a circular section of waveguide. The precise rotation angle of the vane controls the attenuation. Usually, a calibrated dial is installed to provide a direct readout of the attenuation over a range of 0 to 50 dB. The insertion loss experienced in a precision attenuator is typically about 0.5 dB at 35 GHz and 1.0 dB at 95 GHz [8].

Fixed attenuators are comprised of a resistive card attached to a wall within a length of flanged waveguide. Such attenuators are available in values of 3, 6, 10, 20, 30, and 40 dB. with accuracies on the order of 0.1 dB [8].

Phase shifters

Variable phase shifters are flanged waveguide sections with dielectric vanes inserted into the guide. Insertion of the dielectric is accomplished with the aid of a multiturn dial mechanism. Typically, a maximum phase shift of up to 180° is achievable, with an insertion loss of about 0.2 dB at 35 GHz and 0.5 dB at 95 GHz. Direct-reading phase shifters are also available, and they have a half-wave rotating dielectric plate in a circular waveguide section. Readout accuracy on the order of 3° to 4° can be achieved. The insertion loss is about 1 dB at 35 GHz and 1.4 dB at 95 GHz [7, 8]. Figure 10.17 is a photograph of several MMW phase shifters. The larger ones are precision types and the smaller ones are uncalibrated, which are used for phase trimming.

Fig. 10.17 Examples of MMW phase shifters.

10.4.1.2   Ferrite Components

Isolators

Ferrite isolators developed for MMW frequencies work in much the same way as they do for lower frequencies. Y-junction types (E- or H-plane), which employ a matched load in one arm, are available. In addition, isolators that use the Faraday rotation principle are also available. These isolators can typically handle more power than the Y-junction types. Table 10.2 shows some of the more pertinent parameters for these devices at 35 and 95 GHz. A trade-off must generally be made between bandwidth and isolation or loss performance. Some manufacturers offer more isolation at the expense of insertion loss, while some others offer a lower loss at the expense of isolation. The design engineer must determine which set of parameters best fits the system's needs.

Table 10.2   Isolator Parameters.

Parameter	Typical Values	
	35 GHz	95 GHz
Isolation	18 dB	18 dB
Insertion Loss	0.7 dB	1.2 dB
Average Power-Handling Capability	12 W	1.5 W
Bandwidth	2%	2%

Switchable and Fixed Circulators

Y-junction circulators work in much the same way as Y-junction isolators, except that no internal load is incorporated at the third port. This port is used as the isolated port for its application in the radar system. The parameters are very similar to those for the isolator in terms of isolation, insertion loss, and power-handling capability. As with isolators, a trade-off can be made between performance and bandwidth.

The Faraday rotation principle is used to produce a switching action in switchable circulators. Three- and four-port devices are available. As with the ferrite switches to be discussed below, the switching speeds are on the order of 5 to 10 μs, due to the requirement to develop a large current (about 100 MA) in a coil. Table 10.3 shows some of the typical parameters for these devices at 35 and 95 GHz.

Table 10.3   Switchable Circulator Parameters.

Parameter	Typical Values	
	35 GHz	95 GHz
Isolation	18 dB	18 dB
Insertion Loss	1 dB	1.4 dB
Average Power-Handling Capability	12 W	1.5 W

Switches

The Faraday rotation principle is used to produce a switching action in a ferrite switch. Rotation of the field is controlled by current applied to a coil surrounding the guide section. Without current, the rotation (nominally, 0°) is such that the incident power is dissipated in an internal load. When the proper current is applied to the coil the rotation (nominally, 45°) is such that the incident power is allowed to propogate to the output. These switches do not switch as fast as *pin* diode switches, which are to be described below, but they have a higher on/off ratio. Table 10.4 shows some typical parameters at 35 and 95 GHz. Figure 10.18 is a photograph showing a 95 GHz ferrite switch and a 35 GHz ferrite isolator. For contrast, a 95 GHz *pin* diode switch is shown.

Table 10.4   Ferrite Switch Parameters.

Parameter	Typical Values	
	35 GHz	95 GHz
Isolation	40 dB	35 dB
Insertion loss	1.1 dB	1.8 dB
Switching Speed	5-10 μs	5-10 μs
Bandwidth	2 GHz	3 GHz

10.4.1.3   Millimeter-Wave Integrated Circuits

Development of integrated circuit technology for applications at MMW frequencies is being done in anticipation of the need for high-volume production of small, high-performance radar systems. Because of the small-size restrictions posed by the missile environment and the high production quantities required, it is most likely that the first type of system to have the need for IC technology will be the seeker. Due to the fact that most of the developmental work on MMW radar systems is in the range from

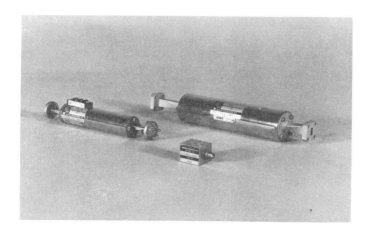

Fig. 10.18 MMW electronic switches.

30 to 100 GHz, this is the region in which most of the IC work is being done.

The areas in which IC technology can improve the size and reproducibility of systems include power amplifiers, oscillators, mixers, low-noise amplifiers, and control circuits, such as phase shifters and switches. The bulk of the research in the area of developing IC forms for the various radar functions involves implementing the building blocks on a piece-by-piece basis. Except for some private industry research and development programs to integrate whole system functions (e.g., an entire receiver), the research has been directed toward improving the performance of the individual elements, with the goal of eventually being able to integrate these functions into a subsystem.

The receiver-related functions given the most attention in IC development are switches and mixers. Development of a 60 GHz amplifier, having a gain of 5 dB and a noise figure of 7.1 dB, has been reported by Watkins *et al.* [9]. Tahim [10] has reported the development of a frequency doubler, which produces a W-band LO signal from a Q-band oscillator. The doubler exhibits a loss of less than 6 dB over a 1.5 GHz bandwidth.

The bulk of the developmental work in active devices is in the area of mixers. The reports of MMW mixers developed by Honeywell [11] and Alpha Industries [12] demonstrate noise figures on the order of 6 to 7 dB at 95 GHz. Figure 10.19 shows an integrated 94 GHz transceiver front-end, including the Gunn LO, receiver protector, duplexer-isolator, and the pulsed IMPATT transmitter. The assembly was developed by Hughes Aircraft Company for a seeker application.

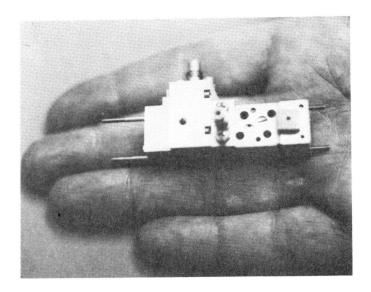

Fig. 10.19 Integrated 94 GHz transceiver front-end. From left to right: Gunn LO, mixer-preamplifier, receiver protector, duplexer-isolator, pulsed IMPATT (courtesy of Hughes Aircraft Company).

10.4.2 Active Components

10.4.2.1 Mixers and Mixer-Preamplifiers

The earliest type of system used at MMW frequencies for the detection of the received signal in a superhetrodyne receiver was the single-ended mixer arrangement. The mixing action occured when a LO signal and the received RF were applied to the diode through a coupler arrangement such as that shown in Figure 10.20. This arrangement was difficult to implement because the conversion loss and noise figure were very poor, and it was difficult to generate ample LO power for properly biasing the diode due to the loss experienced in the coupler.

Later, whisker-contact GaAs Schottky diodes were used in a balanced mixer arrangement. Until recently, this provided a suitable and relatively low-noise approach to the mixer technology. More recently, the use of beam-lead GaAs diodes has greatly improved the reliability of mixers over that experienced with the whisker types.

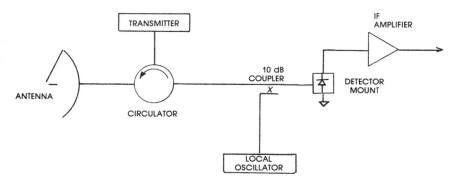

Fig. 10.20 Single-ended mixer arrangement.

Applications for receivers above 95 GHz has resulted in the development of a special class of mixers. Since it is difficult to generate the LO power required to bias the mixer diodes at the higher MMW frequencies, a technique is often employed to multiply the output from a lower frequency LO. A 47 GHz LO and doubler were used in a 95 GHz monopulse instrumentation radar (Scheer and Britt [13]), having four receiver channels. A similar LO multiplier was used for a 140 GHz mixer (Putnam, et al. [14]), exhibiting a 6 dB double sideband (DSB) noise figure.

An extention of the LO multiplier approach was the introduction of the subharmonically pumped mixer, wherein a LO of a lower frequency is used. The multiplying action occurs in the mixer diode itself. A 225 GHz coherent receiver using the fourth harmonic of a 55 GHz LO, having a DSB noise figure of 8.5 dB, has been developed (Forsythe [15]).

10.4.2.2   Down-Converters

The most typical use of a mixer in a receiver subsystem is as a down-converter. The received signal is mixed with a LO signal to produce an IF. Table 10.5 lists the more important parameters associated with this function, and the typical values that are attainable from the various vendors. Figure 10.21 is a photograph of an ensemble of modern balanced mixer-preamplifier assemblies for use at nominal frequencies of 22, 40, and 140 GHz.

Table 10.5 Mixer-preamplifier parameters.

Parameter	Typical values	
	35 GHz	95 GHz
Noise Figure (DSB)	3.5 dB	5.0 dB
LO Drive	+7 dBm	+7 dBm
RF-IF Gain	20–25 dB	20–25 dB
Output Power (1 dB Compression)	0 dBm	0 dBm
LO-IF Isolation	20 dB	20 dB

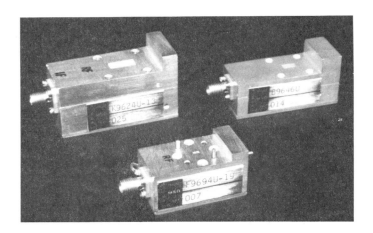

Fig. 10.21 Modern mixer-preamplifiers (courtesy of Alpha Industries).

10.4.2.3 pin Diode Devices

Switches

The *pin* diode switches are typically faster than ferrite switches. The *pin* switches come in a variety of configurations, with or without *transistor-transistor logic* (TTL) drivers, fast switching circuits or conventional switching circuits, and a variety of isolation options. Typical values for the parameters of interest at 35 and 95 GHz are given in Table 10.6. The values in parentheses indicate the typical parameters for a unit specifically designed for fast switching times.

Table 10.6 *pin* Switch Parameters.

Parameter	Typical Values (Special Design)			
	35 GHz		95 GHz	
FULL-BANDWIDTH TYPES				
Isolation	28 dB	(20 dB)	15 dB	(15 dB)
Loss (max.)	1.3 dB	(1.0 dB)	1.8 dB	(1.8 dB)
Loss (typ.)	0.9 dB	(0.6 dB)	1.4 dB	(1.4 dB)
Switching Speed				
(10–90% RF)	150 ns	(50 ns)	150 ns	(50 ns)
(90–10% RF)	30 ns	(25 ns)	30 ns	(25 ns)
NARROW-BAND TYPES				
Bandwidth	6%		3 GHz	
Isolation	60 dB	(30 dB)	15 dB	
Loss (max.)	2.5 dB	(1.5 dB)	2.5 dB	
Switching Speed	5 ns	(1 ns)	5 ns	(1 ns)

Phase Shifters

The *pin* diode switches are used for implementing discrete phase shifters at MMW frequencies in much the same way as at microwave frequencies. Reflection, switchable transmission, and loaded-line phase shifters have all been implemented (Bhartia and Bahl [16]). In general, the insertion loss is greater than that experienced at microwave frequencies, which limits the application of this kind of device. Most of the development of *pin* phase shifters has been done by private industry for their own applications, and so the component catalogs do not always include them.

Attenuators and Modulators

A variation of the use of a *pin* diode as a switching device involves its use as a modulator or variable attenuator. Since the impedance of the diode switch varies with bias current, a variable attenuation will result from impressing a controlled current through the diode.

10.4.2.4 Local Oscillators

Generally, one of two types of oscillators is used as a LO in a MMW radar system: a Gunn or an IMPATT oscillator. IMPATT sources provide up to about 500 mW CW, but they are quite noisy and produce a phase noise that limits the performance of a coherent radar system. Gunn oscillators, although they do not provide as much power as IMPATTs, are considerably quieter, exhibiting less phase noise. IMPATT phase noise is somewhat reduced by phase-locking to a "quiet" source, but not to the same extent as noise from a Gunn.

There are basically two catagories of LOs which are employed in MMW radars: coherent and noncoherent oscillators. Coherent LO signals are usually derived from the transmitter-exciter subsystem, since that function requires the stable LO signal from which to derive the RF. Oscillators, of either the Gunn or IMPATT types, are usually phase-locked or injection-locked to provide a signal that is phase coherent with a reference oscillator. Injection-locked sources are basically oscillators that have a low-level signal injected into the cavity to prompt the device to oscillate at a given frequency and phase. Some systems employ a low-frequency source (100 MHz), which is multiplied up to a MMW frequency. The LO source is then operated in a phase-locked loop.

Table 10.7 shows some of the more pertinent parameters that we must consider when selecting a local oscillator for 35 and 95 GHz.

Table 10.7   Oscillator Parameters.

Parameter	Typical Values	
	35 GHz	95 GHz
Peak power ($<$ 100 ns pulse)	10 W	5 W
Average power	500 mw	100 mw
Efficiency	$\approx 1\%$	$\approx 1\%$

Figure 10.12 shows the phase noise characteristics of a Gunn oscillator used in a phase-locked-loop configuration. Such phase-locked sources usually involve the use of two phase-locked loops. The first is at a relatively low frequency, such as 6 GHz. The second is at the desired MMW frequency, such as 95 GHz. Excellent phase noise can be achieved with this approach. The phase noise is largely determined by the phase noise of the reference oscillator. Figure 10.22 is a photograph of such a phase-locked Gunn oscillator.

If higher power is required than that which is achievable from a phase-locked oscillator, then an injection-locked approach is used. In this case, an oscillator is biased to its operating point, and a lower power reference signal is injected into its cavity, which is usually done by means of a circulator as shown in Figure 10.23. Both IMPATT and Gunn am-

Fig. 10.22 MMW phase-locked oscillator (courtesy of Alpha Industries).

plifiers are available in this configuration, depending on the AM noise and power requirements. The IMPATT device will provide about 500 mW at 35 GHz and 200 mW at 95 GHz. The Gunn device will provide about 200 mW and 30 mW at 35 and 95 GHz, respectively.

Noncoherent applications for LOs call for the use of free-running IMPATT or Gunn oscillators, or voltage-tuned or mechanically tuned versions of these devices. A tuning range on the order of 200 MHz is offered by several vendors at the most common MMW frequencies.

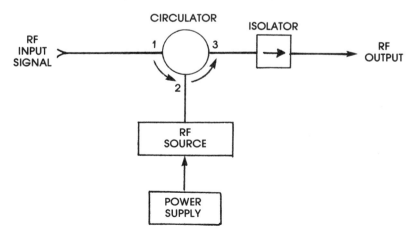

Fig. 10.23 Injection-locked oscillator block diagram.

REFERENCES

1. L.V. Blake, *Radar Range Performance Analysis*, Artech House, Norwood, MA, 1986, pp. 14–15.
2. M.I. Skolnik, *Introduction to Radar Systems*, McGraw-Hill, New York, 1980, p. 345.
3. D.K. Barton, *Radar System Analysis*, Artech House, Norwood, MA, 1976, p. 21.
4. L.V. Blake, "Prediction of Radar Range," Chapter 2, p. 23, and J.W. Taylor, Jr., and J. Mattern, "Receivers," Chapter 5, p. 33 in M.I. Skolnik, ed., *Radar Handbook*, McGraw-Hill, New York, 1970.
5. RHG Electronics Laboratories, Inc., Catalog No. 300, Deer Park, NY, p. 26.
6. S.R. Kurtz, "Mixers as Phase Detectors," Watkins-Johnson, *Tech-Notes*, Vol. 5, No. 1, January-February 1978.
7. Hughes Aircraft Company, Millimeter-Wave Products Catalog, 1986.
8. Alpha Industries, Inc., Millimeter Waveguide Components Catalog (undated).
9. E.T. Watkins, "A 60 GHz FET Amplifier," *IEEE MTT-S International Microwave Symposium Digest*, June 1983, pp. 145–147.
10. R.S. Tahim, "High Efficiency Q-to-W Band MIC Frequency Doubler" *Electronics Letters*, March 1983, pp. 219-220.
11. C.R. Seashore, *et al.*, "Millimeter Wave ICs for Precision Guided Weapons," *Microwave Journal*, Vol. 26, No. 6, June 1983, pp. 51–65.
12. A.G. Cardiasmenos, "Practical MICs Ready for Millimeter Receivers", *Microwave Systems News*, pp. 37-51, August 1980.
13. J.A. Scheer, and P.P. Britt, Solid State, 95 GHz Tracking Radar System, *Microwave Journal*, Vol. 25, No. 10, October 1982, pp. 59–65.
14. J. Putman, *et al.* "140 GHz All Solid State Receiver with System Noise Figure Less Than 6 dB DSB," *IEEE MTT-S International Microwave Symposium Digest*, June 1980, pp. 17–18.
15. Ronald E. Forsythe, "A Coherent, Solid State, 225 GHz Receiver," *Microwave Journal*, Vol. 25, No. 7, July 1982, pp. 64–71.
16. P. Bhartia and I.J. Bahl, *Millimeter Wave Engineering and Applications*, John Wiley and Sons, 1984, p. 469.

Chapter 11
MMW Antennas

D.G. Bodnar

Georgia Institute of Technology
Atlanta, Georgia

11.1 INTRODUCTION

This chapter discusses antenna types that have been used or may find use in MMW radars. Antenna types that are primarily used at lower frequencies (e.g., HF, UHF, and microwave) are not covered here. The interested reader is referred to several excellent references [1, 2, 3] on design techniques for lower frequency antennas.

Millimeter-wave antennas often resemble those at lower frequencies. In fact, many lower frequency (especially microwave) antenna types can also be used at millimeter wavelengths. What then makes MMW antennas different or special? The decrease in size, increase in bandwidth, increased losses, and tighter tolerances set MMW antennas apart from their lower frequency counterparts. Some of these features represent system design opportunities, while others represent difficult challenges. The reduced physical volume required for the antenna is especially attractive for airborne, mobile, and missile applications. The smaller wavelengths involved in MMW antennas creates unique problems including greater losses and the need for more precisely built components. Section 11.2 covers unique MMW antenna characteristics along with general MMW antenna terminology and design concepts. Commonly used MMW antennas, including horns, lenses, and reflectors, as well as other types are discussed in Section 11.3. The availability of millimetric RF components is discussed in Section 11.4, since the lack of a full range of MMW RF components may impede the assembly of certain types of MMW antennas. Two major requirements of a radar are first to find (search for) targets in some region of space and then to track the target once it has been found. These operations require the movement of the antenna beam in some prescribed manner. Achieving this motion is more difficult at millimeter wavelengths because of increased waveguide ohmic losses, lack of rotary joints, and tighter fabrication tolerances. Techniques for MMW antenna beam scanning are discussed in Section 11.5.

11.2 MM ANTENNA THEORY AND TERMINOLOGY

11.2.1 Antenna Terminology

The function of an antenna-on-transmit is to accept the energy from the transmitter, to concentrate said energy in a predetermined beam shape, and to point this beam in a particular direction in space. The antenna-on-receive selectively gathers transmitter energy that has been reflected from targets located in different directions in space, and sends this energy to the receiver. If the receiving and transmitting patterns of the antenna are identical, then the antenna is said to be reciprocal and only one antenna pattern needs to be specified. Both patterns must be specified if the antenna is not reciprocal (for example, if it contains nonreciprocal phase shifters). In the following discussion, a single pattern will be assumed for the antenna. The following discussion can be applied to each pattern separately if the transmitting and receiving patterns are different.

In general the shape of an antenna pattern changes with both range (distance from the antenna) and look direction. Two regions of space about an antenna can be delineated, based on the form of this pattern shape change [4]: the *near-field region* (or Fresnel zone) and the *far-field region* (or Fraunhaufer zone). The pattern changes shape in a complicated manner in the near field of an antenna, while in the far field the shape of the pattern does not change with range. In the far field, the field components (i.e., electric and magnetic fields) vary as $f(\Theta, \Phi) \exp(-jkR)/R$ and the power varies as $g(\Theta, \Phi)/R^2$, where R is the distance from the antenna, Θ and Φ are angular coordinates of the observation point, $k = 2\pi/\lambda$, λ is the wavelength of operation, and f and g are functions of only the angular coordinates, not of range. The antenna pattern has settled into its far-field form when we observe the antenna at ranges greater than the far-field distance; namely, $2D^2/\lambda$ where D is the diameter of the antenna aperture. Only the far-field region of the antenna will be considered in this chapter. However, operation in the near field sometimes occurs in MMW antennas that have very narrow beamwidths and must operate at short as well as long ranges. Hansen [4] should be consulted in these cases to help ascertain the shape of the pattern *versus* range.

Ideally, we would want to produce an antenna pattern that has the shape of a rectangular pulse, and is zero everywhere else. The antenna would see targets equally well in the pulse region (main beam) and have no response to targets outside of this region. However, such a pattern is theoretically impossible unless the antenna is infinitely large. Real antennas, of course, are finite in size and have patterns with (1) a round-top main beam, (2) a main beam that slowly falls to low levels, and (3) contain secondary lobes (sidelobes) in addition to the main beam region. Figure 11.1 shows a typical measured MMW antenna pattern. Notice that item (1) above reduces the pulse integration efficiency of the radar, item (2) tends to smear out the location of the target, while item (3) can produce false target locations if the sidelobes are too large. Pattern parameters that are of interest to the radar designer are the peak gain of the antenna (relative to that of an isotropic radiator), the half-power (usually referred to as the 3 dB) width of the main beam, the peak or highest sidelobe level (specified as the level of the sidelobe relative to the level of the main beam), and the average or rms sidelobe level. Specifying an antenna to achieve desire values for many of these parameters will be discussed in Section 11.2.5. Definitions of these and many other antenna terms can be found in [6].

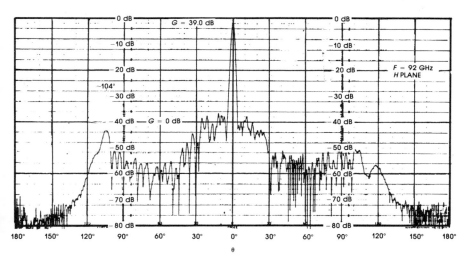

Fig. 11.1 Typical measured MMW antenna pattern (from Dybdal [5], © 1983 IEEE).

11.2.2   Size and Bandwidth Considerations

Two of the major reasons for operating radars at millimeter wavelengths are the reduction in size of RF components, especially the antenna, and the increase in RF component bandwidth as compared to lower frequencies. The size reduction can be understood by considering the half-power beamwidth BW and the gain G (sometimes called power gain to distinguish it from directivity, which is sometimes called directive gain) of a circular aperture antenna, which for radar antennas are typically given by (see Section 11.2.5 for a more detailed discussion of BW and G):

$$BW = 70 \, \lambda/D \quad \text{(in degrees)} \tag{11.1}$$

$$G = 20 \log_{10}(\pi D/\lambda) + \eta_{dB} \quad \text{(in dBi)} \tag{11.2}$$

where

λ = wavelength of operation,
D = diameter of the antenna (in the same units as λ),
dBi = decibels relative to that of an isotropic radiator,
η_{dB} = efficiency of the antenna in dB.

Equations (11.1) and (11.2) show that the beamwidth and gain of an antenna will remain constant with frequency as long as the size of the antenna remains fixed in terms of wavelengths. For gain, the losses must also remain fixed with frequency for this conclusion to hold. Thus, a MMW antenna will be physically smaller than a microwave antenna of the same beamwidth and gain. This size reduction is a desirable feature from a packaging point of view, which allows MMW radars to be made smaller and to be operated in smaller vehicles. Alternatively, a narrower beamwidth and a higher gain can be obtained from the same physical aperture by going up in frequency to millimeter wavelengths. For example, a one-foot diameter antenna at 95 GHz will have a 0.72° beamwidth and 46.6 dBi of gain as compared with one operating at 9 GHz, where it would have a 7.6° beamwidth and 26.2 dBi of gain (3 dB of loss was assumed for both antennas).

The bandwidth of RF components is typically determined as a percentage of the center frequency of operation. Thus, millimetric components will have greater bandwidth than microwave components because of their higher frequency. For example, a 10% bandwidth device will have 9500 MHz of bandwidth at 95 GHz as compared with 900 MHz of bandwidth at 9 GHz. This increased bandwidth can be put to use for improving signal-to-clutter ratio, ECCM performance, or some other important operating characteristic.

11.2.3 Losses

The standard method of interconnecting millimetric RF components is by using a rectangular waveguide. Ohmic losses in a waveguide increase as the cross-sectional area of the guide is decreased, and as its wall surfaces become electrically rougher and more lossy. MMW waveguide is smaller in cross section, as required by the smaller wavelength of operation, and any surface imperfections in the waveguide's walls are larger in terms of wavelengths than for microwave waveguide. Consequently, the losses in MMW rectangular waveguide are relatively high, as can be seen from Table 11.1, even when the guide is made of silver. The losses, even for perfectly smooth walls, increase with frequency due to a decrease in skin depth. For example, the theoretical loss in silver waveguide (using $\sigma = 6.17 \times 10^7$ mhos/m) is 0.16 dB/ft at 35 GHz and 0.76 dB/ft at 95 GHz. Consequently, the length of all waveguide runs must be kept short in order to reduce these losses. Typically, the RF components of the transmitter and receiver are mounted at the antenna, and individual receiver RF components are placed very close together. In addition to keeping the waveguide runs short, the inside surface of the guide must be smooth (Tisher [7, 8]). Otherwise, additional wall losses will occur due to the waveguide currents having to travel over longer paths caused by the surface roughness. This effect can increase the wall losses in the waveguide by 50% at millimeter wavelengths.

Alternative RF transmission methods have been used at millimeter wavelengths for reducing waveguide losses. Quasioptical or beam waveguide techniques (Goldsmith [9], Swanberg and Paul [10], Kay [11]) utilize dielectric lenses or elliptical reflectors to focus energy from a horn at one location onto a horn at another location. Low-loss transmission is possible when the lenses are in each other's near field, and the lenses are many wavelengths in diameter. Researchers at Hughes Aircraft Company [10], for example, used a lens horn approach to develop a mixer operating at 217 GHz. The main disadvantage of this transmission technique is the extra volume required for the quasioptical feed. Another problem of conventional dominant-mode rectangular waveguide is that its power-handling capability is relatively low (e.g., 6 kW peak at 95 GHz). Hence, overmoded waveguide structures are required to handle high power from such devices as gyrotron tubes (Thumm, *et al.* [12]). Overmoded waveguide structures can also be used for reducing losses in long waveguide runs.

Dielectric waveguide, silicon waveguide, and other guiding techniques have been used in building RF components at millimeter wavelengths. In antennas, the main application of these techniques is for integrating antenna and RF components as a monolithic unit, as discussed in Section 11.3.4.

Table 11.1 Theoretical Attenuation in Silver Rectangular Waveguide.

Waveguide Designation	ID Size (inches)	Frequency (GHz)	Waveguide Band Designation	Free-Space Wavelength (inches)	Theoretical Attenuation (dB/100 ft)
WR-28	0.280 × 0.140	26.5–40	K_a	0.445–0.295	21.9–15.0
WR-22	0.224 × 0.112	33–50	Q	0.358–0.236	30.9–21.0
WR-15	0.148 × 0.074	50–75	V	0.236–0.157	57.3–39.2
WR-10	0.100 × 0.050	75–110	W	0.157–0.107	101–70.8
WR-7	0.065 × 0.0325	110–170	—	0.107–0.069	213–135
WR-4	0.043 × 0.0215	170–260	—	0.069–0.045	376–250
WR-3	0.034 × 0.0170	220–325	—	0.054–0.036	510–357

11.2.4 Tolerance Effects

The required fabrication tolerances for a given type of antenna are fixed in terms of wavelengths. Consequently, the shorter wavelengths of MMW radars make the construction of MMW antennas and related RF components more difficult. For example, the deviation of a reflector antenna surface from the desired theoretical shape causes the aperture phase distribution to be different from the desired one. Generally, these surface and hence phase errors are random in nature and must be treated in a statistical manner. Ruze [13] has solved this problem for reflector antennas. His general results can be simplified when the errors are relatively small in both amplitude and extent. Under these circumstances, the on-axis gain of the antenna is

$$G = G_0 \exp[-(4\pi\epsilon/\lambda)^2] \tag{11.3}$$

where

G = gain of antenna with errors,
G_0 = gain of the antenna without errors,
   = $\eta_0(\pi D/\lambda)^2$,
η_0 = antenna efficiency,
D = diameter of antenna,
λ = wavelength of operation,
ϵ = rms surface error.

Figure 11.2 shows the results of evaluation of (11.3) for millimeter wavelengths. Note, for example, that an rms surface error of $\lambda/10$ produces a gain loss of 0.5 dB.

In addition to loss of gain, the antenna sidelobe levels are raised, since energy that is lost from the main beam appears as new sidelobe energy. The nature of the sidelobe perturbation depends on the extent (correlation interval) of the surface error and the level of the original sidelobes in addition to the magnitude of the errors [14]. Typical curves of sidelobe degradation due to random surface effects are shown in Figures 11.3 and 11.4. Notice from Figure 11.3 that these errors produce a wide-angle perturbing pattern that tends to mask the theoretical sidelobe performance of the antenna. From Figure 11.4, we can see that the effect of these errors is less for larger antennas. The study [14] indicates that the actual surface should not deviate from the theoretical surface by more than $\lambda/60$ rms in order to limit the gain loss to 0.2 dB or less, and to maintain sidelobes levels in the -25 to -30 dB range. The rms error has been found experimentally to be approximately one-sixth of the peak-to-peak surface error, which is more commonly measured by shop personnel.

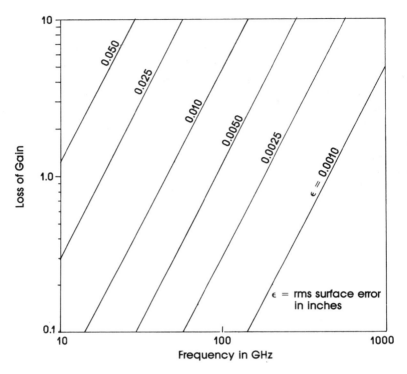

Fig. 11.2 Loss of gain produced by reflector surface roughness.

Thus, the peak-to-peak surface tolerance required on typical reflector antennas is approximately $\lambda/10$ for single-reflector antennas such as a paraboloid, and $\lambda/14$ for double-reflector antennas such as a Cassegrain. At 95 GHz, for example, the actual paraboloidal surface must be held to within 0.002 inches rms of the theoretical surface. This is a reasonable requirement for moderately sized antennas. However, it is difficult to hold such tolerances for large antennas, especially under harsh environmental conditions. Typical numerically controlled milling machines are capable of accuracies of ±0.001 inch peak, while more conventional machining processes often produce errors that are five or ten times as large. In addition to random errors, periodic errors can also result, for example when joining panels of a reflector together. These errors can cause perturbations of the sidelobe levels that are as important as the random ones (Tripp [15]), and so must be controlled during manufacture.

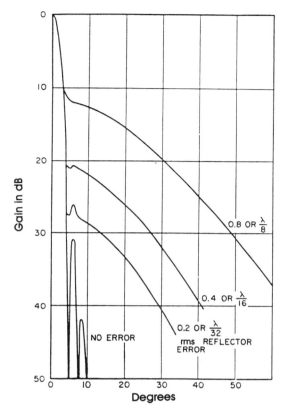

Fig. 11.3 Average system pattern for a circular paraboloid with cosine-squared illumination; correlation interval $c = \lambda$ calculated for a reflector 24λ in diameter (from Johnson and Jasik [14], © 1984 McGraw-Hill Book Company).

11.2.5 Antenna Pattern Characteristics [16]

Both the initial design of an antenna and the validation of measured antenna data are greatly facilitated by the use of theoretically derived pattern characteristics. The patterns produced by a variety of one- and two-dimensional aperture distributions have appeared in the literature over the years. The task of engineers is to determine which theoretical aperture distribution best fits their physical situations so that they can assess antenna

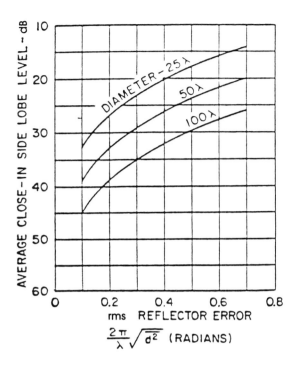

Fig. 11.4 Spurious radiation of paraboloid with cosine-squared illumination; correlation interval $c = \lambda$ (from Johnson and Jasik [14], © 1984 McGraw-Hill Book Company).

performance from the theoretical data. This can be done by measuring or calculating the actual aperture distribution, and then comparing it with a variety of theoretical distributions (Hansen [17]). The three primary factors of concern to antenna designers are gain, beamwidth, and sidelobe level. This section presents data on these three parameters for both line-source and circularly symmetric distributions. These aperture shapes are considered because rectangular and circular apertures are the most common shapes in practice. For rectangular apertures, a line source is used to represent the energy distribution in each of the principal directions.

The gain of an antenna is usually compared to that of an antenna with the same aperture dimensions, but having a constant phase and amplitude distribution (i.e., uniformly illuminated). The power gain G of the antenna then is

$$G = G_0 \eta_0 \tag{11.4}$$

where

$G_0 = 4\pi A/\lambda^2$
A = antenna aperture area
λ = wavelength,
$\eta_0 = \eta\eta_L$ = antenna efficiency,
η = aperture illumination efficiency,
η_L = all other efficiency factors.

The aperture illumination efficiency represents the loss of gain resulting from tapering the aperture distribution in order to produce sidelobes lower than those achievable from a uniform illumination. For a circular aperture η is given by a single number. For separable rectangular aperture distributions, η equals the product (sum in decibels) of the efficiencies in each of the two aperture directions.

The half-power beamwidth BW of an antenna is related to the beam-width constant β by (Silver [18]):

$$BW = 2\sin^{-1}\left(\frac{\beta\lambda}{2L}\right) \approx \beta\frac{\lambda}{L} \quad \text{(linear aperture)} \tag{11.5a}$$

$$BW = 2\sin^{-1}\left(\frac{\beta\lambda}{2D}\right) \approx \beta\frac{\lambda}{D} \quad \text{(circular aperture)} \tag{11.5b}$$

where L is the length of the linear aperture and D is the diameter of the circular aperture. The small-argument approximation for the arcsine is typically used for calculating BW. Values of β and η will be given as a function of sidelobe level for a number of distributions in the following subsections.

Continuous Line-Source Distributions

The problem of determining the optimum pattern from a line source has received considerable attention. The optimum pattern is defined as the one that produces the narrowest beamwidth measured between the first null on each side of the main beam with no sidelobes higher than the stipulated level. Dolph [19] solved this problem for a linear array of discrete elements by using Chebyschev polynomials. If the number of elements becomes infinite, while element spacing approaches zero, the Dolph pattern becomes the optimum continuous line-source pattern (Taylor [20]), called the *Chebyschev pattern*, given by

$$E(u) = \cos\sqrt{u^2 - A^2} \tag{11.6}$$

where

$u = (\pi L / \lambda) \sin\Theta,$
$\lambda = $ wavelength,
$\Theta = $ angle from the normal to aperture,
$L = $ aperture length.

The beamwidth constant β, in degrees, and the parameter A are given by

$$\beta = \frac{360}{\pi^2} \sqrt{[\text{arccosh } (R)]^2 - \left[\text{arccosh } \left(\frac{R}{\sqrt{2}}\right)\right]^2} \qquad (11.7)$$

where

$A = $ arccosh (R),
$R = $ main-lobe-to-sidelobe voltage ratio.

This Chebyschev pattern provides a useful basis for comparison, although it is physically unrealizable, since the remote sidelobes do not decay in amplitude. In fact, all sidelobes have the same amplitude in the Chebyschev pattern. The aperture distribution which produces the Chebyschev pattern has an impulse at both ends of the aperture (Sherman [21]), and produces very low aperture efficiency. In addition, the pattern is very sensitive to errors in the levels of these impulses.

Taylor [20] developed a method for avoiding the above problems by approximating the Chebyschev pattern arbitrarily closely with a physically realizable pattern. Taylor approximated the Chebyschev uniform sidelobe pattern close to the main beam, but let the wide-angle sidelobes decay in amplitude. Taylor used a closeness-parameter $\bar{n}$ in his analysis. As $\bar{n}$ becomes infinite, the Taylor distribution approaches the Chebyschev distribution. By using the largest $\bar{n}$ that still produces a monotonic aperture distribution, we obtain the beamwidth constant and aperture efficiency shown in Figures 11.5 and 11.6. Notice that the beamwidth from this Taylor distribution is almost as narrow as that from the Chebyschev distribution, while still producing excellent aperture efficiency.

Several other common distributions (Silver [18] and Harris [22]) are also listed in Figures 11.5 and 11.6. The advantage of the $\cos^n(\pi x/L)$ distribution and the $\sin (\sqrt{u^2 - B^2})/\sqrt{u^2 - B^2}$ pattern is that both the distribution and the pattern may be obtained in closed form. This mathematical convenience is obtained at the expense of poorer beamwidth and efficiency performance as compared with the Taylor distribution.

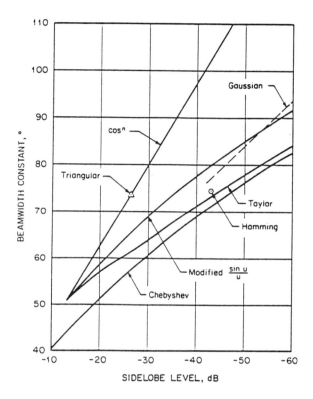

Fig. 11.5 Beamwidth constant *versus* sidelobe level for line-source aperture distributions (from Bodnar [16], © 1984 McGraw-Hill Book Company).

Continuous Circular-Aperture Distributions

The Chebyschev pattern of the preceeding section can also be shown to be optimum for the circular aperture. Taylor has generalized his line-source distribution to the circular case [23, 24], and the pattern approaches the Chebyschev pattern as his closeness-parameter $\bar{n}$ for the circular aperture approaches infinity. The beamwidth constant and aperture efficiency shown in Figures 11.7 and 11.8 are obtained by using the largest $\bar{n}$ that still produces a monotonic aperture distribution (Ludwig [25]).

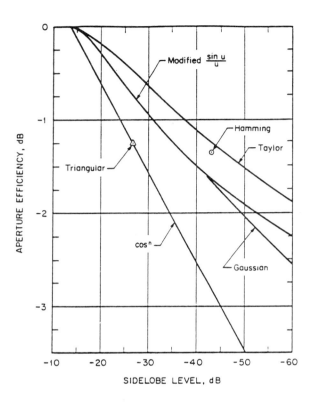

Fig. 11.6 Aperture efficiency *versus* sidelobe level for several line-source
aperture distributions (from Bodnar [16], © 1984 McGraw-Hill
Book Company).

The Bickmore-Spellmire distribution [26] is a two-parameter distribution, which can be considered a generalization of the parabola-to-a-
power distribution [21]. The Bickmore-Spellmire distribution $f(r)$ and pattern $E(u)$ are given by

$$f(r) = p \left[1 - \left(\frac{2r}{D} \right)^2 \right]^{p-1} \Lambda_{p-1} \left[jA \sqrt{1 - \left(\frac{2r}{D} \right)^2} \right] \tag{11.8a}$$

$$E(u) = \Lambda_p \left(\sqrt{u^2 - A^2} \right) \tag{11.8b}$$

where p and A are constants that determine the distribution, Λ is the
lambda function, and

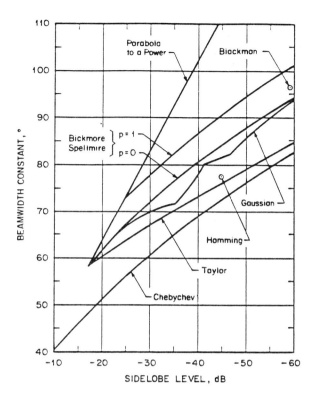

Fig. 11.7 Beamwidth constant *versus* sidelobe level for several circular-aperture distributions (from Bodnar [16], © 1984 McGraw-Hill Book Company).

$$u = \frac{\pi D}{\lambda} \sin \Theta \qquad (11.9)$$

The Bickmore-Spellmire distribution reduces to the parabola-to-a-power distribution when $A = 0$, and to the Chebyschev pattern when $p = -\frac{1}{2}$.

A Gaussian distribution [25] produces a no-sidelobe Gaussian pattern only as the edge illumination approaches zero. In general, the aperture distribution must be numerically integrated to obtain the far-field pattern. The second sidelobe of this pattern is sometimes higher than the first, which accounts for the erratic behavior of β and η in Figures 11.7 and 11.8.

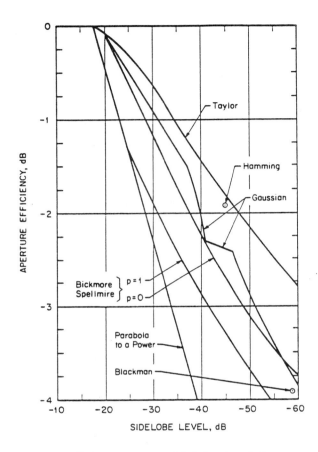

Fig. 11.8 Aperture efficiency *versus* sidelobe level for several circular-aperature distributions (from Bodnar [16], © 1984 McGraw-Hill Book Company).

Blockage

The placement of a feed in front of a reflector results in blockage of part of the aperture energy. In the geometric optics approximation, no energy exists where the aperture is blocked, and the undisturbed aperture distribution persists outside the blocked region (Hannan [27]). Using a line source of length L, with a $\cos^n$ aperture distribution and a centrally located blockage of length L_b, produces the loss of gain and the resulting

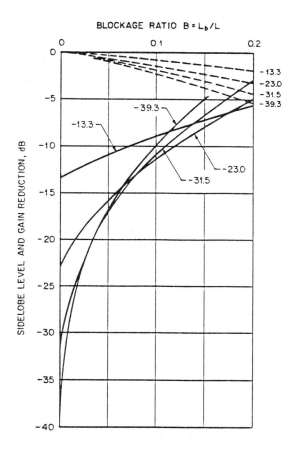

Fig. 11.9 Loss of gain (dashed curve) and resulting sidelobe level (solid curve) for a centrally blocked line-source distribution having specified unblocked sidelobe level (from Bodnar [16], © 1984 McGraw-Hill Book Company).

sidelobe level shown in Figure 11.9. The corresponding changes are shown in Figure 11.10 for a circular aperture of diameter D and a parabola-to-a-power distribution blocked by a centrally located disk of diameter D_b. Notice that the line-source blockage affects the pattern much more rapidly than in the circular case, since it affects a larger portion of the aperture. Calculation of strut-blockage effects is given by Gray [28].

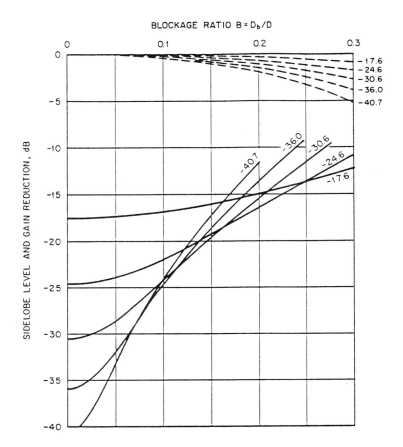

Fig. 11.10 Loss of gain (dashed curve) and resulting sidelobe level (solid curve) for a centrally blocked circular-aperture distribution having a specified unblocked sidelobe level (from Bodnar [16], © 1984 McGraw-Hill Book Company).

11.3   Types of MMW Antennas

A unique feature of MMW antennas is that high gain and narrow beamwidth can be obtained from physically small devices. Application of small, narrow-beamwidth antennas for missiles and remotely piloted vehicles is obvious. Alternatively, a much narrower beamwidth can be obtained by keeping the aperture size fixed and moving down to millimeter wavelengths. For example, the beamwidth would narrow by a factor of 10 and the gain would increase by 20 dB in going from 9.5 to 95 GHz for the same sized aperture.

Many antenna types that are common at lower frequencies are not used in the MMW region due to the small size, tight tolerances, or high losses associated with them at millimeter wavelengths. Yagi-Uda arrays, spirals, and helical antennas are examples of infrequently used antennas. In contrast, horn, lens, and reflector antennas are common types of MMW antennas. These antenna types will be discussed in the following sections.

11.3.1 Horn Antennas

Millimeter-wave horn antennas are useful when low (15 dBi) to moderate (25 dBi) gain is required, since they are relatively easy to fabricate, relatively inexpensive, rugged, and commercially available. Horn antennas are used in laboratory experiments and short-range field tests, and as feeds for reflector antennas. A horn antenna is produced by gradually flaring the walls of a waveguide to the desired aperture size, as shown in Figure 11.11. From (11.5b), we can see that the beamwidth of a horn is inversely proportional to its aperture size in wavelengths. Hence, a larger horn produces a narrower beamwidth. However, the half flare angle θ_f (see Figure 11.11) must be kept small so that the phase variation of the spherical wavefront produced by the input waveguide over the horn aperture does not deviate appreciably from a constant phase front. Phase variation over the aperture reduces the gain of the horn from its maximum value. The aperture phase variation will be acceptably small when (Glover [29]):

$$s = \frac{\Delta}{\lambda} = \frac{d^2}{8 \lambda L} \tag{11.10}$$

where

Δ = maximum path length error over horn aperture,
d = size of the horn aperture,
L = slant length of the horn.

Equation (11.10) reveals that a long flare length is required if a large horn aperture is desired.

Four popular forms of horn antennas are shown in Figure 11.12, and they are the pyramidal, diagonal, conical, and corrugated horn (Love [30]). The pyramidal horn is popular at microwave frequencies and below, since it can be made cheaply from flat pieces of metal that are cut along straight lines. However, it becomes difficult at shorter mm wavelengths (95 GHz and above) to hold accurately the individual pieces for soldering. In addition, the E-plane sidelobes of the pyramidal horn are high (about −13 dB) as compared with the lower H-plane sidelobes (about −23 dB). Low sidelobes are desirable to reduce clutter and interfering signals. A diagonal

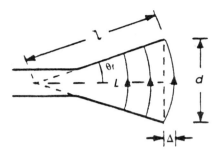

Fig. 11.11 Phase error due to wavefront curvature in the horn aperture (from Love [29], © 1984 McGraw-Hill Book Company).

horn can be used to achieve roughly −23 dB sidelobes in both E-plane and H-plane. However, the diagonal horn suffers from the same fabrication problems as the pyramidal horn. At millimeter wavelengths, it is easier to machine a conical horn on a lathe as a complete unit, including the circular waveguide input section, than it is to assemble the small, individual pieces of a pyramidal or diagonal horn. A taper from rectangular to circular waveguide is required to connect the conical horn to a standard rectangular waveguide. The sidelobe levels of a conical horn are similar to those of a pyramidal horn, being high in the E-plane and low in the H-plane. The sidelobes can be made low in both planes by machining grooves into the inside walls of the conical horn, thus producing a corrugated conical horn (sometimes referred to as a *scalar feed*). The corrugated conical horn has a rotationally symmetric pattern with nearly constant and coincident phase centers, all of which are very desirable characteristics to have when a horn is used as a feed for a reflector antenna. In addition, the corrugated horn has superior far-out and near-in sidelobes as compared with the other horn types. The improved performance of the corrugated horn is obtained at the additional expense of making the grooves in the horn. A comparison of measured patterns of a conical and a corrugated conical horn is given in Figure 11.13. Notice the substantial improvement in E-plane sidelobe performance obtained with the corrugated horn. Corrugated conical horns have been built at frequencies well above 200 GHz.

11.3.2 Lens Antennas

The length of a horn antenna becomes excessive when a high-gain antenna is required. A more compact structure is obtained by combining a dielectric lens with a small horn as shown in Figure 11.14a. The phase

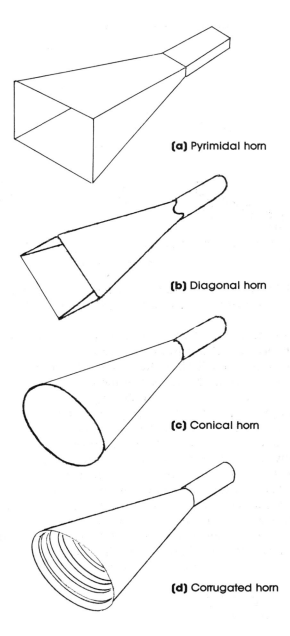

(a) Pyrimidal horn

(b) Diagonal horn

(c) Conical horn

(d) Corrugated horn

Fig. 11.12 Common types of horn antennas.

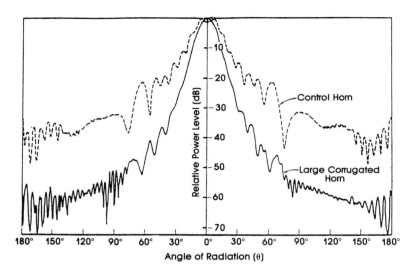

Fig. 11.13 Normalized E-plane patterns of the corrugated horn and the control horn (from Love [30], © 1976 IEEE).

center of the horn is placed at the focal point of the lens. The lens converts the spherical wavefront from the feed into a planar one at the output of the lens, while the fall-off of the horn pattern toward the edges of the lens produces a tapered illumination of the lens aperture distribution. Pattern control can also be achieved by shaping the lens surface (Gunderson and Holmes [31]). In addition to being more compact than a horn antenna, lens antennas also have advantages over reflector antennas. First, blockage of the antenna's energy by the feed does not occur in lens antennas. Secondly, about four times less accuracy is required in lens fabrication than in reflector fabrication. Lens antennas, however, have lower gain than reflector antennas, caused by losses due to reflections at the air-dielectric interface and absorption of energy in the lens.

A planar convex lens, as shown in Figure 11.14a, is the most commonly employed lens shape for converting the spherically diverging wave (on transmit) from the horn into a collimated wavefront on the planar side of the lens. The equation for this lens is

$$r = \frac{(n - 1) f}{n \cos\Theta - 1}$$

(11.11)

where

$$y = \sqrt{(n^2 - 1)x^2 + 2(n - 1)fx}$$

(11.12)

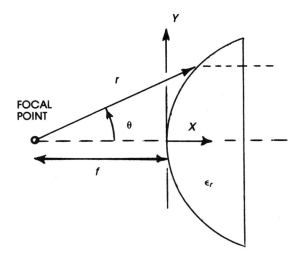

(a) Conventional Lens

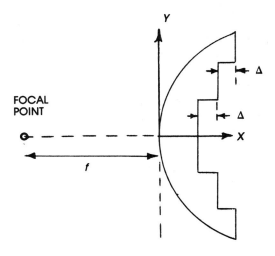

(b) Grooved lens

Fig. 11.14 Lens-horn geometry.

f is the focal length, and n is the refractive index of the lens material. If the lens is too thick or heavy, it may be reduced in thickness by cutting out steps of thickness $\Delta = \lambda_0/(n - 1)$ on the planar side, as shown in Figure 11.14(b). This *zoning* of the lens results in a well formed beam at the center frequency of operation, but makes the lens frequency-sensitive, which can be a problem in very wideband operation. Feed energy that is reflected off the convex side of the lens can be further reflected off metal objects located behind the lens. This reflected energy may then pass through the lens and degrade the antenna sidelobes, as well as enter the feed and increase the antenna VSWR. An enclosure lined with RF absorbing material can be placed over the back of the antenna to absorb the reflected energy and prevent this problem. In addition, quarter-wave or simulated quarter-wave transformer matching surfaces can be applied to the lens surfaces to help reduce the reflections (Morita and Cohn [32]).

Geodesic Luneburg lenses have been used at millimeter wavelengths in addition to previously discussed dielectric lenses. A geodesic Luneberg lens is a parallel-plate structure, usually in the shape of an army helmet, which collimates MMW energy in one plane (Johnson [33, 34]) (Figure 11.15). As shown in Figure 11.16, energy radiated by a feed horn located on the perimeter of the lens travels between the parallel plates and arrives at the diametrically opposite side of the lens as a collimated (constant phase) wavefront. This collimated line source can then, for example, be used to feed a parabolic cylinder reflector to achieve collimation in the other plane. Ring switch electromechanical scanners have been developed for these lenses [35], which permit rapid scanning (e.g., 50 scans/s) over wide angular sectors (e.g., $\pm 45°$). Low-loss geodesic lens have been built up to 70 GHz, and higher frequency operation is feasible. Two of these lenses oriented at 90° with respect to each other could produce two orthogonally scanning fan beams for a track-while-scan (TWS) radar.

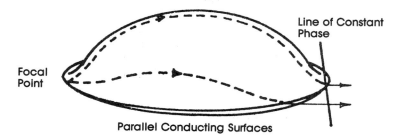

Fig. 11.15 Focusing of a point on one side of the lens to a constant phase front on the opposite side by means of a parallel-plate (metal surface) geodesic Luneberg lens (from Johnson [34], © 1964 Academic Press).

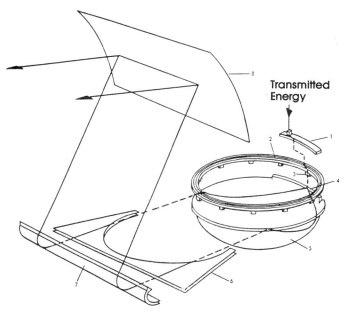

Transmitted
Energy

Key: 1—ring-switch input section   5—geodesic Luneberg lens
2—ring-switch rotating   6—flat-plate extension
section   7—reflector feed assembly
3—active lens feed-horn   8—reflector
4—geodesic lens input-lip
section

Fig. 11.16 Exploded view of a typical geodesic lens scanning antenna (from Hansen [4], © 1964 Academic Press).

11.3.3 Reflector Antennas

Reflector antennas are the most common form of high-gain MMW antenna because they are simple, relatively easy to build, and rugged. In addition, front-fed paraboloidal and rear-fed Cassegrain reflector antennas can be purchased from a number of commercial sources. The front-fed paraboloid is the least expensive to make and the least difficult to align. When the phase center of a small feed horn is placed at the focal point of the reflector, energy radiated by the feed is reflected off the reflector onto the aperture plane, as shown in Figure 11.17(a). All path lengths from the focal point to the aperture plane are equal, and so a collimated beam (i.e., equal phase front aperture distribution) is produced. This antenna is an inherently broadband device, since the constant path length feature is a geometrical characteristic of the paraboloid. However, waveguide losses from the feed to the transmitter and receiver, which must be located behind the reflector, can be high due to the high losses in guide. See Figure 11.18 for a typical installation. If the receiver or transmitter is placed at the feed, instead of behind the reflector, then antenna energy is blocked by these components, causing a loss of gain and an increase in sidelobe levels, as discussed in Section 11.2.5.

To circumvent these problems of a center fed paraboloid, a Cassegrain geometry (see Figure 11.17(b)) can be used, which allows the transmitter and receiver components to be mounted behind the antenna so that

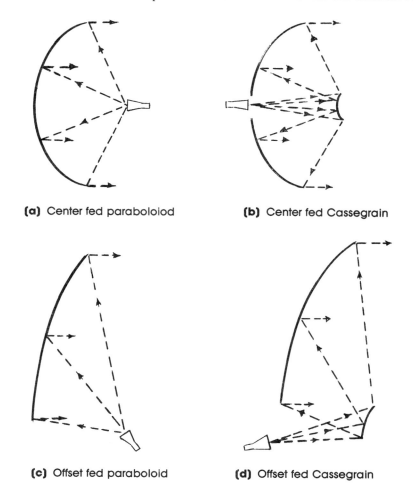

(a) Center fed paraboloiod   **(b)** Center fed Cassegrain

(c) Offset fed paraboloid   **(d)** Offset fed Cassegrain

Fig. 11.17 Reflector antenna geometries.

Fig. 11.18 Front-fed paraboloidal dish MMW (35 GHz) antenna.

they do not create aperture blockage (see Figure 11.19). Energy from a small feed horn, having a phase center that is at the primary focal point, will be first reflected off the hyperboloid in such a manner that it appears to be coming from the focal point of the paraboloid. Finally, the energy reflected by the subreflector is collimated by the paraboloidal main reflector. Like the paraboloid, the Cassegrain reflector is an inherently broadband antenna. The Cassegrain antenna is more expensive to build, since there are two reflectors to construct, instead of one. In addition, it is more difficult to align the two reflectors and the feed than it is to align a single reflector and a feed. In addition, the subreflector produces aperture blockage. An offset section of a paraboloidal reflector (see Figure 11.17(c)), or an offset Cassegrain reflector (see Figure 11.17(d)), can be used to avoid aperture blockage. The feed, feed mounts, transmitter, and receiver are all located below the energy path from the reflectors (Coleman, *et al*. [37]).

A center-fed reflector is usually preferred at microwave frequencies and below, because spun, symmetric reflectors can be obtained relatively cheaply, and they require a relatively inexpensive mount. In contrast, the

Fig. 11.19 Cassegrain 95 GHz antenna. Such a system permits short waveguide runs and may minimize aperture blockage.

backup structure required to hold an offset reflector in its correct shape
is an appreciable portion of the cost of an offset reflector at microwave
frequencies and below. At millimeter wavelengths, however, the cost of
an offset reflector may not be much different from that of a center-fed
reflector, since both surfaces typically must be machined to achieve the
required surface accuracy (see Section 11.2.4).

A hood of RF absorbing material is sometimes placed around the
back and side of an antenna to attenuate wide-angle radiation from the
antenna. This hooded antenna approach is very successful in producing
low sidelobes (about -75 dB [5]) in the back hemisphere of the antenna,
as is usually required for "quiet" radars (see Figure 11.20).

A slice of a paraboloid sandwiched between two parallel plates is
called a *pillbox* antenna. Collimation of the beam in one plane is produced
by the pillbox, while collimation in the other plane is typically produced
at millimeter wavelengths by a parabolic cylinder reflector. An 800 λ

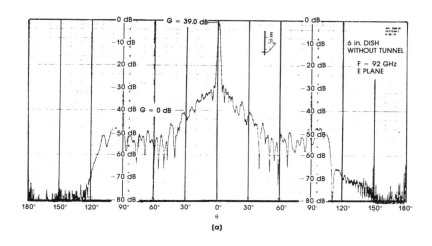

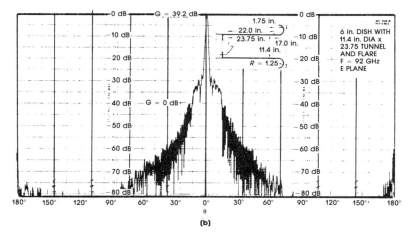

Fig. 11.20 92 GHz measurements of a 6-inch reflector, with and without an absorber-lined tunnel: (a) H-plane without tunnel; (b) E-plane with tunnel (from Dybdal [5], © 1983 IEEE).

aperture pillbox antenna is described by Bodnar [38], operating at 95 GHz. The antenna shown in Figure 11.21 produces a 0.1° to 1.5° fan beam that maintains −26 dB sidelobes over −20° to +70° F and over 2.5 GHz of bandwidth.

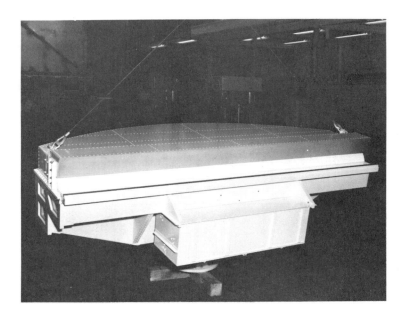

Fig. 11.21 A 95 GHz pillbox antenna, producing a 0.11° × 1.5° beam (photo courtesy of the Georgia Institute of Technology).

11.3.4 Other Antenna Types

Waveguide slotted arrays have been built at millimeter wavelengths, but the losses in the guide, the small sizes of the guide and slots, and the tight tolerances involved make their fabrication difficult. Numerically controlled milling machines currently available are capable of producing slotted arrays up to 95 GHz. Typical MMW slotted arrays are shown in Figure 11.22. Strider [39] has reported on a 20-element array at 34 GHz, a 1040-element array at 56 GHz array, and a 576-element array at 60 GHz, built at Hughes. Excellent agreement between predicted and measured sidelobe levels of −25 dB were obtained at 60 GHz, and the feasibility of corporate feed networks at 60 GHz was also demonstrated.

Although not as common as the antenna types discussed in the previous sections, steerable phased arrays have also been built at millimeter wavelengths. Aerojet-General developed a dual-polarized, phased array at 37 GHz for radiometer application (Pascalar [40]), which was launched on the NIMBUS-F satellite. This antenna uses crossed slots in 109 square waveguides as the aperture. An orthomode transducer feeds each square

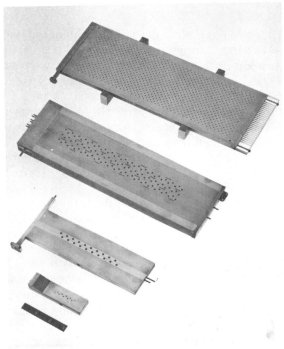

Fig. 11.22 Typical MMW slotted array antennas (courtesy of Hughes Aircraft Company).

guide and separates the two polarization components. Full electronic steering in the cross-track direction is accomplished by using separate phase shifters in each arm of each transducer. Wokurka reported results of a 44 GHz phased array feed for a Cassegrain reflector that was built by Rockwell International [41]. Thirty-one ferrite phase shifters permit beam scan of ±2° in any direction. The highest frequency phase-shifter scanned array ever built appears to be a 16-element linear phased array by TRG at 52 GHz [11]. This array scanned ±30° in the plane of the array, had a switching time of 3 μs, and was used for communications tests.

Printed circuit (also called *microstrip* or *patch*) antennas have been developed at millimeter wavelengths. Microstrip arrays offer a very attractive low-loss approach for combining active devices and radiating elements into one structure. In addition, they have a low profile and can be made conformal to a vehicle, which are often important features in military applications. The main difficulty in their use are maintaining tolerances on the small lines required at these frequencies, reducing the high line losses, and determining accurate design techniques. Weiss [42] built and tested 4 × 4 element arrays at both 37 and 57 GHz, and measured losses were 1.5 dB for each array. These results indicate that the losses in a microstrip array will be essentially independent of frequency for a fixed beamwidth. Menzel [43] reported measured data on a 4 × 24 element 40 GHz microstrip array with a gain of 19 dBi. A 2% bandwidth was obtained with 14 dB sidelobes. A low efficiency of − 6 dB was reportedly caused by radiation of the feed lines and ohmic losses in the conductors. These results suggest that a microstrip array is a viable approach at millimeter wavelengths for low-to-medium gain application. Current results indicate that these antennas can be built at 95 GHz. Larger arrays may be possible by the use of distributed amplifiers in the array.

One of the major problems with microstrip arrays is the high loss that occurs as the array becomes large. These losses can be reduced by the use of leaky-wave dielectric rod or strip antennas, as proposed by Itoh [44], or dielectric waveguide without side walls as proposed by Inggs [45]. Periodic notches, metal strips, or slots are used in the former antenna to couple energy out of the guiding structure. For the latter, uniformly spaced dielectric bars are attached to a metallic ground plane to form dielectric waveguides. The radiation structure consists of waveguide slots etched in a singly clad printed circuit board that covers the dielectric bars and forms the top wall of the waveguides. These types of antennas have been used throughout the MMW region, although the launching and line losses may be higher than desirable.

Considerable research is being performed on monolithic MMW antennas (Niekirk [46]). Such technology combines active components and radiating elements all integrated onto one substrate and promises to produce inexpensive, high performance MMW array antennas. For example, Jain and Bansal [47] utilize semiconductor MMW dipole antennas having radiation characteristics that can be controlled during either operation or manufacture.

11.4 RF COMPONENTS

The lack of availability of certain RF components may hamper the construction of radar antennas at the shorter millimeter wavelengths. However, many components, such as hybrids, *tees*, orthomode transducers, switches, horns, and circulators are commercially available. Typical performance for these components is shown in Tables 11.2 and 11.3. Rotary joints and polarizers are not readily available. However, Strider [39] describes a 60 GHz rotary joint for sector-scanning that has a 3% bandwidth and a 0.6 dB maximum insertion loss.

Table 11.2 Characteristics of Typical Commercial Short-Slot Hybrids.

	Frequency (GHz)		
	75–110	90–140	140–220
Waveguide	WR-10	WR-8	WR-5
Bandwidth (%)	6	6	—
Isolation (dB), min.	18	17	—
VSWR	1.5	1.5	—

Table 11.3 Characteristics of Typical Commercial Ring Hybrids.

	Frequency (GHz)		
	75–110	90–140	140–220
Waveguide	WR-10	WR-8	WR-5
Bandwidth (%)	5	5	5
Isolation (dB), min.	20	20	20
Insertion loss (dB), max.	0.5	0.8	1.1
Power Unbalance (dB), max.	0.5	0.5	0.5
VSWR, max.	1.25	1.25	1.27

Monopulse comparators have been built at 35, 53, 60, 70, and 95 GHz (Waineo and Konieczny [48]). It appears that one could be built at 140 GHz, but it is questionable whether this could be done at 220 GHz. Ferrite switches have been built at 220 GHz, but the losses are a bit too high.

11.5 SCANNING

Scanning of the radar antenna beam is more difficult to accomplish at millimeter wavelengths because of the high transmission-line losses and the lack of rotary joints. Techniques that have been used for performing beam scanning at millimeter wavelengths include mechanical, electromechanical, and electronic, and these techniques are discussed next.

Conceptually the easiest way to point the beam is by moving the entire antenna by using a pedestal or gimbal. This technique, which is commonly used at microwaves and lower frequencies, requires a rotary joint for each axis of beam motion if the transmitter is to be fixed. Losses in these rotary joints, and their lack of availability at 95 GHz and above, make this approach difficult to use at millimeter wavelengths. To alleviate these problems the RF components of the transmitter and receiver can be directly mounted to the antenna structure. Only dc and low-frequency signals then need to be passed through slip rings in the pedestal, which is a much more manageable job. The additional mass of these RF components mounted on the antenna makes it difficult to have rapid movement of the antenna, and hence its beam. Rapid beam motion, however, is usually required to track most military targets. Conical scanning and reflector tilting are all usable techniques for rapid beam motion at millimeter wavelengths.

Conical scanning is the most widely used tracking technique at millimeter wavelengths because of its low loss and ease of implementation. One approach to conical scanning uses a Cassegrain reflector with a fixed feed and main reflector. The subreflector is tilted at a slight angle and rotated by a motor, as shown in Figure 11.23. Notice that no RF rotary joint is required and the RF components can be mounted behind the reflector, thus eliminating blockage by them. This approach can be used throughout the MMW region.

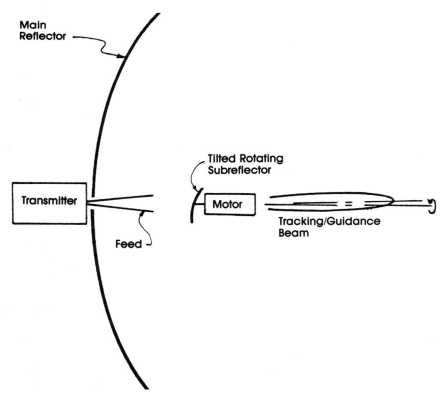

Fig. 11.23 Schematic illustration of the tracking antenna. This is a Cassegrain configuration in which conical scanning is accomplished by rotating a tilted subreflector.

A quasioptical approach to conical scanning utilizes a rotating dielectric wedge in conjunction with a dielectric lens and feed combination [49], as shown in Figure 11.24. The beam from the feed is bent when passing through the wedge. This bending appears as a displaced feed to the lens, which results in a squinting of the lens antenna pattern by

$$s = (d/f)\,(n - 1)a$$

where

s = beam pointing angle,
d = feed to wedge separation,
f = lens focal length,
n = index of refraction of the wedge,
a = wedge vertex angle.

Rotating the wedge about the axis of the system produces conical scanning.

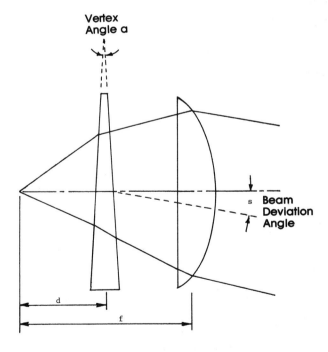

Fig. 11.24 Geometric optics deviation of a beam by a dielectric wedge. The source is at the focal point of the lens, a distance d to the left of the wedge, which has an index of refraction n and a vertex angle a. The net beam deviation angle is s.

Beam scanning accomplished by reflector motion is also possible at millimeter wavelengths. The offset reflector geometry shown in Figure 11.25 has been used at 98 GHz in a helicopter-mounted radiometer system (Wilson, *et al.* [50]). Energy reflected off the flat mirror is directed toward the paraboloid, then focused on the feed. The flat mirror is tilted both to point the beam and to compensate for aircraft motion. The scanner is capable of moving the reflector ±22.5° cross-track at a 4 Hz rate.

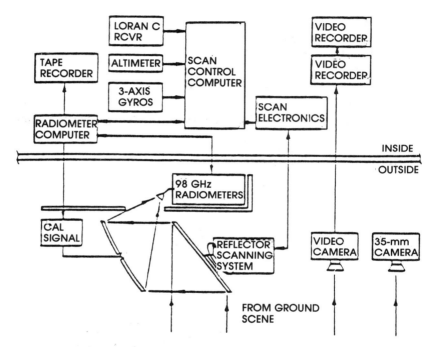

Fig. 11.25 MMW imaging sensor system (from Wilson, *et al.* [50], © 1986 IEEE).

Optical techniques such as multielement lenses have been proposed (Clarke and Dewey [51]) for wide-angle scanning MMW antennas. Houseman [52] uses a *twist reflector* and *transreflector* at 94 GHz to scan a beam by tilting the flat twist reflector, as shown in Figure 11.26. A frequency-scanned silicon waveguide array was built at 60 GHz (Klohn, *et al.* [53]), utilizing metal strips on the silicon rod to produce radiation. A novel focal-point scanning technique has been proposed by Britt [54], wherein the reflector is moved in such a way that the stationary feed is always at the focal point of the paraboloid.

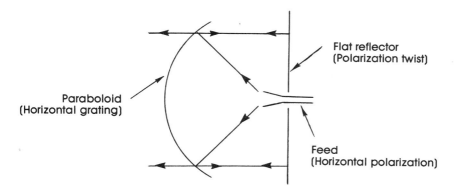

Fig. 11.26 Schematic of a polarization-twist antenna.

11.6 RADOME EFFECTS ON ANTENNA PERFORMANCE

A radome is required over an antenna in order (1) to protect the antenna from sand, dust, water; (2) to provide an aerodynamic shape over the antenna for airborne platforms; (3) to protect the antenna from high thermal loads during supersonic flight or re-entry into the atmosphere. All discussions thus far have neglected the effects of a radome on the antenna's performance. The effect of radomes on MMW antenna performance is discussed in Chapter 12.

REFERENCES

1. J.D. Kraus, *Antennas*, McGraw-Hill, New York, 1950.
2. R.C. Johnson and H. Jasik, ed., *Antenna Engineering Handbook*, McGraw-Hill, New York 1984.
3. A.W. Rudge, K. Milne, A.D. Olver, and P. Knight, *The Handbook of Antenna Design*, Vols. 1 and 2, Peter Peregrinus, London, 1986.
4. R.C. Hansen, *Microwave Scanning Antennas*, Vol. I, Academic Press, New York, 1964, Chapter 1.
5. R.B. Dybdal, "Millimeter Wave Antenna Technology," *IEEE Journal on Selected Areas in Communications*, Vol. SAC-1, pp. 633–644, September 1983.
6. Institute of Electrical and Electronics Engineers, *IEEE Standard Definitions of Terms for Antennas*, IEEE Standard 145-1983, New York, 1983.

7. F.J. Tischer, "Excess Conduction Losses at Millimeter Wavelength," *IEEE Transactions on Microwave Theory and Techniques*, Vol. MTT-24, No. 11, pp. 853–858, November 1976.

8. F.J. Tischer, "Experimental Attenuation of Rectangular Waveguides at Millimeter Wavelengths," *IEEE Transactions on Microwave Theory and Techniques*, Vol. MTT-27, No. 1, pp. 31–37, January 1979.

9. P.F. Goldsmith, "Quasioptical Techniques Offer Advantages at Millimeter Frequencies," *Microwave System News*, Vol. 13, pp. 65–84, December 1983.

10. N.E. Swanberg and J.A. Paul, "Quasi-Optical Mixer Offers Alternative," *Microwave Systems News*, Vol. 9, , pp. 58–60, May 1979.

11. A.F. Kay, "Millimeter Wave Antennas," *Proceeding of the IEEE*, Vol. 54, No. 4, pp. 641–647, April 1966, .

12. M. Thumm, *et al.*, "Very High Power mm-Wave Components in Oversized Waveguides," *Microwave Journal*, Vol. 29, pp. 103–121, November 1986.

13. J. Ruze, "Antenna Tolerance Theory—A Review," *Proceedings of the IEEE*, Vol. 54, No. 4, pp. 633–640, April 1966.

14. R.C. Johnson, Chapter 1 in R.C. Johnson and H. Jasik, eds., *Antenna Engineering Handbook*, 2nd Ed., McGraw-Hill, New York, 1984.

15. V.K. Tripp "A New Approach to the Analysis of Random Errors in Aperture Antennas", *IEEE Transactions on Antennas and Propagation*, Vol. AP-32, No. 8, pp. 857–863, August 1984.

16. D.G. Bodnar, "Materials and Design Data," Chapter 46, *Antenna Engineering Handbook*, 2nd Ed., R.C. Johnson and H. Jasik, eds., McGraw-Hill, New York, 1984. (Reprinted with permission.)

17. R.C. Hansen, *Microwave Scanning Antennas*, Vol. I, Academic Press, New York, 1964, Chapter 1.

18. S. Silver, ed., *Microwave Antenna Theory and Design*, MIT Radiation Laboratory Series, Vol. 12, McGraw-Hill, New York, 1949.

19. C.L. Dolph, "A Current Distribution for Broadside Arrays Which Optimizes the Relationship between Beamwidth and Side-Lobe Level," *Proceedings of the IRE*, Vol. 34, No. 6 pp. 335–348, June 1946.

20. T.T. Taylor, "Design of Line-Source Antennas for Narrow Beamwidth and Low Side Lobes," *IRE Transactions on Antennas Propagation*, Vol. AP-3, pp. 16–28, January 1955.

21. J.W. Sherman, III, "Aperture-Antenna Analysis," Chapter 9 in M.I. Skolnik, ed., *Radar Handbook*, McGraw-Hill, New York, 1970 Chapter 9.

22. F.J. Harris, "On the Use of Windows for Harmonic Analysis with the Discrete Fourier Transform," *Proceedings of the IEEE*, Vol. 66, No. 1, pp. 51–83, January 1978.

23. T.T. Taylor, "Design of Circular Apertures for Narrow Beamwidth and Low Sidelobes," *IRE Transactions on Antennas Propagation*, Vol. AP-8, No. 1, pp. 17–22, January 1960.

24. R.C. Hansen, "Tables of Taylor Distributions for Circular Aperture Antennas, "*IRE Transactions on Antenna Propagation*, Vol. AP-8, No. 1 pp. 23–26, January 1960.

25. A.C. Ludwig, "Low Sidelobe Aperture Distributions for Blocked and Unblocked Circular Apertures," RM 2367, General Research Corporation, April 1981.

26. W.D. White, "Circular Aperture Distribution Functions," *IEEE Transactions on Antenna Propagation*, Vol. AP-25, No. 9, pp. 714–716, September 1977.

27. P.W. Hannan, "Microwave Antennas Derived from the Cassegrain Telescope," *IRE Transactions on Antennas Propagation*, Vol. AP-9, No. 3, pp. 140–153, March 1961.

28. C.L. Gray, "Estimating the Effect of Feed Support Member Blockage on Antenna Gain and Side-Lobe Level," *Microwave Journal*, pp. 88–91, March 1964.

29. A.W. Love, "Horn Antennas," Chapter 15, *Antenna Engineering Handbook*, Second Edition, R.C. Johnson and H. Jasik, eds., McGraw-Hill, New York, 1984.

30. A.W. Love, *Electromagnetic Horn Antennas*, IEEE Press, New York, 1976.

31. L.C. Gunderson and G.T. Holmes, "Lenses Improve Horn Antennas," *Microwaves*, pp. 32–35, August 1971.

32. T. Morita and S.B. Cohn, "Microwave Lens Matching by Simulated Quarter-Wave Transformers, *IEEE Transactions on Antennas and Propagation*, Vol. AP-4, No. 1, pp. 33–39, January 1956.

33. R.C. Johnson, "The Geodesic Luneberg Lens," *Microwave Journal*, pp. 76–85, August 1962.

34. R.C. Johnson, Optical Scanners," Chapter 3, *Microwave Scanning Antennas*, Vol. 1, R.C. Hansen, ed., Academic Press, New York, 1964.

35. R.C. Johnson, A.L. Holliman, and J.S. Hollis, "A Waveguide Switch Employing the Offset Ring-Switch Junction," *IRE Transactions on Microwave Theory and Techniques*, Vol. MTT-8, No. 1, pp. 532–537, September 1960.

36. K.J. Button and J.C. Wiltse, *Infrared and Millimeter Waves*, Academic Press, New York, 1981, pp. 68–73.

37. H.P. Coleman, R.M. Brown, and B.D. Wright, "Paraboloidal Reflector Offset Fed with a Corrugated Conical Horn," *IEEE Transactions on Antennas and Propagation*, Vol. 23, No. 11, pp. 817–819, November 1975.

38. D.G. Bodnar and C.C. Kilgus, "A 95 GHz Surface-Effect Vehicle Antenna and Radome," *International IEEE AP-S Symposium Digest*, Georgia Institute of Technology, Atlanta, GA 10-12 pp. 256–259, June 1974.

39. C.A. Strider, "Millimeter-Wave Planar Arrays," Paper B-6, *Millimeter Waves Techniques Conference*, San Diego, CA, 26-28 March 1974.

40. H.G. Pascalar, "Millimeter Wave Passive Sensing From Satellites", *Millimeter Wave Techniques Conference*, 26-28 March 1974, Naval Electronics Laboratory Center, San Diego, CA, paper F5.

41. J. Wokurka, "A Q-Band Electronically Scanned Cassegrain Antenna," *IEEE International AP-S Symposium Digest*, Quebec, pp. 15–18, 2-6 June 1980.

42. M.A. Weiss, "Microstrip Antennas For Millimeter Waves," Ball Aerospace Corporation, Boulder, CO, ECOM-76-0110-F, October 1977.

43. W. Menzel, "A 40 GHz Microstrip Array Antenna," *IEEE MTT-S International Microwave Symposium Digest*, Washington, DC, pp. 225–226, 28-30 May 1980.

44. T. Itoh and B. Adelseck, "Trapped Image Guide Leaky-Wave Antenna for Millimeter-Wave Application," *IEEE International Antennas and Propagation Symposium Digest*, Quebec, pp. 19–22, 2-6 June 1980.

45. M.R. Inggs, M.T. Birand, and N. Williams, "Experimental 30 GHz Printed Array with Low Loss Insular Guide Feeder," *Electronics Letters*, Vol. 17, No. 3, pp. 146–147, 5 February 1981.

46. D.P. Niekirk, D.B. Rutledge, M.S. Muha, H. Park, and C. Yu, "Progress in Millimeter-Wave Integrated Circuit Imaging Antenna Arrays," *Proceedings of the Society of Photo-Optical Engineers,* Huntsville, AL, Vol. 317, pp. 206–210, November 16-19, 1981.

47. F.C. Jain and R. Bansal, "Monolithic MM-Wave Antennas," *Microwave Journal*, Vol. 27, No. 7, pp. 123–133, July 1984.

48. D.K. Waineo and J.F. Konieczny, "Millimeter Wave Monopulse Antenna with Rapid Scan Capabilities," *IEEE AP-S International Symposium Digest*, Seattle, WA, pp. 477–480, 18-22 June 1975.

49. P.F. Goldsmith and G.J. Gill, "Dielectric Wedge Conical Scanned Gaussian Optics Lens Antenna," *Microwave Journal*, Vol. 29, No. 9, pp. 202–212, September 1986.

50. W.J. Wilson, R.J. Howard, A.C. Ibbott, G.S. Parks, and W.B. Ricketts. "Millimeter-Wave Imaging Sensor," *IEEE Transactions on Microwave Theory and Techniques*, Vol. MTT-34, No. 10, pp. 1026–1035, October 1986.

51. J.A. Clarke and R.J. Dewey, "Millimeter Wave Imaging Lens Antennas," *International Journal of Infrared and Millimeter Waves*, Vol. 5, No. 1, pp. 91–101, 1984.
52. E.O. Houseman, "A Millimeter Wave Polarizer Twist Antenna" *1979 IEEE Antennas and Propagation Society Symposium Digest*, University of Maryland, College Park, MD, pp. 51–54, May 15-19, 1979.
53. K.L. Klohn, R.E. Horn, H. Jacobs and E. Freibers, "Silicon Waveguide Frequency Scanning Linear Array Antenna," *IEEE Transactions on Microwave Theory and Techniques*, Vol. MTT-26, No. 10, pp. 764–773, October 1978.
54. P.P. Britt, "A Focal Pivot Parabola: A Surprising Millimeter Wave Radar Antenna," *Abstracts of the 23rd United States Air Force Antenna Symposium*, Robert Allerton Park, IL, 27-29 April 1977.

Chapter 12
MMW Radomes

E.B. Joy

Georgia Institute of Technology
Atlanta, Georgia

12.1 INTRODUCTION—RADOME REQUIREMENTS

The term *radome* is a contraction of the words "radar dome," and is a subset of the more general term "electromagnetic window," which includes other types of antenna systems, frequency ranges, and shapes, including flat panels and conformal shapes (Tice [1] and Walton [2]). Radomes are designed to protect the enclosed antenna from adverse environmental conditions, while having minimal effect on the electromagnetic performance of the antenna. The antenna and radome are usually designed separately, assuming the interaction between the two is insignificant. A necessary, but not sufficient, requirement for low radome-antenna interaction is the maintenance of a minimum separation between the antenna and radome of at least one wavelength. A special class of electromagnetic windows is the *radant* (Timms and Kmetzo [3]), a contraction of the words "radome antenna," where the radome is a significant part of the antenna performance, and the two are often combined into a single structure. The separation between the antenna and radome in a radant is commonly much less than one wavelength, and the radome thus becomes an integral part of the antenna.

The radome may have requirements in addition to its being an environmental barrier and a transparent electromagnetic window. The exterior shape may have to be streamlined with respect to aerodynamic or hydrodynamic forces, be camouflaged, or architecturally acceptable in appearance. The radome may need additional strength to allow for internal

561

pressurization to increase the dielectric strength for high-power applications or to provide for high-pressure cooling gasses in high-temperature applications. The radome may be required to produce a specified reflection back to the enclosed antenna for impedance-matching purposes. The antenna within the radome may need to gimbal with respect to the radome, which requires the radome to have good electromagnetic performance over a range of aspect angles. These and other requirements for the radome usually result in lowering of the radome's electromagnetic performance. This chapter primarily addresses the electromagnetic aspects of radome requirements, performance, design, and measurement at MMW frequencies.

Electromagnetic performance requirements for MMW radomes are application-dependent. Four primary requirements are identified and related to application. These requirements are high transmission, low boresight error, broadband, and low reflection.

High Transmission

Long-range radar and point-to-point communication systems require high radome transmission efficiency. Such radomes are designed for minimum loss at the peak of the main beam and usually involve a radome having fixed orientation with respect to the antenna, or a radome having a relatively flat face parallel to the aperture of the antenna. It is shown that the transmission performance of a radome wall is highly dependent on the angle between the direction of propagation of the incident electromagnetic energy and the normal to the radome wall, which is termed the *angle of incidence*. Flat radomes or those with large radius of curvature encounter a very limited range of angles of incidence near normal to the radome wall. The transmission properties of a radome wall can be optimized in such walls with limited angle of incidence. The gain of the antenna may even be increased by use of a radome. The radome may act as a lens to correct a known phase front abnormality of the antenna, and thus increase its gain. Transmission loss at MMW frequencies is especially acute because the radome walls are generally electromagnetically thicker than lower frequency designs. The minimal physical thickness of a radome wall is dependent on the forces that the radome must support, and the thickness is independent of frequency. This resulting thickness, when measured in terms of the millimeter wavelength, is often larger than desirable for low electromagnetic loss. Thus, high transmission and low loss are prime objectives of all MMW radome designs.

Low Boresight Error

Directional guidance radar systems have a prime requirement for low boresight error and boresight error slope. High-speed aircraft and missile guidance radar systems can become destabilized due to high boresight error slopes. Boresight error is the difference between the mechanical and the electromagnetic pointing direction of the antenna, as determined by the radar system. The radome contributes to this error, and can be the only source of this error if the antenna boresight error without radome has been corrected. Boresight error slope is the rate of change of the boresight error as the antenna is gimballed with respect to the radome, and is usually expressed as degrees of boresight error per degree of antenna gimbal. Radome-induced boresight errors are caused by asymmetries in the amplitude and phase of the transmitted or received electromagnetic field over the surface of the radome. If each point of the radome surface transmitted each component of the electromagnetic field equally well, there would possibly be a loss of gain, but no boresight error or boresight error slope. Maximum asymmetry, and thus boresight error, for pointed radomes usually occurs when one-half of the antenna pattern is directed toward the tip region of the radome and the other half is pointed toward the relatively flat side wall of the radome.

Broadband

Electronic warfare, electronic countermeasures, electronic counter-countermeasures, multifrequency radar, and communication systems typically require that the radome be electromagnetically transparent over a broad range of frequencies, not just one narrow band of frequencies. The prime requirement is usually minimum loss of gain over a broad range of frequencies and antenna gimbal angles.

Low Reflection

High-power systems and low-sidelobe antenna systems require low reflection of the incident electromagnetic energy as it passes through the radome. Even a small percentage of a high-power field reflected from the radome interior surface and returned to the transmitting antenna could cause the high-power source to change frequency or reduce power output. Higher levels of reflected signal could cause dielectric breakdown in the high-power feed system. Levels for low-sidelobe antennas will be increased if energy from the main beam of the antenna is reflected from the radome into sidelobe regions. The radome is designed to have low radome wall reflection for such applications.

Specifications and Standards

United States Military Specification MIL-R-7705B, "General Specification for Radomes," (14 January 1975) gives general design and performance requirements for radomes used in flight vehicles, surface vehicles and fixed ground installations. Within this specification is the following list of Federal, Military, and Industrial specifications and standards that are directly applicable to radomes and radome materials.

SPECIFICATIONS

FEDERAL

L-P-383	Plastic Material, Polyester Resin, Glass Fiber Base, Low Pressure Laminated

MILITARY

MIL-B-5087	Bonding, Electrical, and Lightning Protection, for Aerospace Systems
MIL-C-7439	Coating System, Elastomeric, Rain Erosion Resistant and Rain Erosion Resistant with Anti-Static Treatment, for Exterior Aircraft and Missile Plastic Parts
MIL-R-7575	Resin, Polyester, Low Pressure Laminating
MIL-C-8073	Core Material, Plastic Honeycomb, Laminated Glass Fiber Base, for Aircraft Structural Applications
MIL-C-8087	Core Material, Foamed in-Place Polyester Diisocyanate Type Interchangeability and Replaceability of Component Parts for Aircraft and Missile
MIL-M-8856	Missiles, Guided: Strength and Rigidity: General Specification for
MIL-A-8860	Airplane Strength and Rigidity, General Specification for
MIL-A-8869	Airplane Strength and Rigidity, Special Weapons Effects
MIL-S-9041	Sandwich Construction, Plastic Resin, Glass Fabric Base, Laminated Facings and Honeycomb Core for Aircraft Structural Applications
MIL-C-9084	Cloth, Glass, Finished, for Polyester Resin Laminates
MIL-R-9299	Resin, Phenolic, Laminating
MIL-R-9300	Resin, Epoxy, Low-Pressure Laminating

MIL-P-9400	Plastic Laminate Materials and Sandwich Construction, Glass Fiber Base, Low Pressure Aircraft Structural, Process Specification Requirements
MIL-Q-9858	Quality Program Requirements
MIL-F-18264	Finishes, Organic, Weapon Systems, Application and Control of
MIL-C-22750	Coating, Epoxy-Polyamide
MIL-P-23377	Primer Coating, Epoxy Polyamide, Chemical and Solvent Resistant
MIL-R-25042	Resin, Polyester, High Temperature Resistant, Low Pressure Laminating
MIL-S-25392	Sandwich Construction, Plastic Resin, Glass Fabric Base, Laminated Facings and Polyurethane Foamed-in-Place Core for Aircraft Structural Applications
MIL-P-25395	Plastic Materials, Heat Resistant, Low Pressure Laminated Glass Fiber Base, Polyester Resin
MIL-P-25421	Plastic Materials, Glass Fiber Base-Epoxy Resin, Low Pressure Laminated
MIL-R-25506	Resin, Silicone, Low Pressure Laminating
MIL-P-25515	Plastic Materials, Phenolic-Resin, Glass Fiber Base, Laminated
MIL-P-25518	Plastic Materials, Silicone Resin, Glass Fiber Base, Low Pressure Laminated
MIL-C-27315	Coating Systems, Elastomeric, Thermally Reflective and Rain Erosion Resistant
MIL-C-81773	Coating, Polyurethane, Aliphatic, Weather Resistant
MIL-C-83231	Coatings, Polyurethane, Rain Erosion Resistant for Exterior Aircraft and Missile Plastic Parts
MIL-C-83286	Coating, Urethane, Aliphatic Isocyanate, for Aerospace Applications
MIL-Y-83370	Yarn, Roving and Cloth, High Modulus, Organic Fiber
MIL-A-83377	Adhesive Bonding for Aerospace Systems, Guidelines for

STANDARDS

FEDERAL

FED-STD-5	Standard Guides for Preparation of Item Identification
FED-STD-102	Preservation, Packaging and Packing Levels
FED-STD-406	Plastics, Methods of Testing

MILITARY

MIL-STD-100	Engineering Drawing Practices

MIL-STD-129	Marking for Shipment and Storage
MIL-STD-130	Identification Marking of US Military Property
MIL-STD-210	Climatic Extremes for Military Equipment
MIL-STD-401	Sandwich Constructions and Core Materials; General Test Methods
MIL-STD-490	Specification Practices
MIL-STD-794	Parts and Equipment, Procedures for Packaging and Packing of
MIL-STD-810	Environmental Test Methods

NAVAL AIR SYSTEMS COMMAND

| NAVAIR 01-1A-22 | Maintenance Instruction Manual—Aircraft Radomes and Antenna Covers |

AEROSPACE INDUSTRIES ASSOCIATION

| ATC Report ARTC-4 | Electrical Test Procedures for Radomes and Radome Materials |

AMERICAN SOCIETY FOR TESTING AND MATERIALS

ASTM C-177	Methods of Test for Thermal Conductivity of Materials of the Guarded Hot Plate
ASTM C-373	Test for Water Absorption, Bulk Density, Apparent Porosity, and Apparent Specific Gravity of Fired Porous Whiteware Products
ASTM C-407	Compressive Crushing Strength of Fired Whiteware Materials
ASTM C-674	Flexural Properties of Ceramic Whiteware Materials
ASTM E-228	Linear Thermal Expansion of Rigid Solids with a Vitreous Silica Dilatometer

The outer and inner surface shapes of a radome affect all aspects of its electromagnetic performance, and shapes are therefore discussed in the following section.

12.2   RADOME SHAPE

The shape of the radome contributes to the electromagnetic, mechanical, thermal, and architectural performance of the radome. The exterior shape must also interact with the environment, and thus has many nonelectromagnetic requirements, as discussed above. The interior shape of a radome and associated radome wall thickness are designed to meet or surpass the mechanical and thermal requirements, and then are tailored for optimum electromagnetic performance.

The preferred radome shape is a wall of large radius of curvature and constant thickness, with curvature matching the constant phase fronts of the enclosed antenna. Matching of the radome curvature and the phase front results in a normal incidence field to the radome at all points on the radome. The constant wall thickness allows these normal incidence fields to be transmitted with the same amplitude and phase. After having selected the radome material, the thickness of the wall can be designed for maximum transmission or minimum reflection for optimum performance, as described below. The preferred separation between the radome and the antenna is greater than 10 wavelengths. This minimal separation ensures that the radome is in the radiating near field of the antenna and not in the reactive near field, where other forms of electromagnetic coupling between the antenna and the radome wall are possible. Often, a spherical radome shape with large radius centered on the antenna can approximate the preferred radome design.

Radome shapes that have a radius of curvature which is less than the radius of curvature for the antenna phase fronts cause the angle of incidence to vary over the radome surface, resulting in unequal transmission for a constant wall thickness. Non-normal angle of incidence and radome curvature also combine to produce refraction of the field passing through the radome wall. Figure 12.1 shows this effect (Tice [1]). The electromagnetic field represented by the arrow A is incident on the outer surface of a radome at an incidence angle Θ. The electromagnetic field leaves the radome wall as arrow A'. The incident and transmitted fields have been refracted due to the constant thickness of the wall and the non-normal angle of incidence. The amount of refraction, δ, can be determined by using the figure. An example for a radius of curvature equal to 100 times the wall thickness and for an incidence angle of 60° for a wall with a dielectric constant $\epsilon = 6$, yielding a refraction of 0.7°, is also shown in the figure. The radome wall thickness could be varied such that the refraction due to nonparallel wall thickness would cancel the refraction due to wall curvature. This cancellation, however, would only occur for an incident angle of 60° at that one point on the radome.

Aircraft and missile radomes must have aerodynamically shaped outer surfaces for minimum drag, heating, and rain erosion. These shapes tend to be pointed, introducing a discontinuity in the surface shape. Figure 12.2 shows five such axially symmetric shapes with the same axial length and base diameter (Walton [2]). The five shapes shown in the figure are the cone, Von Karman, tangent ogive, L.V. Haack, and the 60° log spiral. Other shapes not shown are the nontangent ogive, paraboloid, three-quarter power, ellipsoidal, and numerically generated. Radome shapes may also be nonaxially symmetric. Each shape has its advantages and disadvantages. For a given length and diameter, some shapes produce minimum drag at a specific velocity, some produce minimum heating, some

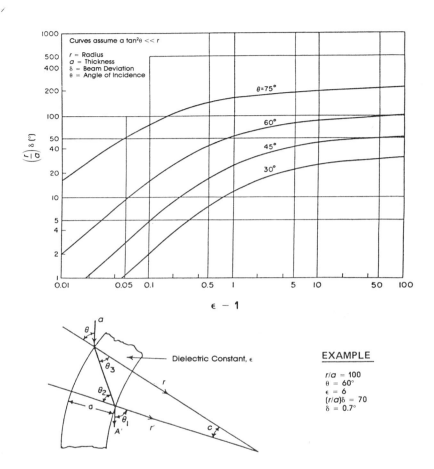

Fig. 12.1 Refraction due to radome wall curvature as a function of wall dielectric constant for several angles of incidence (from Tice [1]).

have the best electromagnetic performance for a given application, and antenna. Each of the shapes has a parameter called the *fineness ratio,* which is the ratio of length to diameter. The performance of each shape is also a function of its fineness ratio. Figure 12.3 shows the boresight error slope of a half-wavelength wall, constant-thickness, tangent ogive radome using Pyroceram™ material with a dielectric constant of $\epsilon = 5.5$ *versus* the fineness ratio of the shape (Yost, *et al.* [4]). This figure shows the strong dependence of an electromagnetic performance indicator *versus* fineness ratio. Electromagnetic performance of the radome is also affected by the antenna characteristics, as shown in Figure 12.4 (Kilcoyne [5]). The

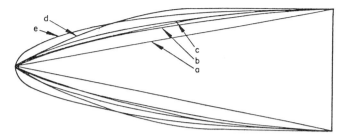

a. CONE
b. VON KARMAN
c. TANGENT OGIVE
d. HAACK
e. 60° LOG SPIRAL

Fig. 12.2 Five streamlined radome shapes, each having the same axial length to base diameter ratio of 2.5 (from Walton [2]).

boresight error of a fixed radome and fixed size antenna *versus* antenna gimbal angle, here called *look angle*, can be seen as a function of the type of antenna aperture distribution. A cosine-squared distribution, which produces a lower gain, wider beamwidth pattern, can be seen to produce larger boresight errors than the higher gain, narrower beamwidth uniform distribution. This difference in boresight error is probably due to the wider range of incident angles, and associated wider range of transmission values, produced by the wider beamwidth distributions.

A thorough analysis of the range of incident angles encountered for a specific radome, antenna and antenna location, and gimbal angle usually requires computer assistance. Figure 12.5 shows the results of such a computer-aided analysis of incident angle distributions (Joy, *et al.* [6]). The figure shows the 5% and 95% incident-angle probability distribution points for a tangent ogive radome with fineness ratio of 2.0 for a uniformly illuminated antenna located near the base of the radome. The figure illustrates how the incident angle distribution function changes with antenna look angle in the azimuth plane. Note that for zero look angle, the antenna is pointed through the tip of the radome and the incident angle distribution for the sum pattern is very narrow with a mean value of approximately 74°. As the antenna scans, the distribution widens and 90% of the energy encompasses incident angles ranging from 12° to 70° at a 60° scan angle. A similar computer analysis could be conducted at each point of the radome surface to show the range of incident angles encountered at each point.

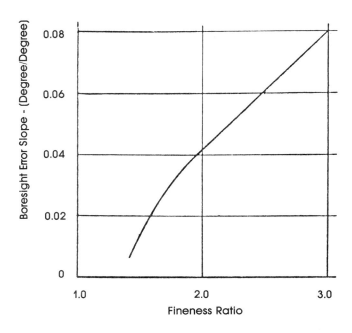

Fig. 12.3 Radome-induced boresight error slope averaged over a 10° look angle *versus* fineness ratio for a tangent ogive, tuned, half-wavelength thickness, Pyroceram® radome (from Yost, Weckesser, and Mallalieu [4]).

The wall thickness at each point could then be optimally designed, based on the selected electromagnetic performance criterion. Other electromagnetic effects of radome shape, in addition to gain loss, boresight error, boresight error slope, sidelobe level increase, and reflected power in the antenna and feed system, include main beam distortion, beamwidth change, polarization change, and decreased difference-pattern null depth. The next section outlines the procedure and tools used for the design of MMW radomes.

12.3 RADOME DESIGN

Radome design begins with radome performance specifications, measured antenna near-field or far-field radiation patterns, antenna geometry and location, antenna gimballing geometry, gimballing scenarios, environmental specifications, and often the size, shape, and location of the

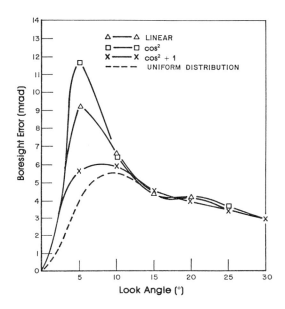

Fig. 12.4 Calculated radome-induced boresight error *versus* antenna look angle within the radome for several antenna aperture distribution functions (from Kilcoyne [5]).

exterior surface of the radome. Mechanical, thermal, material, environmental, and electromagnetic designs must interact to achieve all performance specifications. Often, computer-based analyses will be used for the mechanical, thermal, and electromagnetic parts of the design. A flow chart of the typical design process for an aerodynamic radome is shown in Figure 12.6 (Cary [7]). This figure illustrates the interrelationships among the various disciplines, and between man and computer. The electromagnetic parts of the design are discussed further, and the mechanical and thermal analysis techniques are presented elsewhere (Walton [8]).

The radome wall is initially chosen to be of a constant thickness with a material and thickness that conservatively meet the mechanical, thermal, and environmental requirements. The thickness is then electromagnetically refined by using a computer-based incident angle analysis. The power-weighted, average angle of incidence is computed for each point on the interior of the radome wall by using a transmitting formulation. The two-dimensional plane-wave spectrum of the antenna is calculated from the two-dimensional near-field or far-field antenna patterns. The plane-wave

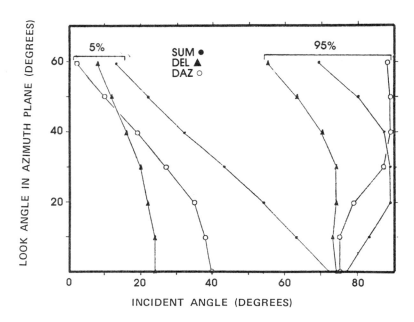

Fig. 12.5 The 5% and 95% points of the power-weighted incident angle
probability distribution function for an tangent ogive radome
with a fineness ratio of 2.0 *versus* antenna look angle in the
azimuth plane for the sum (SUM), elevation difference (DEL),
and azimuth difference (DAZ) patterns of a monopulse radar
antenna (from Joy, Huddleston, Bassett, and Wheeler, [6]).

spectrum is the complex amplitude of each of the plane waves that con-
stitute the radiation pattern of the antenna. A grid of points is established
on the interior surface of the radome. The antenna is gimballed to a fixed
orientation, and transmits its spectrum of plane waves from its aperture,
with each transmitted plane wave spanning the entire aperture. The incident angle and power density of the perpendicular and parallel components of each plane wave are determined and recorded at each interior
radome surface point. Note that not all interior points are illuminated by
all transmitted plane waves. The average incident angle may then be computed at each interior point. The power-weighted incident angle may also
be calculated for each polarization. This process is then repeated to find
averages for other antenna patterns, such as sum and difference patterns
for radar antennas, and for other gimbal orientations, and the averages
could be weighted by the probabilities associated with the gimballing scenario. The thickness of the radome wall is then chosen for maximum
transmission, minimum reflection, or constant insertion phase delay, based

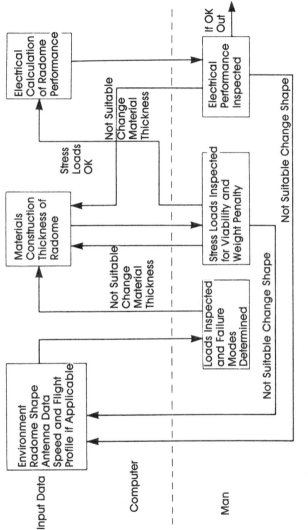

Fig. 12.6 Iterative radome design process block diagram showing the interaction between man and computer aided analyses (from Cary [7], © 1983 Peter Peregrinus).

on the power-weighted average incidence angle. The chosen thickness is a compromise between the best for each polarization. This thickness distribution is often called a *prescription,* which is an optical term.

The next step in the electromagnetic design is an entire radome analysis to ascertain the cumulative effects of imperfect transmission, reflection, and insertion phase delay over the entire radome surface. This analysis will culminate in the determination of overall radome electromagnetic performance, including such parameters as loss of gain, boresight error, and boresight error slope *versus* gimbal angle. Several radome analysis techniques are in use [9–14]. The two most popular computer-based techniques are called *surface integration* (Park [11]) and *plane-wave spectrum* analyses (Joy and Huddleston [13]). Each has several forms, some optimized for transmitting or receiving, or for the calculation of a single parameter such as boresight error. The equivalent aperture, plane-wave spectrum, transmitting formulation, radome analysis technique is briefly described as an example [13].

The plane-wave spectrum analysis begins with the same information required for the incident angle analysis described above plus the exact inner surface or wall thickness description. A square grid of points is established over a planar surface parallel to the antenna aperture and separated from the antenna by one to two wavelengths. It is convenient if this is the surface over which near-field measurements of the antenna were made. The points are spaced approximately one-quarter of a wavelength. The antenna near-field or far-field pattern is expressed in terms of a plane wave spectrum. Each plane wave of the spectrum is represented by samples located at the planar surface grid points. Note that the complex vector sum of all the plane waves in the spectrum evaluated at these grid points would form the near-field distribution of the antenna. Each sample of each plane wave of the antenna is propagated in the direction of propagation of that plane wave to the inner surface of the radome. The normal to the radome wall, the plane of incidence, the angle of incidence, and the perpendicular and parallel components of the plane-wave sample are determined at the inner surface intersection point. The complex perpendicular and parallel transmission and reflection coefficients for the possibly multilayered wall (Richmond [15]), and the directions of reflection and refraction are calculated. The perpendicular and parallel transmitted plane-wave components are propagated through the wall to the exterior surface, where the plane wave is again refracted. Note that curvature of the wall will change the direction of propagation of the transmitted plane wave. The reflected plane wave is propagated to an intersection with another interior point of the radome, where the above process is repeated. Multiple reflection can be indefinitely continued. Typically, only one reflection is considered. The transmitted plane-wave samples are then propagated in the direction opposite to the refracted direction, back to the original planar surface, with the radome removed to form an equivalent plane wave. The equivalent plane wave has been changed point by point for the effects of the radome wall, including change in amplitude, phase, polarization, and direction of propagation. This process is repeated for all samples of all plane waves of the antenna, yielding an equivalent near-field distribution for the antenna. Note that the planar surface area is made much greater than the antenna aperture area to accommodate the backward propagated reflected plane wave samples. The far-field pattern is then calculated for the equivalent near-field distribution by using the fast Fourier transform to ascertain the electromagnetic effects of the radome. Analysis of the computed results may suggest further tapering of the radome wall, both longitudinally and circumferentially, to achieve the desired radome electromagnetic performance. This radome analysis technique, although computationally intense, is faster than the surface integration technique due to the use of the fast Fourier transform in place of the two-dimensional surface integrations required by the surface intergration technique.

Additional radome analysis tools, which are often required, are models for scattering by metallic objects such as radome tips and lightning diverter strips (Waterman [16]), and models for trapped and traveling waves on the radome surface. The next section describes radome wall configurations and their electromagnetic characteristics.

12.4 RADOME WALL CONFIGURATIONS

The radome wall is designed for highest electromagnetic transmission amplitude, lowest reflection amplitude, and lowest insertion phase delay over the expected range of incident angles and frequencies. In addition to good electromagnetic performance, the radome wall should be designed for minimum weight and maximum strength. Figure 12.7 shows the common configurations for radome walls of constant thickness. The two basic groups are the monolithic (solid) walls and the layered walls. These wall groups are discussed separately in the following sections.

12.4.1 Monolithic Walls

The monolithic wall configuration is shown in Figure 12.7 with various thicknesses. The thin wall is most attractive because it is lightweight and operates with low loss for all frequencies where the wall is less than approximately 1/20 of a wavelength (in the material) thick. The thin wall is

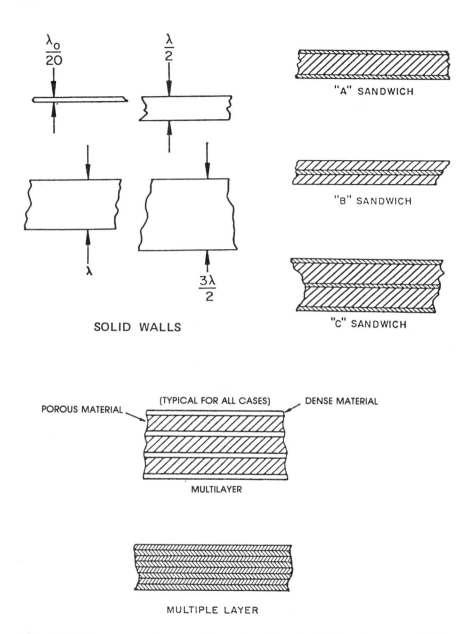

Fig. 12.7 Common radome wall construction, including the monolithic, sandwich, and multilayer configurations (from Tice [1]).

a naturally broadband wall. Thin walls, however, are rarely useful at millimeter wavelengths due to the resulting low strength of such a thin structure. A thin wall configuration at 90 GHz, using a material with dielectric constant of $\epsilon = 4$, would have a maximum thickness of only 3.28 thousandths of an inch. The most popular of the monolithic wall thicknesses is the one-half wavelength (in the material) wall. This wall exhibits 100% normal incidence transmission for lossless materials, and the thickness can be optimized for maximum transmission at any desired incident angle. Normal incidence transmission of 100% may also be achieved for any integer multiple, N, of one-half wavelength. Maximum transmission at any desired incident angle, Θ, is achieved, for a monolithic wall, by choosing the thickness of the wall, d, as follows:

$$d = N\lambda/2 \,(1/\sqrt{\epsilon - \sin^2\theta}), \; N > 0 \text{ (integer)} \qquad (12.1)$$

where

ϵ = the relative dielectric constant of the wall material,
λ = the free-space wavelength.

This equation shows that the required wall thickness for maximum transmission increases with increasing incident angle and decreasing dielectric constant. The transmission through any flat panel wall is polarization-sensitive. An arbitrarily polarized, incident plane-wave field, E_i, can be decomposed into two orthogonal components with respect to the plane of incidence. The plane of incidence is defined as the plane containing the direction of propagation of the incoming plane wave, $\hat{k}$, and the normal to the flat panel wall, $\hat{n}_R$, as shown in Figure 12.8 [13]. The two components are the perpendicular and parallel, and they are the components of the incident field which are perpendicular and parallel to the plane of incidence, respectfully. The equations in the figure allow determination of the perpendicular and parallel directions, perpendicular and parallel components of the incident electric field, and the angle of incidence, given the complex amplitude, polarization direction and propagation direction of the incident field, and the normal to the flat panel wall. The transmitted field is also shown in the figure, which has the same direction of propagation as the incident field. The perpendicular and parallel components of the transmitted field can be seen as equal to the perpendicular and parallel components of the incident field, respectively multiplied by the complex perpendicular and parallel transmission coefficients of the wall. The total transmitted field will generally not be parallel (i.e., not have the same polarization) to the incident field, unless the complex transmission coefficients are equal. The two coefficients are equal at normal incidence, for all incident polarizations, the electric field is totally parallel to the surface of the panel. Figure 12.9 shows the magnitude of the transmitted power (in percentage) for the two polarizations of a low-loss dielectric flat panel with a relative dielectric constant of $\epsilon = 9$ and a loss tangent of $\delta = 0.0004$ *versus* thickness of the panel measured in free space wavelengths for various incident angles (Jasik [17]). Notice the great differences between the transmission properties of two polarizations especially at an incident angle of 70°. The parallel polarization exhibits high transmission for all wall thicknesses, whereas the perpendicular polarization exhibits low transmission, except at the wall thickness specified in the above equation for maximum transmission thickness. This high parallel polarization transmission effect is called the *Brewster effect*. The Brewster angle, θ_B, for maximum parallel polarization transmission of lossless flat panels with relative dielectric constant of ϵ, is given by

$$\theta_B = \tan^{-1}(\sqrt{\epsilon}) \tag{12.2}$$

The Brewster angle for a flat panel wall with a relative dielectric constant of $\epsilon = 9$ is 71.6°. Also notice in the figure that the maximum transmission at other than the Brewster angle depends on both the thickness of the wall and the angle of incidence. This figure shows that for a fixed wall thickness it is difficult to achieve high transmission over a wide range of incident angles. Figure 12.10 compares the normal incidence power transmission for flat panels of various dielectric constants *versus* the thickness of the panel given in free-space wavelengths (Yost, *et al.* [4]). This figure illustrates several benefits of a low dielectric wall material. First, the shaded region of the figure shows that the low dielectric constant material has high transmission over a much larger range of wall thicknesses, or for a fixed physical thickness, over a much larger range of frequencies. The low dielectric wall is more broadband than the high dielectric wall. The shape of the transmission curves repeat for the larger multiples of one-half wavelength (in the material) wall thickness. This results in a decrease in percentage bandwidth by the same multiple. Second, the low dielectric constant wall is thicker than the high dielectric constant wall at the maximum transmission thickness, which often means a stronger wall. A wall with a dielectric constant of $\epsilon = 2$ is twice as thick as a wall with a dielectric constant of $\epsilon = 8$ for maximum transmission. The effect of a nonzero loss tangent on normal incident power transmission is shown in Figure 12.11 *versus* wall thickness, measured in wavelengths in the material, for several values of loss tangent (Jasik [17]). One can see that for thin wall radomes, with wall thickness less than 1/20 of a wavelength in the material, loss tangents as large as 0.1 result in minimal extra loss as compared with the lossless case. Walls which are multiples of one-half wavelength in the material, however, are very sensitive to loss tangent, and the figure shows the importance of low loss tangent if high-power transmission is a design objective. Often, low loss and high strength can be obtained by using a wall composed of layers as discussed in the following section.

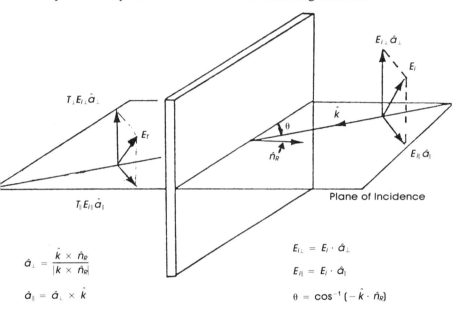

$$\hat{a}_\perp = \frac{\hat{k} \times \hat{n}_R}{|\hat{k} \times \hat{n}_R|}$$

$$\hat{a}_\parallel = \hat{a}_\perp \times \hat{k}$$

$$E_{i\perp} = E_i \cdot \hat{a}_\perp$$

$$E_{i\parallel} = E_i \cdot \hat{a}_\parallel$$

$$\theta = \cos^{-1}(-\hat{k} \cdot \hat{n}_R)$$

Fig. 12.8 Illustration of the plane of incidence and incidence angle including equations for the determination of the perpendicular and parallel directions and the field components (adapted from Joy and Huddleston [13]).

12.4.2 Layered Walls

Figure 12.11 shows several different configurations of flat panel walls constructed of more than one monolithic layer. The most common of the layered wall constructions is the *A-sandwich wall*. The A-sandwich wall is composed of two outer skins of thin but strong material, and a low loss, low dielectric constant, and usually low strength inner core material with thickness of approximately one-quarter wavelength in the material for

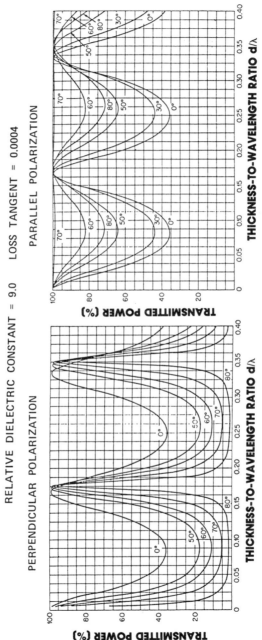

Fig. 12.9 Perpendicular and parallel polarization power transmission coefficients for a flat panel with a dielectric constant of 9.0 and loss tangent of 0.0004 *versus* panel thickness for several incidence angles (from Kay [17], © 1961 McGraw Hill Book Company).

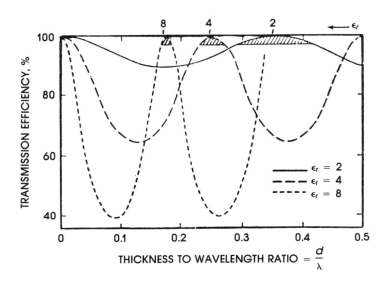

Fig. 12.10 Normal incidence power transmission coefficient for a lossless flat panel *versus* panel thickness-to-wavelength ratio for several panel dielectric constants. The shaded region shows transmission in excess of 97% (from Yost, Weckesser, and Mallalieu [4]).

good, normal incidence transmission. The core material is often a foam or a honeycomb material, which is largely air or void and typically has a dielectric constant of $\epsilon < 1.3$. The skin materials are typically chosen to withstand stress, temperature, or erosion due to the environment, and may have a dielectric constant or loss tangent higher than desired. Figure 12.12 shows the perpendicular and parallel power transmission factor and insertion phase delays for such an A-sandwich wall, where the skins have a dielectric constant of $\epsilon = 4.0$ and a loss tangent of $\delta = 0.02$, and the core has a dielectric constant of $\epsilon = 1.2$ and a loss tangent of $\delta = 0.005$ [17]. The figure shows four different core thicknesses designed to maximize power transmission at four associated incident angles. Notice that for the 50° incident angle optimization case, there is good tracking between the perpendicular and parallel polarization transmission amplitudes and insertion phase delays *versus* incident angle. Such good tracking normally results in minimum depolarization of the transmitted field. Care must be taken in selection of adjoining materials in a layered radome wall, especially if a large temperature change is expected. The materials must have

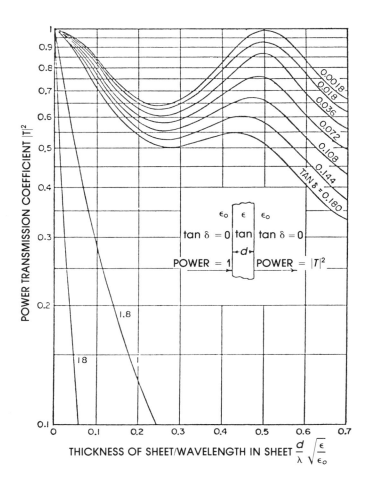

Fig. 12.11 Normal incidence power transmission coefficient for a flat plane *versus* normalized panel thickness for various panel loss tangent values (from Kay [17], © 1961 McGraw Hill Book Company).

similar thermal expansion rates or be compressible such that delamination will not occur. The electromagnetic performance of a layered wall is often affected by small layers with poor loss tangents, such as glue, which holds the layers together, outer surface paint, which is used for rain erosion or ultraviolet protection, and sealants, which are used to prevent the entrance of moisture. These organic substances typically have a dielectric constant between 2 and 4, but often have loss tangents in excess of 0.05. These layers should be considered in the design and analysis of radome walls.

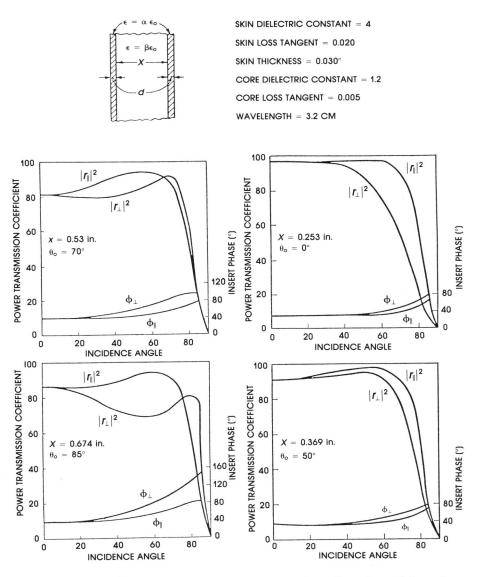

Fig. 12.12 Power transmission coefficients and insertion phase delays for an A-sandwich radome wall construction *versus* incidence angle for four design angles and corresponding core thicknesses (from Kay, [17], © 1961 McGraw Hill Book Company).

Similar layered wall constructions are possible by using ceramic materials. The most successful layering is that which uses different densities of the same basic material. Silicon dioxide and reaction sintered silicon nitride have been successfully layered. The dielectric constant and strength of the material increases with the density. An A-sandwich construction is obtained by using a low density for the core, yielding a low dielectric, low strength core with good electromagnetic properties, and by using a high-density formulation for the skins to obtain high strength. Radome walls designed to have high transmission over a wide frequency range are called *broadband walls*, and these are presented in the next section.

12.4.3 Broadband Walls

There is often a need for a radome wall to be highly transmissive at more than one frequency. A system may operate over several widely spaced narrow bands, or may require a large continuous bandwidth. Both applications might be called *broadband*. Several wall configurations have proved to be effective in achieving broadband radome performance. These are (a) thin wall, (b) low dielectric constant wall, (c) multiresonant monolithic wall, (d) multilayer bandpass wall, (e) reactively loaded wall, and (f) grooved wall.

The thin wall has been discussed above and shown to operate with low loss over all frequencies up to the frequency where the thickness of the wall is approximately 1/20 of a wavelength in the material. The loss of the thin wall radome becomes excessive above this frequency, except at frequencies where the thickness of the wall is an integer multiple of one-half wavelength of the material in the wall. If the material is low loss, high transmission performance can be achieved at these increasingly narrow bands in addition to its wideband low-frequency performance.

The low dielectric constant monolithic wall is the ultimate broadband radome wall configuration. Should the material have a dielectric constant of $\epsilon = 1.0$ and a low loss tangent, it could operate over all frequencies for any thickness of the material. A lossless material with a dielectric constant of $\epsilon = 2.0$ has greater than 90% normal incidence transmittance for all frequencies, as shown in Figure 12.10. Only a nonzero loss tangent and associated additional losses limit the thickness of such a material. PTFE, with a dielectric constant of $\epsilon = 2.1$, is an example of such a material.

The multiresonant monolithic radome wall, as discussed above, can provide narrow windows of high transmission at widely spaced frequencies. The frequencies may be integer multiples of one another, or an integer multiple of some lower frequency. A monolithic wall can be designed to have high transmission at 60 and 90 GHz by having a thickness corresponding to two and three half-wavelengths, respectively, of a 30 GHz half-wave wall. The wall will have high transmittance at 30 GHz as well as at the desired 60 and 90 GHz. The loss tangent must be low because three half-wavelengths must be traversed at 90 GHz. The bandwidth of the 60 GHz band is one-half of the 30 GHz bandwidth, the bandwidth of the 90 GHz is one-third of the 30 GHz bandwidth, and the range of incident angles for high transmission is also reduced as the integer multiple increases.

A multilayered wall can also be designed with similar low dielectric constant material of comparable thicknesses to form a bandpass filter. The exact thickness of each layer is optimized to produce high transmission over a band of frequencies. A practical limit for the number of layers is approximately five. Little extra bandwidth is achieved with additional layers. The analogy to transmission-line bandpass filters can be made for normal incidence design. All materials must have relatively low dielectric constant for a successful broadband design.

Radome walls can also include conductors for reactive loading of the wall. The conductors can be shaped and oriented with respect to the incident electric field to produce either capacitive or inductive loading of the wall. The effect is either to increase or decrease, respectively, the effective dielectric constant of the wall. Capacitive reactance can be achieved with spherical conductors of dimensions much less than a wavelength, or with wires oriented perpendicular to the incident electric field. Inductive reactance can be achieved with a wire oriented parallel to the incident electric field. Reactive loading allows more degrees of freedom in the design of the radome wall's electromagnetic performance. Reactive loading is often applied to increase bandwidth, but normally can be achieved for only one polarization with respect to the conductors. Figure 12.13 shows the transmission loss for a quartz fiber resin laminated monolithic wall with dielectric constant of $\epsilon = 3.15$ and loss tangent of $\delta = 0.004$, which is 1.7 mm thick and contains parallel wires (Carey [7]). The wall without the wires is less than one-half wavelength thick, even at 45 GHz. The inductive wires compensate for the capacitance of the thin wall, and thus decrease the required thickness of the wall and produce a broadband structure. The structure is only broadband for incident electric fields that are in the same plane as the conductors.

A grooved wall, as shown in Figure 12.14, has exhibited broadband properties. Several panel configurations of this structure have been fabricated and electromagnetically measured at Georgia Tech (Bodnar and Bassett [18]). Panels were made with grooves on both sides, as shown in the figure, and with grooves on only one side of the panel, which is more

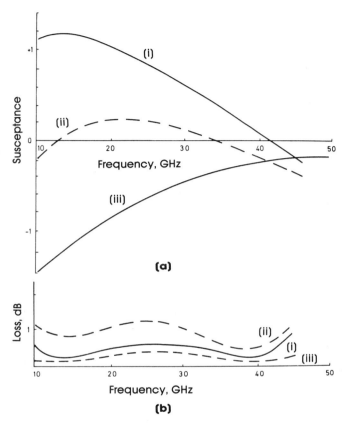

Fig. 12.13 Susceptance and transmission loss for a flat panel of quartz fiber
resin laminate with dielectric constant of 3.15, loss tangent of
0.004, and thickness of 1.7 mm loaded with thin parallel wires
versus frequency: (a) normal incidence susceptance for (i) the
dielectric only, (ii) dielectric and wires, and (iii) the wires only;
(b) transmission loss for (i) normal incidence, (ii) perpendicular
component at an incidence angle of 45°, and (iii) parallel component at and incidence angle of 45°. The electric field is in the
plane of the wires in all cases (from Cary [7], © 1983 Peter
Peregrinus).

appropriate for aerodynamic radome applications. The doubly grooved
panels provide better transmission and lower depolarization properties
than the singly grooved panels. The lower frequency limit of the structure

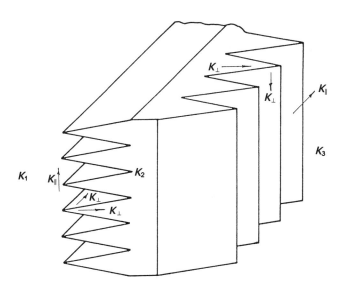

Fig. 12.14 Doubly grooved flat panel radome wall for broadband performance (from Bodnar and Bassett [18], © 1975 IEEE).

occurs when the groove depth is less than approximately one free-space wavelength. The upper frequency limit occurs when the grooves are spaced greater than approximately one-half a free-space wavelength (Rope and Tricoles [19]). Doubly grooved structures have been constructed with greater than 90% transmission over the frequency range from 5 to 40 GHz for incident angles from 0° to 50°. Gradient dielectric radome walls can also provide broadband properties (Loyet [20]), but such walls are electromagnetically thick and often difficult to manufacture.

12.5 RADOME MATERIALS

Materials used for MMW radome construction are primarily selected for good MMW transmission and high structural strength, including flexural strength, thermal shock, and rain erosion rates, at the highest operating temperature. Good electromagnetic transmission at MMW frequencies is normally achieved with a wall construction that is a multiple of one-half wavelength (in the material) thick, as discussed above. The electromagnetic properties of materials that are beneficial for such wall constructions are low dielectric constant and low loss tangent—the lower the better!

Radome materials are divided into two primary categories: *organic* and *inorganic* materials. Organic materials are usually large molecule polymers used in applications where the temperature is less than 700°F. Inorganic materials are usually ceramic materials, and can be used for temperatures up to 4000°F. Table 12.1 presents an overview of the two types of radome materials. Organic material radomes are preferred due to the low cost of the material and associated ease of manufacture. Inorganic material radomes must be used when the temperature or strength requirements cannot be met by use of the organic materials.

Table 12.1   Organic *versus* Inorganic Radomes* (from Joy, Hudleston, Bassett, and Wheeler [6]).

Consideration	*Organic*	*Inorganic*
RF Transmission	Good	Good
Mechanical Strength	Low	High
Usage	Aircraft and Missiles	Missiles
Max. Operating Temperature	300–700°F	1500–4000°F
Thermal Shock Resistance	Good	Poor
Rain and dust Damage Resistance	Low—Needs Protection	High
Cost	Low	High
Handling Breakage	Low	High

The following are several sources which list the electromagnetic or structural properties of radome materials: Chapter 46, *Antenna Engineering Handbook*, Jasik and Johnson (eds.), McGraw-Hill, 1984 [21]; *Radome Engineering Handbook: Design and Principles*, Walton (ed.), Marcel Dekker, 1970 [8]; *Dielectric Constant and Loss Data*, Westphal and Sils, AFML-TR-72-39, 1972 [22]; *Dielectric Materials and Applications*, Von Hipple, MIT Press, 1954 [23]; *Optical Properties of Advanced Infrared Missile Dome Materials*, Bennett, *et al.*, NWC-TP-6284, July 1981 [24]. The dielectric constant and loss tangent of some of the more common radome materials are given in Table 12.2 [25]. The weight of radome materials is important, especially in physically large radomes or missile applications. The thickness of a radome wall is approximately inversely proportional to the square root of the dielectric constant. Thus, a weight parameter for radome materials equal to the density of the material divided

*These are general observations and certainly do not apply for all materials in that category.

by the square root of the dielectric constant is more relevant for weight comparisons of radome materials. Table 12.3 lists some common organic and inorganic radome materials, their density, dielectric constant, and weight parameters [4]. As an example, although dielectrically loaded PTFE has a density that is almost a factor of two lighter than alumina, a dielectrically loaded PTFE radome of the same electromagnetic thickness as an alumina radome would weigh more.

Table 12.2 Dielectric Constant and Loss Tangents for Some Common Radome Materials (adapted from Huddleston and Bassett [25], © 1984 McGraw Hill Book Company).

Material	Dielectric Constant	Loss Tangent
ORGANIC		
Thermal plastics		
Lexan	2.86	0.006
Teflon	2.10	0.0005
Noryl	2.58	0.005
Kydox	3.44	0.008
Laminates		
Epoxy-E glass cloth	4.40	0.016
Polyester-E glass cloth	4.10	0.015
Polyester-quartz cloth	3.70	0.007
Polybutadiene	3.83	0.015
Fiberglass laminate polybenzimidazole resin	4.9	0.008
Quartz-reinforced polyimide	3.2	0.008
Duroid 5650™ (loaded PTFE)	2.65	0.003
INORGANIC		
Aluminum oxide	7.85	0.0005
Alumina, hot-pressed	10.0	0.0005
Beryllium oxide	6.62	0.001
Boron nitride, hot-pressed	4.87	0.0005
Boron nitride, pyrolytic	5.12	0.0005
Magnesium aluminate (spinel)	8.26	0.0005
Magnesium aluminum silicate (cordierite ceramic)	4.75	0.002
Magnesium oxide	9.72	0.0005
Pyroceram 9606	5.58	0.0008
Rayceram 8	4.72	0.003
Silicon dioxide	3.82	0.0005
Silica-fiber composite (AS-3DX)	2.90	0.004
Slip-cast fused silica	3.33	0.001
Silicon nitride	5.50	0.003

Table 12.3   Radome Materials and Weight Parameters (from Yost,
Weckesser and Mallalieu [4])

Material	Density (ρ) lb/in^3	Dielectric Constant (ϵ)	Weight Parameter $(\rho/\sqrt{\epsilon})$ lb/in^3
Epoxy Laminate			
EPON 838 (ECC-181)	0.070	4.64	0.032
Phenolic Laminate			
CTL-91LD (ECC-181-114)	0.067	5.25	0.029
Polyester Laminates			
Selectron 5065 (ECC-181)	0.072	4.79	0.033
Vibrin 135 (ECC-181)	0.078	4.16	0.038
Silicone Laminate			
30% dc 2106 (ECC-181)	0.061	4.35	0.029
PTFE Laminate (ECC-112)	0.083	2.97	0.042
Alumina (99.5%)	0.137	9.60	0.044
Beryllia (99.5%)	0.105	6.60	0.041
Cordierite	0.085	4.80	0.039
Fused Silica (Slip Cast)	0.069	3.30	0.038
Pyroceram 9606™	0.094	5.60	0.040
Duroid 5870™	0.078	2.35	0.051
Duroid 5650™	0.079	2.38	0.051
Hot Pressed Boron Nitride	0.045	4.10	0.022
Isotropic Pyrolytic			
Boron Nitride	0.046	3.20	0.026
Reaction Sintered			
Silicon Nitride	0.090	5.60	0.038
Hot Pressed			
Silicon Nitride	0.116	8.82	0.039

Almost gem-quality materials are available for millimeter and submillimeter frequencies. The materials include silica mullite, germanium
mullite, boron aluminate, zinc germanate, thorium germanate, sapphire,
magnesium flouride, spinel, pollucite, nafnium titanate, aluminum nitride,
calcium lanthanate sulfide, zirconite, and zirconia-tougheneol oxide (Bennett, *et al.* [24])

A-sandwich organic material radome walls and other multilayered
radome wall constructions may use a lightweight, low dielectric constant,
and usually low strength material as a core layer. Honeycomb, foam, and

syntactic foam materials have been developed for this application. Honeycomb material is shaped into a honeycomb pattern that forms hexagonal voids. Honeycombs are commonly made from nylon, phenolic, Nomex℠, or polyimide resin reinforced materials. The dielectric constant of the material varies with polarization, and typically ranges between 1.1 and 1.3. Loss tangents for honeycombs typically range from 0.001 to 0.005, and increase with increasing incidence angle. Polyurethane closed cell foam is popular for low-temperature applications. The dielectric constant can be varied by varying the density of the foam. Dielectric constants from 1.05 to 1.2 and loss tangents from 0.0005 to 0.002, respectively, are possible. A syntactic foam is a mixture of glass or ceramic microballoons in an organic resin. The syntactic foam has the advantage of the uniform electromagnetic properties of the foams with the higher strength and temperature performance of the honeycombs. Dielectric constant and loss tangent of the syntactic foams are dependent on the resin used, but typical values range from 1.8 to 2.1 and 0.005 to 0.01, respectively.

12.6 ENVIRONMENTAL EFFECTS

Radomes are designed to protect the enclosed antenna from the effects of the environment. Thus, the radome must withstand this environment, and not be subject to electromagnetical or structural degradation. Military Specifications MIL-STD-210B and MIL-STD-810, method 516.2, specify environmental conditions and vibration and shock requirements, respectively. Terrestrially based systems must endure temperature extremes, humidity, rain, ice, snow, hail, lightning, wind, blowing dust and sand, salt spray, soot and smog accumulation, ultraviolet radiation, and seismic vibration. Flight systems must also endure aerodynamic heating and thermal shock, aerodynamic drag forces, acceleration and vibration, and high-speed impact with precipitation and airborne particles. Some of these environmental conditions primarily affect the required strength of the radome wall, such as the force of the wind, the weight of rain, snow, and ice, and the temperature. Other environmental conditions primarily affect the outer surface of the radome wall, such as humidity, rain, lightning, salt spray, and ultraviolet radiation. Often, it is not possible to find a single radome material able to withstand all of the above environmental conditions. It is thus necessary to apply outer layers of special materials to a radome for environmental protection purposes. Examples of such layers are ultraviolet blocking, moisture sealant, antistatic, lightning diversion (metallic strips or buttons), hydrophobic, rain erosion, and heat ablation (Letson, *et al.* [26]).

Heating of a radome causes the radome material to change its physical thickness, dielectric constant, and loss tangent. These effects combine to change the electromagnetic transmission and reflection properties of the radome wall. Figure 12.15 shows the effect of temperature on the dielectric constant and loss tangent of several popular inorganic radome materials [8]. Changing wall properties can have an effect on boresight error for radar systems. Figure 12.16 shows boresight error for a radome, with and without a temperature gradient [5]. Other boresight error measurements, using the solar furnace in France, show even greater boresight error shifts when the temperature gradient is from radome tip to base as well as from exterior surface to interior surface. (Frazer [27]).

Water on the surface of a radome or moisture within the radome wall has a significant effect on the electromagnetic performance of the radome. The real and imaginary components of the dielectric constant of liquid water as a function of frequency are displayed in Figure 12.17. The loss tangent of a material is equal to the imaginary part of its dielectric constant, ϵ'', divided by the real part of its dielectric constant, ϵ'. The relative dielectric constant can be seen as greatly different than typical low dielectric, low loss tangent, radome materials. Water is also shown to be very lossy at MMW frequencies. Transmission loss through layers and drops of water on the exterior surface of a 0.04-inch thick, PTFE coated, woven glass fabric (RAYDEL) radome is displayed in Figure 12.18 (Joy, *et al.* [28]). This figure shows the high loss at MMW frequencies of layers of water, and illustrates the advantage of hydrophobic radome outer surfaces.

12.7   ELECTROMAGNETIC TESTING OF RADOMES

The electromagnetic testing involves testing of the finished radome and the constitutive materials used in the radome. The electromagnetic design and subsequent performance of a radome depends on the electromagnetic properties of the materials used to fabricate it. These electromagnetic properties are often functions of the material or radome manufacturing process, frequency, and temperature. It was shown above that moisture absorption can greatly effect the electromagnetic properties of materials at MMW frequencies. The dielectric constant and loss tangent are the material properties that are measured for homogeneous and isotropic materials. The following section presents the more common electromagnetic testing methods for radome materials.

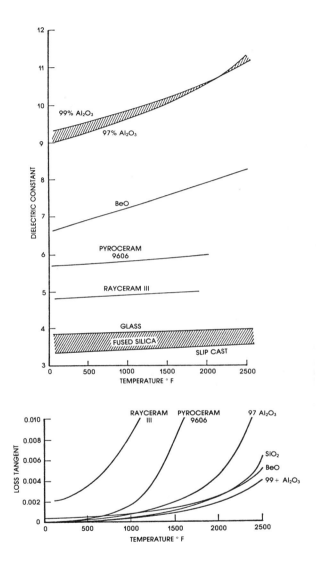

Fig. 12.15 Measured dielectric constants and loss tangents for several popular inorganic radome materials *versus* temperature (from Walton [8], © 1970 Marcel Dekker, Inc.).

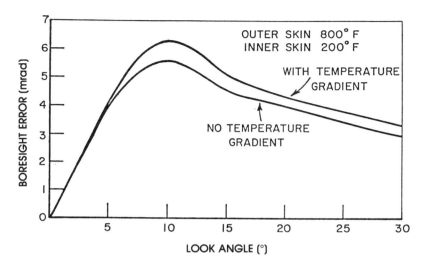

Fig. 12.16 Radome induced boresight error *versus* antenna look angle within the radome showing the effect of a temperature gradient within the radome wall (from Kilcoyne [5]).

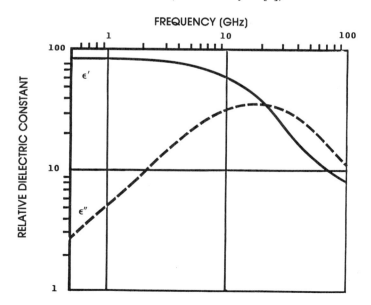

Fig. 12.17 The relative dielectric constants of fresh liquid water *versus* frequency. The dielectric constant is given by ϵ' and the loss tangent given by ϵ''/ϵ' (from Joy, Wilson, Effenberger, Strickland, and Punnett [28]).

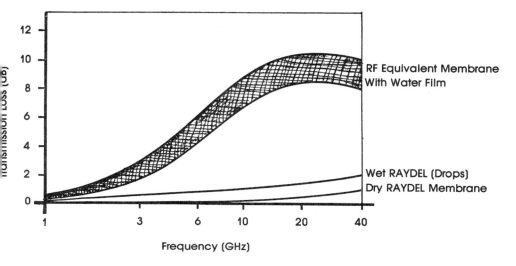

Fig. 12.18 Transmission loss of hydrophobic (RAYDEL) and nonhydro-
phobic radome membrane subjected to rain *versus* frequency
(from Joy, Wilson, Effenberger, Strickland, and Punnett [28]).

12.7.1 Radome Materials Testing

There is no direct method for the measurement of the dielectric
constant and loss tangent of radome materials that is independent of the
geometry of the material sample. Thus, the accurate measurement of the
electromagnetic properties of the materials depends on accurate measurement of the sample geometry. If it is desired to measure the electromagnetic properties of the material over a range of temperatures, the geometry
of the material must likewise be measured over the same range of temperature, and then used in the extraction of the dielectric constant and
loss tangent of the material from the measurements.

A flat panel transmission measurement system is shown in Figure
12.19 [1]. The system is composed of a high-gain transmitting antenna and
a high-gain receiving antenna, coaxially located and separated by sufficient
distance to allow the rotation of the panel and to minimize the multiple
reflections between the antennas without the panel in place. The effect of
multiple reflections between the two antennas can be quantified by performing repeated transmission measurements without the panel in place,
while varying the separation distance in small increments of a wavelength.
A sinusoidal amplitude variation with period of one-half wavelength of
separation indicates the presence of multiple reflections. The multiple

reflections can be reduced by better matching of the antennas to their source or receiver and increased separation of the antennas. The amplitude and phase of the transmitted signal is measured without the panel in place. Next, the panel is inserted in the transmission path and the amplitude and phase of the transmitted signal are measured *versus* rotation angle of the panel. If the axis of panel rotation is perpendicular to the plane formed by the polarization of the antennas and the axis of transmission, then the parallel transmission is being measured. If the axis of panel rotation is parallel to the polarization of the antennas, then the perpendicular transmission is being measured. The measurements are carried out for both polarization rotations. The panel material must be homogeneous and of constant thickness, since this measurement is an average over a large number of square wavelengths of the material. The transmission amplitude and insertion phase *versus* incident angle measurements are compared to calculated values in an iterative fashion to determine the dielectric constant and loss tangent of the material. A good estimate of both the dielectric constant and loss tangent can be made by using the Brewster effect. The incidence angle of peak transmission (other than normal incidence) of the parallel polarization measurement (the Brewster angle) is equal to the inverse tangent of the square root of the dielectric constant for a lossless material. The transmission loss at the Brewster angle is simply related to the loss tangent and thickness of the panel material. These estimates are then refined by using the complete data sets. Scattering from the edge of the panel sample will produce a ripple in the transmission data *versus* rotation angle as the scattering adds in phase and out of phase due to rotation. This edge scattering may be minimized by reducing the illumination level at the panel edge. This can be accomplished by using large panel sizes, large aperture antennas, or absorbing materials at the edge of the panel. Phase curvature of the incident field will produce a smearing of the data *versus* incident angle, since not all of the transmitted energy will be passing through the material at the same angle. This effect may be reduced by using large aperture (high directivity) antennas.

Another flat panel technique to measure dielectric constant for low-loss materials is shown in Figure 12.20 (Walton [2]). This technique measures the normal incidence reflection of a flat panel of the material, placed at the mouth of a well matched, large aperture antenna such as a horn. The aperture size of the antenna must be many wavelengths in diameter such that the wave impedance approaches that of free space. The measurement system may be a frequency-agile network analyzer, reflectometer, or VSWR measurement system. Higher accuracy may be obtained if the phase as well as the amplitude of the reflection is measured. By varying the frequency of the incident field, a minimum reflection condition

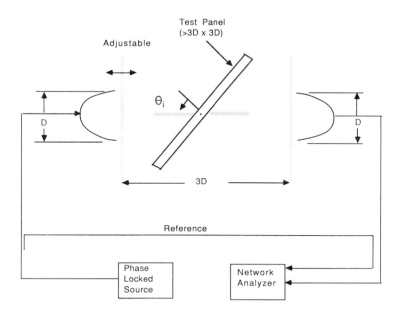

Fig. 12.19 Flat panel transmission test system using a network analyzer (adapted from Tice [1]).

is found, indicating that the panel thickness is electromagnetically equal to an integral number of half-wavelengths. Figure 12.21 shows such measurements for a variety of dielectric constants [17]. The peak amplitude of reflectivity, compared to the return from a metal plate, is also an estimate of the dielectric constant, as shown in the figure. The peak amplitude value and frequency of the peak return nearest the minimum return are used to determine which integer multiple of one-half wavelength is being measured at minimum return. This measurement is usually designed to measure the one-half wavelength electromagnetic thickness, based on an approximation of the dielectric constant of the material. In this case the electromagnetic thickness of one-half wavelength is equal to the physical thickness measured in free-space wavelengths at the frequency of minimum reflection divided by the square root of the dielectric constant. A panel with thickness of 1.875 mm and dielectric constant of $\epsilon = 4$ should show a minimum of reflection at 40 GHz and a maximum at 60 GHz. If the measurement system has phase measuring capability, the mismatch of the antenna without sample may be removed from the measurement, thus yielding higher accuracy.

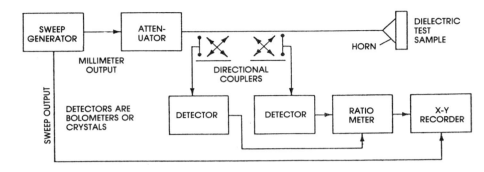

Fig. 12.20 Normal-incidence flat panel reflection test system (from Walton
[2]).

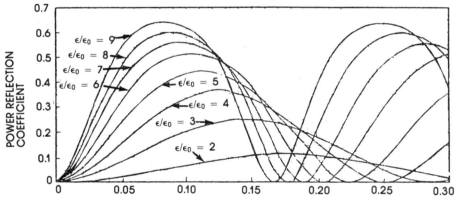

SAMPLE THICKNESS/FREE SPACE WAVELENGTH

Fig. 12.21 Lossless flat panel normal-incidence power reflection coefficient
versus panel thickness for several panel dielectric constants.
(from Kay [17], © 1961 McGraw Hill Book Company).

    The shorted waveguide measurement technique [8] is useful at lower
MMW frequencies. A rectangular or cylindrical sample of the radome
material must be accurately machined or molded into the open end of a
dominant-mode rectangular or cylindrical waveguide, respectfully, and the
open end of the waveguide fitted with a flat shorting plate in contact with
the radome material sample. The VSWR patterns in the air-filled section
of the waveguide, with and without the sample, are compared to determine
the dielectric constant and loss tangent of the material at the frequency

of excitation. The VSWR patterns are measured with a slotted waveguide probe. Such a system is shown in Figure 12.22. A material with a dielectric constant greater than air will cause the VSWR null to shift toward the material as the material causes the wavelength to shorten in the material. The shift of the null location is easily related to the dielectric constant of the material, and the depth of the null is related to the loss tangent of the material. Once the dielectric constant is known, highly accurate loss tangent measurements can be made for low-loss materials by using long sample lengths. An ambiguity exists in this measurement concerning the number of half-wavelengths of material length. This ambiguity can be resolved with measurement of several lengths of material, or measurement of the same material at several frequencies.

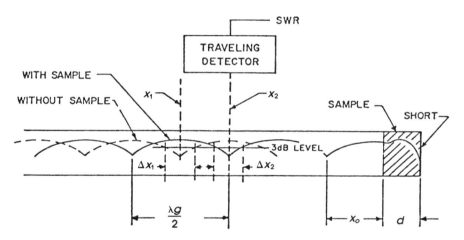

Fig. 12.22 Shorted waveguide system for the determination of dielectric constant and loss tangent of a sample of radome material (from Walton [8], © 1970 Marcel Dekker, Inc.).

Correction factors are presented in Von Hippel [23] for the effect of gaps existing between the sample and the walls of the waveguide. This technique is difficult to apply at the higher MMW frequencies due to the small dimensions of dominant-mode waveguide and associated required sample geometrical tolerances.

The above materials testing techniques can be modified for measurements at elevated temperatures. Two examples are presented here. Figure 12.23 shows a mechanism developed at Georgia Tech for the transmission measurement of a flat panel at normal incidence of a ceramic radome material (Bassett [29]). Two elliptical reflector antennas with a

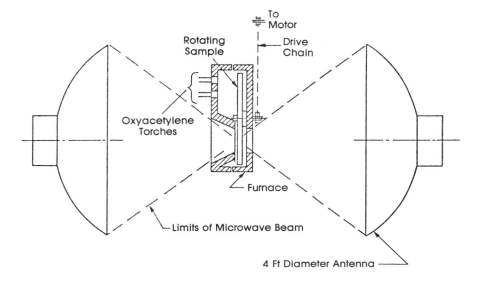

Fig. 12.23 System for transmission measurements of heated radome material samples. The system employs two ellipsoidal reflector antennas with common exterior focal point located at the sample position and an oxyacetylene furnace (from Bassett [29].

common outer focal point are coaxially aligned such that the power transmitted by one is passed through the common focal point and received by the other, as shown in the figure. Not shown in the figure is the inner focal point of each elliptical reflector, where a small horn feed is located. A rotating plate with several circular openings holds several samples of ceramic radome material and one opening is left open as the reference. The rotating samples are alternately measured for transmission loss and heated with oxyacetylene torches. An optical pyrometer is used to measure the temperature of each sample as the transmission through the sample is being measured.

A second high-temperature material testing apparatus is shown in Figure 12.24 [8]. This apparatus employs the shorted waveguide technique to measure the dielectric constant and loss tangent of the radome material, and uses electric resistance coils to heat the sample. The top of the cylindrical waveguide is water-cooled to protect the measurement equipment. Four models of this apparatus were constructed to cover the frequency range from 8.6 to 50 GHz. The dimensions for each of the four models are given in the figure. Temperatures up to 600° F can be achieved with

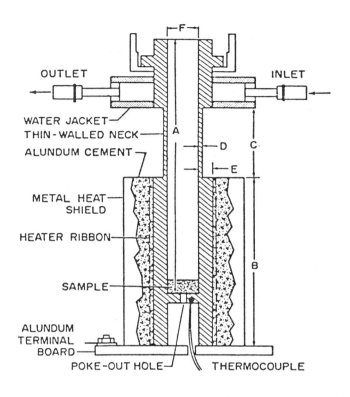

NOMINAL FREQUENCY	8.6 X 10⁹	1.4 X 10¹⁰	2.4 X 10¹⁰	5 X 10¹⁰
A. INSIDE LENGTH	15.0 CM	7 CM	5.3 CM	3.1 CM
B. LENGTH HEATED	10.0 CM	4.2 CM	3.6 CM	2.4 CM
C. LENGTH OF NECK	3.6 CM	2.1 CM	1.0 CM	0.8 CM
D. NECK THICKNESS	0.030 IN.	0.012 IN.	0.010 IN.	0.008 IN.
E. WALL THICKNESS	1/8 IN.	1/8 IN.	1/16 IN.	1/16 IN.
F. INSIDE DIAMETER	1.0 IN.	5/8 IN.	3/8 IN.	11/64 IN.

Fig. 12.24 Heated shorted waveguide system for the determination of dielectric constant and loss tangent of a sample of radome material. The system employs a cylindrical waveguide and electric resistance heating (from Pentecost [8], © Marcel Dekker, Inc.).

this apparatus. The thermal expansion of both the material and the waveguide must be carefully examined such that, at the measurement temperature, the size of any air gap between the sample and the waveguide will be known and ideally small, and the sample may be inserted and removed at room temperature. The next section presents several techniques in common use for the measurement of overall radome performance.

12.7.2 Radome Testing

Radomes are electromagnetically tested to determine the degree of pattern distortion, gain loss, boresight error, and backscatter due to the radome. Pattern distortion, gain loss, and boresight error measurements are made on conventional or modified antenna measurement ranges. The amount of backscatter is measured indirectly by measuring the input VSWR of the antenna, with and without radome in place.

Pattern distortion due to radome enclosure of an antenna is measured on conventional antenna ranges. Antenna ranges can be categorized into three major types and subdivisions as follows:

- Far-Field Ranges
  Elevated
  Ground Reflection
  Slant
  Celestial Source
  Anechoic Chamber

- Compact Ranges
  Point Source
  Line Source
  Dual Cylindrical
  Subreflector

- Near-Field Ranges
  Planar Surface
  Cylindrical Surface
  Spherical Surface
  Source Synthesis

The most popular of these ranges for MMW radome measurement are the elevated far-field range, the anechoic chamber far-field range, and the point source compact range. Some recent work has been reported on the use of near-field measurements for the detailed diagnostics of radome defects.

There are several important differences between antenna measurements and radome-enclosed antenna measurements. First, the size of the radome must be considered in determining the required far-field measurement distance on far-field ranges and the size of the quiet zone for compact ranges. The standard far-field distance, R, for far-field measurements is given by (Hollis, *et al.* [30]):

$$R = 2D^2/\lambda \qquad\qquad (12.3)$$

where

D = the diameter of the minimum sphere, centered on the center of rotation of the range positioner, which completely encloses the antenna and radome for all range positioner motion;

λ = the wavelength of the frequency of measurement.

Note that this criterion is sufficient for antennas with first sidelobe levels greater than or equal to 20 dB below the peak of the main beam. Larger far-field distances are needed for lower sidelobe level antennas. Likewise, the diameter of the quiet zone for the compact range must be greater than or equal to D, as defined above. Second, two positioners are required to perform pattern measurements if the antenna is gimballed within the radome. The antenna must be gimballed with respect to the radome, and then the antenna-radome assembly is rotated for pattern measurement. This second degree of freedom, the antenna look angle or scan angle with respect to the radome, must be accommodated through the use of a second positioner. Often the antenna's own gimballing system can be used to rotate the antenna with respect to the radome, and then the whole assembly is rotated with the range positioner. If it is not possible to use the antenna's gimballing system, then the antenna is mounted to the range positioner, and a special radome rotator is also mounted to the range positioner, which allows the rotation of the radome with respect to the antenna. Let an antenna and radome system be operating at 90 GHz and the radome extend 12 inches beyond the center of the range rotator, giving a diameter D of 24 inches. The required far-field range distance by using the above equation is 732 feet, and the required quiet zone size for compact range measurements is two feet. These sizes compare favorably with existing far-field range and compact range capabilities. Most far-field ranges are shorter than 2000 feet, and most compact ranges have a quiet zone less than four feet. The example also illustrates that anechoic chambers with typical lengths of less than 100 feet are only useful for smaller antenna-radome systems or at lower frequencies. The accuracy for far-field ranges at high frequencies is limited by range length and ground reflections. Ground reflections can be minimized through the use of low-sidelobe-level range antennas and diffraction fences. The accuracy limitation for compact ranges at high frequencies (above 40 GHz) is the surface roughness of the range reflectors. The accuracy limitation for the anechoic chamber far-field range at high frequencies (above 40 GHz) is the reflectivity of the chamber absorbing material. Near-field antenna-radome measurements are becoming common as more near-field measurement facilities are being constructed. Figure 12.25 is a diagram of Georgia Tech's spherical surface near-field and anechoic chamber far-field facility, which is used for radome measurement. Measurements are made in amplitude and phase over a sphere enclosing the antenna and radome under test. The radius of the sphere is two to ten wavelengths greater than the minimum sphere enclosing the antenna and radome. These near-field measurements must then be transformed to calculate and display the associated far field of the antenna and radome. The near-field measurements may also be "backward transformed" to obtain the fields on the surface of the radome, and thus identify the location of scattered field sources. Figures 12.26 and 12.27 respectively show the amplitude of the near-field measurements of a simple nongimballed, open-ended, cylindrical waveguide antenna, with and without radome enclosure [31]. The measurements with radome show the effects of radome scattering as rippling of the amplitude of the near-field measurements. The phase measurements show a similar effect.

Measurements of loss in gain are performed as pattern comparison measurements, with and without the radome in place, and normally include the effect of antenna mismatch differences for gimballed systems. For nongimballed systems, the antenna matching should be adjusted for measurements both with and without radome. The range source level and receiver attenuation settings are held constant for this comparison, and are normally of interest only at the peak of the main beam for each relative orientation of the antenna with respect to the radome.

Boresight error measurement can also be conducted as a pattern comparison measurement, with and without the radome in place. However, whenever greater precision is required, special boresight measurement apparatus is typically used on far-field measurement ranges. The most common of these systems is the difference-pattern null seeker system [2] shown in Figure 12.28. Such systems are used to measure the boresight error of the difference patterns central null for a radar system. The null seeker is a small aperture antenna mounted on a planar positioner at the required far-field distance. The antenna without radome is mechanically aligned such that the null seeker antenna is in the center of the planar positioner and is located in the central null of the difference pattern of the antenna under test. The radome is then placed on a radome positioner and located with respect to the antenna under test. The seeker antenna is then moved throughout the measurement plane to seek a minimum field (null) location. The distance from the center of the plane is recorded and translated into a boresight error, based on the range length. The radome

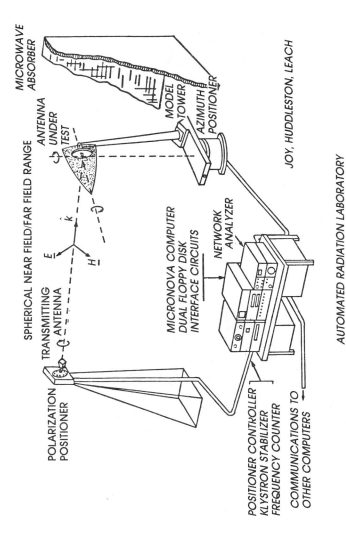

Fig. 12.25 Georgia Tech's spherical surface, near-field and far-field, antenna-radome measurement range (from Joy, Wilson, Caraway, Hill, and Edwards [31]).

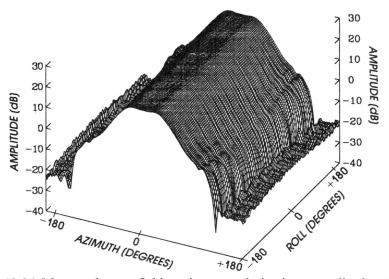

Fig. 12.26 Measured near-field, primary polarization, amplitude of a choked, cylindrical, open-ended waveguide without radome (from Joy, Wilson, Caraway, Hill and Edwards [31]).

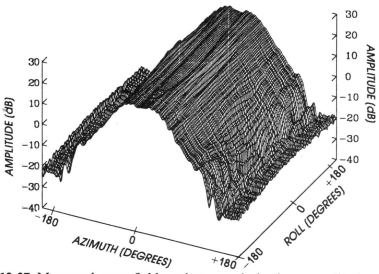

Fig. 12.27 Measured near-field, primary polarization, amplitude of a choked, cylindrical, open-ended waveguide with a tangent ogive, 2.5 fineness ratio, slip-cast fuzed silica radome in place (from Joy, Wilson, Caraway, Hill and Edwards [31]).

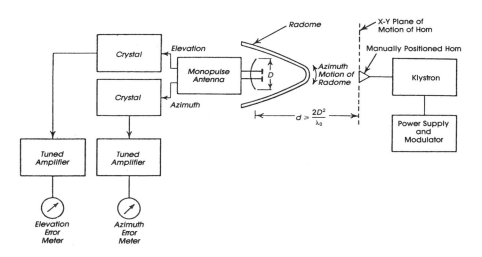

Fig. 12.28 Far-field null seeker, boresight error measurement system (from Tice [1]).

is then rotated to a new relative position with respect to the antenna and the null seeking process is repeated for all radome gimbals. The accuracy of this technique is highly dependent on the magnitude of ground reflections and other range scattering. Figure 12.29 gives the required suppression of range scattering *versus* desired boresight measurement accuracy for various sizes of phase monopulse radar antennas [30].

Variation in the electromagnetic properties of a finished radome wall can be assessed by using a pair of small antennas that are axially aligned and held a constant distance apart, as shown in Figure 12.30. The radome wall is inserted between the two antennas, and the insertion phase of the wall is measured at each point of the radome wall. Radome wall variations and defects can be identified and corrected by adding or subtracting wall material. Near-field measurements of the radome-enclosed antenna, as discussed above, may also be able to provide this type of localized information about the electromagnetic performance of a radome.

Additional information on radome design can be found in Chapter 44, "Radomes," by Huddleston and Bassett, in *Antenna Engineering Handbook,* Johnson and Jasik (eds.), McGraw-Hill Book Company, 1984 [21]; Chapter 14, "Radomes," by Cary, *Handbook of Antenna Design,* Rudge, Milne, Olver, and Knight (eds.), 1983 [7]; *Radome Engineering Handbook,* Walton (ed.), Marcel Dekker, 1968 [8]; "Techniques for Airborne Radome Design," AFAL-TR-66-391, Vol. 1, Tice (ed.), 1966 [1], and Vol. 2 Walton (ed.), 1966 [2], and the proceedings of the biennial Electromagnetic Windows Symposia, which are published in even-numbered years by the Georgia Institute of Technology.

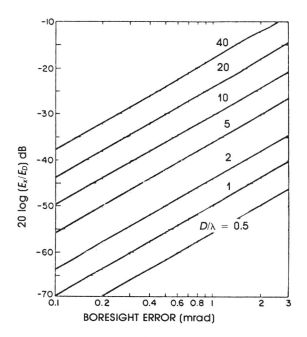

Fig. 12.29 Boresight measurement error *versus* ratio of extraneous range field, (E_x) to the direct range field (E_D) for several antenna effective aperture diameters (D) (from Hollis, Lyon, and Clayton [30], © 1970 Scientific-Atlanta, Inc.).

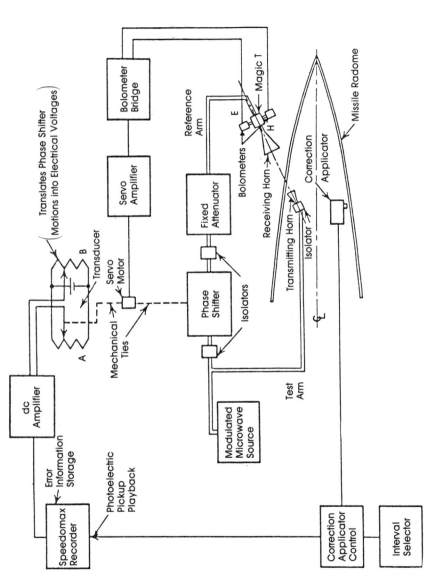

Fig. 12.30 Automated system for the measurement and correction of the insertion phase delay of a finished radome (from Tice [1]).

REFERENCES

1. T.E. Tice (ed.), "Techniques for Airborne Radome Design," AFAL-TR-66-391, Vol. 1, December 1966.
2. J.D. Walton, Jr. (ed.), "Techniques for Airborne Radome Design," AFAL-TR-66-391, Vol. 2, December 1966.
3. R. Timms and J. Kmetzo, "Traveling Wave Radants-Applications, Analysis, and Models," *Proceedings of the 9th USAF-Georgia Tech Electromagnetic Windows Symposium*, June 1968.
4. D.J. Yost, L.B. Weckesser, and R.C. Mallalieu, "Technology Survey of Radomes for Anti-Air Homing Missiles," JHU/APL Report FS-80-022, Contract N00024-78-C-5384, March 1980.
5. N.R. Kilcoyne, "An Appropriate Calculation of Radome Boresight Error," *Proceedings of the 9th USAF-Georgia Tech Electromagnetic Windows Symposium*, June 1968.
6. E.B. Joy, G.K. Huddleston, H.L. Bassett and S.R. Wheeler, "Multipurpose Missile (MPM) High Performance Radome Tradeoff and Development Study," Final Report on Martin Marietta Aerospace Division, Contract ZDZ/283105, April 15, 1975.
7. R.H.J. Cary, "Radomes," Chapter 14 in A.W. Rudge, K. Milne, A.D. Olver, and P. Knight (eds.), *Handbook of Antenna Design*, Vol. 2, Peter Peregrinus, London, 1983.
8. J.D. Walton, Jr. (ed.), *Radome Engineering Handbook: Design and Principles*, Marcel Dekker, New York, 1970.
9. G.P. Tricoles, "Radiation Patterns and Boresight Error of a Microwave Antenna Enclosed in an Axially Symmetric Dielectric Shell," *Journal of the Optical Society of America*, Vol. 54, No. 9, September 1964, pp. 1094-1101.
10. M. Travis, "A Three-dimensional Ray Tracing Method of the Calculation of Radome Boresight Error and Antenna Pattern Distortion," Report TOR-0059 (56860)-2, Air Force Systems Command, May 1971.
11. D.T. Paris, "Computer-Aided Radome Analysis," *IEEE Transactions on Antennas and Propagation*, Vol. AP-18, No. 1, January 1970, pp. 7-15.
12. D.C.F. Wu and R.C. Rudduck, "Plane Wave Spectrum-Surface Integration Technique for Radome Analysis," *IEEE Transactions on Antennas and Propagation*, Vol. AP-22, No. 3, May 1974, pp. 497–500.
13. E.B. Joy and G.K. Huddleston, "Radome Effects on Ground Mapping Radar," Contract DAAH01-72-C-0598, AD-778 203/0, US Army Missile Command, March 1973.

14. E.B. Joy, R.E. Wilson, D.E. Ball, and S.D. James, "Comparison of Radome Electrical Analysis Techniques," *Proceedings of the 15th Symposium on Electromagnetic Windows*, Atlanta, GA, June 18-20, 1980, pp. 25-29.

15. J.H. Richmond, "Calculation of Transmission and Surface Wave Data for Plane Multilayers and Inhomogeneous Plane Layers," OSU Antenna Laboratory Report 1751-2, Contract AF33 (615)-1081, October 1963.

16. S.W. Waterman, "Calculation of Diffraction Effects of Radome Lightning Protection Strips," *Proceedings of the 4th International Electromagnetic Windows Conference*, Direction des Contructions et Armes Navales de Toulon, Toulon, France, June 1981.

17. H. Jasik (ed.), *Antenna Engineering Handbook*, McGraw-Hill, New York, 1961.

18. D.G. Bodnar and H.L. Bassett, "Analysis of an Anisotropic Dielectric Radome," *IEEE Transactions on Antennas and Propagation*, Vol. AP-23, No. 4, November 1975, pp. 179-185.

19. E.L. Rope and G.P. Tricoles, "Anisotropic Dielectrics: Tilted Grooves on Flat Sheets and on Axially Symmetric Radomes," *IEEE AP-S International Symposium Digest*, June 1979, pp. 610–611.

20. D.L. Loyet, "Broadband Radome Design Techniques," *Proceedings of the 13th Electromagnetic Windows Symposium*, Georgia Institute of Technology, GA, 1976, pp. 169–173.

21. R.C. Johnson and H. Jasik (eds.), *Antenna Engineering Handbook*, McGraw-Hill, New York, 1984.

22. W.B. Westphal and A. Sils, "Dielectric Constant and Loss Data," Report No. AFML-TR-72-39, April 1972.

23. A.R. Von Hipple (ed.), *Dielectric Materials and Applications*, MIT Press, Cambridge, MA, 1954.

24. H.E. Bennett, *et al.*, "Optical Properties of Advanced Infrared Missile Dome Materials," Technical Report NWC TP 6284, Naval Weapons Center, July 1981.

25. G.K. Huddleston and H.L. Bassett, "Radomes" *Antenna Engineering Handbook*, Chapter 44, R.C. Johnson and H. Jasik (eds.), McGraw-Hill, New York, 1984.

26. K.N. Letson, *et al.*, "Rain Erosion and Aerothermal Sled Test Results on Radome Materials," *Proceedings of the 14th Electromagnetic Windows Symposium*, Georgia Institute of Technology, Atlanta, GA, June 1978, pp. 109-116.

27. R.K. Frazer, "Duplication of Radome Aerodynamic Heating Using the Central Receiver Test Facility Solar Furnace," *Proceedings of*

the 15th Electromagnetic Windows Symposium, Georgia Institute of Technology, Atlanta, GA, June 1980, pp. 192-197.

28. E.B. Joy, R.E. Wilson, J.A. Effenberger, R.R. Strickland, and M.B. Punnett, "The Electromagnetic Effects of Water on the Surface of a Radome," Proceedings of the 18th Electromagnetic Windows Symposium, Georgia Institute of Technology, Atlanta, GA, September 1986.

29. H.L. Bassett, "A Free-Space Focused Microwave System to Determine the Complex Permittivity of Materials to Temperatures Exceeding 2000°C," Review of Scientific Instruments, Vol. 42, No. 2, February 1971, pp. 200-204.

30. J.S. Hollis, T.J. Lyon, and L. Clayton, Jr. (eds.), Microwave Antenna Measurements, Scientific-Atlanta, Inc., Atlanta, GA, 1970.

31. E.B. Joy, R.E. Wilson, W.D. Caraway, C.E. Hill, and S.J. Edwards, "Near-field Measurement of Radome Performance," Proceedings of the 18th Electromagnetic Windows Symposium, Georgia Institute of Technology, Atlanta, GA, September 1986.

PART IV
SYSTEMS APPLICATIONS

Chapter 13

MMW Radar Design Considerations

J.A. Scheer

Georgia Institute of Technology
Atlanta, Georgia

13.1 INTRODUCTION

In Chapter 2, an introduction to MMW radar technology is presented to give the reader some of the design philosophy and rationale for selection of the MMW band for a particular radar application. This chapter presents a set of radar system design considerations that allow the design engineer to make the decisions associated with selection of the various radar system design parameters. The material presented here demonstrates the application of many of the design considerations discussed in other chapters of this book.

Although a type of design methodology is presented here, there is no "cookbook" approach to the MMW radar design process. Several interrelationships exist in developing the system parameters, such as the relationships among PRF, unambiguous Doppler, and unambiguous range.

The use of the radar range equation is presented again, this time from a designer's point of view. Antenna, transmitter, receiver, and processing topics are covered as they relate to the development of the specified performance of these functions. The details of implementation of the transmitting, receiving and antenna functions are presented in their respective chapters. Various signal processing concepts and their relationship to the overall system performance are also presented.

The sequence of thought processes through which a radar system designer might progress is presented here as an example of this design process. The waveform desired is often relatively easy to select. Table 13.1 gives a list of several salient radar system performance specifications

and lists the key radar parameters driving each particular performance
parameter. The discussions in this chapter provide the designer with an
understanding of the detailed relationships between the desired parameters
and the required radar specifications.

Table 13.1   Key Radar System Relationships.

Given Specification	Radar Parameter	Implied Performance
Range Resolution	Pulsewidth	Bandwidth
		Signal-to-Interference
		Performance
Range (max.)	PRF	Integration Gain
Target Velocity	PRF	Integration Gain
(Doppler Detection)		
Cross-Range Resolution	Beamwidth	Antenna Gain
Detection Probability	SNR	Range of Detection
Target Evaluation	Processing	Resolution, Coherence

Following this chapter are several chapters on systems applications
of MMW radars, each including a design example associated with achieving
a set of specified system performance parameters for the given radar application. Seeker technology, as discussed in Chapter 14, poses a set of
problems associated with detecting targets immersed in ground clutter,
and includes the added complexity of system constraints in the areas of
size, weight, and power requirements. All radar functions, including signal
processing and target detection, must be autonomously performed within
the airborne seeker. Airborne mapping systems, as described in Chapter
15, often have added requirements of coherence to detect slowly moving
ground targets in the presence of clutter. These coherence requirements
for detection of slowly moving targets often surpass the requirements of
most other applications of MMW radars. Ground-based low-angle tracking
systems are discussed in Chapter 16. While these systems have the luxury
of relatively large size and weight limits, they require target tracking precision which is unmatched in the aforementioned systems. Each of the
system applications described requires a different MMW radar system
design as far as the waveform and associated signal processing are concerned.

13.2   THE RADAR EQUATION

The most basic measure of the performance of a radar system is its
signal-to-interference ratio. Chapter 3 presented an analysis of the detection of targets in interference. The discussion here addresses the analysis

of the detection of targets in the presence of clutter and noise from a more practical point of view, primarily for the benefit of the system designer.

13.2.1 Interference Due to Noise

The single-pulse signal power returned from a target may be accurately predicted from the radar range equation:

$$P_r = \frac{P_t G^2 \lambda^2 \sigma}{(4\pi)^3 L_s L_{atm} R^4} \tag{13.1}$$

where

P_r = received power,
P_t = peak transmitted power,
G = antenna gain,
λ = wavelength,
σ = target cross section,
L_s = system loss (>1.0),
L_{atm} = atmospheric loss (>1.0),
R = range to the target.

The noise power, P_n, with which the target signal must compete is also predictable from

$$P_n = kTF_n B_n \tag{13.2}$$

where

k = Boltzmann's constant (1.38×10^{-23} J/K),
T = receiver temperature (290K),
F_n = receiver noise figure,
B_n = receiver noise bandwidth.

The resulting signal-to-noise ratio is found from

$$\text{SNR} = \frac{P_t G^2 \lambda^2 \sigma}{(4\pi)^3 kTF_n B_n L_{atm} L_s R^4} \tag{13.3}$$

A more precise form of the expression for the signal-to-noise ratio, although similar to the one given above, was developed in Chapter 3, which involved the expression for target energy, instead of power. The received target energy, E_{rt}, is found by

$$E_{rt} = P_r \tau \tag{13.4}$$

where τ is the transmitted pulse length. For the target-to-noise energy ratio to be unitless, the expression for the noise power spectral density must be used in the denominator, which results in the expression:

$$E_{rt}/N_0 = \frac{P_t G^2 \lambda^2 \sigma \tau}{(4\pi)^3 k T F_n L_{atm} L_s R^4} \tag{13.5}$$

Expressions (13.3) and (13.5) are equivalent when $B_n = 1/\tau$.

Figure 13.1 is a curve showing the SNR as a function of range for a set of typical parameters for a 35 GHz radar system. Notice that the slope is -12 dB per decade, plus the effects of atmospheric attenuation. The -12 dB part of the roll-off is predictable from the R^4 term in the denominator of the radar equation. Table 13.2 lists the parameters used in determining the performance of this system. Figure 13.1 and Table 13.2 also give the set of parameters for a hypothetical 95 GHz radar system having the same signal-to-noise performance as the 35 GHz system at 2.5 km. This is shown to demonstrate the severe effect of the atmospheric attenuation on the 95 GHz system relative to that for the 35 GHz system.

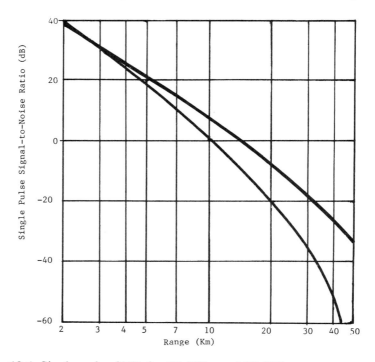

Fig. 13.1 Single-pulse SNR for 35 GHz and 95 GHz systems.

Table 13.2 Radar Parameters for SNR Analysis.

Parameter	35 GHz Value	95 GHz Value
P_t	1000 W	1000 W
G	38 dB	43 dB
λ	8.57 mm	3.16 mm
σ	10 m^2	10 m^2
B_n	1 MHz	1 MHz
F_n	8 dB	8 dB
L_s	7 dB	7 dB
L_{atm}	0.3 dB/km	1 dB/km

13.2.2 Signal-to-Noise Requirements

A system designer needs to know the relationship between the SNR and the performance of a radar system. There are relatively sophisticated techniques for determining these relationships, some of which have been reduced to plots that can be used to relate the SNR to the probabilities of detection and false alarm. Although there are only a limited variety of target and interference conditions for which the plots are developed, they provide a very good first-level approximation of system performance (Blake [1], Mayer and Meyer [2]).

A system designer can determine the expected SNR by using the radar range equation and the generally accepted rules of thumb for approximating the various terms. A key remaining question concerns the SNR required for a given system application. The basic system performance parameters affected by the SNR are the probability of false alarm and the probability of detection. The requirements for these parameters are normally determined from mission analysis. A system is usually set up for a given false alarm rate, often established by using a constant false alarm rate (CFAR) function. Once the CFAR is established, the probability of detection (P_d) is a direct function of the SNR; that is, as the SNR increases, so does the P_d. The relationships between P_d and SNR are somewhat complex, depending on the probability density functions and the fluctuation statistics of the target and interference signals. Fortunately, there are several references that plot a variety of conditions, which represent typical applications. Blake [1] shows curves for several conditions, which may represent a designer's scenario. A more complete set of curves has been generated by Mayer and Meyer [2]. By using these curves, a designer can determine the SNR requirements for a wide range of false alarm rates and target fluctuation conditions, with and without noncoherent integration processing. Improvement on the basis of coherent integration gain is not plotted due to the simplicity of manual calculation.

A simplification of a plot from Mayer and Meyer [2] is shown in Figure 13.2. The sample given here represents a false alarm number of 6×10^5, which is equivalent to a probability of false alarm of 1.16×10^{-6}. The target is nonfluctuating in this example, as signified by the Swerling case "0" notation. The example shows a family of curves, which provides the probability of detection, from 0.01% to 99.99%, *versus* number of pulses noncoherently integrated, from 1 to 10,000, for a range of input SNR values.

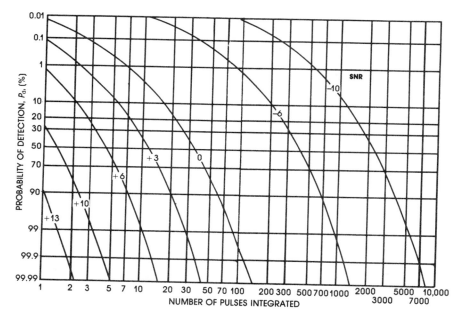

Fig. 13.2 Probability of detection *versus* number of pulses noncoherently integrated for several input SNR values for a nonfluctuating target and a false alarm number of 6×10^5 (adapted from Mayer and Meyer [2], © 1973 Academic Press, Inc.).

The curves are used in several ways. One of the most common involves determination of the required output SNR for a given P_d. This is accomplished by determining which SNR line passes through the desired P_d point on the leftmost vertical axis (number of pulses = 1). In the example shown, a P_d of 90% is achieved with a SNR of 13 dB.

The curve can also be used to determine the input SNR required for a desired P_d if some noncoherent integration is employed. If, for example, a P_d of 90% is desired and noncoherent integration of 64 pulses can be achieved, the junction of these two coordinates falls on the 0 dB SNR line, indicating that the input SNR must be at least 0 dB for the specified P_d performance.

Note that if coherent integration were implemented in the example above, only 20 pulses would have to be integrated to achieve the required 13 dB output SNR, whereas 64 pulses are required for noncoherent integration. This demonstrates the reduced efficiency of noncoherent integration. Integration loss is about 5 dB in this case. As the input SNR increases, so does the efficiency of noncoherent integration.

The example given is for a nonfluctuating target. Generally, this is the case requiring the lowest SNR for a given P_d and P_{fa} performance. For fluctuating targets, the required SNR approaches 20 dB for typical cases requiring P_d of 90%. The plots given by Mayer and Meyer [2] are for random Gaussian noise. Other distributions require a different analysis. Chapter 3 provides a more in-depth look at these conditions.

13.2.3 Interference due to Clutter

Two forms of natural interference restrict the performance of a radar system that is designed to detect targets. The most obvious form is the receiver noise, which is highly predictable; the other is clutter backscatter. Terrestrial (ground), sea, and atmospheric clutter have a strong influence on the performance of MMW radar systems, and as such have a strong influence on the design parameters.

Received clutter power is a function of range in much the same fashion as is target power. The radar cross section of clutter is a function of how much clutter is illuminated and received in a given range gate. Although ground clutter is not usually a homogeneous area scatterer, for purposes of first-order analysis, it is often treated as such, the cross section depending on the area defined by the beam shape and pulsewidth. This treatment of ground clutter characterizes it in terms of average radar cross-section per unit of illuminated area. At MMW frequencies, typical backscatter radar cross section values for ground clutter range from -10 to -30 dB (sm per sm), with extremes ranging from $+5$ to -50 dB for special cases.

The radar cross section, σ_c, from an area clutter cell is found from:

$$\sigma_c = A\sigma^\circ \tag{13.6}$$

where

A = illuminated area,

σ° = clutter reflectivity factor.

The illuminated area is found by defining the range extent of the clutter cell (based on either the pulse length or the beam shape) and cross-range extent (based on the antenna azimuth beamwidth). For ground (or sea) clutter, the area illuminated is either nearly an ellipse, defined by the azimuth and elevation beamwidths, or a rectangle (or a close approximation thereto), defined by the azimuthal beamwidth and the pulsewidth. The range resolution, R_τ, along the direction of the radar beam, is determined by

$$R_\tau = c\tau/2 \tag{13.7}$$

However, for a projected area on the ground of a range dimension determined by the pulse length, the resolution is found from

$$R_\tau = c\tau/2 \cos\Phi \tag{13.8}$$

where Φ is the depression angle.

For the beam-limited case, the range dimension R_b is found from

$$R_b = \frac{R\Theta_{EL}}{\sqrt{2} \sin\Phi} \tag{13.9}$$

where Θ_{EL} is the 3 dB elevation beamwidth, and $\Theta_{EL}/\sqrt{2}$ is an approximation for the effective elevation beamwidth.

In either case, the cross-range dimension, D_c, is found from

$$D_c = \frac{R\Theta_{AZ}}{\sqrt{2}} \tag{13.10}$$

which provides expressions for the effective area of

$$A = \frac{R\Theta_{AZ}c\tau}{2\sqrt{2} \cos\Phi} \tag{13.11}$$

for the pulse-limited case, and

$$A = \frac{\pi R^2 \Theta_{AZ} \Theta_{EL}}{8 \sin\Phi} \tag{13.12}$$

for the beam-limited case.

The $\pi/4$ term accounts for the fact that the area is an ellipse, where the remainder of the terms define a rectangle of dimensions equivalent to the beamwidth and pulsewidth.

The boundary for these two regions is found by equating the two expressions and solving for the antenna depression angle. When doing so, the boundary can be found from

$$\Phi = \arctan\frac{\pi R \Theta_{EL}}{2\sqrt{2}c\tau} \tag{13.13}$$

When substituting the expression for clutter reflectivity in place of the target cross section in the radar equation, we see the nature of the relationship between range and SNR changes. In particular, as contrasted with $1/R^4$ variation for received target power, clutter received power varies approximately as $1/R^3$ (except for the usual decrease in σ with decreased depression angle) for the pulse-limited case and approximately as $1/R^2$ for the beam-limited case. This has major significance for MMW radars operating in a "look-down" mode when searching for stationary targets in ground clutter, since the ground clutter, not thermal noise, will often prove to be the dominant factor limiting slant-range performance.

Atmospheric clutter such as rain is a volumetric scatterer, depending on the volume defined by the azimuth and elevation beamwidths and the pulsewidth. The volume defined by these parameters is found from

$$V = \pi R^2 \Theta_{AZ} \Theta_{EL} c\tau/16 \tag{13.14}$$

When examining the range dependence of received power from volumetric clutter, we find that the power varies with R^2.

The reduced range dependence for area and volumetric clutter means that the signal-to-clutter ratio decreases with increasing range. For this reason, in many cases, the radar slant-range performance, in terms of detecting a one-square-meter target, becomes limited by clutter interference before being limited by thermal noise.

13.2.4 Antenna Related Parameters

Chapter 11 provides a comprehensive review of antenna technology supporting the development of millimeter wave radar systems; however, there are some very simple techniques typically used to determine the expected antenna performance, which provide the designer with a first-level approximation of the related parameters. Two antenna parameters are of primary importance to the radar designer, beamwidth and gain.

The radian beamwidth of a pencil-beam antenna can be estimated from

$$\text{BW (radians)} = 1.26\lambda/D \tag{13.15}$$

where λ is the wavelength and D is the antenna diameter in the same units. To estimate the beamwidth in degrees, we find

$$\text{BW (degrees)} = 72\lambda/D \tag{13.16}$$

For most applications and antenna designs, an estimated antenna efficiency factor of 60% will suffice to describe the antenna performance. Based on an efficiency factor of 60%, the gain of an antenna may be estimated from

$$G = 28{,}000/(\Theta_{AZ}\Theta_{EL}) \tag{13.17}$$

where

Θ_{AZ} = azimith beamwidth, in degrees;
Θ_{EL} = elevation beamwidth, in degrees.

For a radar with a scanning antenna, the number of pulses that can be integrated to improve the SNR, or processed to determine Doppler characteristics, is limited by the number of pulses transmitted while the antenna is pointed at the target of interest. The dwell time, T_d, is found from

$$T_d = \Theta_{AZ}/\Omega \tag{13.18}$$

where Ω is the scanning rate, and the resulting number of related pulses, n, is

$$n = T_d\text{PRF} \tag{13.19}$$

Because the antenna gain normally specified is that at the center of the beam and some of the pulses integrated occur at times when the effective gain is less than the peak, there is a scanning loss factor on the order of 1.45 (1.6 dB), which reduces the overall integration gain of the system (Barton [3]).

13.3 WAVEFORM SELECTION

Three basic elements of a radar system that the designer has at his or her disposal are (1) the technology which is available, (2) the choice of waveform to implement, and (3) the signal processing which will follow the detection process. This book features a series of chapters that deal with the technology available to the MMW radar system designer. The following sections of this chapter provide a discussion of the basic waveforms that are at the designer's disposal and the general class of processing techniques that can be employed with these waveforms to achieve a desired level of system performance. Among the waveforms available, those most commonly employed are pulse, pulse compression, FMCW (frequency-modulated continuous wave), interrupted FMCW, and stepped-frequency waveforms.

13.3.1 Pulsed Systems

Often, the first waveform to be considered is the pulse waveform because of its simplicity and ease of implementation. In this case, a burst of RF energy is transmitted, and it propagates to the target, reflects from the target, and propagates back to be received as a much attenuated burst of energy, the strength of which is related to the target RCS and range. The time delay, t_d (relative to the time the pulse was transmitted), is directly related to the range to the target, as given by

$$t_d = 2R/c \tag{13.20}$$

The pulsed radar implementation often allows the highest level of peak transmitted power, and thus provides the highest level of single-sample SNR. This is particularly important in systems that cannot afford a long dwell time on the target.

Range Resolution

The range resolution of a pulsed radar system is related to the pulse length by (13.7). A convenient rule of thumb for the relationship between range and time is that 100 ns of time delay represents 15 m of slant range, both in describing the slant range to a target and in dealing with range resolution.

PRF Selection

The range of possibilities for the PRF of a radar system is very large. The designer needs a methodology for determining an optimum PRF for a given application. Two specific radar performance factors generally help to establish the desired PRF. Although the requirements are sometimes conflicting, they are not always so, and for those cases in which they are, they at least bound the problem and provide the system designer with an initial estimate.

Usually, the highest PRF possible is used to allow faster antenna scans without missing target "hits," and also to produce the highest average power available from the transmitter. The higher the PRF, the more integration that can be performed. There is, however, a natural maximum PRF which can be employed.

Generally, it is desired, although not always possible, to avoid transmitting a pulse until after the signal resulting from the previous pulse has had sufficient time to reflect from the target and return to the radar system. If sufficient time is not provided, then it is not known whether a given returned signal is a result of the reflection of the most recent pulse from a nearby target or the reflection of an earlier pulse from a more distant target. If this condition cannot be met, the radar is said to exhibit *range ambiguities*. The delay time (interpulse period) required to allow unambiguous "ranging" to a target at range R is found by employing (13.20).

If t_{min} is the time required for the signal to come back from a target at the longest range of interest, then this represents the minimum time required between transmitted pulses to ensure unambiguous range operation. The maximum PRF allowable in this case is thus the reciprocal of the minimum interpulse time, which is found by

$$\text{PRF}_{max} = 1/t_{min} = c/(2R_{max}) \tag{13.21}$$

A more exact analysis considers the fact that the pulse length (τ) must be added to the delay time before the reciprocal is taken, yielding

$$PRF_{max} = \frac{1}{\dfrac{2R}{c} + \tau}$$

(13.22)

Another factor influenced by the PRF is the maximum Doppler frequency that can be unambiguously measured. The Doppler frequency shift due to a moving target is found from

$$F_d = 2V/\lambda$$

(13.23)

where

V = relative radial velocity, in meters per second;
λ = wavelength, in meters.

A pulsed radar operates as a sampled data system, which can reconstruct data having a bandwidth up to half the sample rate. Therefore, the minimum PRF which can be employed for unambiguous measurement of a Doppler frequency (e.g., distinguish closing targets from departing targets) is $2F_d$, or

$$PRF_{min} = 4V/\lambda$$

(13.24)

Sometimes, the maximum PRF desired due to maximum range considerations is less than the minimum PRF desired due to Doppler considerations; in these cases, a compromise must be made. Either the Doppler measuring capability or the range measuring capability of the system will be ambiguous for a given PRF. For some PRFs, both of these parameters may be ambiguous. Systems having a PRF high enough to be always unambiguous in Doppler are called *high-PRF systems*. Systems which are always unambiguous in range are called *low-PRF systems*, and systems that are ambiguous in both range and Doppler are *medium-PRF systems*. In medium-PRF systems, the range and Doppler can be generally resolved by using various combinations of PRF staggering techniques and associated processing to resolve the ambiguities (Schrader [4]).

Bandwidth

The desired bandwidth of the radar receiver is dictated by the expected bandwidth of the received signal, which is approximately equivalent to that of the transmitted signal. The spectrum of a rectangularly shaped pulse has a sin(*x*)/*x* form, where the first null is at the frequency represented by 1/τ and the null-to-null separation is 2/τ. The transmitted pulse shape is generally not purely rectangular, however, and the shape of the receiver bandpass is such that the optimum 3 dB bandwidth, to optimize the SNR, is approximated by 1/τ (Van Voorhis [5]). Therefore, the receiver bandwidth required for a 100 ns transmitted pulse (providing 15 m range resolution) is 10 MHz; 1.5 m resolution requires 100 MHz bandwidth; 0.3 m resolution requires 500 MHz, *et cetera*. In general, the required bandwidth may be found from

$$BW(MHz) = 150/R_{res}(m) \tag{13.25}$$

13.3.2 Pulse Compression

In some system applications, due to limited dwell time on the target, it is neither possible nor practical to employ pulse-to-pulse integration to achieve the required SNR. In these cases, a technique referred to as *pulse compression* may be used to increase the single-pulse SNR. Pulse compression may be used by itself or in conjunction with pulse-to-pulse integration to achieve SNR improvement superior to that which can be achieved independently with either of the two techniques.

The pulse compression mode involves transmitting a pulse of relatively long duration, which has some form of intrapulse coding. Most often, this coding is a linear swept-frequency chirp. Other popular, but more complex, codes are phase coding and polarization coding. The received signal from a point target will exhibit the same coding as the transmitted signal, and when it is applied to a matched filter, the filter transforms the swept-frequency long pulse to a fixed-frequency short pulse. The amount of SNR gain achieved is approximately equivalent to the pulse time-bandwidth product, $B\tau$.

Most pulse compression systems use *surface acoustic wave* (SAW) technology to implement the pulse expansion and compression functions. The maximum $B\tau$ product which is readily achievable with current technology is about 1000. The technology exists at frequencies up to about 500 MHz. Therefore, for a MMW radar that implements this mode, the waveform is generally generated at IF and up-converted to the desired RF for transmission.

As an example of a typical pulse compression system, a wide pulse (6.5 μs) that has a linear frequency sweep of 20 MHz is transmitted. This pulsewidth limits the minimum range to beyond 1000 m. The chirp is developed at IF by exciting a SAW delay line with an impulse that has frequency components beyond 20 MHz. The chirp is developed at an IF of 500 MHz, and then up-converted to the RF for amplification and transmission.

The received signal is down-converted to 500 MHz, where it is dechirped (compressed) by individual SAW filters in each channel. The receiver (compression) filters are the reciprocal of the transmitter (expansion) filters. The resulting pulse out of the compression network is about 50 ns wide. Figure 13.3 is a pictorial description of the waveforms at the various stages of this process.

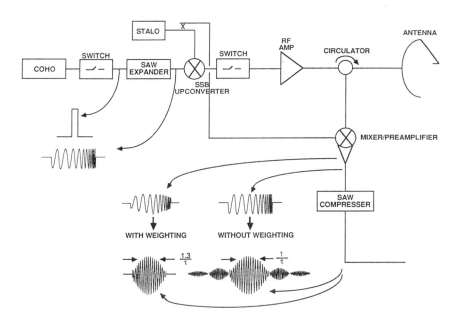

Fig. 13.3 Pulse compression system block diagram.

The received pulse is applied to a compression filter, which produces a $\sin(x)/x$ time response for a chirped input. The –13 dB time sidelobes associated with such a response are unacceptable if high range resolution is required of the system. The sidelobes can be reduced by implementing a weighting function on the received signal as part of the compression process (Harris [6]). This weighting can reduce the time sidelobes to about 40 dB below the main lobe response at the expense of some range resolution. Main lobe spreading on the order of approximately 30% will be experienced with such weighting. In this case, a pulse compression bandwidth of 1.25 to 1.3 times the real pulse's required bandwidth is needed. For a required range resolution of 0.3 m, this represents a required bandwidth of 625 to 650 MHz, as opposed to the 500 MHz theoretically required for a simple pulsed radar system.

13.3.3 Stepped Frequency Waveform

When using a real or compressed pulse waveform to achieve high range resolution, there is a challenge to acquire the received data at the rate, or bandwidth, at which it is received. In particular, for 0.3 m resolution, the received single bandwidth is about 500 MHz, which is consistent with the 2 ns pulsewidth required to achieve that resolution. The output of a pulse compression network likewise needs to be 2 ns video for the equivalent resolution. This requires a data acquisition system that is capable of processing data at a peak rate of at least 500 MHz, for a period of time at least equivalent to that required for instrumentation of the range extent of interest. Whether the video is applied to a digitizing network, a signal processor, a recorder, or simply a display, advanced circuit technology is required for a video amplifier and sample-and-hold networks as well as sophisticated A/D converter and buffer memory functions. In fact, except for a converter with only a few quantum levels, the A/D converter required for this bandwidth is beyond the current technology. Several converters would have to be multiplexed to digitize the data. The data rate burden then falls on the buffer memory, which would have to acquire data at rate of at least 500 megabytes per second.

One waveform technique, which may be used to avoid the data acquisition problems associated with the real pulse mode, involves the pulse-to-pulse stepped frequency waveform. The bandwidth required for a given range resolution is still the same as determined in Section 13.3.1. However, in the stepped frequency waveform mode, this bandwidth is not achieved instantaneously (i.e., within a single pulse). The individual bandwidth of each pulse is much less than the total radar bandwidth.

An example of the stepped frequency waveform is one in which a 100 ns pulse is transmitted. The frequency of each transmitted pulse is stepped relative to an adjacent pulse by 5 MHz. The received signal is coherently detected and recorded or stored in memory for subsequent processing. A single range sample will provide data for a 15 m range extent.

The disadvantage of the frequency stepped waveform is associated with the fact that, since the bandwidth required for a given resolution is not achieved in a single pulse, an ensemble of pulses must be stored and processed to develop the high resolution range profile. This process takes time, requiring the system instability to remain within limits during the waveform interval. The specific process is described below.

Referring to the parameters depicted in Figure 13.4, the phase, Φ_1, of the returned signal associated with a transmitted pulse of frequency F_1 is found from

$$\Phi_1 = 4\pi F_1 R/c \tag{13.26}$$

for a nonmoving target at range R. If a pulse at a different frequency, F_2, is transmitted, the phase, Φ_2, of the signal returning from a nonmoving target would be

$$\Phi_2 = 4\pi F_2 R/c \tag{13.27}$$

For a sequence of pulses, where each pulse is equally spaced from adjacent pulses by a frequency step of δF, there is a predictable pulse-to-pulse phase shift of $\delta\Phi$ of the target, which is found by

$$\delta\Phi = \Phi_2 - \Phi_1 = 4\pi R\delta F/c \tag{13.28}$$

This pulse-to-pulse phase shift manifests itself within the radar as an apparent Doppler frequency, which is a function of the range to the target. A spectral analysis of the sequence of received. A range profile of the scattering elements of the target. This analysis often takes the form of a fast Fourier transform (FFT) of the ensemble of received pulses. Figure 13.5 is a pictorial representation of the transmitted sequence of pulses and the action of the FFT in developing the high resolution range profile. The total extent of the range swath cannot be any longer than that which is equivalent to the range covered by the original pulsewidth.

The total unambiguous range, after the FFT process has been performed, is $c/2\delta F$, where δF is the pulse-to-pulse frequency step size. The range resolution of the output array is $C/2F_{total}$, where F_{total} is the total frequency excursion of the transmitted waveform. For a sequence of N pulses, each separated by δF, the total frequency excursion is $N\delta F$. A restriction on the required δF as a function of the transmitted pulsewidth is that undersampling of the apparent Doppler (induced by the frequency stepping) does not occur. This requires that δF be no larger than $1/\tau$ for a purely rectangular pulse. The suggested δF is $1/2\tau$, which is adequate for a Gaussian pulse (Einstein [7]).

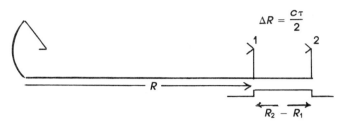

$$\phi_1 = \frac{4\pi R_1}{\lambda} = \frac{4\pi R_1 F}{c}$$

$$\phi_2 = \frac{4\pi R_2}{\lambda} = \frac{4\pi R_2 F}{c}$$

$$\delta\phi_1 = \frac{4\pi}{c} R_1 \Delta F$$

$$\delta\phi_2 = \frac{4\pi}{c} R_2 \Delta F$$

$$\delta\phi_2 - \delta\phi_1 = \frac{4\pi}{c} \Delta F (R_2 - R_1)$$

$$= \frac{4\pi}{c} \Delta F \left(\frac{c\tau}{2}\right)$$

$$\delta\phi_2 - \delta\phi_1 = 2\pi\tau\Delta F$$

To avoid phase ambiguity on pulse-to-pulse basis, for pulse length τ,

$$\delta\phi_2 - \delta\phi_1 < 2\pi$$

$$2\pi\tau\Delta F < 2\pi$$

$$\Delta F < \frac{1}{\tau}$$

Fig. 13.4 Phase relationships for two targets using stepped-frequency waveform.

The above conditions apply for a nonmoving target, which exhibits no Doppler frequency shift. If the target has a radial component of relative velocity, then the Doppler effects will shift the apparent range to the target, since frequency maps to range as a result of the FFT process.

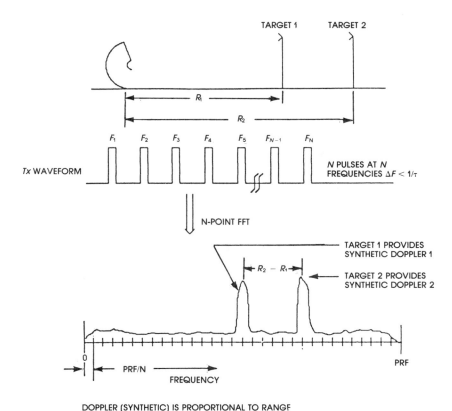

Fig. 13.5 Depiction of stepped-frequency waveform processing.

13.3.4 CW Waveforms

Another commonly used class of waveforms is a CW signal. This type of system typically has a lower peak power than the pulsed system, due to transmitter limitations. As with the pulsed system, the returned signal strength is related to the target reflectivity and its range from the radar, and the delay is related to the range, but the delay time to a point target cannot be easily determined for a simple CW waveform because of the continuous nature of the returned signal making it difficult to establish a time reference. Because of this, for cases in which information regarding the range to the target is desired, the waveform must somehow be coded to provide a means to determine the delay between the transmitted waveform and the received waveform.

FMCW Systems

CW systems are often coded with a frequency ramp (or chirp), as is the case with FMCW systems. Figure 13.6(a) is a simplified block diagram of an FMCW system. In this case, at any instant of time, the received waveform is at a different frequency from the signal being transmitted by an amount related to the range to the target and the chirp rate. Figure 13.6(b) depicts the operation of the FMCW waveform, showing the transmitted signal and the received signal for a given set of targets. Typically, the received signal is mixed with a sample of the transmitter signal. As can be seen in the figure, the signal received from a target is at a different frequency from the transmitter frequency because the transmitter has moved in frequency during the time that the signal propagated from the radar to the target and back to the radar. This results in a constant-difference (or beat) frequency for a particular target at a specific range. A spectral analysis of received signal after it is down-converted is required to determine the range to the targets being illuminated. As with the stepped frequency pulsed waveform, any Doppler frequency on the target return will be manifest as an error in determining the range to the target.

Each target produces a signal at a different beat frequency. The resolution with which targets can be separated in range is a function of the total bandwidth of the transmitted signal and is also directly related to the linearity of the transmitted ramp. Ramp nonlinearity causes a spreading of the spectrum of the received signal from a point target, thus contaminating the degree to which its range can be determined. If two targets are at nearly the same range, this spreading will cause the two targets to merge and to appear as one.

The ramp slope, as shown in Figure 13.6(b), is given by $\delta f / \delta t$. As with the pulsed system, the delay T_r for a target at range R is $2R/c$. The frequency offset is found from

$$F_{\text{offset}} = T_r \delta f / \delta t \qquad (13.29)$$

Therefore, for a range resolution of δR, the required frequency resolution δF_r is found from

$$\delta F_r = (\delta f / \delta t)(2\delta R / c) \qquad (13.30)$$

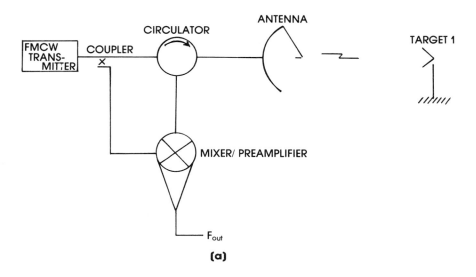

Fig. 13.6(a) FMCW radar block diagram.

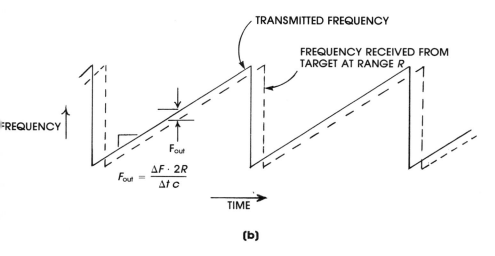

Fig. 13.6(b) FMCW operation.

For a spectral resolution of δF_r, the minimum dwell time, T_d, required for analysis is $1/\delta F_r$, thus giving

$$T_d = (c/2\delta R)1/\delta f/\delta t \tag{13.31}$$

At a ramp rate of $\delta f/\delta t$, the total sweep-frequency excursion is the product of the sweep rate and the dwell time. After cancellation of terms, this reduces to

$$F_{\text{total}} = c/2\delta R \qquad\qquad (13.32)$$

Notice that the total frequency excursion required for a given range resolution is consistent with the bandwidth required for a given range resolution for a pulsed system.

FMCW and Interrupted FMCW

Chapter 14 provides a discussion of the advantages and disadvantages of the FMCW waveform relative to a pulse waveform. The FM CW waveform offers a simpler RF design, which uses a single RF source for the transmitter and LO, high average power, and good short-range performance. This waveform demands careful design to maximize front-end isolation and to minimize front-end VSWR and reflections. A highly linear frequency sweep is required for high range resolution.

IFMCW

The *interrupted FMCW* (IFMCW) waveform involves interrupting the CW transmitted signal to eliminate the requirements for transmitter-to-receiver isolation during the reception time and extremely low antenna VSWR. One IFMCW approach involves a transmission time matched to the round-trip propagation time, followed by an equal reception time. The resulting duty factor is 0.5, which reduces the average power by 3 dB relative to FMCW. The improved isolation may be expected to reduce the system interference by more than 3 dB which improves the overall SNR. A high speed-modulation switch is required for IFMCW operation at short range. Since the target cannot be blanked by the transmitted signal, the range to the target must be known in order to decide when to transmit and to receive. A range acquisition scheme must be implemented to find the target range for this type of system.

13.4 SIGNAL PROCESSING CONCEPTS

13.4.1 Integration

Typically, MMW systems do not exhibit the same SNR performance as microwave systems. This is because there is generally less power available from the transmitter, due to reduced sizes in the resonant structures, *et cetera*, poorer achievable receiver noise figure, higher waveguide and device losses, and higher atmospheric loss, even in the so-called "windows" of the absorption curves. Systems designers are continually being taxed to provide as good a system performance as possible from MMW systems.

The signal-to-noise performance of a system can almost always be improved relative to the single-sample SNR by a process called *integration*. This is the case if the dwell time on a cell is longer than the single sample time. Two kinds of integration can be incorporated, coherent and noncoherent integration. Coherent integration provides an improvement in SNR of n, where n is the number of pulses coherently integrated. Noncoherent integration is not as efficient as coherent integration, providing an improvement of n^m, where m is generally somewhere between 0.5 and 0.9, depending on the input SNR (Barton [8]). The higher the input SNR is, the more efficient noncoherent integration becomes. Noncoherent integration is sometimes referred to as *postdetection integration* because it is accomplished by merely adding consecutive detected video signals. Coherent integration, however, requires vectorial addition of coherently detected signals.

Sometimes both coherent and noncoherent integration can be employed in a system to improve its detection performance. Coherent integration, although more effective than noncoherent integration, is sometimes limited by the coherence of the system. The coherence may persist for only a limited time, depending on the stability of the oscillators and the velocity and acceleration characteristics of the target itself. As an example, suppose that an oscillator has the required stability for 100 ms for effective coherent integration, but dwell time on the target could allow for up to one second of integration time. In addition to the loss of stability of the oscillator after 100 ms, the target may experience an acceleration, which would change the integration characteristics. In a case such as this, coherent integration could be employed in 100 ms periods, and groups of 10 coherently integrated pulses could thcn be noncoherently integrated.

13.4.2 Coherent Detection and Processing

Several processes require coherent detection of the returned signal. Among them are coherent integration, as discussed above, moving-target-indication (MTI) processing, and Doppler processing. The stepped frequency process, resulting in high-resolution-range profiles, also requires coherent detection. Coherent detection of the received signal implies that the phase of the transmitted signal is known and that a reference signal exists, which is phase locked to the transmitted signal. The received signal can then be detected in the form of a vector quantity, the phase of which is known as well as its amplitude, as opposed to noncoherent detection, which provides only amplitude information.

Figure 13.7 shows a very basic block diagram of a coherent system. The stable local oscillator (STALO) and coherent oscillator (COHO) combine to generate the transmitted signal, and the received signal is downconverted by using the STALO as the first LO and the COHO as the reference for the synchronous detectors. The in-phase (I) and quadrature (Q) outputs represent the two orthogonal components of the received signal vector. The I and Q components are defined by the following expressions:

$$I = A \cos(\Phi) \tag{13.33}$$

$$Q = A \sin(\Phi) \tag{13.34}$$

where A is the signal amplitude and Φ is the phase.

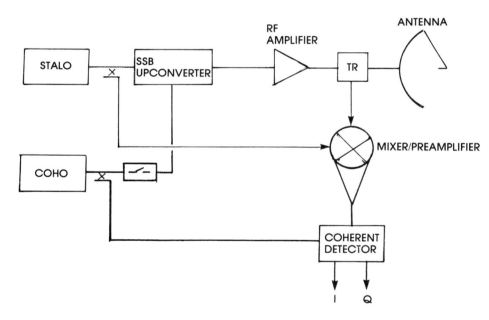

Fig. 13.7 Coherent radar block diagram.

Figure 13.8 shows a block diagram of a simple I/Q detector network, which generates these components of the received vector.

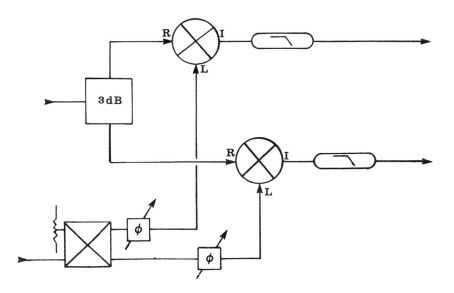

Fig. 13.8 Coherent detector block diagram.

MTI Processing

The use of MTI processing provides the operator with a radar system output that ideally includes only targets which are moving, and thus eliminates stationary targets and clutter from the display. Two kinds of MTI processes are employed, noncoherent MTI and coherent MTI. Noncoherent MTI is sometimes termed *clutter-reference MTI* because it uses the stationary clutter as the phase reference for the detection of moving targets. Coherent MTI requires coherence in the radar system, but it allows for detection of moving targets in a clutter-free environment, and often has much better performance than a noncoherent MTI.

In noncoherent systems, the phase of the received vector is a randomly varying quantity, and provides no information concerning the actual target phase. Instead, the phase of the vector for a coherent system provides much information about the target. For example, slight motion of the target will cause a pulse-to-pulse variation of the phase due to the fact that there will be a change in the radial component of range to the target. In general, radar systems do not measure the absolute range to the target in terms of wavelengths, but a change can be detected in the number of fractional wavelengths on a pulse-to-pulse basis. The fineness with which this change can be detected is a function of the stability of the reference oscillators. If the oscillators themselves exhibit phase scintillation, then the target phase will be scintillating, although the target may not be moving.

This phase shift is exploited in several ways. For cases in which it is merely desired to determine whether a target is moving, an MTI processor may be employed. This process is essentially a high-pass filter, which, for a pulsed system, has a repeating frequency response. Targets that exhibit a Doppler characteristic will pass the filter, and targets with no Doppler, such as stationary targets and ground clutter, will not pass the filter.

Most MTI systems take the form of a single delay, double delay, or double delay feedback canceller. Figure 13.9 shows the block diagram of these three configurations. The delay can be implemented in one of several ways.

In the early configurations of MTI processors, the delay was implemented by using a quartz delay line. The delay was made precisely equivalent to the interpulse period so that the input to the subtractor network would be the response from two consecutive pulses. Since those early days, digital technology has been used to perform the delay as well as the cancellation (subtraction) function.

If there is any target motion, the two consecutive signals will be different, due to the change in phase, and there will be a resulting signal out of the subtractor. The output is full-wave rectified (because the difference may be plus or minus) and applied to some display device. Stationary targets and stationary clutter are "cancelled," and not applied to the display. Moving targets pass the canceller and show up on the display.

Obviously, there is no abrupt threshold between the declaration of moving and nonmoving targets. The frequency response of the canceller (filter) is a function of how many pulses are used in the cancellation process and the feedback configuration of the canceller. Figure 13.10 shows the frequency response of a single delay canceller, a double delay canceller and typical response for a double delay feedback canceller. The system designer needs to know the spectrum of the clutter, the spectrum of the phase instabilities in the radar system and the expected range of Doppler frequencies from the target to optimize the MTI performance of the radar system.

Several considerations need to be addressed here to complete even a cursory discussion concerning MTI processing. First, phase and amplitude instabilities in the radar itself will make stationary targets appear to be moving when processed in such a fashion. It is therefore important to

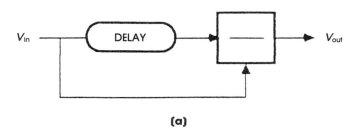

(a)

Fig. 13.9(a) Single delay canceller.

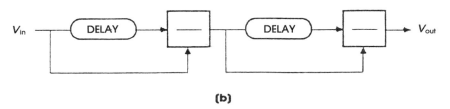

(b)

Fig. 13.9(b) Double delay canceller.

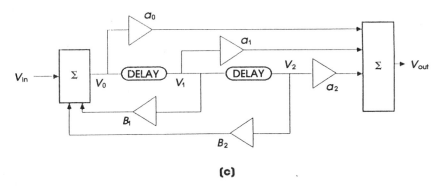

(c)

Fig. 13.9(c) Double delay feedback canceller.

minimize the phase and amplitude instabilities, and to design the filter to optimize the target-to-residue performance around the existing phase noise.

Second, it is important to point out that, since the radar senses the radial component of velocity, the Doppler effects of a moving target will

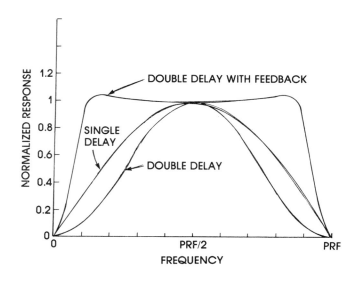

Fig. 13.10 Frequency response of MTI (high-pass) filters.

be reduced by a factor related to the off-radial angle. Additionally, the radar cannot differentiate target motion from platform motion. Therefore, if the radar itself is moving, stationary targets and clutter will exhibit Doppler effects, and hence appear to be moving. If this is to be avoided, motion compensation needs to be employed. That is, a phase correction is required before the signal is applied to the MTI canceller. The phase correction needed on a pulse-to-pulse basis is related to the platform velocity, and both the speed and direction relative to the radar antenna pointing angle are considered.

Doppler Processing

An extension of the MTI process involves processing of the Doppler frequency exhibited by a moving target. The Doppler frequency is the difference between the transmitted frequency and the received frequency of a signal reflected from a moving target. The effect is essentially the same as may be experienced with a train whistle as the train passes by an observer. The apparent frequency of the whistle is higher than the actual frequency as the train approaches, and the apparent frequency is lower as it departs. The amount of Doppler frequency offset is dependent on the speed of the train. Likewise, the frequency of the wave reflected from a moving target is shifted by an amount related to the radial component of velocity, as given in (13.23).

Processing of the received signal to exploit the Doppler characteristics involves a spectral analysis of the sampled signal, where the sampling occurs at the PRF. In modern radar systems, this analysis is done by employing the FFT technique on the sampled and digitized data. As with the coherent MTI system, the analysis of the Doppler characteristics depends on the coherence of the radar itself. Figure 13.11 shows the frequency response of a typical bank of Doppler filters, implemented by means of an FFT process (Stimson [9]). The number of filters is exactly the same as the number of FFT points processed, and the total passband goes from dc to the PRF.

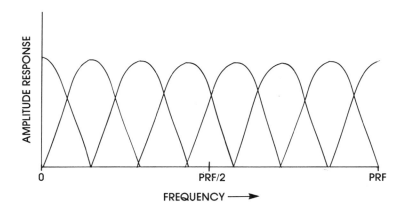

Fig. 13.11 Doppler filter bank.

The FFT peforms a service equivalent to coherent integration of the received signal. No matter what the Doppler frequency is, the signal will integrate up in one of the Doppler bins associated with the spectral analysis performed by the FFT. This is different from simple noncoherent integration from the point of view that the signal will not integrate up monotonically if the signal is not of fixed phase on a pulse-to-pulse basis. In other words, if the N vectors are not all in the same direction, when they are added, the result will not be N times the original vector. The FFT analysis can be thought of as the addition of a series of N vectors for a variety of pulse-to-pulse phase corrections, one of which will be correct for optimal summation. Each FFT or Doppler bin represents a different pulse-to-pulse phase correction. The bin in which the target integrates up to the highest value represents the bin of the Doppler frequency due to the target velocity.

Because the received signal is sampled at the PRF, it is treated as a sampled data system. Most digital signal processing techniques involve Fourier transform analysis, in which case several characteristics apply. For example, the frequency resolution, F_{res}, is as fine as that allowed by the dwell time, T_d, as determined by

$$F_{res} = 1/T_d \tag{13.35}$$

which, for a scanning antenna system, can be found by using (13.18).

As an example, for a system with a 1° beamwidth, and a scanning rate of 60°/s, the dwell time is 16.6 ms, resulting in a Doppler resolution of 60 Hz. At 95 GHz this represents a velocity resolution of about 0.09 m/s. This fine velocity resolution is important in applications involving target identification, where the target Doppler signatures are applied to pattern recognition algorithms. By contrast, an X-band system having the same dwell time, or coherence time, would have a velocity resolution of about 1 m/s.

Sometimes the maximum processing time is not limited by the dwell time, but rather by the period of time over which the received signal is coherent, which is limited by either the stability of the radar system components, or the target motion. Target velocity manifests itself as a Doppler frequency, or pulse-to-pulse phase shift of the received signal. Doppler processing, typically in the form of an FFT, provides a means to analyze the spectral characteristics of the target. As long as the target velocity is constant, the components will fall into their respective bins. If the target is experiencing an acceleration, the Doppler will smear across several bins, the extent to which depends on the acceleration and the processing interval.

The Doppler resolution is found from (13.35), and the Doppler spread is found from

$$\delta F_d = 2\delta V/\lambda \tag{13.36}$$

Equating (13.35) and (13.36), and solving for T_d, we define the limit on the processing time as a function of wavelength and acceleration, A, by

$$T_d = \sqrt{\lambda/2A} \tag{13.37}$$

For example, an acceleration, A, of 1 G (9.8 m/s^2) will result in the target Doppler slewing through about six Doppler bins during one processing period in the preceding example in which the Doppler resolution is 60 Hz.

Another limitation to the processing interval is the stability of the radar system. The phase noise of the STALO, for example, can limit the coherent processing time, as described simplistically in Section 10.2.8.

It is often difficult to avoid Doppler and range ambiguities. Using (13.21) and (13.24), and eliminating the PRF term, the relationship between unambiguous range, R_u, and unambiguous velocity, V_u is found to be

$$R_u V_u = c\lambda/8 = 118.5 \times 10^3 \text{ (at 95 GHz)} \tag{13.38}$$

This means that for a system having a PRF consistent with 5 km range, the maximum unambiguous velocity is about 23.7 m/s. This is adequate for many ground targets but not for most airborne targets.

13.5 SUMMARY

Several design considerations are presented here, including an example of the solution to the radar range equation and several waveform considerations. Also, some of the basic processing techniques have been introduced. The chapters to follow provide more specific examples of MMW system applications, employing the techniques introduced here as well as others.

REFERENCES

1.   L.V. Blake, "Prediction of Radar Range," Chapter 2, pp. 19–25, in M.I. Skolnik, ed., *Radar Handbook,* McGraw-Hill, New York, 1970.
2.   D.P. Meyer and H.A. Mayer, *Radar Target Detection,* Academic Press, New York, 1973, pp. 156–157.
3.   D.K. Barton, *Radar System Analysis,* Artech House, Dedham, MA, 1979, p. 146.
4.   W.W. Shrader, "MTI Radar," Chapter 17, p. 38, in M.I. Skolnik, ed., *Radar Handbook,* McGraw-Hill, New York, 1970.
5.   S.N. Van Voorhis, *Microwave Receivers* (MIT Rad. Lab. Series), McGraw-Hill, New York, 1948.
6.   F.J. Harris, "On the Use of Windows for Harmonic Analysis with the Discrete Fourier Transform," *Proceedings of the IEEE,* Vol. 66, No. 1, January 1978.
7.   T.H. Einstein, "Generation of High Resolution Range Profiles and Range Profile Auto-Correlation Functions Using Stepped Frequency Pulse Trains," Project Report TT-54, MIT Lincoln Laboratory, October 1984, p. 28.

8.   D.K. Barton, *Radar System Analysis,* Artech House, Dedham, MA, 1979, p. 26
9.   G.W. Stimson, *Introduction to Airborne Radar,* Hughes Aircraft Company, El Segundo, CA, 1984.

Chapter 14
MMW Seekers

S.O. Piper

Georgia Institute of Technology
Atlanta, Georgia

14.1 MISSIONS AND SCENARIOS

Millimeter-wave seekers are suited for a wide variety of missions and scenarios, including air-to-surface, surface-to-air, and air-to-air missiles. The maturation of MMW technology makes MMW seekers attractive solutions for "fire-and-forget" precision-guided weapons (PGW). This chapter will introduce basic missile guidance nomenclature, identify major trade-off issues, and project performance for several candidate seeker designs.

Millimeter-wave seekers have beamwidths comparable to those of microwave radars, but in smaller and missile-compatible volume. Millimeter-wave seekers offer superior adverse weather performance relative to infrared (IR) or electro-optical (EO) technologies for precision-guided weapons. The autonomous capabilities of the MMW seekers permit fire-and-forget operation, which extends the stand-off range of the launch platform and enhances its survivability. Fire-and-forget operation also increases potential fire power by enabling multiple engagements on a single pass.

Table 14.1 lists the major subsystems and components of a typical missile. The MMW seeker is part of the missile guidance and control, along with the inertial reference components such as gyroscopes for stabilization and navigation. The autopilot controls the missile, and the prime electrical power is typically provided by a thermal battery. The airframe includes the body of the missile, the wings, which may fold out to provide lift, and the control surfaces or fins with actuators. Control may alternatively be provided by steering jets or control rockets. The armament includes the warhead, which may be unitary, shaped charge, or self-forging, the safe and arm mechanism, and the fuze. If the missile is powered, then the propulsion will include the propellant, ignitor, and safety mechanism.

Table 14.1   Missile Subsystems and Components.

Subsystem	Components
Guidance and Control	Target Seeker
	Electronics
	Inertial Reference, Gyroscope
	Computer, Autopilot
	Prime Electrical Power
Airframe	Body
	Lift Surfaces
	Control Surfaces
Armament	Warhead
	Unitary
	Shaped Charge
	Self Forging
	Safe and Arm
	Fuzes
Propulsion	Propellant
	Ignitor
	Safety Mechanism

14.1.1   Air-to-Surface Seekers

Millimeter-wave seekers are under development for a number of air-to-surface missions, including antiarmor and high-value target attack (see [19, 23, 35, 39, 40, 44]). This chapter emphasizes the antiarmor mission.

Antiarmor Systems

Table 14.2 lists a number of antiarmor MMW seeker development programs. Armor targets include tanks, armored personnel carriers (APC), mobile artillery, and air defense units (ADU). Note that the missiles may be launched from fixed or rotary wing aircraft, larger carrier missiles, or artillery pieces. An aircraft-launched missile with a MMW seeker may be cued or given the target position prior to launch. The aircraft fire control computer may compensate for wind and aircraft manuevers at launch to improve the accuracy of the missile flight path. Submissiles or submunitions launched from a larger carrier missile will be more suitable for attacking a general target area without specific prior information on target position. Figure 14.1 shows a developmental submissile. Because the missile platform is consumed in use, economic considerations will limit the cost, and consequently the precision, of the on-board inertial guidance sensors. Artillery-launched weapons may also attack target areas. These weapons are designated *sensor-fuzed munitions* or *target-activated munitions* because the seeker triggers the warhead.

The artillery-launched sensor-fuzed munitions are deployed in groups of several submunitions. Typically, a parachute is used to control both the descent and spin of the body that performs a circular spiral scan on the ground. The antenna is mounted without gimbals so that it leads the warhead axis to provide processing time for the seeker. The seeker functions as an open-loop fire control system that fires the warhead when it detects targets of opportunity. With the self-forging warhead, only the warhead itself actually impacts the target.

Table 14.2   Antiarmor MMW Seeker Development Programs.

Aircraft Launched
Advanced Millimeter-Wave Seeker (AMMWS)
WASP [1, 2, 8, 10, 14, 28, 33, 43, 52, 57]
Millimeter-Wave Contrast Guidance Demonstration (MCGD) [16]
Millimeter-Wave Semiactive Guidance Program (MSGP)

Missile Launched
Multiple Launch Rocket System/Terminally Guided Weapon
   (MLRS/TGW) [45]
Terminally Guided Submunition (TGSM) [60]
Assault Breaker [15, 21, 32]
Cyclops [60]
Millimeter Radiometric Seeker Subsystem (MRSS)

Artillery Launched
Sense and Destroy Armor (SADARM) [22, 31, 57, 27, 60]
Advanced Indirect Fire Seeker (AIFS) [57]
Extended Range Antiarmor Munition (ERAM) [57, 60]
Smart Target-Activated Fire and Forget (STAFF) [27]
Copperhead

Fig. 14.1 Developmental submissile (photo courtesy of General Dynamics, Valley Systems Division).

High-Value Targets

For high-value targets such as airfields, bridges, power stations, petroleum oil lubricant (POL) facilities, or fixed ADUs, the MMW seeker provides an image of the target to the signal processor, which recognizes the target for detection and chooses the aim point. Two-dimensional correlation of the seeker data with the target reference data may be employed. Data from a MMW imaging radar or radar altimeter may also be correlated with prestored reference terrain reflectivity or elevation data to obtain position fixes for long-range missile systems (Roeder and Teti [42]).

14.1.2 Surface-to-Air Systems

Millimeter-wave seekers can be employed for surface-to-air missions, including both air defense and ballistic missile defense [3, 4]. In these missions, the MMW seeker offers better angular resolution than microwave seekers, better adverse weather capability, and less overall susceptibility to countermeasures than IR or EO sensors. Millimeter-wave seekers are being considered for re-entry vehicle interceptors for exoatmospheric operation. An erectable or "pop-out" type of antenna increases aperture for narrower beamwidth and higher gain in the exoatmosphere, while fitting within the diameter of the kill vehicle in the storage mode [34].

14.1.3 Air-to-Air Systems

Millimeter-wave seeker technology is also a candidate for air-to-air missions when high angular resolution and adverse weather capability are important. The narrow antenna beamwidth also reduces the countermeasures susceptibility of the MMW seeker.

14.2 TYPES OF SEEKERS

Table 14.3 lists five major types of MMW seekers and several dual-mode sensor combinations.

Table 14.3 MMW Seeker Types.

MMW Seekers
Active Monostatic Radar
Semiactive Bistatic Radar
Passive Home-on-Radiation Seeker
Active Radiometer or Noise Illumination Bistatic Radar
Passive Radiometric

Dual-Mode Combinations
Millimeter-Wave/Infrared (MMW/IR)
Millimeter-Wave/Antiradiation Missile (MMW/ARM)
Millimeter-Wave/Electro-optical (MMW/EO)

14.2.1 Active

Active MMW seekers employ monostatic radar techniques with both the transmitter and receiver in the seeker. The active seeker is fully autonomous and has many operational advantages. It also presents the greatest design challenges. This chapter emphasizes active MMW seekers.

14.2.2 Semiactive

Semiactive MMW seekers detect and track the signals reflected from targets that are illuminated by a separate illuminator. The illuminator often has fewer cost, size, and weight constraints than the seeker because the illuminator is reusable and it is mounted on the launch platform. The semiactive approach can be used to extend effective range, for remote target designation, and for precision attack in the vicinity of friendly units. The illuminator must track the target throughout the missile flight, which reduces overall firepower, since only one target is engaged at a time, and also exposes the illuminating platform to return fire.

14.2.3 Passive ARM

Passive MMW seekers home on the tranmissions from threat systems. Of course, the threat must operate at MMW. By receiving and tracking the threat radiation, passive seekers provide guidance for antiradiation missiles (ARM). Radio-frequency interferometer techniques are often used for precise angle tracking.

14.2.4 Active Radiometer

Active radiometer MMW seekers are bistatic radar systems, which use a wideband noise waveform for the illuminator. The seeker operates much as a radiometric seeker, only with inverted track polarity to exploit the contrast enhancement that results from the noise illumination. The broadband illumination smooths target glint and scintillation to provide good aim-point data [20].

14.2.5 Radiometric

Radiometric MMW seekers exploit the apparent thermal contrast at MMW frequencies between the background clutter at the ambient temperature and the cold sky temperature reflected from metallic targets. Radiometric technology is covered in detail in Chapter 18. Because the radiometric aim point exhibits much less glint and scintillation than that for the radar aim point, some seekers include radiometric tracking for terminal tracking (Harrop, Stump, and Teti [18]). The wide RF bandwidth and low contrast make radiometric operation susceptible to countermeasures.

14.2.6 Dual-Mode Combinations

By combining complementary sensor technologies, the more complex dual-mode seeker exploits the best features of each. Target discrimination techniques can compare the information from the two sensors for performance which is better than that available for either sensor alone. The dual-mode seeker can switch modes when countermeasures are encountered.

In addition to active MMW seekers with a radiometric terminal phase, other dual-mode seekers combine MMW and another sensor such as IR to improve performance. For example, a dual-mode MMW/IR seeker offers good area search and adverse-weather capability from the MMW sensor while the high resolution IR sensor offers good target discrimination and terminal aim-point performance. A common-aperture MMW/IR sensor presents many challenges in antenna and radome design because of the large difference in wavelengths for MMW and IR [12, 29, 55, 56]. Figure 14.2 shows a developmental dual-mode MMW/IR seeker with radome removed. By combining a MMW sensor with an EO sensor, such as a low-light-level television, is also feasible, but an IR sensor offers better tactical performance than a TV sensor. Another interesting combination is an antiarmor missile with an RF ARM interferometer seeker combined with a MMW seeker. The MMW sensor measures range to the target and is able to guide the missile to the target in the event that the interferometer guidance is lost due to shutdown of threat system radiation.

14.3 GUIDANCE

14.3.1 Lock-On Before Launch (LOBL)

For lock-on before launch (LOBL), search and detection is done by a separate target-acquisition system, which will hand over the target position to the MMW seeker prior to launch. The seeker acquires the target and maintains track throughout the missile flight. Launch transients in velocity and attitude impose stringent dynamic requirements on the seeker range and angle tracking control loops. The seeker may require a reacquisition capability to recover in the event of a loss of target track. The

Fig. 14.2 Developmental dual-mode MMW/IR seeker with radome removed (photo courtesy of General Dynamics, Valley Systems
Division).

LOBL provides shorter stand-off range because it is limited by the maximum seeker tracking range. Operator participation enhances target discrimination capability, but reduces the rate of fire.

14.3.2 Lock-On After Launch (LOAL)

For the lock-on after launch (LOAL) approach, the missile flight
includes both a midcourse phase and terminal guidance phase. The expendable hardware cost is greater for LOAL than for LOBL due to the
midcourse guidance requirements. The LOAL extends the platform stand-
off range to enhance survivability.

Midcourse Phase

During the midcourse phase, the missile flies to the target area without seeker guidance input. The missile may follow a simple ballistic trajectory during the midcourse phase, or the autopilot may utilize the on-
board inertial sensors to navigate toward the cued target position or to
achieve a desired dispersion for multiple missiles in an area attack. The

missile could alternatively be guided from the launch platform or other location by using a spatially encoded beam in a beam-rider approach (McMillan, *et al.* [30], Wallace [53]), or with a command guidance data link.

Terminal Guidance Phase

At the end of the midcourse phase, the seeker begins the target area search. After autonomous target detection, discrimination, possible classification and ranking of priority targets, and acquisition, the seeker initiates target track and provides guidance to the missile autopilot.

14.3.3 Guidance Techniques

Candidate guidance techniques include proportional navigation, pursuit navigation, and optimal control. Most guidance and autopilot control loops use proportional navigation (Goedeke [17]).

Proportional Navigation

Proportional navigation requires the missile autopilot to hold the target line-of-sight (LOS) angle constant. If the seeker employs a type 1 antenna servo loop, then the loop error signal is proportional to the target LOS rate. Proportional navigation is sensitive to errors in the target position error signal slope, but not to absolute position errors such as LOS bias. The seeker provides azimuth and elevation gimbal errors to the autopilot. This requires maintaining the angle track control-loop bandwidth constant, even as range and target RCS change.

Pursuit Navigation

In pursuit navigation, the missile autopilot attempts to align the missile velocity vector with the target location. Pursuit navigation is more appropriate for stationary targets. Pursuit navigation will be sensitive to measurement errors in the seeker-to-target angle.

Optimal Control

Optimal control techniques apply modern control theory to missile guidance. These techniques exploit the knowledge of expected target and missile dynamics to improve performance. Optimal control techniques can anticipate target glint and minimize the resultant tracking degradation. Range information may also be utilized by the autopilot to shape the terminal trajectory for best warhead effectiveness.

14.4   SEEKER SUBSYSTEMS

Figure 14.3 is a simplified seeker block diagram, which shows the major seeker subsystems. This block diagram is similar to that for any radar system, except that there is no display. No switches or indicators are needed because there is no operator. The only seeker outputs are guidance data to the missile autopilot. The seeker will also include a compact, high-efficiency power supply to condition the prime power for the seeker electronics. Figure 14.4 is a photograph of a developmental MMW seeker. Figure 14.5 shows the front end, including the transmitter, mixer preamplifier, and antenna for this developmental seeker.

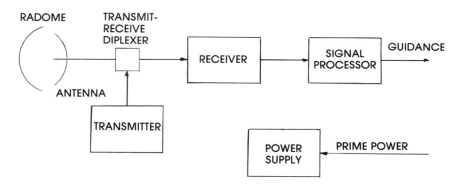

Fig. 14.3 Simplified seeker block diagram.

Fig. 14.4 Developmental MMW seeker (photo courtesy of Hercules Defense Electronics Systems).

Fig. 14.5 Developmental MMW seeker front end (photo courtesy of Hercules Defense Electronics Systems).

14.4.1 Transmitter

The transmitter subsystem provides the transmitted waveform at the desired power level to the antenna [7, 13, 54, 58, 59]. This subsystem will include the pulse modulator for a pulsed waveform, or the frequency-sweep linearizer for an FMCW waveform. Thermal stabilization electronics and frequency control are also typically included. Chapters 8 and 9 cover transmitter subsystems in more detail.

14.4.2 Receiver

The receiver subsystem accomplishes the down-conversion of the RF return signals to IF and to video with the required filtering and amplification. The transmit-receive diplexer, which may be a circulator or a transmit-receive (T/R) switch, isolates the receiver from the high-power transmission signal. The mixer preamplifier mixes the RF return signals with the LO to produce the IF signals. The mixer preamplifier is typically located adjacent to the diplexer to minimize losses for the best system noise figure. Highly integrated transmitter-receivers, or *transceivers,* offer good performance along with small size and low mass, which are especially important for applications where the RF components are gimbal mounted [9, 36, 47]. Figure 14.6 shows an integrated MMW transceiver. Chapter 10 presents the receiver discussion.

Fig. 14.6 Developmental MMW integrated transceiver (photo courtesy of
Hercules Defense Electronics Systems).

14.4.3 Antenna

The antenna subsystem is very important. Both the electrical and
mechanical characteristics of the MMW antenna are critical to performance. Key electrical characteristics include antenna power gain, beamwidth, sidelobe levels, and whether the antenna employs conical scan or
monopulse techniques for generating the track error signal. The antenna
gimbals, motors, linkage, backlash, and moment of inertia are important
mechanical characteristics. The antenna is a major contributor to important
seeker features, such as target-to-clutter ratio, target-to-noise ratio, search
scan rate, and prime power consumption. Chapter 11 is devoted to antenna
subsystems.

14.4.4 Radome

The radome must satisfy electrical, mechanical, and aerodynamic
requirements. For example, a radome with hemispherical shape and half-
wavelength thickness will offer good electrical performance, but the aerodynamic drag may be too high. Electrically, the radome introduces loss,
boresight errors, and reflections. Low reflection and VSWR improve transmit-receive isolation, which is particularly important for FM CW waveforms. The radome will also contribute boresight error and boresight slope
error, which degrade tracking performance. Aerodynamically, the radome

protects the antenna from aerodynamic loading and weather effects, and provides an aerodynamic fairing for the front of the missile to reduce drag. Radomes are covered in greater detail in Chapter 12.

14.4.5 Signal Processor

The MMW seeker signal processor subsystem accomplishes the mode sequence, does the calculations required for control loops, and performs the operations needed to extract target information from the seeker signals. Because these functions must be done in near-real-time and involve computations of at least moderate complexity, the signal processor is typically a relatively powerful digital computer (Cowan [11]). The seeker control loops, including antenna servo loops, range and angle tracking control loops, and automatic gain control (AGC) loops include relatively low-bandwidth digital filters. The target detection and discrimination signal processing may include matrix operations and Fourier transforms as part of algorithms for selecting the target in ground clutter. Increasingly complex seeker signal processing techniques are being developed, including artificial intelligence (AI) concepts. A sophisticated seeker signal processor may accept environmental and target information from the target acquisition system, and then, based on its assessment of the situation including clutter, weather, and ECM conditions, adjust its algorithms and parameters for best performance in that situation. Chapter 6 provides additional description of MMW radar signal processing techniques.

14.5 SEEKER MODE SEQUENCE

Figure 14.7 shows the mode sequence for a typical seeker. The seeker requires a full set of modes, comparable with that for an aircraft fire control radar plus automatic sequence and recovery. Note that the major phases include initialization, range track, target acquisition, target track and fuzing, and warhead detonation.

14.5.1 Initialization

The seeker initially performs a built-in confidence test and begins thermal stabilization of the transmit source. The seeker may accept mission data as well as guidance and control input from the autopilot. The seeker may also measure range to the ground in this phase to aid missile stabilization and navigation.

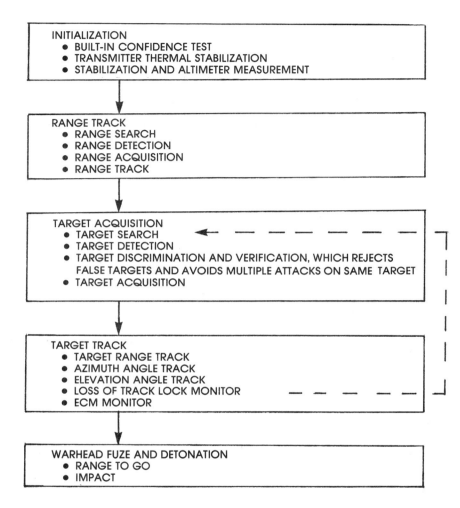

Fig. 14.7 Typical seeker mode sequence.

14.5.2 Range Track

Prior to the target-acquisition phase the seeker must position the range bins at the slant range corresponding to the antenna-beam ground interception area. This requires range search, detection, acquisition, and track. The ground clutter-to-noise ratio is the important factor in establishing this range track. Back and forth azimuth scans with an azimuthal outer gimbal and elevation inner gimbal will result in arcs on level ground at constant slant range. In this case, the range track control loop will only have to respond to terrain elevation variation. Azimuth scans with an azimuthal inner gimbal will yield straight lines on level terrain with slant range variation, which must be accommodated by the range track loop.

14.5.3 Target Acquisition

As shown in Figure 14.7, the target acquisition phase includes target search, detection, discrimination, and acquisition. The seeker target detection and discrimination techniques may include target return power, target Doppler frequency shift for MTI, target range or height extent, target angular extent, scan-to-scan correlation, polarization characteristics, and spatial distribution of apparent target returns. The seeker will attempt to reject false targets and to avoid multiple attacks on the same target. Target acquisition, which is the transition to target track, requires the antenna to stop searching and slew back to the target angle to begin target range and angle track.

14.5.4 Target Track

During the target track phase, the seeker provides guidance data to the missile autopilot. For proportional guidance and type 1 angle track control loops, the angle error signal, which is proportional to the LOS rate, is sent to the autopilot.

Care must be exercised in design of not only the azimuth and elevation angle tracking control loops, but also of the range track and AGC loops to maintain the angle loop gain constant so that the angle error signals, and consequently the LOS rate signals, maintain the proper scale factor. The changing range, target, and clutter reflectivity variations make this difficult. During target track, the seeker will employ centroid guidance techniques to minimize effects of glint and scintillation to guide the missile to the optimal approach.

The missile needs guidance data for a long enough time to permit manuevering to the target. Typically, guidance data acquired after the missile is closer than 100 m to the target is of little value, except to initiate warhead detonation.

The automatic seeker mode sequencer must also provide for coasting during short interruptions in track, in addition to recovery logic for loss of track lock. Note that if the seeker loses target track lock, it must return to target search to reacquire the target or acquire another. The mode sequencer will also monitor the ECM environment and modify the seeker operation to minimize ECM degradation (Johnston [25]).

14.5.5 Warhead Fuze and Detonate

The final steps prior to impact are for the warhead to fuze and to detonate. The seeker may be used to initiate detonation at the optimum range for the expected impact angle. A shaped warhead may be triggered by the radome contact with the target. The seeker may also be used to provide a proximity signal, such as "10 m to go," based on target range and missile velocity estimates to the safe and arming mechanism.

14.6 SEEKER SYSTEM TRADES

This section includes preliminary performance projections for a 94 GHz pulsed waveform seeker, a 35 GHz pulsed waveform seeker, and a 94 GHz FM CW waveform seeker system.

14.6.1 Constraints

Millimeter-wave seeker design is challenging because of the many constraints which bound the design. Table 14.4 lists some of the major constraints. Cost is most likely the major factor because the seeker is consumed in use. The antenna aperture diameter is bounded by the missile diameter because an oversized missile nose presents aerodynamic problems. The missile size also constrains the volume available for the seeker front end and the electronics. The dense packaging requirements aggravate the cooling problems. Packaging is further complicated by the requirement to distribute the seeker's mass to satisfy the overall missile's weight and balance requirements. A shaped charge warhead may also mean that the radome, antenna, and other RF components ahead of a shaped charge warhead may have to be frangible to maintain warhead effectiveness. Prime power, typically provided by a thermal battery, is often limited. The seeker must be very reliable, even after a long shelf life. The frequency allocations also constrain the MMW frequency selection (Johnston [26]). The environmental conditions experienced in both storage and operation must be considerable in all aspects of the design [24, 37, 38, 41, 46, 48, 51].

14.6.2 Illustrative Example

This section includes an illustrative seeker design example for an air-to-ground antiarmor seeker. Table 14.5 lists a set of typical seeker re-

Table 14.4 MMW Seeker Design Constraints.

Cost
Aperture Diameter
Volume
Cooling
Physical Mass and Distribution of Mass
Frangibility
Prime Power
Reliability
Frequency Allocations
Environmental Conditions

Table 14.5 Illustrative Seeker Requirements.

Parameter	Value	Units
Antenna Aperture Diameter	150	mm
Detection Range	1	km
Target RCS	20	m^2
Missile Velocity	200	m/s
Inclement Weather Requirement		
(maximum rain rate)	4	mm/hr
Missile Altitude	340	m
Depression Angle	20	degrees
Cross-Range Search Swath Requirement	350	m
Minimum Range Requirement	100	m
Maximum Target Velocity	15	m/s
Antenna Aperture Efficiency	0.5	—

quirements for this example. These requirements are representative and reasonable, but they do not apply to any specific mission. This example is based on catalog component capabilities, which will be lower in cost, more producible, and of lower risk than developmental components.

14.6.3 Baseline 94 GHz Pulsed Waveform Seeker Design

Based on the requirements in Table 14.5, a baseline seeker design is envisioned to operate at 94 GHz by using a pulsed waveform.

Peak Target Return Power

Table 14.6 shows the calculation of peak target return power of -72.7 dBm, referenced to the antenna port for the baseline seeker.

Table 14.6   Target Return Power Calculation for 94 GHz Pulse Seeker.

$$P_r = \frac{PG^2\lambda^2\sigma}{(4\pi)^3R^4\alpha} \tag{14.1}$$

P = Transmitting Power = 5000 mW =	$+$ 37.0 dBm
G^2 = (Antenna Power Gain)^2 = $(10{,}901)^2$	$+$ 80.8 dBi
λ^2 = (Wavelength)^2 = $(3.2$ mm$)^2$	$-$ 49.9 dBsm
σ = Target RCS = 20 m^2 =	$+$ 13.0 dBsm
α = Atmospheric Attenuation =	$(-)$ $+$ 0.6 dB
0.3 dB/km $\times$ 1.0 km $\times$ 2 =	
$(4\pi)^3$ = 1984 =	$(-)$ $+$ 33.0 dB
R^4 = (Slant Range)^4 = $(1000$ m$)^4$ =	$(-)$ $+120.0$ dBm^4
P_r = Target Return Power =	$-$ 72.7 dBm

Transmitting Power

The transmitting power is based on a catalogued 94 GHz IMPATT oscillator with 5 W peak power and 50 ns minimum pulse length [20]. The 150 mm antenna aperture diameter provides a 0.0177 m^2 antenna aperture area. The antenna power gain, G, is given by

$$G = \frac{4\pi\eta Af^2}{c^2} \tag{14.2}$$

where

   η = antenna aperture efficiency,
   A = antenna aperture area,
   f = RF,
   c = propagation velocity = 3.0×10^8 m/s (Barton [6]).

For an antenna aperture efficiency of 0.5, the 150 mm aperture yields a gain of 10,901 or 40.4 dBi at 94 GHz. The 20 m^2 target RCS used here is a conservatively low average value for armored vehicle targets. The one-way atmospheric attenuation at 94 GHz for clear weather is 0.3 dB/km.

In the inclement weather condition of 4 mm/hr rainfall rate, the one-way atmospheric attenuation increases to 4.5 dB/km, which reduces the target return power to −81.1 dBm. Chapter 4 provides a more complete discussion of MMW atmospheric attenuation.

Noise Power

Table 14.7 shows the receiver noise power calculation. The 20 MHz IF or predetection bandwidth, is matched to the 50 ns pulse length, and the 8.0 dB single-sideband noise figure is representative for 94 GHz receivers. The noise figure contribution of the antenna to the total system noise figure is negligible, and so ignored here. The −93.0 dBm noise power yields a +20.3 dB target-to-noise power ratio in clear weather, and a +11.9 dB target-to-noise ratio under 4 mm/hr rainfall rate conditions.

Table 14.7 Noise Power Calculation for 94 GHz Pulse Seeker (Assumes Receiver Terminated at 290 K).

$$N = kTBF \qquad (14.3)$$

k = Boltzmann's Constant = 1.38×10^{-20} mW/Hz/K =	− 198.6 dBm/Hz/K
T = Reference Temperature = 290 K =	+ 24.6 dBK
B = Receiver Equivalent Noise Bandwidth = 20 MHz =	+ 73.0 dBHz
F = Receiver Noise Figure = 6.3 =	+ 8.0 dB
N = Noise Power Referenced to the Antenna Port =	− 93.0 dBm

Ground Clutter RCS

The RCS of the ground or terrain clutter is projected in Table 14.8. The 6.0 m² clutter RCS means that the target-to-clutter ratio is approximately +5.2 dB. The peak power of the clutter return referenced to the antenna port will be approximately −77.9 dBm in clear weather, and −86.3 dBm in 4 mm/hr rainfall rate. (The resolution cell area dependence on depression angle is not included in this top level calculation).

Table 14.8   Clutter RCS Calculation for 94 GHz Pulse Seeker.

$$\text{Clutter RCS} = \sigma°(\pi/4)\Theta R\Delta R/\sqrt{2} \tag{14.4}$$

$\sigma°$ = Nominal Clutter Reflectivity = 0.05 =	− 13.0 dB
$\pi/4 = 0.785$ =	− 1.0 dB
Θ = Antenna Beamwidth = 1.6° = 28 mr =	− 15.5 dB
R = Slant Range = 1000 m =	+ 30.0 dBmeter
ΔR = Range Resolution = 7.5 m	+ 8.8 dBmeter
$\sqrt{2} = 1.414$ =	(−)+ 1.5 dB
Clutter RCS = 6.0 m^2 =	+ 7.8 dBsm

The 0.05 or − 13.0 dB clutter reflectivity is representative for moderate ground clutter, such as cultivated fields at 94 GHz, for moderate depression angles. Chapter 5 presents a more complete discussion of clutter reflectivity. The 40.4 dBi antenna power gain at 94 GHz corresponds to a one-way half-power antenna beamwidth of approximately 1.6°. The $\pi/4$ factor is needed because the resolution cell area is an ellipse. Division by $\sqrt{2}$ reduces the one-way antenna beamwidth to approximately the two-way beamwidth, which determines the azimuthal extent of the resolution cell. The 7.5 m range resolution, ΔR, results from the following expression:

$$R = c\tau/2 \tag{14.5}$$

where

c = propagation velocity = 3.0×10^8 m/s,
τ = pulse length = 50 ns.

Rain RCS

The RCS for volumetric rain with a 4 mm/hr rainfall rate is calculated in Table 14.9. Rain RCS is calculated as the product of the rain reflectivity, which is normalized by volume, and the resolution cell volume. The resolution cell volume is a right circular cylinder with diameter equal to the two-way antenna beamwidth at the range of interest and height equal to the range resolution. At the range where the antenna LOS intercepts the ground, half of this cylinder is below the ground and does not contain rain. Figure 14.8 shows a side view of the resolution volume.

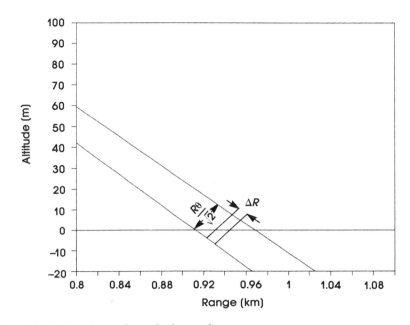

Fig. 14.8 Side view of resolution volume.

Table 14.9 Rain RCS Calculation for 94 GHz Pulse Seeker.

$$\text{Rain RCS} = \eta(1/2)(1/2)/(\pi/4)\Theta^2 R^2 \Delta R \qquad (14.6)$$

η = Volume Reflectivity of Rain = 7.0 × 10^{-5} m²/m³ =	−41.5 dBsm/m³
1/2 = Allowance for Half of Resolution Cell Volume = 0.5 =	− 3.0 dB
1/2 = Allowance for Two-Way Antenna Beamwidth = $(1/\sqrt{2})^2$ = 0.5 =	− 3.0 dB
$\pi/4$ = 0.785 =	− 1.0 dB
Θ^2 = (Antenna Beamwidth)² = (1.6°)² = (28 mr)² =	−31.1 dB
R^2 = (Slant Range)² = (1000 m)² =	+60.0 dBsm
ΔR = Range Resolution = 7.5 m =	+ 8.8 dBmeter
Rain RCS = 0.08 m² =	−10.8 dBsm

This calculation indicates that the rain RCS is approximately 23.8 dB less than the target RCS.

94 GHz Pulse Seeker Assessment

Table 14.10 lists the target to clutter ratio, the target-to-noise ratio and the target-to-rain ratio plus the peak target, clutter, rain, and noise power levels for both clear and 4 mm/hr rainfall rate weather conditions. No transmitting or receiving losses are considered in this top level analysis. Figure 14.9 shows return power as a function of slant range for target, clutter, and noise for clear weather. Figure 14.10 shows power as a function of range for target, clutter, rain, and noise under 4 mm/hr rainfall rate conditions.

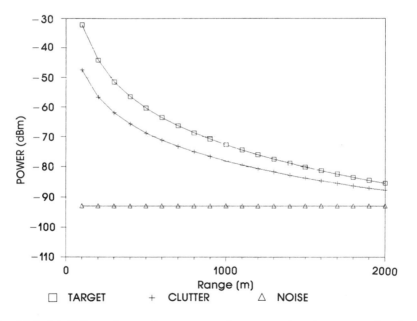

Fig. 14.9 94 GHz pulse seeker target, clutter, and noise power in clear weather as a function of slant range.

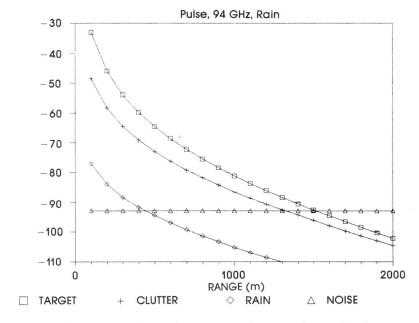

Fig. 14.10 94 GHz pulse seeker target, clutter, rain, and noise power in rain as a function of slant range.

Table 14.10 94 GHz Pulse Seeker Summary.

	Clear	*4 mm/hr Rain*
Target-to-Clutter Ratio	+ 5.2 dB	+ 5.2 dB
Target-to-Noise Ratio	+20.3 dB	+ 11.9 dB
Target-to-Rain Ratio	—	+ 23.8 dB
Target Return Power	−72.7 dBm	− 81.1 dBm
Clutter Return Power	−77.9 dBm	− 86.3 dBm
Rain Return Power	—	−104.9 dBm
Noise Power	−93.0 dBm	− 93.0 dBm

The target-to-noise ratios of 20.3 dB and 11.9 dB for clear and rain conditions, respectively, are sufficient for most detection and tracking, while the 5.2 dB target-to-clutter ratio is marginal and will require significant postdetection processing. The 23.8 dB target-to-rain ratio means that the seeker performance is limited by the ground clutter, even under the rain condition. For example, a single-scan SNR of 10 dB yields a probability of detection of 0.5 and a probability of false alarm of 0.0001 for a Swerling case 1 target (Skolnik [49]). These are predetection power levels and power ratios. Postdetection integration and filtering will further improve the ratios of target to interference power. Because the power spectral densities for noise, ground clutter, and rain are different, the effect of integration and other signal processing techniques will result differently for each. For example, the nonfluctuating component of clutter will not be reduced by averaging. The rain return power can be decreased by employing a lower RF and by improving the range resolution. There is little that can be done to improve the azimuthal resolution without increasing the aperture.

Pulse Repetition Frequency

A PRF of 100 kHz, which is the maximum capability of the IMPATT source [20], yields a pulse repetition interval (PRI) of 10 μs. The corresponding unambiguous range is 1.5 km.

Unambiguous Velocity

The 100 kHz PRF also provides unambiguous Doppler frequencies from −50 kHz to 50 kHz, which corresponds to relative velocities of −80 m/s to +80 m/s (±290 km/hr or ±180 mi/hr). The center frequency of the scatterers in the antenna-beam ground interception area will be shifted up by approximately 125 kHz due to the 200 m/s missile velocity. By shifting the Doppler frequency of the main beam clutter down to zero hertz, the full ±80 m/s unambiguous velocity region is available for MTI and other target detection or discrimination signal processing.

Duty Factor

The 100 kHz PRF together with the 50 ns pulse length yield a 0.005 duty factor. The average transmitting power is the product of this duty factor and the 5 W peak power, which yield 25 mW average power. For matched-filter range bin bandwidth, the average target-to-noise ratio is approximately 23 dB less than the peak target-to-noise ratio.

Scan Geometry

Millimeter-wave seekers typically scan back and forth as the missile flies down range to cover the search area. Figure 14.11 shows a side view of the antenna beam. Note the 340 m altitude, 20° depression angle relative to horizontal, and 1000 m slant range along the antenna LOS for level terrain. The slant-range swath for the 1.1° (20 mr), two-way, antenna-beam ground interception area is 54 m. The seeker will require eight range bins with 7.5 m slant-range resolution to cover this swath. This 54 m slant-range swath corresponds to a 58 m ground-range swath at the 20° depression angle. To maintain full ground coverage, the antenna search scan period must be shorter than the time required for the missile to move 58 m forward. At the 200 m/s velocity, this is 0.29 s, which corresponds to a 3.45 Hz scan frequency. The 350 m or ±175 m cross-range swath requirement corresponds to a ±9.9° search scan amplitude. This ±9.9° amplitude at 3.45 Hz corresponds to a 137°/s average scan rate. Because the antenna scan must decelerate and accelerate at the scan extremes, the scan rate must be somewhat higher between extremes.

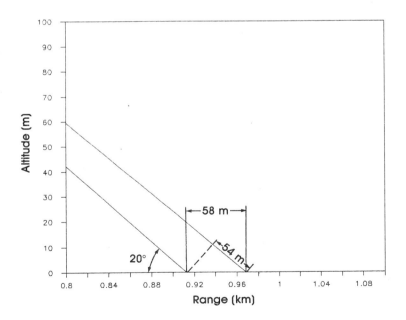

Fig. 14.11 Side view of 94 GHz antenna-beam ground interception.

Dwell Time

If we assume the average 137°/s scan rate and a 1.1° two-way antenna beamwidth, then the resulting target dwell time is 8.0 ms so that the video bandwidth must be at least 125 Hz. Thus, the maximum postdetection integration gain is approximately 26 dB. Wider video bandwidths to accommodate shorter dwell times due to higher scan rates, or for other reasons, will reduce the postdetection integration gain. The effective gain of video filtering will be less for constant or slowly fluctuating interference components.

14.6.4 Frequency

Table 14.11 lists some of the major frequency trade-off issues in terms of reasons to increase or to decrease frequency. Increased frequency yields narrower antenna beamwidth, higher antenna power gain, and greater Doppler frequency-shift sensitivity. A narrower antenna beamwidth will reduce the ground and volumetric clutter, and improve the angular resolution. For very high velocity targets, this higher Doppler sensitivity may be a disadvantage due to Doppler ambiguities. Lower frequency results in lower atmospheric attenuation and lower rain reflectivity. Higher source power and lower noise figure receivers are also available at lower frequencies. In addition, the greater tolerances for lower frequencies help to lower costs and increase producibility.

Table 14.11   Frequency Trade-Off Issues.

Reasons to Increase Frequency

Narrower Antenna Beamwidth
Higher Antenna Power Gain
Greater Doppler Sensitivity
Smaller Volume Components

Reasons to Decrease Frequency

Lower Atmospheric Attenuation
Lower Clutter Reflectivity
Lower Rain Reflectivity
Lower Receiver Noise Figure
Greater Transmitting Power Available
Greater Component Tolerances
Lower Cost Components

Target Return Power

This section considers a 35 GHz version of the 94 GHz baseline seeker.

Table 14.12 shows that, for a 35 GHz pulse seeker, the target return power is 5.0 dB lower at -77.9 dBm than for the 94 GHz system. The 10 W peak power used here is based on current catalog component specifications. Increases in peak power at 35 GHz relative to 94 GHz greater than the 3 dB assumed here are feasible. The 3.0 dB increase in available transmitting power, 8.6 dB decrease in antenna gain, 4.3 dB increase in wavelength, and 0.25 dB/km decrease in clear weather atmospheric attenuation yield 5.1 dB lower power relative to the antenna port than for the 94 GHz seeker. The target RCS is assumed to remain 20 m^2 at the lower frequency. For a 4 mm/hr rainfall rate the target return power for the 35 GHz seeker will decrease to -79.7 dBm due to the 1.0 dB/km one-way attenuation at 35 GHz. This is 1.4 dB greater power than for the 94 GHz seeker under rain conditions.

Table 14.12 Target Return Power Calculation for 35 GHz Pulse Seeker.

$$P_r = \frac{PG^2\lambda^2\sigma}{(4\pi)^3 R^4 \alpha} \qquad (14.7)$$

P = Peak Transmitting Power = 10000 mW =	+40.0 dBm
G^2 = (Antenna Power Gain)^2 = $(1511)^2$ =	+63.6 dBi
λ^2 = (Wavelength)^2 = $(8.6 \text{ mm})^2$ =	−41.3 dBsm
σ = Target RCS = 20 m^2 =	+13.0 dBsm
α = Atmospheric Attenuation = 0.05 dB/km	dB
$\times$	−0.1
1.0km × 2 =	
$(4\pi)^3$ = 1984 =	(−) +33.0 dB
R^4 = (Slant Range)^4 = $(1000 \text{ m})^4$	(−) +120.0 dBm^4
P_r = Target Return Power =	−77.8 dBm

Noise Power

The noise power will be approximately 1.0 dB lower because of the lower noise figure mixer preamplifiers, which are available at 35 GHz.

Clutter RCS

The clutter power will be lower at 35 GHz because, while the wider antenna beamwidth of 4.4° will increase the resolution cell area by approximately 4.4 dB, the normalized clutter reflectivity is approximately 7.0 dB lower at 35 GHz. Table 14.13 shows the clutter RCS calculation.

Table 14.13   Clutter RCS Calculation for 35 GHz Pulse Seeker.

$$\text{Clutter RCS} = \sigma_0 \, (\pi/4) \, \Theta R \Delta R / \sqrt{2} \qquad (14.8)$$

σ^0 = Nominal Clutter Reflectivity = 0.01 =	−20.0 dB
$\pi/4$ = 0.785 =	−1.0 dB
Θ = Antenna Beamwidth = 4.4° = 76 mr =	−11.2 dB
R = Slant Range = 1000 m =	+30.0 dBmeter
ΔR = Range Resolution = 7.5 m =	+8.8 dBmeter
$\sqrt{2}$ = 1.414 =	$(-) + 1.5$ dB
Clutter RCS = 3.2 m^2 =	+5.1 dBsm

Rain RCS

The rain return power will be slightly higher for the 35 GHz seeker. As described above, the resolution cell area will increase by 4.4 dB, and the rain reflectivity decreases by 8.5 dB to −50 dBsm/m^3. Table 14.14 shows the 35 GHz rain RCS calculation.

35 GHz Pulse Seeker Assessment

Table 14.15 lists the target-to-clutter ratio, the target-to-noise ratio, and the target-to-rain ratio plus the peak target, clutter, rain, and noise power levels for the 35 GHz pulse seeker. Figure 14.12 and Figure 14.13 show the power as a function of range for clear weather and 4 mm/hr rainfall rate conditions, respectively.

Table 14.14   Rain RCS Calculation for 35 GHz Pulse Seeker.

$$\text{Rain RCS} = \eta \, (1/2) \, (1/2) \, (\pi/4) \, \Theta^2 R^2 \Delta R \qquad (14.9)$$

η = Volume Reflectivity of Rain = 1.0×10^{-5} m²/m³ =	−50.0 dBsm/m³
1/2 = Allowance for Half of Resolution Cell Volume = 0.5 =	−3.0 dB
1/2 = Allowance for Two-Way Antenna Beamwidth = 0.5 =	−3.0 dB
$\pi/4$ = 0.785	−1.0 dB
Θ^2 = (Antenna Beamwidth)² = (4.4°)² = (76 mr²) =	−22.3 dB
R^2 = (Slant Range)² = (1000 m)² =	+60.0 dBsm
ΔR = Range Resolution = 7.5 m =	+8.8 dBmeter

Rain RCS = 0.09 m² =	−10.5 dBsm

Table 14.15   35 GHz Pulse Seeker Summary.

	Clear	4 mm/hr Rain
Target-to-Clutter Ratio	+8.0 dB	+8.0 dB
Target-to-Noise Ratio	+16.2 dB	+14.3 dB
Target-to-Rain Ratio	—	+23.5 dB
Target Return Power	−77.8 dBm	−79.7 dBm
Clutter Return Power	−85.8 dBm	−87.7 dBm
Rain Return Power	—	−103.2 dBm
Noise Power	−94.0 dBm	−94.0 dBm

Scan Geometry

Figure 14.14 shows a side view of the seeker-to-ground geometry for the wider beamwidth at 35 GHz. The altitude, depression angle, and slant range are the same as for the 94 GHz system. The 3.1° (54 mr) two-way antenna beamwidth yields an antenna-beam ground interception area with 148 m slant-range extent. Note that 20 range bins with 7.5 m resolution

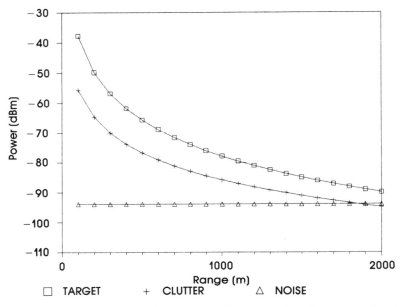

Fig. 14.12 35 GHz pulse seeker target, clutter, and noise power in clear weather as a function of slant range.

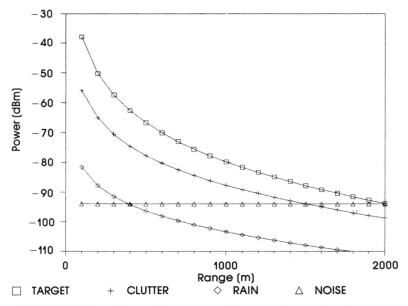

Fig. 14.13 35 GHz pulse seeker target, clutter, rain, and noise power in rain as a function of slant range.

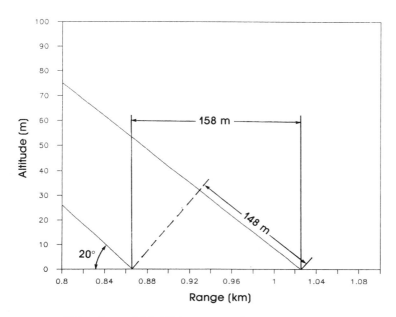

Fig. 14.14 Side view of 35 GHz antenna-beam ground interception.

are required to cover this slant-range extent. The corresponding ground-range extent is 158 m, so the maximum antenna scan period is 0.79s for 1.27 Hz minimum scan frequency. The $\pm 9.9°$ scan amplitude yields a 50°/s average scan rate, which is significantly slower than the 137°/s required for the 94 GHz seeker.

Dwell Time

If we assume the average 50°/s scan rate and a 3.1° two-way antenna beamwidth, the resulting target dwell time is 62 ms so that the video bandwidth must be at least 16 Hz. Thus, the maximum postdetection integration gain is approximately 30.5 dB.

Increased Scan Rate

Because of the greater slant-range extent, this 35 GHz seeker is processing over twice as many range bins as the 94 GHz seeker. The signal processing requirement for the 35 GHz seeker can be reduced by decreasing the number of range bins that are processed. This will demand an increase in the scan rate, but doubling the scan rate to 100°/s is not unreasonable and would permit halving the number of range bins that are processed. Doubling the scan rate will halve the target dwell time to 31 ms, double the minimum video bandwidth to 32 Hz, and reduce the maximum postdetection integration gain to 30 dB. Alternatively, the seeker can continue to process the full slant-range swath and exploit the additional looks to each point in the search area. These additional looks will enable scan-to-scan correlation processing to help compensate for the lower target-to-clutter ratio for the 35 GHz seeker relative to the 94 GHz seeker.

14.6.5 Waveform

The baseline 94 GHz and 35 GHz seekers analyzed in the preceeding sections employed a pulsed waveform. This section presents the FMCW waveform, which is employed in some MMW seekers.

FMCW

Table 14.16 lists some of the major advantages and disadvantages of the FMCW waveform relative to a pulsed waveform. The FMCW waveform offers high average power and good short-range performance. This waveform demands careful design to maximize front-end isolation and to minimize front-end VSWR and reflection. A highly linear frequency sweep is required for high range resolution.

FMCW Equation

Figure 14.15 shows both the transmitting and receiving frequencies for a linear FMCW triangular waveform as a function of time. Note that by similar triangles, the ratio of the frequency deviation to the beat frequency is equal to the ratio of the sweep time to the round-trip propagation time. The FMCW equation may be written as

$$\Delta F/F_b = T/t = T/(2R/c) \tag{14.10}$$

where

ΔF = RF deviation,
F_b = beat frequency between the transmitting and receiving frequencies,
T = frequency sweep time,
t = round-trip propagation time,
R = slant range.

Table 14.16   Advantages and Disadvantages of FMCW Waveform
Relative to Pulsed Waveform.

Advantages of FMCW Relative to Pulsed Waveform

Low Peak Power, Greater Covertness
Small Size
Simple Mechanization
Single RF Source for Transmitter and LO
Narrowband IF Processing
Easier to Operate at Extremely Short Range
Less Likely to Have Range Ambiguity Problems

Disadvantages of FMCW Relative to Pulsed Waveform

Transmitter-Receiver Isolation
Front-End Reflections-VSWR
Requires Highly Linear Frequency Sweep
Range Resolution May Be Increasingly Coarse
    with Increasing Range
Requires Low FM/Phase Noise Source
Requires Waveform Weighting to Reduce Range Sidelobes
Saw-toothed FMCW and Noncoherent MTI—No Doppler
    Compensation
Triangular FMCW Requires Doppler Compensation
    based on Velocity Estimation

Note that as the slant range changes, one or more of the other parameters, including frequency deviation, sweep time, and beat frequency, will change.

FMCW Waveform Trade-offs

As indicated by Table 14.17, the FM CW waveform design is a trade-off between many competing factors.

Frequency Deviation

Increase of the frequency deviation will result in greater clutter smoothing, lower RF power spectral density, and higher theoretical range resolution capability. The wider bandwidth source will have lower quality factor (Q), which will reduce the output power and increase the phase noise. The wider bandwidth required for the other RF components will

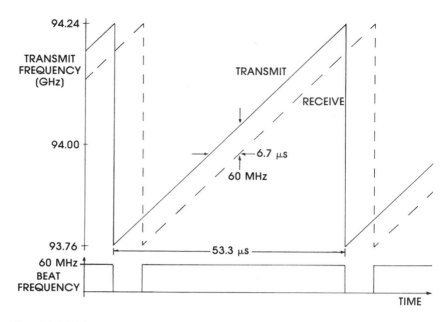

Fig. 14.15 Transmitted frequency and beat frequency as a function of time for linear FMCW.

Table 14.17 FMCW Waveform Trade-off Issues.

Issue	Reasons to Decrease	Reasons to Increase
Frequency Deviation (ΔF)	Higher Powered Source Available	Greater Clutter Smoothing
	Narrower Band RF Components	Lower RF Power Spectral Density
	Lower Source Phase Noise	
	Easier to Linearize	
Beat Frequency (F_b)	Narrower Bandwidth for IF Range Bin Filters—SNR	Lower Phase Noise Further from Carrier
$\Delta F / F_b$ Ratio		Longer Sweep Time More Time in Band

increase cost. The wider frequency deviation will also be more difficult to linearize.

For a linear FMCW waveform, the theoretical time resolution is equal to the inverse of the frequency deviation (Bartlett, Couch, and Johnson [5]). For example, 480 MHz frequency deviation provides potential time resolution of 2.1 ns, and corresponding range resolution of 0.3 m. Frequency deviation beyond the minimum required for a given range resolution is often used to provide clutter smoothing. Typically, the frequency deviation is held constant, and either the sweep time or the beat frequency is adjusted to accommodate range changes.

Range Resolution Requirements

The range resolution achieved for the FMCW waveform depends on satisfying transmission bandwidth, receiver frequency resolution, and frequency-sweep linearity requirements. Generally, the nonlinearities are the limiting factor. Receiver frequency resolution may be matched to the expected beat-frequency bandwidth for a target at the maximum range. Transmission bandwidth which exceeds that required for the achievable range resolution helps to smooth target and clutter returns.

Saw-Toothed Versus Triangular Waveform

The triangular waveform permits measuring Doppler frequency shift. This is a disadvantage because the seeker must compensate for platform motion. For the saw-toothed waveform, during the transit-time interval after the sweep recovery, the beat frequency is much greater and will be outside of the receiver bandwidth.

Beat Frequency

Higher beat frequency is favorable because the source phase noise decreases as the separation from the center frequency increases. To accommodate range changes, FMCW systems either can adjust the sweep time to maintain a constant beat frequency or have a variable beat frequency for a constant sweep time. For a constant-sweep-time system, the range resolution will be constant. For a constant-beat-frequency system, the range bin bandwidth is a fixed percentage of the nominal beat frequency, and therefore the bandwidth is a constant percentage of the nominal range. Thus, for a constant-beat-frequency or variable-sweep-time system, increasing the beat frequency means increasing the bandwidth and, consequently, the noise. Fractional bandwidth of the range bin filters will drive the selection of range bin filter technology (Solie and Wohlers [50]). Figure 14.16 illustrates how frequency-sweep nonlinearities produce a spread in beat frequency for a target.

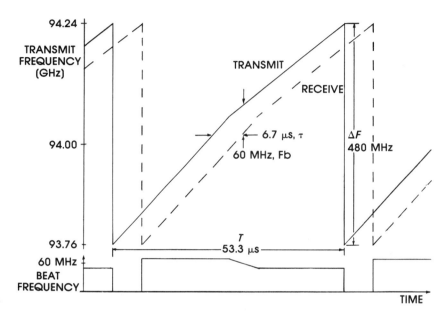

Fig. 14.16 Transmitted frequency and beat frequency for nonlinear frequency sweep.

Ratio of Frequency Deviation to Beat Frequency

Low frequency deviation and high beat frequency are attractive, but a high ratio of frequency deviation to beat frequency is also important. An increase of this ratio increases the frequency sweep time and decreases the modulation frequency, which make linearization easier. Higher frequency deviation to beat frequency ratios also decrease the percentage of the modulation period that the beat frequency is outside of the receiver bandwidth following frequency-sweep turnarounds.

Interrupted FMCW

The interrupted FMCW (IFMCW) waveform involves interrupting the transmission signal to improve isolation during the reception time. One IFMCW approach matches the transmit time to the round-trip propagation time, followed by an equal reception time, as shown in Figure 14.17. The 0.5 duty factor for the IFMCW waveform reduces the average transmitted power by approximately 3 dB relative to FMCW. The improved isolation may be expected to reduce the system noise by more than 3 dB, which improves the overall SNR. A high-speed modulation switch is required for IFMCW operation at short range. For most efficient operation of the IFMCW waveform, the ratio of frequency deviation to beat frequency must be an even integer. The minimum typical ratio is 6, which has out-of-band loss of 0.8 dB.

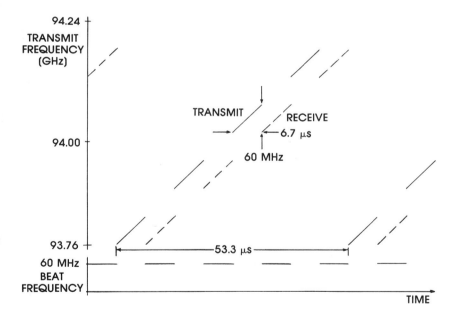

Fig. 14.17 IFMCW waveform.

Candidate FMCW Waveform

A candidate FMCW seeker might have a 480 MHz frequency deviation and 60 MHz beat frequency. This 480 MHz frequency deviation is only 0.5% of the 94 GHz center frequency, and it provides a $\Delta F / F_b$ ratio of 8 for a 0.6 dB out-of-band loss. The 60 MHz beat frequency is well separated from the carrier for low phase noise. At 1 km slant range, the required sweep time is 53.3 μs for a 18.75 kHz modulation frequency. For range bin filters at fixed frequency, the sweep time must decrease to 5.3 μs or 187.5 kHz modulation frequency at the 100 m minimum range. Alternatively, with fixed sweep time, the range bin filtering must shift down to 6 MHz to handle the 100 m minimum range requirement.

Range Resolution

The range resolution in FMCW is typically limited by the frequency-sweep linearity. Figure 14.16 shows a nonlinear sweep along with the resulting beat frequency. This illustrates how frequency-sweep nonlinearities will result in a distribution of beat frequencies for the target return. The distributed target return bandwidth will limit the range resolution. The bandwidth of beat frequencies for a target is proportional to the range resolution.

For a system that achieves RF sweep linearity of 300 kHz, which is 0.00032% (or 3.2 parts per million (ppm)) relative to the 94 GHz center frequency, and 0.06% (or 625 ppm) relative to the 480 MHz frequency deviation, and 0.5% (or 5000 ppm) relative to the 60 MHz nominal beat frequency, the range resolution will be 0.5% of range or 5 m at the 1000 m nominal search range. The thermal noise power in this 300 kHz range bin filter for an 8.0 dB receiver noise figure will be only -111.2 dBm. Front-end reflections and leakage will contribute additional noise, which typically exceeds the thermal noise component.

Target Return Power

Table 14.18 lists the target return power calculation for a 94 GHz FMCW seeker. The 25 mW average transmitting power is equal to the average power of the 5 W IMPATT oscillator with 0.005 duty factor. Gunn oscillators are often used in FMCW seekers because of their lower AM and FM noise. This yields a 15.5 dB target-to-noise ratio in clear weather, and 7.1 dB target-to-noise ratio in 4 mm/hr rainfall rate. The improved 5 m range resolution will improve the target-to-clutter and target-to-rain ratios by 1.8 dB relative to the 7.5 m range resolution.

Table 14.18 Target Return Power Calculation for 94 GHz FMCW
Seeker.

$$P_r = \frac{P\,G^2\lambda^2\,\sigma}{(4\pi)^3\,R^4\alpha} \qquad (14.11)$$

P = Transmitting power = 25 mW =	$+14.0$ dBm
G^2 = (Antenna Power Gain)^2 = $(10{,}901)^2$ =	$+80.8$ dBi
λ^2 = (Wavelength)^2 = $(3.2\text{ mm})^2$	-49.9 dBsm
σ = Target RCS = 20 m^2 =	$+13.0$ dBsm
α = Atmospheric Attenuation = 0.3 dB/km $\times$	
1.0 km $\times$ 2 =	$(-)+0.6$ dB
$(4\pi)^3$ = 1984 =	$(-)+33.0$ dB
R^4 = (Slant Range)^4 = $(1000\text{ m})^4$	$(-)+120.0$ dBm^4
P_r = Target Return Power =	-95.7 dBm

Table 14.19 summarizes the target-to-clutter ratio, target-to-noise
ratio and the target-to-rain ratio plus the average target, clutter, rain, and
noise power levels for both clear and 4 mm/hr rainfall rate weather conditions. Figure 14.18 shows the target, clutter, rain, and noise power as a
function of range for the 94 GHz FMCW seeker in clear weather. Figure
14.19 shows the corresponding values under 4 mm/hr rainfall rate conditions.

Table 14.19 94 GHz FMCW Seeker Summary

	Clear	*4mm/hr Rain*
Target-to-Clutter Ratio	$+7.0$ dB	$+7.0$ dB
Target-to-Noise Ratio	$+15.5$ dB	$+7.1$ dB
Target-to-Rain Ratio	—	$+25.6$ dB
Target Return Power	-95.7 dBm	-104.1 dBm
Clutter Return Power	-102.7 dBm	-111.1 dBm
Rain Return Power	—	-129.7 dBm
Noise Power	-111.2 dBm	-111.2 dBm

The FMCW is the waveform of choice for sensor-fuzed munitions
because of the short range requirement. The FMCW is less attractive for
applications requiring a complex radome, such as might be required for
stringent aerodynamic specifications, because of the higher expected reflection and VSWR. The IFMCW may be employed in these cases.

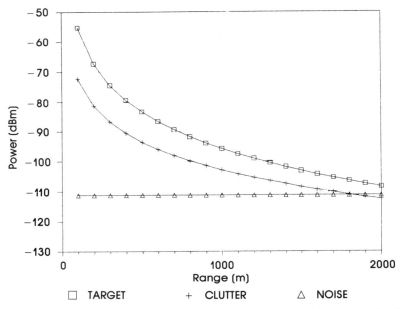

Fig. 14.18 94 GHz FMCW seeker target, clutter, and noise power in clear weather as a function of slant range.

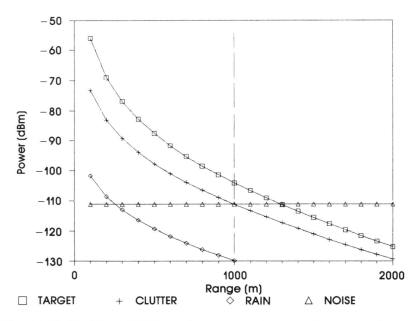

Fig. 14.19 94 GHz FMCW seeker target, clutter, rain, and noise power in rain as a function of slant range.

Scan Geometry

The scan geometry for the FMCW 94 GHz seeker is similar to that for the 94 GHz pulsed seeker. The finer range resolution of 5 m will require 11 range bins to cover the 54 m slant-range swath. The 8 ms dwell time and 125 Hz video bandwidth are unchanged. The narrower 300 kHz IF bandwidth reduces the maximum postdetection integration gain to approximately 17 dB.

14.6.6 Antenna

The antenna is a particularly critical component in a MMW seeker. Chapter 11 discusses antennas, but there are several trade-off issues of special interest in MMW seeker systems. Important seeker parameters, such as target-to-clutter ratio, target-to-noise ratio, and tracking performance, are directly dependent on the antenna. Target-to-clutter ratio is inversely proportional to antenna beamwidth because of the angular resolution dependence. Target-to-noise ratio is proportional to the square of the antenna power gain. Target tracking performance depends on both the antenna beamwidth and the angle sensing technique, such as conical scan or monopulse. The polarization of the antenna affects target and clutter reflectivity, and dual polarization capability is required for polarimetric target discrimination processing. Antenna sidelobe levels are important to ECM susceptibility. In addition, antenna shortcomings are difficult to compensate in other areas. For example, increasing transmitting power does nothing to improve low target-to-clutter ratio, since both target and clutter power increase equally, and for an FMCW seeker, the noise due to isolation and reflection will increase. Increasing range resolution will help compensate for low angular resolution, but requires wider receiver bandwidth.

The antenna rotational inertia about the gimbal axes is important to scan capability, target tracking rate capability, and motor torque requirements and, consequently, motor power requirements with implications for overall prime power or thermal battery demands. Minimizing the moment of inertia about gimbal axes is complicated by the fact that the transmitter source and receiver-mixer preamplifier often must be mounted on the gimbals to minimize losses.

14.6.7 Search Technique

For area attack, the seeker search area must be large enough for the desired overlap between adjacent submissiles. In the cued approach, the search area must cover the span within which the target may manuever during the missile flight as well as the missile dispersion and navigation errors. Of course, there is no point in searching an area larger than that within which the missile can manuever.

Increasing the search area also increases the number of false alarms. The circular or spiral search of sensor-fuzed munitions is appropriate for vertical trajectories. The antenna LOS is at a constant angle relative to vertical, so the search area covered decreases as the munition descends, which matches the area within which the warhead is effective. Table 14.20 lists the major seeker search trade-off issues for a horizontal trajectory.

Table 14.20 Seeker Search Area Trade-Off Issues.

	Reasons to Decrease	*Reasons to Increase*
Depression Angle	Lower Clutter Reflectivity	Narrower Slant-Range Swath
Slant Range	Better Target-to-Clutter Ratio Better Target-to-Noise Ratio	Greater Reaction Time Greater Area of Authority
Search Width	Lower Scan Rate Longer Dwell Time Fewer False Alarms	Greater Area of Authority
Search Frequency	Longer Dwell Time	Multiple Looks

ACKNOWLEDGEMENT

The author wishes to acknowledge the contributions of Mr. John J. Teti, Jr., Mr. Alan M. Gauzens, and Mr. Robert S. Roeder of Hercules Defense Electronics Systems, Inc., in Clearwater, Florida, and Mr. James R. Gallivan of General Dynamics Valley Systems Division, in Rancho Cucamonga, California, for chapter review and photographs. The author was previously associated with the millimeter-wave engineering group at Hercules Defense Electronics Systems (formerly Sperry Microwave Electronics).

REFERENCES

1. "Millimeter Wave Guidance Under Study," *Aviation Week and Space Technology,* July 16, 1979, pp. 59–65.
2. "Flight Tests Under Way on Wasp Missile Program," *Aviation Week and Space Technology,* April 19, 1982, p. 73.
3. "Industry Observer," *Aviation Week and Space Technology,* June 4, 1984, p. 13.
4. "Army/LTV Missile Intercepts Reentry Vehicle," *Aviation Week and Space Technology,* July 14, 1986.
5. M.C. Bartlett, L. W. Couch, and R.C. Johnson, "Ambiguity Function Measurement," Report HDL-CR-75-039-6, June 1975.
6. D.K. Barton, *Radar System Analysis,* Artech House, Dedham, MA, 1976.
7. F.J. Bernes, R.S. Ying, and M. Kaswen, "Solid State Oscillators Key to Millimeter Radar," *Microwave Systems News,* May 1979, Vol. 9, No. 5, pp. 79–86.
8. "The New Defense Posture Missiles, Missiles and Missiles," *Business Week,* August 11, 1980, pp. 76–81.
9. Apostle Cardiasmenos, "Design Considerations Examined for Millimeter Wave Defense Electronics Systems," *Microwave System Designer's Handbook,* 1986, 4th Ed., pp. 280–296 [MSN&CT, July 1986].
10. Robert Carroll, "Smart Weapons Changing Tactical Warfare," *Defense Electronics,* Vol. 13, No. 6, June 1981, pp. 37–40.
11. A.E. Cowan, "Brilliant Munitions Must Merge Hardware and Software," *Defense Electronics,* July 1983, pp. 60–70.
12. Patrick E. Crane, UBC, Inc., "Dual Mode Infrared-Millimeter Sensor Design Study," Contract No. DAAG29-76-D-0100, Battelle Columbus Laboratories, Columbus, OH, November 1981.
13. D.R. Decker, "Are MMICs a Fad or Fact?," *Microwave Systems News,* July 1983, pp. 84–92.
14. "Wasp: New Missile That Thinks for Itself," *Defense Electronics,* Vol. 12, No. 6, June 1980, pp. 44–45.
15. "Flight Tests Evaluate Antitank Assault Breaker," *Defense Electronics,* Vol. 13, No. 4, April 1981, pp. 41–43.
16. Albert N. DiSalvio, USAF Millimeter-Wave Seeker Developments, *International Defense Review,* January 1980, pp. 42–46.
17. R.C. Goedeke, "Characterization of RF Sensor/Seekers," *GACIAC SOAR 81-02,* December 1981.

18. Joseph D. Harrop, Ronald Stump and John J. Teti, Jr., "Surface Navy Applications of Millimeter Wave Sensors," Vol. 1, Design Trade-Off Study, Final Report NSWC/DL TR-3748, November 1977.

19. S.A. Hovanessian, "Detection of Non-Moving Targets by Airborne MM-Wave Radars," *Microwave Journal*, Vol. 29, No. 3, March 1986, pp. 159–166.

20. Hughes Aircraft Company, Microwave Products Division, Millimeter-Wave Products Catalog, 1986, p. 48.

21. *International Defense Review*, "Assault Breaker," January 1980, pp. 91–94.

22. *Jane's Defense Weekly*, "SADARM Moves Ahead," October 25, 1986, p. 938.

23. *Jane's Defense Weekly*, "Marconi Seeker to Upgrade Hellfire," November 1, 1986, p. 1017.

24. S.L. Johnston, ed., *Millimeter Wave Radar*, Artech House, Dedham, MA, 1980.

25. S.L. Johnston, "MM-Wave Radar: The New ECM/ECCM Frontier," *Microwave Journal*, Vol. 27, No. 5, May 1984, pp. 265–271.

26. S.L. Johnston, "Radar Frequency Management and the New MM-Wave Radar Operating Frequencies," *Microwave Journal*, Vol. 27, No. 12, December 1984, pp. 36–44.

27. T. Kennedy, "Advances in Smart Munitions," *Defense Science and Electronics*, Vol. 5, No. 10, October 1986, pp. 63–67.

28. P.J. Klass, "Air to Surface Weapons Sensors Advance," *Aviation Week and Space Technology*, March 9, 1981, pp. 199–202.

29. A.C. Levitan, "MM-Wave and Infrared," Military *Electronics/Countermeasures*," February 1982, pp. 100–103.

30. R.W. McMillan, R.G. Shackelford, and J.J. Gallagher, "Millimeter Wave Beamrider and Radar System," *Proceedings of the SPIE Technical Symposium 259*, Huntsville, AL, September 1980, pp. 166–171.

31. "The Front End," *Microwaves and RF*, May 1985, p. 31.

32. "Assault Breaker Stays on Target," *Microwave Systems News*, Vol. 10, No. 6, June 1980, p. 28.

33. "Wasp Seeker Enters Testing," *Microwave Systems News*, Vol. 11, No. 9, September 1981, p. 38.

34. "SDIO Seeks MM-Wave Radar from Italy," International Report, *Microwave Systems News and Communications Technology*, Vol. 16, No. 11, October 1986, p. 49.

35. W. Perry, "PGM's Are Crucial," *Defense Electronics*, Vol. 13, No. 6, June 1981, pp. 45–51.

36. R.A. Phaneuf, "A Sensible Approach to Realizing MMW Seeker Systems," *Microwave Journal,* Vol. 26, No. 6, June 1983, pp. 109–112.

37. E.K. Reedy and G.W. Ewell, "Millimeter Radar,"*Infrared and Millimeter Waves,* Chapter 2, *Millimeter Systems,* Vol. 4, K.J. Button and J.C. Wiltse, eds., Academic Press, New York, 1981, pp. 23–94.

38. D.D. Rhodes and R.S. Roeder, "Automatic Test Techniques—Next Generation Millimeter Wave Guidance Systems," *AUTOTESTCON '85 Symposium,* New York, 1985.

39. R.S. Roeder, "Millimeter Wave Technology Advances the Science of Force Multiplication," *Military Electronics/Countermeasures,* Vol. 8, No. 12, December 1982, pp. 24–26.

40. R.S. Roeder and A.H. Green, Jr., "Millimeter Technology for Air to Ground Guidance," *Military Microwaves,* 1984.

41. R.S. Roeder and A.H. Green, Jr., "MMW Seekers for Terminal Homing — Manufacturing Methods and Technology Delvelopment," *ADPA 1984 Symposium on Avionics,* October 30–31, 1984.

42. R.S. Roeder and J.J. Teti, Jr., "Millimeter Wave Passive and Active Sensors for Terrain Mapping, *SPIE Proceedings,* Vol. 791, May 22, 1987.

43. R.P. Ropelewski, "Wasp Antiarmor Tests Slated," *Aviation Week and Space Technology,* May 10, 1982, pp. 63–65.

44. J.B. Schultz, "Airland Battle 2000: The Force Multiplier," *Defense Electronics,* December 1983, pp. 48–67.

45. J.B. Schultz, "International Competition Heats Up For Millimeter Wave Munition," *Defense Electronics,* June 1984, pp. 100–110.

46. C.R. Seashore, "Missile Guidance," in *Infrared and Millimeter Waves,* Chapter 3, *Millimeter Systems,* Vol. 4, K.J. Button and J.C. Wiltse, eds., Academic Press, New York, 1981, pp. 95–150.

47. C.R. Seashore and D.R. Singh, "Millimeter-Wave ICs for Precision Guided Weapons," *Microwave Journal,* Vol. 26, No. 6, June 1983, pp. 51–65.

48. C.R. Seashore, "MM-Wave Sensors for Missile Guidance," *Microwave Journal,* Vol. 26, No. 9, September 1983, pp. 134–144.

49. M.I. Skolnik, *Introduction to Radar Systems,* 2nd Ed., McGraw-Hill, New York, 1980.

50. L.P. Solie and M.D. Wohlers, "Use of an SAW Multiplexer in FMCW Radar System," *IEEE Transactions on Microwave Theory and Techniques,* Vol. MTT-29, No. 5, May 1981.

51. R.A. Sparkes, "Missile Guidance Electromagnetic Sensors," *Microwave Journal,* Vol. 26, No. 9, September 1983, pp. 24–32.

52. Edgar Ulsamer, "Smart and Standing Off," *Air Force Magazine,* November 1983, pp. 56–62.

53. H. Bruce Wallace, "140 GHz Capture Antenna Multipath Experiment," Memorandum Report ARBRL-MR-02855, No. ADA059712, August 1978.

54. J.B. Winderman and G.N. Hulderman, "Solid State MM-Wave Pulse Compression Radar Sensor," *Microwave Journal,* Vol. 20, No. 11, November 1977, pp. 45–50.

55. J.B. Winderman, "Common Optics Millimeter Wave and Infrared Seeker," paper presented at *5th Annual KRC Symposium on Ground Vehicle Signatures,* Houghton, MI, August 1983.

56. J.B. Winderman, "Captive Flight Demonstration of a Common Optics Millimeter Wave and Infrared Seeker," paper presented at *National IRIS,* May 1984.

57. James C. Wiltse, "Millimeter-Wave Radar Features Unique Characteristics and Designs," *Microwave Systems News,* Vol. 14, No. 5, May 1984, pp. 58–76.

58. R.S. Ying, "Recent Advances in Millimeter-Wave Solid State Transmitters," *Microwave Journal,* Vol. 26, No. 6, June 1983, pp. 69–76.

59. George Ziff and Paul Schwartz, "Implementation of Millimeter Wave Technology Offers Improved Missile Guidance Performance," *Microwave Systems News,* November 1983, pp. 65–75.

60. R.T. Pretty, ed., *Jane's Weapon Systems 1980–81,* Jane's, London, 1980.

Chapter 15
MMW Airborne Mapping Radar

G.V. Morris

Georgia Institute of Technology
Atlanta, Georgia

15.1 MAPPING RADAR TECHNIQUES

Millimeter waves were first applied to airborne mapping radars to improve the resolution within the antenna size constraints imposed by aircraft installation. Improved resolution was desired to provide better detection and recognition of stationary military targets. Millimeter-wave radar is a compromise between the low atmospheric loss and poor resolution of lower frequency systems and the high resolution and clear-weather-only operation of optical systems. Earth resources applications, such as geological mapping and crop monitoring, were the first to utilize MMW radars. The need to derive further information from the radar signal led to application of synthetic array resolution improvement and multiple polarization techniques. Today, the classes of MMW airborne mapping radar in use include:

- Side-Looking Airborne Radar
- Scanning Radar
- Synthetic Aperture Radar

These three types will be addressed in this chapter.

15.2 SIDE-LOOKING AIRBORNE RADAR (SLAR)

The term *side-looking airborne radar* (SLAR) usually refers to a radar system employing a long antenna, typically 100 to 200 wavelengths, mounted with its long dimension parallel to the longitudinal axis of the aircraft. The radar beam is nominally perpendicular to the aircraft's flight path. The scanning of the antenna is provided by the aircraft motion as shown in Figure 15.1. The aircraft maps an area by flying a number of straight-line segments. The elevation pattern of the antenna is shaped from the radar horizon down to a depression angle of 45° to provide a wide mapping swath. Usually, two antennas are installed so that mapping can be performed on the right or the left of the aircraft or simultaneously on both sides.

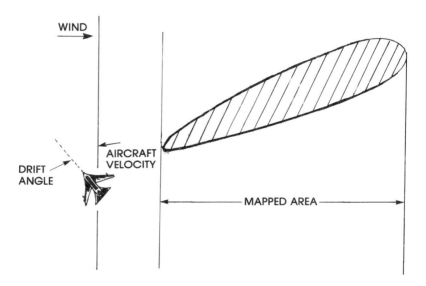

Fig. 15.1 SLAR mapped area.

The azimuth or cross-range resolution is achieved by the narrow beamwidth of the physical antenna without using the synthetic aperture techniques, which will be discussed later. The beamwidth of the antenna, β, in radians, is given by

$$\beta = \frac{k\lambda}{L} \tag{15.1}$$

where

λ = wavelength;
L = antenna length;
k = a factor on the order of 1.3, which is a function of amplitude taper used to reduce azimuth sidelobes.

From (15.1) it is obvious that millimeter wavelengths make it possible to achieve a given resolution with a shorter antenna than is necessary at X band. Table 15.1 compares some characteristics of an X-band and a MMW SLAR.

Table 15.1   Comparison Between X and K_a Band SLAR.

Characteristic	X–Band	K_a–Band
Frequency (GHz)	9.3	34
Beamwidth (mrad)	8	1.7
Array length (ft)	16	20

Figure 15.2 is a simplified block diagram of a SLAR. The magnetron transmitter generates a narrow pulse of high peak power, which establishes the system range resolution. The amplitude of the radar return *versus* time (range) is stored for every radar sweep. A point target on the ground may be illuminated by several hundred radar pulses as it passes through the radar beam. The integration function improves the SNR and reduces the amount of redundant data that must be stored. Since the antenna scan is provided by the aircraft moving along its flight path, several minutes may be required before enough data are gathered to make a recognizable map. Therefore, the storage element is vital to having a usable display.

Frequently, SLAR systems include a noncoherent MTI processor. Two maps are simultaneously produced: a *fixed-target* (FT) *map* and a *moving-target* (MT) *map*. One figure of merit for an MTI system is the slowest detectable radial velocity. Low velocity detection performance is limited by the spectral width of the ground clutter. For a SLAR, the clutter spectral width f_{DC} can be approximated as

$$f_{DC} = \frac{2V}{\lambda} \beta \qquad\qquad (15.2)$$

where

$V =$ aircraft velocity;
$\lambda =$ wavelength;
$\beta =$ beamwidth, in radians.

Substituting the value of β from (15.1) into (15.2) yields

$$f_{DC} = \frac{2Vk}{L} \qquad\qquad (15.3)$$

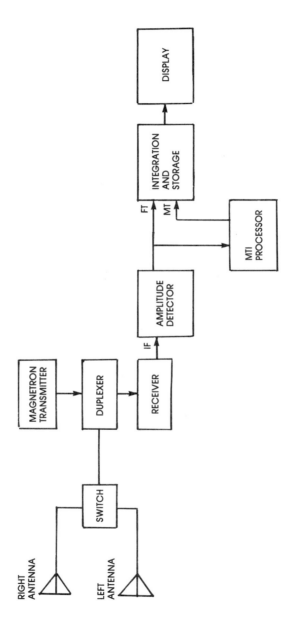

Fig. 15.2 SLAR block diagram.

Equation (15.3) shows that the low velocity detection performance is independent of wavelength, and depends on the physical antenna length. Therefore, a MMW SLAR with the same length of antenna as an X-band system will have the same low velocity detection performance, but proportionally higher resolution in both the MT and FT maps.

One of the pioneer MMW SLAR systems was the AN/APQ-97 system built by Westinghouse. Derivatives of the AN/APQ-97 were used for commercial radar mapping. The characteristics of the commercial mapping radar are shown in Table 15.2. Figure 15.3 is an example of the imagery produced.

Table 15.2 Characteristics of MMW SLAR.

Parameter	Value
Peak Power Output	100 kW
Nominal Transmitted Frequency	34.85 GHz
Transmitted PRF	2088 pps
Pulsewidth	0.07 s
Effective Antenna Resolution	0.1°
Receiver IF	60 MHz
Receiver Bandwidth	30 MHz
Antenna Polarization	Dual
Antenna Roll Compensation	±4.0°
Antenna Pitch Compensation	±1.5°
Antenna Yaw Compensation	±1.5°
Optimized Radar Slant Range	16 km
Optimized Antenna Altitude	20,000 ft
Ground Speed Range	130 to 300 knots
Slant-Range Accuracy, 2σ	30 m
Azimuth Accuracy, 2σ	40 m
Range Resolution at 16 km Slant Range	12 m
Azimuth Resolution at 16 km Slant Range	21 m
Input Signal Dynamic Range	50 dB
Power Consumption, 3-Phase, 400 Hz	3000 VA
Power Consumption, 28 Vdc	700 W

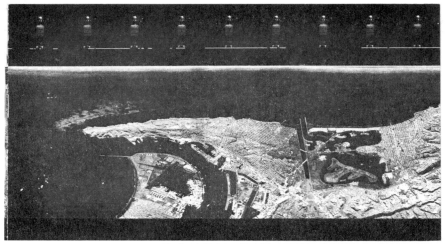

Standing out with photographic clarity . . . such details as the kelp beds off
Point Loma (upper left) . . . piers and ships at Coronado and Mission Bay . . .

Fig. 15.3 MMW SLAR image of Point Loma, California (photo courtesy
of Westinghouse Electric).

15.2.1 Antenna Considerations

Elevation Beam Shaping

Two idealized elevation-gain shapes are often discussed. The gain
function, which provides a constant return power from a point target of
constant cross section, independent of slant range, assuming a flat earth,
is

$$G(\epsilon) = C_1 h^2 \csc^2\epsilon \tag{15.4}$$

where

$G(\epsilon)$ = gain,
C_1 = constant function of maximum antenna gain,

and h and ϵ are defined in Figure 15.4.

If the backscatter coefficient of the ground clutter is $\sigma°$, then the
elevation gain function that provides a uniform return from ground clutter
with constant γ ($\gamma = \sigma° \sin\epsilon$) is:

$$G(\epsilon) = C_2 h^{3/2} \csc^2 \epsilon \sqrt{\cos\epsilon} \tag{15.5}$$

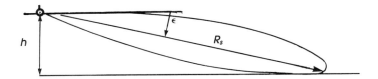

Fig. 15.4 Shaped elevation beam.

Beam shaping is typically achieved from 5° to 45°. From a practical standpoint, it is difficult to have close control of the beam shape, so it matters very little which of these idealized functions is used as the design goal. For example, at 45° depression angle, where the gain is on the order of 20 dB below the peak gain, the gain difference between the two is 0.7 dB. Note from (15.4) and (15.5) that the ideal elevation beam shaping is a function of altitude above ground level. If the aircraft is flown at a different altitude than the design altitude, then the shaping is no longer optimum.

Stabilization

Antenna stabilization is often provided to permit ground mapping in the presence of air turbulence. The antenna is movable over a range of about 5° in one or more axes—yaw, roll, or pitch. This stabilization is not intended to compensate for aircraft maneuvers or heading changes.

Yaw stabilization provides a uniform scan across the ground, and uniform dwell time for all points on the ground at the same range. Non-uniform dwell time would result in striations in the ground map image perpendicular to the flight line.

Roll stabilization maintains the proper orientation of the shaped elevation beam. The gain slope of the elevation pattern is on the order of 3 dB per degree, two-way. Uncompensated roll would cause the map from one side of the aircraft to be of higher intensity than it should be, and the map from the other side would be less intense.

Pitch stabilization maintains the plane of the main-beam vertical and preserves the geometric fidelity of the map.

15.2.2 Signal Processing

Several signal processing techniques are used to improve the image quality and make it easier to identify the terrain features of interest. The radars generally include the processing techniques described below for geometrically correcting the image and optimally using the available dynamic range of the display medium. Noncoherent MTI processing is also included in some radars. In some systems, the data are recorded for non-real-time processing to enhance the contrast between the features of interest and the background, and to remove artifacts introduced by the radar system (e.g., nonuniform antenna gain). The image-enhancement techniques are similar to those used in processing photographic images (Wiener and Hull [1], Mannos [2]).

Dynamic Range Compression

The dynamic range of returns from various types of terrain at the input to the antenna can exceed 100 dB. The dynamic range processed by the radar prior to the display is generally on the order of 60 to 80 dB. The dynamic range of monochrome cathode-ray tube (CRT) displays and film is generally on the order of 20 to 25 dB. Different transfer functions are required to highlight various types of terrain (e.g., desert, cultivated lands).

Slant-Range Correction

If a simple linear time sweep is used in the display, then the cross-track dimension of the imagery is slant range. At shallow depression angles, as occur at long ranges or during low altitude flight, the distortion may be negligible. However, the SLARs are frequently designed to operate at altitude of 20,000 feet, and the antenna patterns are shaped to provide illumination down to 45° below the horizon to minimize the unmapped area. A nonlinear time sweep can be used to correct the imagery to represent true ground range. The altitude above the terrain must be known.

Roll Compensation

If the antenna is not mechanically roll-stabilized, then roll effects cause uneven illumination of the ground, resulting in striations in the imagery. Electronic roll compensation can be used over about a ±3° range. The receiver gain is dynamically changed during the range sweep to adjust for the error in two-way antenna gain due to roll.

Drift Angle Correction

Wind can cause drift angles (see Figure 15.1) of up to 20°. SLAR antennas are typically 8 to 20 feet long. Providing drift angle compensation by mechanically moving the antenna ±20° is difficult, and not usually necessary. The CRT sweep can be rotated in azimuth by the drift angle to rectify the map.

Moving-Target-Indication Processing

As discussed earlier, millimeter wavelengths provide a higher resolution MTI ground map, but no particular benefit in low velocity target detection performance relative to microwave frequencies. However, the long antennas used for SLAR result in a very narrow clutter patch and excellent MTI performance. Noncoherent or clutter-referenced MTI is most often employed.

15.2.3 Display

Film has been historically used to produce SLAR imagery. The image of Figure 15.3 was recorded directly on film. Film continues to be an important medium for SLAR because it performs the signal processing function of integration, provides the archival storage, and can be viewed directly. Digital storage is widely used in recent applications, but film continues to be an economical method of presenting large synoptic scenes. Some SLARs expose the film in flight, and process the image on the ground. A few SLARs process the film when airborne, and it is viewed by the operator while passing over a back-lighted viewing surface.

The CRT displays have a limited number of pixels (e.g., 400 × 400) compared with the resolution elements available from the radar. For example, the radar described in Table 15.2 requires approximately 1000 × 2000 pixels to display a 20 km² map. Therefore, zoom magnification, scrolling features, and a digital data base are used. The digital data base also makes it possible to apply an operator-selectable amplitude to pseudocolor transformation. Pseudocolor expands the effective dynamic range of the display, and can enhance subtle contrast differences.

15.3 SCANNING RADARS

Scanning airborne ground-mapping radars are primarily used as navigational aids. In addition to providing a map display, the scanning radar is frequently used for terrain following, terrain avoidance, moving ground target detection, and air-to-ground ranging. The principal operational benefits of applying MMW are a result of the higher angular resolution relative to the K_u-band (J-band) systems that are the most widely used today. Some of the specific benefits are as follows:

    a. The radar map is easier to interpret for navigation because it more closely resembles the visual image (e.g., an aerial reconnaissance photograph).

    b. The safety of low-level flight is increased through the ability to detect small obstacles such as wires and poles.

    c. Smaller stationary ground objects can be detected and identified.

    d. The resolution improvement is achieved without the increase in complexity necessitated by other techniques, such as large antennas or synthetic array processing, and is therefore suitable for lightweight aircraft and helicopters.

    e. The higher carrier frequency facilitates achieving the wide signal bandwidth necessary for higher range resolution.

15.3.1 Antenna

A shaped elevation beam is used to illuminate a broad ground swath as the antenna is scanned in azimuth. An approximation to csc^2 shaping is employed, rather than the more accurate shaping used with SLAR systems. The antennas are frequently multimode to provide a narrow pencil beam for tracking and air-to-ground ranging, and to have a wider "spoiled beam" for mapping.

    The antenna scan pattern is stabilized so that the elevation-depression angle (measured from the horizontal plane) is maintained during aircraft maneuvers. This stabilization ensures that the area of the ground that the operator has selected to map continues to be illuminated. In addition, the antenna array is usually mechanically roll-stabilized so that the shaped elevation beam remains vertical. The center of the azimuth scan pattern rotates with the aircraft as it turns.

15.3.2 Signal Processing

    Dynamic range compression and moving-target processing, similar to that described for SLAR, are generally used. Slant-range correction is sometimes used to provide an undistorted map display at short ranges.

15.3.3 Display

    A *plan position indicator* (PPI) is the most common display. Digitally generated displays are supplanting the analog sector-scan presentations that use retentive phosphors. More display features can be provided, such as (1) a nonfading image that can be indefinitely stored; (2) zoom magnification; (3) pseudocolor to enhance the effective dynamic range.

15.3.4 Applications

Several 35 GHz airborne radars have become operational. The technology at 95 GHz is rapidly emerging, but the number of applications is limited. The characteristics of several representative MMW radars are described below to present an overview of the technology. This should not be interpreted as an exhaustive list of MMW radar applications, nor of the organizations involved in MMW research and development.

K_a-Band Operational Radar

The AN/APQ-137 radar, commonly referred to as MOTARDES (*moving target detection system*), built by Emerson Electric, is one example of an operational airborne MMW mapping and tracking radar. It is mounted in the nose of a helicopter as shown in Figure 15.5. A dual-mode antenna contained in a single reflector structure provides a shaped elevation beam to scan wide areas in search and a pencil beam for target tracking.

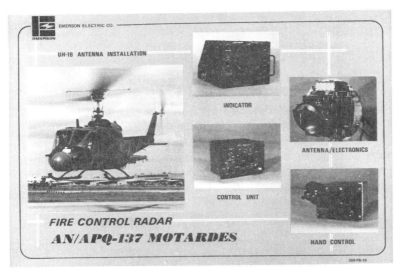

Fig. 15.5 AN/APQ-137 35 GHz Radar (courtesy of Emerson Electric).

MOTARDES uses clutter-referenced airborne moving-target-indication (AMTI) processing, and has demonstrated the ability to detect persons walking and a variety of vehicles. Figure 15.6 is a photograph of the CRT display. The upper portion of the display shows the returns from

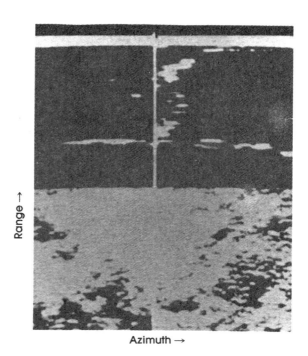

Fig. 15.6 CRT photo showing processed and unprocessed AMTI video (courtesy of Emerson Electric).

the ground without AMTI processing. The lower portion shows the output of the AMTI processor.

In the ground mapping mode, MOTARDES presents a map of prominent terrain features, such as rivers, lakes, hills, and coastlines, as shown in Figure 15.7. A B-scan display is used and results in map distortion of close-in ranges. The characteristics of the MOTARDES are shown in Table 15.3.

Dual-Frequency X and K_a Band Operational Radar

The AN/APQ-122 (V) is a dual-frequency radar developed for the C-130E transport aircraft. The equipment provides ground mapping out to 200 nmi, weather information out to 150 nmi, and beacon interrogation out to 240 nmi, when using the X-band radar. K_a-band frequencies are used to provide short-range and high resolution ground mapping to facilitate accurate aerial delivery of cargo.

Table 15.3 Characteristics of AN/APQ-137.

Frequency (GHz)	34.5
Pulsewidth (μs)	0.25
PRF (pps)	4000
Peak Power (kW)	25
Antenna Beamwidth (degrees)	
Azimuth	1.8
Elevation,	1.8
track	csc^2, 6 to 70
search	depression angle
Antenna Polarization,	
search	V
track	H
Antenna Gain (dB),	
search	33
track	36
Gimbal Limits (degrees)	
Azimuth	45
Elevation	$+11, -45$
AMTI Range Gates	80
AMTI Passband (Hz)	242 to 2000
Subclutter Visibility (dB)	30
Display	5-inch direct-view storage tube
Weight (lbs.)	212
Range Scale Selections (km)	2, 6, 18

Developmental 35 GHz Multimode Radar

The WX-50, built by Westinghouse Electric [3], provides a high resolution forward-mapping mode and a terrain-clearance mode for descending to low-altitude flight. The forward section contains the main scanning antenna, an auxiliary elevation interferometer antenna, three-axis gimballing, and all RF transmitting and receiving equipment. This particular configuration eliminates the need for slip rings and rotary joints. A cross-polarized signal can be selected to improve mapping through rain. The system uses a monopulse antenna and two receivers to implement an azimuthal monopulse resolution improvement technique to improve the azimuth resolution of the 1.7° antenna beamwidth. The characteristics of the WX-50 are shown in Table 15.4.

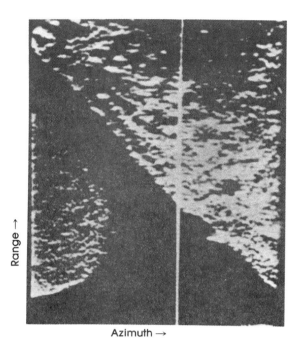

Fig. 15.7 CRT photo of ground map display (courtesy of Emerson Electric).

Table 15.4 WX-50 Characteristics.

Frequency (GHz)	35
Antenna Size (in)	15
Azimuth Beamwidth (degrees)	1.7
Elevation Beam Shape	$\csc^2$
Receive Polarization Selections	V,H
Azimuth Scan Width (degrees)	±35
Peak Transmitter Power (kW)	100
Transmitter Tube	Magnetron
Range sweep selections (nmi)	5, 10, 15, 30
Volume (ft^3)	2.5

Experimental 95 GHz Scanning Radar

A 95 GHz proof-of-principle transmitter-receiver was constructed by the Georgia Tech Research Institute (GTRI) for the General Electric Company, who then integrated it with other subsystems to form a complete detection and tracking radar for the MIT Lincoln Laboratory (Scheer and Britt [4]). The unit was mounted in the nose of a UH-1 helicopter and used for mapping and acquisition of moving ground targets. The radar parameters are shown in Table 15.5. Figure 15.8 is a block diagram of the RF unit.

Table 15.5   Characteristics of 95 GHz Airborne Radar.

TRANSMITTER	
Type	Pulsed IMPATT
Frequency	94 to 95 GHz
Pulse Length	100 ns
Chirp Bandwidth	250 MHz
Power	5 W (Peak)
PRF	20 kHz

ANTENNA	
Type	4-Horn Monopulse Lens
Aperture	10-in Diameter
Beamwidth	1° to 3°
Gain	42 dB

RECEIVER	
Type	Integrated Mixer-Preamp (Linear)
Dynamic Range	40 dB (Instantaneous)
Gain Control (AGC)	120 dB
Bandwidth	500 MHz
Noise Figure	8 dB
Input Losses	3 dB

The antenna lens provides a beamwidth of 1° when focused, but may be defocused to provide up to a 3° beamwidth.

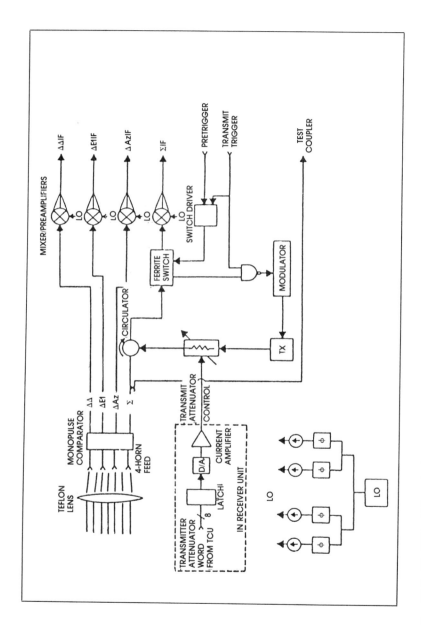

Fig. 15.8 95 GHz RF unit block diagram (from Scheer and Britt [4], © 1982 Horizon House).

Dual Experimental MMW Radars

Two short-range, airborne mapping radars were assembled by the Environmental Research Institute of Michigan (ERIM), assisted by the Georgia Tech Research Institute, to collect data on typical natural background scenes and man-made objects [5]. The characteristics of the radars are shown in Table 15.6.

Table 15.6 Characteristics of 35 and 95 GHz Radars.

Characteristic	35 GHz	95 GHz
PRF	37,500	37,500
Scan Rate (scans per second)	7.32	58.6
Slant Range (m)	300–1200	300–1200
TRANSMITTERS		
Output Stage	Phase-Locked Gunn Diode	EIO
Center Frequency (GHz)	35.24	94.7
Modes	wideband (230 MHz) narrowband (20 MHz)	Frequency Diversity Chirp (160 MHz) Coherent MTI
Peak Power (W)		
wide pulse	1.7	5.0
narrow pulse		7.2
Pulsewidth, Selectable (ns)		
wide	250	1150
narrow	50	850
ANTENNAS		
Type	Parabolic	Rotating Mirror Illuminated Dielectric Lens
Scan Type	Sinusoidal	Linear
One-way (3 dB)		
Beamwidth (mrad)	27.1	10
Resolution at 300 m (m)	5.7	2.1
Scan Angle (degrees)	27	21.6
Transmitted Polarization	V, H, Circular	V, H, Circular
Receiver Polarization	V, H	Parallel and Cross
Gain (dB)	40.8 V 41.6 H	47.5
Cross-Polarization		
Isolation (dB)	22	15

The collision-warning radar built by AEG-Telefunken [6] takes advantage of two of the unique properties of MMW: the high angular resolution provided by a small aperture and the high atmospheric attenuation at certain frequencies. The high angular resolution facilitates the detection of power lines and poles that are extremely hazardous to low-altitude helicopter flight at slant ranges of 200 to 400 m. The high attenuation reduces the probability of interference between groups of similarly equipped aircraft and probability of detection by other systems.

The RCS, σ, of an infinite straight wire (normal to the wire length) is given by

$$\sigma = \frac{\pi d R}{2} \tag{15.5}$$

where d is the wire diameter and R is the range to the wire. The formula is valid for wavelengths that are smaller than the wire circumference. The SNR is proportional to $1/R^3$, instead of $1/R^4$, since the RCS is proportional to R. Figure 15.9 shows the measured cross section (Rembold, et al., [6]).

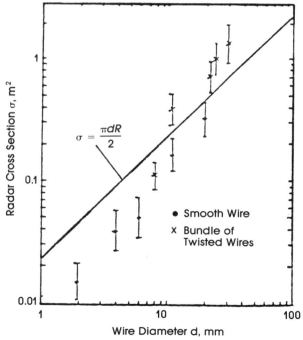

Fig. 15.9 Measured RCS of straight wires with different diameters (from Rombold, *et al.* [6], © 1982 Microwave Exhibitions and Publishers).

60 GHz Collision-Warning Radar

The antenna consists of a fixed parabolic reflector illuminating a rotating plane mirror that produces a 1.6 degree pencil beam. The mirror rotates at 900 revolutions per minute in azimuth and also causes the beam to move from −15 to +15 degrees in elevation each second. The characteristics of the collision warning radar are shown in Table 15.7.

Table 15.7 60 GHz Radar Characteristics.

Frequency (GHz)	59.1
Peak Power (W)	3
Pulse Length (ns)	20
PRFs (kHz)	125, 250
Antenna Gain (dB)	40
Beamwidth (degrees)	1.6
Receiver Noise Figure (dB)	13
Field of View (degrees)	
azimuth	180
elevation	30
Scan Time (s)	1

95 GHz Developmental Multimode Radar

A 95 GHz radar designed by Norden Systems under the joint Army-Air Force program, Tactical Avionics for Low Level Navigation and Strike (TALONS), was configured in a 15-inch diameter pod, and was capable of being carried by a variety of aircraft, including remotely piloted vehicles. The primary modes were terrain-following and terrain-avoidance (TF/TA). Ground-mapping or ground-target tracking modes can be selected by the operator and are interleaved with the TF/TA mode. The ground-mapping mode provides either moving target indication or fixed target enhancement. The fixed target enhancement feature exploits the difference in depolarizing characteristics between ground clutter and man-made objects. Normal and high resolution modes are selectable, which provide 200-foot and 40-foot resolution, respectively. During the tests of the mapping modes, the ability to detect cables (e.g., high-voltage power lines), was demonstrated.

Two antenna elevation beamwidths are provided: a 1.2° beamwidth monopulse pattern is used for tracking; a 6° beamwidth "spoiled beam," which is an approximation to a csc^2 pattern, is used to enlarge the ground area being illuminated in ground-mapping modes. Right-hand circular polarization is transmitted. Signals are received simultaneously from left-hand circular monopulse sum and right-hand circular monopulse sum-and-difference channels. The transmitter tube is an *extended interaction oscillator* (EIO). Table 15.8 lists the characteristics of the radar.

Table 15.8   Characteristics of 95 GHz Multimode Radar.

Frequency (GHz)	95
Peak Power (kW)	1.5
Pulsewidth (ns)	400, 80
Average Power (W)	4.5, 0.9
PRF (Hz)	7500
Antenna Gain (dB)	
monopulse	46
spoiled beam	40.5
Azimuth Beamwidth (degrees)	0.7
Elevation Coverage (degrees)	
monopulse	1.2
spoiled beam	6
Noise Figure (dB)	9.0
Gimbal Limits (degrees)	
azimuth	± 45
elevation	$+10, -15$
roll	± 40
Azimuth Scan Rate (degrees per second)	100

15.4 APPLICATION OF SYNTHETIC APERTURE TECHNIQUES

There have been few applications of MMW synthetic aperture radar (SAR) to date. In fact, MMW real aperture radars have been frequently used as an alternative to lower frequency SAR because the azimuth resolution required by the specific application could be provided by using MMW and an antenna of acceptable physical size. However, experimental MMW SARs are beginning to be used, which may pave the way for broader applications.

Several trade-offs must be considered in the decision to use MMW or microwaves and real aperture or SAR. Several considerations apply to MMW in general, and are not unique to airborne mapping radar applications (e.g., higher antenna gain, but lower available transmitter power at MMW), and such considerations are discussed at length in other chapters. The considerations that are directly applicable to SAR are summarized in this section. As an example, a typical SAR requirement is that the range resolution be matched to the azimuth resolution. To achieve a range resolution of one foot, the bandwidth of the transmitted waveform must be on the order of 500 GHz. This bandwidth is only 0.5% at 95 GHz and is comparatively easy to achieve, but at 2 GHz it is considerably more difficult to implement.

The minimum theoretical linear azimuth resolution, ΔR_c, for an unfocused SAR is given by

$$\Delta R_c = \frac{\sqrt{\lambda R}}{2} \qquad (15.6)$$

where R is range.

The derivation is presented in Dunn, et al. [7]). Clearly, MMW unfocused SAR is capable of better resolution than its lower frequency counterparts.

For a focused SAR, the minimum resolution is given by

$$\Delta R_c = \frac{L}{2} \qquad (15.7)$$

where L is the horizontal length of the physical antenna.

The resolution is theoretically independent of frequency. However, a MMW antenna providing the gain necessary to achieve the required SNR might be physically smaller than that of a microwave system.

The above equations assume that the aircraft flies in a straight line without acceleration. Deviations from linear unaccelerated motion must be compensated. The accuracy required is on the order of $\lambda/4$. Application of SAR at MMW requires a proportionate increase in the accuracy of motion compensation. The motion compensation requirements are mitigated by the fact that the synthetic array length is usually less. Consider the following relationships:

$$\beta R = \frac{k\lambda}{L_R} \qquad (15.8)$$

$$L_s = vT \qquad (15.9)$$

$$(L_s)\text{max} = \beta_R R \qquad (15.10)$$

where

βR = beamwidth of the real aperture,
λ = wavelength,
L_R = length of real aperture,
v = aircraft velocity,
T = aperture time,
L_s = length of synthetic aperture,
k = constant related to antenna illumination design.

Solution of the above equations for T yields the following expression, which shows that the aperture time is proportional to wavelength:

$$T_{\max} = \frac{k\lambda R}{L_R v} \tag{15.11}$$

Another source of phase error, which generally increases with increasing frequency, is the effect of atmospheric refraction. The effects of phase errors on the SAR beam include beam spreading and redistribution of energy from the main lobe to the sidelobes.

35 GHz SAR Instrumentation Radar

A 35 GHz instrumentation radar with a SAR mode is under development by MIT Lincoln Laboratory [8]. The airborne element consists of a pulsed, coherent, dual-polarization, polarization-agile radar sensor. The radar video is recorded in flight. High resolution images are produced in the ground processing element. The radar specifications are shown in Table 15.9.

15.5 MAPPING RADAR DESIGN EXAMPLES

A 35 GHz MMW airborne ground-mapping radar that can directly replace the many radars commonly in use today, which operate mostly in K_u band, is used here as an example. The primary functions are ground mapping, terrain following, terrain avoidance, and air-to-ground ranging. The current radars are noncoherent and use magnetron transmitters. Noncoherent MTI processing provides detection of moving surface targets. We will assume that the MMW antenna is of the same physical size as the current K_u-band antennas. We will show that the increased antenna gain partially offsets the smaller available transmitter power and higher atmospheric attenuation, and yields acceptable mapping range performance plus the advantage of a higher resolution map. Table 15.10 summarizes the assumed operational requirements for the radar.

Table 15.9   35 GHz SAR Specifications.

Frequency (GHz)	33.5
Modes	Sidelooking SAR
	Spotlight SAR
	Forward real aperture
Polarization	
transmitting	H or V
	RHC or LHC
Polarization	
receiving	Simultaneous co- and cross-polarized
Amplitude Balance (dB)	<0.5
Phase Balance (degrees)	<3.0
Absolute Calibration (dB)	<2.0
Swath Width (m)	150
Antenna Depression Angle (degrees)	10 to 50
Aircraft Velocity (m/s)	50 to 110

Table 15.10   Radar Operational Requirements.

Antenna Scan Width (degrees)	±45
Azimuth Scan Rate (degrees per second)	60
Display Range (km)	20
Maximum Aircraft Velocity (knots)	400
Range Resolution (ft)	50
Moving Target Detection Range (km)	10
(5 m² target)	

Antenna

The antenna characteristics are shown in Table 15.11. The apertures are assumed to be elliptical. The elevation gain pattern is shaped to provided relative uniform illumination of the ground from 1 to 20 km in range. The gains and beamwidths were calculated by using the following relations [9]:

$$\Theta_{AZ} = k \left(\frac{\lambda}{w} \right) \tag{15.12a}$$

$$\Theta_{EL} = k \left(\frac{\lambda}{h} \right) \tag{15.12b}$$

$$G_{max} = \frac{41,000}{\Theta_{AZ}\Theta_{EL}} \cdot \epsilon \cdot n \tag{15.13}$$

where

Θ_{AZ} = azimuth beamwidth, in degrees;
Θ_{EL} = elevation beamwidth, in degrees;
ϵ = efficiency = 0.5;
s = peak sidelobe level, dB down from peak gain = 25;
λ = wavelength;
w = antenna width;
h = antenna height;
G_{max} = antenna gain;
n = elevation beam shaping loss = 0.63 (-2 dB);
k = constant.

Table 15.11   Antenna Characteristics.

Width (mm)	600
Height (mm)	400
Elevation Beam Shape	csc^2
Azimuth beamwidth (degrees)	0.92
Gain (dB)	40.1

Transmitter

An important criterion used in the selection of a transmitter tube for an airborne radar is weight. A weight of 10 lbs or less is judged to be suitable for use in a 150 to 200 lb radar. The 100 kW peak power used in the analysis is a modest projection of the technological limit from that which is widely available now.

Mapping Performance

A measure of mapping performance is the SNR produced by low backscatter terrain at the longest mapping range of interest. The RCS, σ_c, of the terrain patch was computed by using

$$\sigma_c = R\Theta_{AZ}\Delta R\sigma° \tag{15.14}$$

where

R = maximum range = 20 km;
Θ_{AZ} = Azimuth beamwidth, in radians;
ΔR = Range resolution, in meters;
$\sigma°$ = Backscatter coefficient = 0.01 (-20 dB).

The SNR was calculated by using

$$\frac{S}{N} = \frac{P\,G^2\lambda^2\sigma_c}{(4\pi)^3 k T_0 B\,\text{NF}\,LR^4} \tag{15.15}$$

where

P = peak transmitter power,
G = antenna gain,
λ = wavelength,
σ_c = radar cross section of clutter,
k = Boltzmann's constant,
T_0 = reference temperature,
B = IF Bandwidth,
L = losses,
NF = noise figure,
R = maximum range.

The values used in the calculations and the results are presented in Table 15.12. The performance greatly exceeds the objective of providing a single-pulse SNR of 3 dB, thereby leaving a margin for other factors, such as lower terrain reflectivity or higher atmospheric attenuation due to increased moisture.

Table 15.12 Values Used and Results of Mapping Performance
Analysis.

Quantity	Units	Value
P	dBW	50.0
G^2	dBi	80.2
λ	mm	8.6
σ_c	dBsm	7.1
kT_0	dBW	-204.0
B	MHz	10
NF	dB	6.0
R	km	20
L		
microwave	dB	4.0
processing	dB	3.0
atmosphere	dB (total)	2.8
	dB/km	0.07
S/N Calculated	dB	28.7
S/N Required	dB	3.0

REFERENCES

1. T.F. Wiener and E.L. Hall, "Digital Processing of Aerial Images," *Proceedings of the Society of Photo-optical Instrumentation Engineers*, Vol. 186, 1979.
2. J.L. Mannos, "Design of Digital Image Processing Systems," *Proceedings of the Society of Photo-optical Instrumentation Engineers*, Vol. 301, 1982.
3. "Compact Radar Adds Punch to Attack Aircraft," *EW Magazine*, January-February 1977.
4. J.A. Scheer and P.P. Britt, "Solid State Tracking Radar," *Microwave Journal*, Vol. 25, No. 10, October 1982, pp. 59-65.
5. *WAAM Target/Background Signature Measurement Program Final Report*, Vol. II, System Description, Environmental Research Institute of Michigan Report 140400-24-T, March 1981, sponsored by US Air Force Armament Laboratory (AFATL/ASR), Eglin AFB, FL.
6. B. Rembold, *et al.*, "A 60 GHz Collision Warning Sensor for Helicopters," *Proceedings of Conference on Military Microwaves* (3rd: London), Microwave Exhibitions and Publishers, Kent, 1982.

7. J.H. Dunn, *et al.*, "Tracking Radars," Chapter 21 in M.I. Skolnik, ed., *Radar Handbook*, McGraw-Hill, New York, 1970.

8. *Advanced Detection Technology Program Specification for Sensor and Processor Development: Data Collection, Processing and Data Base Management*, Massachusetts Institute of Technology, Lincoln Laboratory, RFI 38671, December 1984.

9. *Reference Data for Radio Engineers*, 5th Ed., Howard W. Sams, New York, 1968.

Chapter 16
MMW Low Angle Tracking Radars

J.A. Bruder

Georgia Institute of Technology
Atlanta, Georgia

16.1 INTRODUCTION

The tracking of targets at low angles presents a number of problems to tracking radars if terrain or other objects are within the same antenna beamwidth as the tracked target. Ground clutter and multipath will degrade the ability of the radar to track a target accurately, can cause severe fading of the target signal, and, in certain cases, can cause loss of target track. In the tracking of low-altitude targets, if the antenna beamwidth is sufficiently small, ground clutter and multipath can be kept out of the main lobe of the antenna. Narrow antenna beamwidths can be obtained at millimeter wavelengths, while still maintaining reasonable antenna sizes.

However, even with very narrow antenna beamwidths, it is not always possible to keep multipath from low-altitude targets out of the main lobe of the antenna. One method for rejecting ground clutter is through the use of MTI processing, which for best performance requires that the system be coherent, or, at least, coherent-on-receive. Noncoherent MTI radars have been implemented. However, in order for the target to generate a Doppler beat frequency with this type of radar (which is not coherent-on-receive), clutter must be present in the same range cell as the target. Furthermore, the PRF of the radar must be staggered to avoid target spectra "foldover" for which the target Doppler corresponds to that of the clutter returns. Thus, the cost of the system is considerably increased by incorporating MTI processing. For cases when multipath cannot be eliminated from the main lobe, the use of MTI processing still enables target tracking, but with somewhat degraded performance. The MTI processing will not correct the effects of angular errors or fading due to multipath, but the use of a narrow-beam antenna can minimize the resulting errors to acceptable limits.

16.2 SYSTEM CONSTRAINTS

Since antenna beamwidth is inversely related to antenna size, fractional degree antenna beamwidths require rather large antennas at microwave bands. Physical restraints often limit the practical size of antennas. For fixed-site applications, large antennas require correspondingly large supporting structures and servo motors, thus resulting in a high-cost system. Further, the use of a large antenna size normally limits the slew response of the servo system. For mobile applications, antenna sizes are generally restricted to approximately 4 m in diameter, while for shipboard applications, antennas larger than 4 to 5 m in diameter are prohibitive. From this standpoint, MMW radars offer an attractive alternative in that smaller antenna apertures are required for narrow beamwidths. However, there are factors that provide practical frequency limits to the application of MMW frequencies to low angle tracking radars. Such factors include transmitter power, receiver sensitivity, propagation losses, and device availability.

16.2.1 MMW Radar

Millimeter wavelengths offer an attractive compromise between microwaves, where narrow antenna beamwidths are restricted by practical size limitations, and optical wavelengths, where tracking ranges are often limited due to atmospheric effects.

The continued development of MMW components over the last few years has enabled the manufacture of practical narrow-beamwidth angle tracking radars. For example, Vitro's RIR778 K_a-band range instrumentation radar [1], shown in Figure 16.1, has a diameter of 2.3 m. At an operating frequency of 35 GHz, the antenna beamwidth is 0.28°. The radar has a 135 kW peak-power magnetron transmitter and sufficient sensitivity to track a 15 cm diameter sphere out to 27 km under clear-air conditions. Operating frequencies of MMW tracking radars can range from 30 GHz to potentially as high as 440 GHz. In general, components are readily available from multiple sources at frequencies through 110 GHz, and are available from limited sources for frequencies up to 440 GHz. The development of MMW integrated circuit techniques, such as fin line, provides efficient packaging MMW transmitter-receiver systems (Meier [2]). Such techniques can be used up to frequencies of at least 140 GHz.

Fig. 16.1 Vitro RIR 778 K_a-band radar (photo courtesy of Vitro Corporation).

16.2.2 MMW Radar and Microwave Radar Trade-Offs

In selecting an operating frequency for a low-altitude tracking radar, the factors which must be considered are the desired beamwidth, antenna size, required operating range, search and acquisition time, and component availability, reliability, and ease of maintenance.

The selection of antenna beamwidth is the most important criterion for a low angle tracking radar. Figure 16.2 shows the elevation angle between the target and the earth's surface (or horizon) as a function of target height and range. The plot assumes that the radar's antenna is located 20 m above a smooth surface, and includes the effects of the earth's curvature and nominal radar diffraction. As can be seen from the plot in Figure 16.2, to isolate targets at altitudes of 30 m from sea return, even at ranges as close as 10 km, it is necessary to have an antenna beamwidth of $0.1°$ or less. Beamwidths of $0.5°$ or greater cannot isolate targets, even at heights of 100 m above the sea surface at ranges greater than 10 km. Low-altitude targets at ranges greater than 50 km are likely to be over the horizon, or masked by terrain, thus limiting the range requirements for low-altitude radars. The required antenna heights for a $0.2°$ elevation beamwidth would be 10.5 m at X band and 6.5 m at K_u band, which would prohibit their use for mobile radar applications and most shipboard applications. A K_a band radar, with a $0.2°$ elevation antenna beamwidth, would permit the use of a reasonable antenna height (3 m), while maintaining reasonable range performance. At 95 GHz, a radar with $0.20°$ antenna beamwidth could be even more compact ($\approx$ 1 m antenna height), but the range of a pulsed radar would be severely limited due to low transmitter power and propagation attenuation. However, even narrower beamwidths could be achieved at 95 GHz, and other transmitting waveforms such as chirp could be used to improve the maximum detection range. Frequencies above 95 GHz offer the potential of still smaller antenna sizes, or narrower antenna beamwidths, and could be particularly well suited for use with short-range systems.

16.3 SCENARIO AND ENVIRONMENT

As mentioned previously, angle tracking of low-altitude targets presents severe tracking problems to radars. As can be seen from Figure 16.2, the elevation angle between the target and the earth's surface is only a fraction of a degree for low-altitude targets. An aircraft or missile at an altitude of 10 m will be out of sight for the radar until it is within approximately 30 km from the ship, and when it becomes visible, multipath and

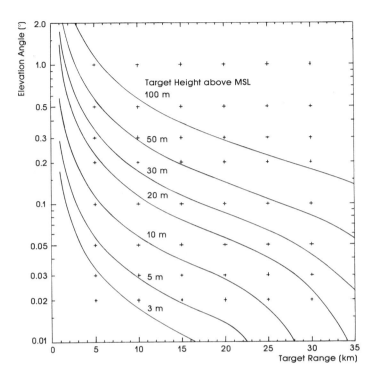

Fig. 16.2 Elevation angle between target and sea surface ($k = 4/3$), radar
antenna height 20 m above mean sea level (MSL).

sea clutter will prevent most radars from accurately tracking the aircraft
or missile until it is within approximately 1 km from the ship. This is clearly
too late to enable effective weapons deployment for the ship's defense.
The same situation is also present for a land-based radar, in which masking
from local terrain and vegetation limits the visible range to the aircraft or
missile. Even with MTI processing, a radar having an antenna beamwidth
of 0.5° or greater will not be capable of accurately tracking the aircraft or
missile as long as multipath is present in the antenna beam.

16.3.1 Search Areas and Volume Constraints

Prior to target acquisition, a radar must either have *a priori* knowledge of target location, or must search a predetermined volume of uncertainty to locate and to acquire the target. The time required to locate a

target depends on numerous factors, including the volume of airspace to be searched, antenna scan rate, radar PRF, antenna beamwidth, and radar processor requirements. The time required for search increases as the square of the inverse of the antenna beamwidth (assuming that the power transmitted is adjusted as a function of antenna beamwidth to maintain constant SNR). Thus, the search time for a radar with a 0.2° pencil beam would be 25 times that required for a 1.0° pencil beam. To require a low angle tracking radar to search over a full hemisphere would demand an inordinate amount of time to complete a single scan. However, even restricting the elevation coverage to 10° would require approximately 90 s for a radar with 0.2° antenna beamwidth to complete a single scan, if it is assumed that 1 ms dwell time is required per beam spot.

Multipath and sea clutter effects on low-altitude target tracking are primarily evident in the elevation plane of the antenna, while the effects on the azimuth track tend to be minimal. A possible antenna configuration, which could enable faster search and acquisition, would be to use a fan-beam antenna with a narrow elevation beam (such as 0.2°), and a relatively broad azimuth beam (such as 1.0°). However, even with this antenna, the time for the complete search cycle could only be reduced to approximately 18 s.

Other options need to be considered to reduce acquisition time with a narrow-beam radar. If the radar is to be used for range instrumentation purposes to track designated targets, then *a priori* information is often available on the target launch point or flight path. This information can then be programmed into the radar's computer to reduce the time required for search and acquisition. A second alternative is to have another radar (with a wider beam) perform the search and acquisition function, and supply the target position information to the narrow-beam radar to facilitate acquisition. For example, in a ship's defense applications, an X-band search radar could be used to provide initial target detection and coarse target location. The narrow-beamwidth tracking radar would then only be required to search a limited sector in order to acquire the target. In addition, the X-band radar would normally have a much greater detection range than a MMW radar, particularly in rain and fog.

16.3.2 Environmental Constraints

The major problem associated with radar tracking of low-altitude targets is that of isolating the target from clutter and multipath. Furthermore, the visible range of the target is limited by masking due to terrain features, the earth's curvature, and propagation losses. The designer of a low-altitude tracking radar must understand the restraints placed on the radar for search, acquisition, and track when the target is close to terrain.

In the search mode, MTI processing can distinguish between target and terrain or sea returns, but cannot eliminate the effects of signal fading and angular errors due to multipath. This can only be achieved by preventing the multipath return from entering the main beam of the tracking radar.

Propagation

The visible range to low-altitude targets is restricted by the environment, and also the maximum visible range of the target is restricted by the earth's curvature. The effect of the earth's curvature can be better described by converting to *spherical earth geometry,* as described by Durlatch [3], and shown in Figure 16.3. The radar beam above the earth's surface actually curves due to atmospheric refraction, so that the distance to the radar horizon is in fact greater than that to the optical horizon. This is approximated by increasing the effective radius of the earth's curvature by a factor k, where k is customarily taken to be 4/3. The actual k factor varies in practice [2], and according to Hunter and Senior [5] the stratification of the atmosphere near the sea surface, on the average, could produce greater radar detection ranges on a low-level target than that predicted by the k factor of 4/3.

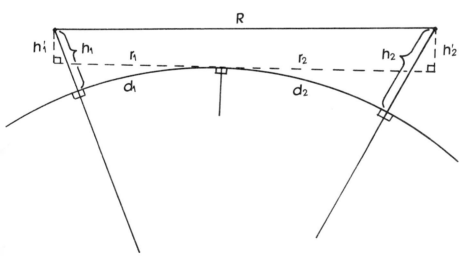

Fig. 16.3 Spherical earth parameters (from Durlatch [3]).

The effect of the earth's curvature is to reduce the apparent height of the radar (h_2) and the target (h_1). The equations for the reduced heights, h_1' and h_2', can be calculated from the equations [3]:

$$h_1' = h_1 - \Delta h_1 = h_1 - (d_1)^2/(2ka) \tag{16.1}$$

$$h_2' = h_2 - \Delta h_2 = h_2 - (d_2)^2/(2ka) \tag{16.2}$$

where

k = assumed refraction index,
a = radius of the earth.

Figure 16.3 can be simplified to the form shown in Figure 16.4 by setting the distance h_1' equal to zero. This enables us to determine the angle α, by computing the height h_2' and using the angle formula:

$$\alpha = \sin^{-1}\left(\frac{h_2'}{R}\right) \tag{16.3}$$

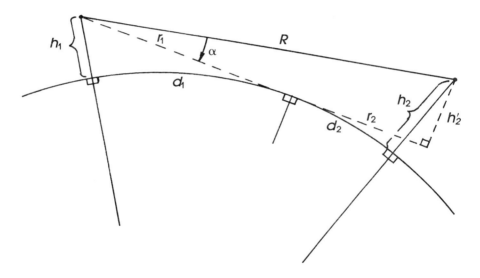

Fig. 16.4 Radar horizon geometry.

Because the radius of the earth (ka) is several orders of magnitude larger than d_1 and d_2, $r_1 \approx d_1$ and $r_2 \approx d_2$. Also, $r_1 + r_2 \approx R$, but, since $h_1 = 0$,

$$h_1 = (d_1)^2/2ka \tag{16.4}$$

Then,

$$h_2' \approx h_2 - \frac{(R - \sqrt{2kah_1})^2}{2ka} \tag{16.6}$$

and

$$\alpha \approx \sin^{-1}\left[\frac{h_2}{R} - \frac{(R - \sqrt{2kah_1})^2}{2Rka}\right] \tag{16.7}$$

for $R \geqslant \sqrt{2kah_1}$.

Thus, for ranges $R = \sqrt{2kah_1}$,

$$\alpha = \sin^{-1}\left(\frac{h_2}{R}\right) \tag{16.8}$$

For ranges $R < 2kh_1$ and small α,

$$\alpha \approx \sin^{-1}\left(\frac{h_2}{R}\right) \tag{16.9}$$

Equation (16.9) can then be used to compute the angle (α) between a low-altitude target and the angle to the sea surface immediately below the target when the distance to the target is less than or equal to the distance to the horizon. Equation (16.7) can be used to compute the elevation angle between the target and the horizon when the distance to the target is greater than the distance to the horizon. These equations have been used to compute the angles plotted in Figure 16.2.

The height of both the radar and the target will determine the maximum visible range (R_{max}) to the target. The maximum range to the target occurs when angle α is equal to zero. For this condition, (16.7) becomes

$$2kah_2 = (R_{max} - \sqrt{2kah_1})^2 \tag{16.10}$$

which further reduces to

$$R_{max} = \sqrt{2ka}\,(\sqrt{h_1} + \sqrt{h_2}) \tag{16.11}$$

For the assumption of $k = 4/3$, (16.11) becomes

$$R_{max} = \sqrt{17}\,(\sqrt{h_1} + \sqrt{h_2}) \tag{16.12}$$

where h_1 and h_2 are in meters, and R_{max} is in kilometers.

Figure 16.5 plots the maximum visible target range over a water path *versus* target height for radar antennas 5, 10, 20, 30, and 50 m above the water. As one can see from the chart, the maximum visible range to a 10 m target height is approximately 31.5 k (for an antenna height of 20 m). For overland tracking situations, the maximum visible range is generally more restricted by local masking due to hills and vegetation than by the earth's curvature. Thus, maximum visible range for acquisition of low-altitude targets is limited not only by the curvature of the earth, but also by the terrain features.

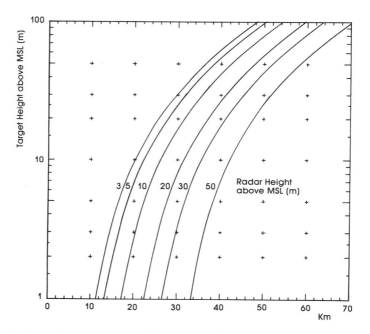

Fig. 16.5 Maximum target visible range ($k = 4/3$).

Propagation losses vary as a function of frequency as discussed in Section 4.1. Selection of frequency for MMW tracking radars is generally restricted to the "windows" [4] at 35, 95, 140, 220, and 440 GHz, where the propagation losses due to atmospheric absorption are relatively low for clear-air tracking. For fog and moderate rainfall rates, the propagation losses increase with frequency, but still permit tracking with major reductions in the maximum radar range. By comparison, optical and laser trackers are totally disabled in fog or rain.

Multipath

Multipath is probably the most difficult problem encountered by a low angle tracking radar, especially in an over-the-water application. A target with a 5 m altitude above the sea surface at 10 km range has an elevation angle of only 0.029° so that even a narrow-beam MMW radar would encounter main lobe multipath. A thorough understanding of the multipath phenomenon and potential solutions are required for low angle radar design.

When a target is situated over a flat surface, such as water or land, specular reflections can occur so that the radar sees not only the direct returns, but also those reflected off the surface. In addition, if the surface is rough, as is often the case (especially at MMW bands), diffuse reflections of the radar signals from an area of the sea or land surface will contribute to the multipath returns. These multipath returns can cause severe tracking problems when both the direct target returns and the multipath returns are within the main beam of the antenna. Depending on the amplitude and phase of the multipath returns with respect to those of the direct returns, multipath can, and often does, result in large amplitude and angular fluctuations in the composite return signals. We can see the effect of multipath in Figure 16.6, which shows the error in the measured elevation angle (Alexander and Ewell [7]) to an aircraft over a body of water. Multipath can also be experienced from low-altitude aircraft flying over land, although the effects of multipath are not as severe as those experienced over water. However, the problem is still serious enough to cause severe angle tracking errors. In general, the major angle tracking errors are experienced in the elevation channel, but azimuth multipath can be caused by reflections from buildings or mountain sides.

A number of techniques have been used to try to eliminate the effects of multipath, particularly in an over-the-water situation, where the problem is most severe. Barton [8], in his article on low angle tracking, thoroughly analyzes sidelobe multipath, main lobe multipath, and horizon region multipath for over-the-sea tracking. He also evaluates a number of potential methods for reducing the angular errors due to multipath, based on both real and analytical data, and a summary of the results of his analysis are shown in Table 16.1 [9]. Computational methods for eliminating the angular errors from multipath have not been too successful because they must assume a fixed configuration of the reflecting surface. With changing wind states, wind directions, and azimuth pointing angles, the properties of the reflecting surface are constantly changing. The use of frequency agility will tend to eliminate the fading, but cannot eliminate the angular errors due to multipath. Instead, the radar boresight will tend to track at

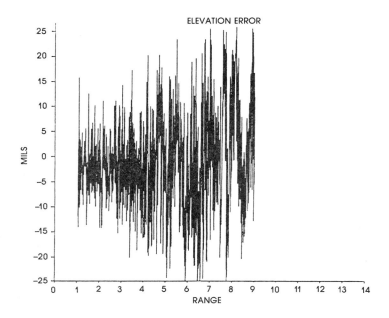

Fig. 16.6 Elevation error as a function of range for a Cessna inbound at an altitude of 500 feet (from Alexander and Ewell [7]).

an elevation angle between the aircraft and the multipath reflection. A technique investigated by Barton [9] involves off-axis tracking. The lowest angle that the monopulse boresight is permitted is 0.8 beamwidths above the horizon. Thus, multipath returns tend to be attenuated more than the direct returns from the target because of the beam shape. Stable tracking can be provided in the main lobe region by using this technique, and it provides stable track in the horizon region with the increasing error until loss of signal. Another effective technique involves the use of a monopulse radar with an asymmetrical monopulse elevation pattern. This technique appears to provide even better performance than off-axis tracking in the horizon region.

Multipath returns in the sidelobes of the antenna pattern can also produce angle tracking errors, but their effects should be small compared to those experienced in the main lobe. The effects of multipath in the sidelobes can be further reduced by good antenna design to minimize the sidelobe response of the antenna.

The best way to eliminate the effects of multipath is to keep the antenna beam narrow enough to prevent the multipath returns from entering the main beam of the antenna. Figure 16.7 shows the severe angular

Techniques	Sidelobe region $\theta_t > 1.5\theta_e$, $\sigma_E < 0.03\,\theta_e$	Mainlobe region, $0.3\,\theta_e < \theta_t < 1.5\,\theta_e$ $0.03\,\theta_e < \sigma_E < 0.3\,\theta_e$	Horizon region, $\theta_t < 0.3\theta_e$ peak error $\to \theta_e$
Narrow beamwidth	Error $= C\theta_e/\sqrt{G_{se}}$	Error and region proportional to θ_e	Region of large error proportional to θ_e
Range resolution with signal band-width B	Effective for $\dfrac{h_r\,h_t}{R}\dfrac{c}{2B}$	Effective only if $h_r\,\theta_e > \dfrac{c}{2B}$	Ineffective
Data smoothing or Doppler resolution	Effective for $\dfrac{2h_r\,\theta_t}{\lambda} > \dfrac{1}{t_o}$	Effective for $2\dfrac{h_r\,\theta_t}{\lambda} > \dfrac{1}{t_o}$	Ineffective
Frequency agility or diversity	Effective for	$\dfrac{h_r\,h_t}{R} > \dfrac{c}{2\Delta f}$	Permits track on centroid at horizon, but no height data
Off-axis mono-pulse tracking	Ineffective	Provide stable track with $\sigma_E \to 0.2\,\theta_e$ as $\theta_t \to 0.3\,\theta_e$	Provides stable track with increasing error until loss of signal
Symmetrical Δ/Σ pattern	Inapplicable	Effective for smooth and medium surfaces	Relatively large diffuse error and deep fades limit applicability
Asymmetrical monopulse	Inapplicable	Effective, giving $0.05 < \sigma_E/\theta_e < 0.1$	Provides stable track with increasing error until loss of signal
Complex angles	Ineffective	Effective for specu-lar reflection at low sites	Possible data over calibrated surface until loss of signal
Multiple-target estimation	Ineffective	Effective for smooth and medium surfaces	Impractical except as implemented with asymmetrical beams
Radar fences	Effective	Effective when main-lobe clears fence	Detrimental: extends horizon to top of fence
Circular polarization	Ineffective unless $\theta_t \geqslant$ Brewster angle; not applicable for narrow beams		
Use of *a priori* altitude	Inapplicable	Poor tracking is better than *a priori* data	Permits continued range and azimuth track until loss of signal

Table 16.1 Effectiveness of Anti-multipath Techniques (from Barton [9], © 1974 IEEE)

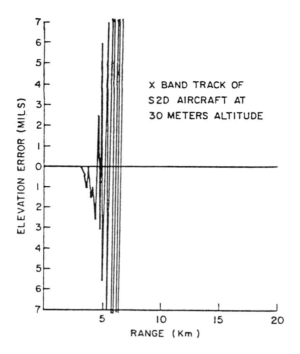

Fig. 16.7 X-band monopulse (1° beamwidth) tracking error *versus* range skin tracking on S2D aircraft at 33 m altitude (from Cross, *et al.* [10], © 1976 Microwave Journal).

error experienced by the TRAKX (Cross, *et al.* [10]) X-band radar, with a 1° antenna beamwidth, while tracking an aircraft at 30 m altitude. However, the TRAKX K_a-band radar, with a 0.22° antenna beamwidth, shows negligible angular error (see Figure 16.8), while tracking a target at the same altitude.

While the best method for eliminating angular errors due to multipath is to keep the multipath out of the main beam of the antenna, this is often not possible when tracking a low-altitude target. The use of a multipath reduction technique can provide rms accuracies for low elevation angle tracking of between 0.05 and 0.1 beamwidths, and, in combination with a narrow antenna beamwidth, this technique can reduce to acceptable limits the angular errors due to multipath.

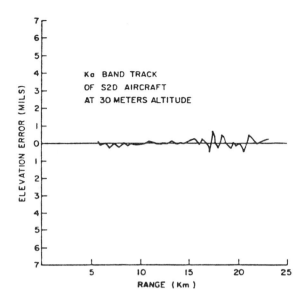

Fig. 16.8 K_a-band monopulse (0.22° beamwidth) tracking error *versus* range skin tracking on S2D aircraft at 33 m altitude (from Cross, *et al.* [10], © 1976 Microwave Journal).

Backscatter

Backscatter from clutter that is at the same range as the target and within the radar beam's footprint will degrade the radar's ability to track the target. For low-altitude tracking, the angular height of the target above the terrain is generally only a few tenths of a degree, so that most tracking radars will have clutter in the main beam while tracking such targets. Clutter returns from land and sea are discussed in Chapter 5, but consider the following example of the possible effect of clutter on target tracking. A radar with a 1° antenna beamwidth, tracking a target at 10 km, will illuminate a terrain patch of approximately 175 m wide by 37.5 m in extent (for a 250 ns pulsewidth). If we assume a $\sigma°$ of -30 dB, the total cross section of the terrain patch, which is in the same range cell as the target, will be approximately 8.2 m^2. This is greater than the cross sections of many targets, and therefore would seriously affect tracking the target. For over-the-sea tracking, clutter spikes at the range of the target could appear as potential targets, which would confuse the radar and might result in the radar tracking a clutter spike. Even if the sea or terrain clutter did not cause loss of track, it would still degrade the radar's ability of accurately tracking the target. The best solution is to keep the radar beamwidth small enough so that the clutter returns remain outside the main lobe of the antenna. However, this is not always possible for a radar tracking low-altitude targets, even a radar with an extremely narrow elevation beamwidth.

The effects of clutter in the main lobe can be greatly attenuated by the use of MTI processing. The clutter viewed by the radar will either be stationary, or have rather low velocity components. The target, on the other hand, will have a large velocity component in the direction of the radar, and hence produce a large Doppler component. An MTI radar produces a notch in the receiver response for low Doppler components, while gating through returns with higher Doppler components. Thus, an MTI radar provides rejection of clutter, thus greatly improving the target-to-clutter ratio. (See Chapter 6.)

16.4 TYPES OF TRACKING SYSTEMS

While numerous types of radar tracking systems exist, this section will focus on those techniques with particular application to low angle tracking. Also, rather than try to provide a comprehensive analysis of the tracking techniques, this section will provide an overview of the techniques and their application to low angle tracking. In each of the sections, references are given to the literature that provides comprehensive and in-depth analysis of the radar techniques.

16.4.1 Range Tracking

Automatic range tracking is virtually a necessity for tracking of low-altitude targets, and providing the accurate range information required for weapons systems. Several types of range tracker implementations are available, including split-gate (or centroid tracker), leading-edge tracker, and trailing-edge tracker. The use of a leading-edge tracker has the potential for eliminating the multipath effects if it can isolate the leading edge of the direct return from that of the delayed multipath return. However, for MMW radars with narrow-beamwidth elevation antennas, the multipath of concern would be displaced only 0.1° or 0.2° from the direct return so that the multipath is likely to be delayed by only 1 to 3 ns from the direct return. Thus, a leading-edge tracker would be impractical for this application. Centroid trackers utilize the entire pulse energy for establishing the range-gate position, and can be implemented as either analog or digital range trackers.

Range tracking accuracy depends on a number of factors, including radar-dependent errors, target-dependent errors, and environment-dependent errors. These error sources are discussed in detail by Barton [11, 12]. One of the prime sources for range tracking errors (especially at low SNR) is that due to thermal noise. The standard deviation of range error due to thermal noise (σ_r) for a single-pulse basis is given by the equation [13]:

$$(\sigma_{r1})_{\text{opt}} = \frac{1}{2B\sqrt{S/N}} \qquad (16.13)$$

for ($B\tau \geqslant 1$).

With most range trackers, the tracked range is averaged by the range servo over a number of pulses, so that (Barton [14])

$$\sigma_r = \frac{1}{2B \sqrt{f_r t_0 (S/N)}} \qquad (16.14)$$

where

f_r = radar PRF,
t_0 = observation time.

16.4.2 Angle Tracking

Of the various types of angle tracking techniques available, monopulse appears to provide the best capability for accurate low angle tracking. This is because an independent angle estimate is established on each radar pulse, and therefore is less sensitive to target return amplitude fluctuations. The greatest source of elevation tracking error occurs when the target and the multipath are both located in the main beam of the antenna, as is frequently the case with a low-altitude target. An evaluation has been performed by Barton [12] of various multipath rejection techniques, and a summary of those techniques is shown in Table 16.1.

Selection of the best angle tracking technique for a particular low angle tracking application depends on the system requirements and the allowable system complexity. Two techniques that appear to provide stable tracking when multipath is present in the main lobe region of the antenna are off-axis tracking and asymmetric elevation monopulse. In off-axis tracking, a symmetric elevation monopulse antenna is controlled so that the elevation boresight is permitted to have elevation angles at 0.8° or higher above the horizon. In asymmetric elevation monopulse, the lower lobe of the antenna has less gain than that of the upper lobe. In both cases, the multipath returns receive less gain than that of the direct target returns due to the shape of the antenna patterns, and thus a measure of multipath rejection is achieved. Near the horizon, the asymmetrical elevation monopulse implementation appears to offer better performance than the off-axis elevation monopulse, but the implementation of an antenna with an asymmetric elevation pattern is somewhat more complex than use of a symmetric monopulse pattern. Selection of an appropriate elevation processing scheme should enable rms angle tracking accuracies to between 0.05 and 0.1 beamwidths. Thus, when designing a low angle tracking radar, the elevation beamwidth should be selected so that the desired tracking accuracy is achieved with

$$\Theta_e \leqslant 20 \, \sigma_e \qquad\qquad\qquad (16.15)$$

where σ_e is the desired maximum rms elevation error.

The tracking accuracy will improve as the elevation angle between the target and the sea surface increases, since the multipath is further attenuated by the antenna pattern.

Azimuth multipath from the sea surface is much less severe than multipath in the elevation channel, and on the average should be symmetrically distributed in the azimuth antenna pattern. Therefore, the rms deviation of azimuth error (σ_a) can be held to less than 0.01 Θ_a.

16.4.3   Moving Target Indication (MTI)

The tracking of low altitude targets often results in sea or land clutter being in the antenna beam, which prevents accurate target tracking. The use of MTI processing enables rejection of the unwanted clutter returns.

The received frequency reflected from a moving target will be shifted from that of the transmitted frequency by a Doppler component due to the motion of the target velocity vector in line with the radar. The absolute value of the frequency of the Doppler shift is given by (Barton [15]):

$$|f_d| = \frac{2 \, v_r}{\lambda} \qquad\qquad\qquad (16.16)$$

where

v_r = target radial velocity with respect to the radar,
λ = transmitter wavelength.

In order to suppress the clutter, the Doppler frequencies resulting from the clutter must be filtered out while passing the Doppler components due to the clutter. The standard deviation of the clutter power spectrum (σ_f) is given by (Shrader [16]):

$$\sigma_f = \frac{2\sigma_v}{\lambda} \qquad (16.17)$$

where σ_v is the rms velocity spread in meters per second.

The MTI improvement factor (IM_1) for a single-delay coherent canceller, without feedback, is given by (Barton [17]):

$$IM_1 = 2 \left(\frac{f_r}{2\pi\sigma_f}\right)^2 \qquad (16.18)$$

where f_r is the radar PRF.

Further clutter rejection can be achieved by using a dual-delay coherent canceller without feedback, where the improvement factor (IM_2) is given by (Barton [18]):

$$IM_2 = 2 \left(\frac{f_r}{2\pi\sigma_f}\right)^4 \qquad (16.19)$$

16.4.4 Pulse Compression

Traditional pulsed (constant frequency) radars rely on a narrow pulsewidth to provide the range resolution (Δr), which is directly proportional to the pulsewidth, or

$$\Delta r \approx ct_p \qquad (16.20)$$

where t_p is the time duration of the radar pulse.

The requirement for high peak power to achieve the desired range coverage often cannot be met with currently available MMW transmitters, which in addition to having narrow pulsewidths, result in a low duty factor that does not effectively utilize the transmitter's capabilities. This is especially true with MMW coherent transmitters, which are likely to rely on either traveling wave tubes (TWTs) or solid-state amplifiers as the final transmitter output stage. The required range resolution can be achieved by modulating the transmitter pulse with a high range resolution waveform and a relatively long transmission pulse. The desired range resolution is then achieved by processing the received pulse to achieve a range resolution (Δr_s) given by (Wehner [19]):

$$\Delta r_s \approx \frac{c}{2\beta} \tag{16.21}$$

where β is the frequency excursion (bandwidth).

Thus, in pulse compression, the frequency excursion is used to compress a relatively long pulse in the receiver into an equivalently narrow, high range resolution pulse. By compressing the pulse, the energy present in the long pulsewidth is preserved in the compression process so that the effective SNR is equivalent to that obtained with a narrow pulse transmitter of much higher peak power.

According to Wehner [20], high range resolution waveforms include: (1) binary phase coding, which provides intrapulse phase modulation of the pulse; (2) discrete frequency coding (contiguous), in which frequency-stepped harmonically related waveforms are superimposed; (3) stretch, which uses an exchange of frequency bandwidth for time; (4) chirp, which provides a linear frequency-modulated transmission pulse and time compression of the received pulse; (5) discrete frequency coding (pulse-to-pulse), in which the transmit frequency is incrementally stepped between transmission pulses over the desired frequency excursion range.

The selection of the type of high range resolution waveform depends on the particular radar application. For a tracking radar with pulse compression waveform, the use of chirp is relatively easy to implement, and the receiver processing provides a real-time compressed waveform. A typical chirp radar implementation utilizes a pair of surface acoustic wave (SAW) delay lines: one delay line in the transmitter stretches a narrow input pulse into a stretched frequency modulated waveform; a second delay line in the receiver provides a delay *versus* frequency characteristic so that the stretched received pulse is compressed into a narrow, high range resolution pulse.

In designing a radar with pulse compression, particular care must be exercised to avoid range sidelobes, which could mask other nearby targets. This can be accomplished by incorporating appropriate weighting (Wehner [21]) (such as Dolph-Chebyschev or Taylor) in the receiver processor. In a system that uses SAW delay lines, the weighting can be incorporated in the SAW receiver delay line.

16.5 DESIGN EXAMPLE

Ships' protection from low-altitude aircraft and missiles is an increasingly important problem, especially when considering the vulnerability of ships demonstrated during the Falklands War (1982). A radar is required that will not only detect a low-altitude aircraft or missile, but one that will also provide accurate tracking to support gunfire or a command guided missile in order to "kill" the intruder. Further, ships' protection is required not only in clear weather, but also in fog or rain. Other radar contaminants which may be encountered are sea birds, chaff left over from antimissile self-protection attempts of other ships or one's own ship, and ECM from the missile or one's own ships. Thus, the scenario encountered in a combat environment places severe requirements on the radar.

Present X-Band MTI radars can adequately provide detection of aircraft or missiles out to ranges of 30 km, provided that the clutter rejection is adequate to prevent the sea clutter from masking a low-RCS, low-altitude target.

However, a typical X-Band radar with a 2° to 3° elevation beamwidth cannot adequately track a target for which the elevation is only a small fraction of a degree above the surface. However, a MMW tracking radar with a fraction of a degree elevation beamwidth can provide good low angle tracking, but in a search mode would require an inordinate amount of time because of the small spot size and the limited detection range (especially in rain). The best solution is to have the X-Band radar perform the search function, and hand off the target to a MMW radar for tracking.

According to Figure 16.2, a target at 10 m MSL at 10 km will have an elevation angle of only 0.057°, and therefore the target will most likely have sea clutter in the main beam of the radar. In order to enable tracking of a low-altitude target for these conditions, it is almost imperative that the tracking radar have MTI processing to suppress the unwanted clutter.

If the assumed worst case is for a target with a cross section of 0.1 m² and traveling toward the ship at wave-top height (3 m above MSL for sea state 5), we further assume that target acquisition range of 10 km is required in clear air, and 5 km when the radar return is corrupted by rain, chaff, or sea birds.

16.5.1 Performance Requirements

Based upon the scenario assumed in Section 16.5, the following system requirements are dictated for the design example.

1. *Tracking accuracy* is 1 m rms in the three coordinates at 5 km (to support gunfire or some hypothetical command guided missile).

2. *Acquisition range* on 0.1 m target at 3 m altitude is 10 km in clear weather, 5 km in rain.
3. *Acquisition time* is 2 s, assuming target designation is given to 1° (3σ).
4. *Environment* is sea state 5, rain to 3 mm/hr, chaff (left over from antimissile self-protection attempts of other ships or own ship), sea birds, and ECM.

16.5.2 Derived System Requirements

An X-Band radar can provide the search function for initially detecting the intruder, and provide the target designation required to enable acquisition with a MMW tracking radar.

The derived requirements for a MMW tracking radar are then as given below.

1. *Angular Accuracy*: At 5 km, 1 m error in elevation (or azimuth) corresponds to an angle of 0.0115°. Therefore,

$$\sigma_\epsilon = 0.0115°\text{rms}$$

2. *Elevation beamwidth*: In order to provide the angular accuracy in elevation,

$$\Theta e_{max} \leqslant 20 \, \sigma_\epsilon = 0.23°$$

Based on this result, an elevation beamwidth of 0.2° was selected.

3. *Azimuth beamwidth*: In order to provide the angular tracking accuracy in azimuth,

$$\Theta a_{max} \leqslant 100 \, \sigma_0 = 1.15°$$

Based on the above result, an azimuth beamwidth of 1.0 degree was selected.

4. *Unambiguous range* (to allow some scan time before target reaches 10 km acquisition range in clear-air conditions), then is

$$R_U > 15 \text{ km}$$

5. *Pulse repetition frequency*: In order to obtain a maximum unambiguous range of 15 km,

$$f_r \leqslant 10 \text{ kHz}$$

6. *First blind speed*: According to Shrader (Figure 17-13 in the *Radar Handbook* [22]), for 30 dB MTI rejection of rain clutter turbulence (with a two-delay canceller) with a $\sigma_v = 1$ m/s, requires a first blind speed of approximately 62 knots, or 32 m/s. Thus,

$$V_b > 32 \text{ m/s}$$

7. *Wavelength*: From Shrader [23], we have

$$\lambda_{min} > \frac{2V_b}{f_r} = 0.0064 \text{ m} \tag{16.22}$$

This corresponds to a frequency of 46.9 GHz. For this design example, the selection of an operating frequency of 35 GHz is more appropriate for the radar, since it is in the middle of an atmospheric propagation window. Thus,

$$\lambda = 0.0086 \text{ m}$$

8. *Antenna dimensions*: For a paraboloidal antenna with tapered illumination across the reflector, the dimension (Γ) of the reflector is given by (Sherman [24]) the relation:

$$\Gamma = \frac{1.2 \, \lambda}{\Theta} \tag{16.23}$$

where Θ is the 3 dB beamwidth expressed in radians.

Thus, the antenna dimensions will be approximately 0.6 m wide by 3.0 m high.

9. *Clutter spectral spread*: The standard deviation of the clutter spectral spread, σ_f, is given by (16.17):

$$\sigma_f = \frac{2\sigma_v}{\lambda} = 233 \text{ Hz} \tag{16.24}$$

10. *Maximum MTI improvement factor*: For a dual-delay coherent canceller, without feedback, (16.19) gives the improvement factor as

$$IM = 2 \ (f_r/2\pi\sigma_f)^4 \tag{16.25}$$

or

$$IM = 4350 = 36.4 \ dB \tag{16.26}$$

This is the improvement factor for a perfect MTI system. If we allow 6 dB of instabilities in the system, a more typical improvement factor would be 30.4 dB.

11. *Signal-to-clutter ratio*: If we assume that the target corresponds to a Swerling type 1 (slowly fluctuating target with no predominant scatterers), then a P_d of 0.9 and a P_{FA} of 10^{-6} requires a signal-to-noise (clutter) ratio of $+15$ dB, with a minimum integration of six pulses. (See Chapters 3 and 13.)

12. *Allowable input clutter*: For an output signal-to-clutter ratio (S/C) of $+15$ dB and a minimum target RCS of 0.1 m rms, the allowable clutter RCS in the antenna footprint is $+5.4$ dBsm, or 3.5 m².

13. *Pulsewidth*: The selection of pulsewidth is limited by two factors. First, 1 m rms tracking accuracy is needed and, second, the amount of clutter in the antenna-pulse length footprint must be limited. This assumes (which is usually the case for low angle tracking) that the length of the footprint due to the pulsewidth is less than that due to the antenna elevation beamwidth.

14. *Maximum pulsewidth due to tracking precision*: For a target with $+15$ dB S/C, the range tracking precision of an optimum range tracker is given by (16.14):

$$\sigma_r = \frac{1}{2B\sqrt{f_r t_0 (S/N)}} \tag{16.27}$$

where B is receiver bandwidth and t_0 is the observation interval.

A range error limit of 1 m rms would require $\sigma_r = 0.0067$ μs. Thus, assuming an observation time ($f_r t_0$) of 0.01 s, the maximum pulsewidth due to range precision is 0.75 μs. However, to allow some leeway for absolute range errors in the system, a pulsewidth of 0.5 μs is a more reasonable assumption.

Pulsewidth limitations due to clutter depend on the average cross section, $\sigma°$, of sea clutter at 35 GHz. Although the data for RCS for sea at very low grazing angles is not readily available, the data published in

Long [3] suggests that a $\sigma° = -30$ dBsm is a conservative number for sea state 5. Thus, at 10 km, a pulsewidth of 133 ns (20 m) would provide a clutter cross section of $+3.5$ m (approximately $+15.4$ dB). Thus, the pulsewidth limit due to the requirement for clutter suppression is more critical than the limit due to the range tracking accuracy requirement. Thus, a pulsewidth of 0.125 μs has been selected to provide the required clutter suppression.

Maximum Range Computation

As determined previously, a P_d of 0.9 (Swerling case 1) and a P_{FA} of 10^{-6} requires a minimum SNR of 15 dB, and six pulses of noncoherent integration. According to Skolnik [26], the maximum range (R_{max}) to a target is given by the equation:

$$R_{max}^4 = \frac{P_t G_t G_r \lambda^2 \sigma}{(4\pi)^3 S_{min}} \tag{16.28a}$$

where

$$S_{min} = kT_0 BF_n \, (S/N)$$

and

P_t = transmitted power, in watts;
G_t = transmitting antenna gain;
G_r = receiving antenna gain;
λ = wavelength of transmitting frequency, in meters;
σ = target RCS (relative to one-meter metal sphere);
k = Boltzmann's constant [4] (1.38×10^{-23} Joule/K);
T_0 = temperature of the receiver relative to absolute zero (nominally 290 K);
B = receiver bandwidth, in Hertz;
F_n = receiver noise figure;
S/N = signal-to-noise ratio (SNR) at the receiver.

For a MMW receiver, there are two additional factors, which must be added to the equation: first, the system losses between the transmitter and antenna, and between the antenna and the receiver; second, the propagation losses due to atmospheric absorption.

Combining these two factors into (16.26) and (16.27) yields

$$R_{\max}^4 = \frac{P_t G_t G_r \lambda^2 \sigma}{(4\pi)^3 k T_0 B F_n \ (S/N) \ L_S L_p} \tag{16.28b}$$

where

L_S = radar system losses,
L_p = propagation path losses (two-way).

A coherent system will provide the best MTI performance, but this requires an amplifier to supply the transmitter pulse power. The Hughes 921H TWT provides 4 kW minimum (typically 5 kW at 35 GHz), with 3–4% duty cycle for the air-cooled version and 5% for the liquid-cooled version. This TWT is a reasonable tube for this application because of its size and power requirements. Higher power TWT amplifiers have been built at this frequency, but would be impractical for such an application. The bandwidth of the receiver is optimized by setting it equal to 8.0 MHz, which effectively matches it to a 0.125 μs transmitted pulsewidth.

The remainder of the parameters assumed for the system are

$G_t = G_r$ = 49 dB,
λ = 0.86 cm (0.0086 m),
σ = 0.1 m^2,
F_n = 8 dB,
S/N = 15 dB,
L_s = 5 dB.

For clear-air conditions, the original design goal was for target detection at 12 km, assuming that 2 s would be required for target acquisition (with a maximum target speed of 1000 m/s). Chapter 4 lists the clear-air attenuation in Table 4.9 as 0.05 dB (one-way); however, immediately above the sea surface, it is expected that there will be higher moisture content so that 0.1 dB is probably a more reasonable number to use for this example. Thus, for a 12 km path, an attenuation (L_S) of 2.4 dB can be expected.

Substituting these parameters into (16.28) results in a computed maximum range of 7.5 km, which is considerably less than the minimum required detection range of the specification. In order to extend the range to 12 km, an effective increase of 8.2 dB of transmitter power is required. This would require a peak transmitter power of over 26 kW, or, alternatively, pulse compression could be used to decrease the effective bandwidth of the system. At the proposed PRF, a pulsewidth of 4 μs would provide a duty cycle of 4%, which is within the duty cycle limits of the tube. With the 4 μs pulsewidth and a 32:1 pulse compression ratio, a 15 dB increase in receiver sensitivity potentially can be obtained. Even if a 3 dB loss is allowed for the weighting required to reduce the range sidelobes and other pulse compression losses, the 12 dB effective increase in receiver sensitivity would provide a detection range of 15 km, which is more than sufficient to meet the required range performance in clear air.

One of the primary advantages of using a MMW radar, rather than an optical or infrared tracker, is that it provides tracking in fog and moderate rain, although with some degradation in range performance. From Table 4.9 in Section 4.4, a 4 mm/hr rain rate will provide a one-way attenuation of 1 dB/km, and therefore a 7 km path will provide 14 dB of attenuation. This attenuation will result in a reduction of the maximum detectable range (with S/N-15 dB) to approximately 7.5 km. If we assume that 2 km maximum is required for acquisition, then acquisition will be achieved in less than 5 km.

Residual chaff from other ships or one's own ship's self-protection attempts could affect the ability of the radar to track the missile through the residual chaff cloud. If the cross-section of the chaff cloud is strong enough, it could obscure the return from the missile or aircraft as it flies through the cloud. Also, the attenuation of the radar signal through the chaff could obscure the target when it is in line with the cloud. From Table 17-2 of Shrader [28], the reflectivity (n) for chaff is given by the expression

$$n = 3 \times 10^{-8}\lambda \tag{16.29}$$

where n is calculated as the chaff reflectivity factor per unit volume.

Thus, for $\lambda = 0.0086$ and at a range of 10 km, the cross section is $\sigma = 3 \times 10^{-5}$ for the compressed range resolution cell. Thus, the cross section of the chaff will be smaller than that of the target, and will certainly be rejected by the MTI.

The attenuation due to the chaff depends greatly on the density and other factors, but assuming a worst-case attenuation of 10 dB/km, and a chaff extent of 100 m, the two-way chaff attenuation would be equal to 4 dB. With this attenuation, the maximum detection range would be reduced to 11.6 km, and the worst-case acquisition to 9.6 km.

For the purpose of this design example, the only type of jammer that will be considered is that of a barrage noise jammer, which generates CW noise spectrum over a large bandwidth. The radar designer would normally be concerned with incorporating ECCM techniques into the radar (for example, frequency hopping and PRF stagger) to defeat more sophisticated jamming techniques, and thus the jammer would have only the use of barrage noise jammming with which to defeat the radar. When encountering jamming, the radar operator is concerned with the burn-through range of the target, which is normally considered to be the range at which the target signal strength in the receiver is equal to that from the jammer.

The received power in the radar due to the target (P_{rt}) is given by the equation (Skolnik [29]):

$$P_{rt} = \frac{P_t G_t \, \sigma \, A_r}{(4\pi)^2 R^4} \tag{16.30}$$

where

$$A_r = \frac{G_r \lambda^2}{4\pi} \tag{16.31}$$

Combining terms yields

$$P_{rt} = \frac{P_t G_t G_r \, \sigma \lambda^2}{(4\pi)^3 R^4} \tag{16.32}$$

The received power in the radar due to the jammer (P_{rj}) is given by the equation:

$$P_{rj} = \frac{P_j G_j G_r \lambda^2}{(4\pi)^2 R^2} \tag{16.33}$$

where

P_j = jammer power,
G_j = jammer antenna gain.

In computing the burn-through range, only the frequency components of the jammer power that are within the bandwidth of the receiver (P_{jB}) will contribute to the obscuration of the target, or

$$P_{jB} = \frac{P_j B}{B_j} \tag{16.34}$$

where

B = receiver bandwidth,
B_j = bandwidth of the jammer spectrum.

Depending on the radar pulse shape and bandwidth, some of the transmitted pulse spectrum may also be outside the receiver bandwidth. However, for a matched receiver, the received power lost due to bandwidth will be small so that the effect on burn-through range will be negligible.

The burn-through range (R_{bt}) is then determined by setting P_{rt} equal to P_{rj}, and substituting P_{jB} for P_j to yield the formula:

$$R_{bt} \approx \sqrt{\frac{P_t G_t \, \sigma \, B_j}{4\pi \, P_j G_j B}}$$
(16.35)

For the purpose of our example, we will assume that the jammer is generating 100 W (CW), spread over a 1 GHz bandwidth, and has a 10 cm diameter ($\approx 5°$) aperture. The effective radar bandwidth for the radar receiver will be 0.25 MHz, since the target signal will compress coherently, while the jammer will not compress, due to the random nature of the noise. With these parameters, the computed burn-through range is 14.1 km, which is approximately equal to the detection range in clear air. Thus, ECM from this type of jammer should not adversely affect the performance of the radar.

It is possible that sea birds can affect the tracking performance of the radar. The cross-section of an individual sea bird is rather small (probably less than -20 dBsm), but a flock of sea birds, particularly if they were at close range, could be mistaken as a potential target by the radar. However, a skilled radar operator could distinguish between a real target and the unwanted birds by observation of the speed and flight pattern of the flock. Also, it is possible that a flock of birds in line with the target could obscure the target from the radar's sight due to the attenuation of the target return. It is unlikely that this condition would remain for an extended period of time, nor would it sufficiently attenuate the target signal when the target range decreases.

This design example shows that the implementation of the radar with currently available components appears to be practical. The addition of MTI to the radar enhances the target detection against a sea clutter background, as is likely to be the case for a low-altitude target. As a result, the use of MMW tracking that uses a narrow elevation beamwidth provides the best solution for low-altitude target tracking.

REFERENCES

1. Automation Industries, Inc., Vitro Services Division, Fort Walton Beach, FL (undated).
2. P.J. Meier, "Integrated Finline: The Second Generation," *Microwave Journal*, Vol. 28, No. 11, November 1985, pp. 31-50; No. 12, December 1985, pp. 30-48.

3. N.I. Durlatch, "Influence of the Earth's Surface on Radar", MIT Lincoln Laboratory Technical Report No. 373, Massachusetts Institute of Technology, January 1965.

4. US Department of Commerce, *A World Atlas of Atmospheric Radio Reflectivity*, ESSA Monograph No. 1, Washington, DC, 1966.

5. I.M. Hunter, and T.B.A. Senior, "Experimental Studies of Sea-Surface Effects on Low Altitude Radars," *Proceedings of the IEE,* Vol. 113, November 1966, pp. 1731-40.

6. K.J. Button, J.C. Wiltse, eds., *Infrared and Millimeter Waves*, Academic Press, New York, 1981.

7. N.T. Alexander, and G.W. Ewell, "Low Altitude Tracking Experiments," Final Report on Contract DAAK40-77-C-0192, Georgia Institute of Technology, Atlanta, GA, September 1978.

8. D.K. Barton, "Low Angle Radar Tracking," *Proceedings of the IEEE*, Vol. 62, No. 6, June 1974, pp. 687-704.

9. D.K. Barton, "Low Angle Radar Tracking," *Proceedings of the IEEE*, Vol. 62, No. 6, June 1974, p. 702.

10. D. Cross, *et al.*, "TRAKX: A Dual Frequency Tracking Radar," *Microwave Journal*, Vol. 19, No. 9, September 1976, pp. 39-41.

11. D.K. Barton, *Radar System Analysis*, Artech House, Dedham, MA, 1976, p. 373.

12. D.K. Barton, "Low-Angle Radar Tracking," *Proceedings of the IEEE*, Vol. 62, No. 6, June 1974, pp. 687-704.

13. D.K. Barton, *Radar System Analysis*, Artech House, Dedham, MA, 1976, p. 42.

14. D.K. Barton, *Radar System Analysis*, Artech House, Dedham, MA, 1976, p. 43.

15. D.K. Barton, *Radar System Analysis*, Artech House, Dedham, MA, 1976, p. 190.

16. M.I. Skolnik, *Radar Handbook*, McGraw-Hill, New York, 1970, p. 17-9.

17. D.K. Barton, *Radar System Analysis*, Artech House, Dedham, MA, 1976, p. 212.

18. D.K. Barton, *Radar System Analysis*, Artech House, Dedham, MA, 1976, p. 219.

19. D.R. Wehner, *High Resolution Radar*, Artech House, Norwood, MA, 1987, p. 101.

20. D.R. Wehner, *High Resolution Radar*, Artech House, Norwood, MA, 1987, p. 103.

21. D.R. Wehner, *High Resolution Radar*, Artech House, Norwood, MA, 1987, p. 129.

22. W.W. Shrader, "MTI Radar," Chapter 17 in M.I. Skolnik, ed., *Radar Handbook*, McGraw-Hill, New York 1970, p. 17-16.
23. W.W. Shrader, "MTI Radar," Chapter 17 in M.I. Skolnik, ed., *Radar Handbook*, McGraw-Hill, New York, 1970, p. 17-7.
24. J.W. Sherman, "Aperture Antenna Analysis," Chapter 9 in M.I. Skolnik, ed., *Radar Handbook*, McGraw-Hill, New York, 1970, p. 9-21.
25. M.W. Long, *Radar Reflectivity of Land and Sea*, 2nd Ed., Artech House, Dedham, MA, 1983, p. 270.
26. M.I. Skolnik, "An Introduction to Radar," *Radar Handbook*, McGraw-Hill, New York, 1970, Chapter 1, p. 1-5.
27. W.W. Shrader, "MTI Radar," Chapter 17 in M.I. Skolnik, ed., *Radar Handbook*, McGraw-Hill, New York, 1970, p. 2-5.
28. L.V. Blake, "Prediction of Radar Range," Chapter 20 in M.I. Skolnik, ed., *Radar Handbook*, McGraw-Hill, New York, 1970, p. 17-10.
29. M.I. Skolnik, "An Introduction to Radar," *Radar Handbook*, McGraw-Hill, New York, 1970, Chapter 1, p. 1-4.

PART V
SPECIAL TOPICS

Chapter 17
MMW Reflectivity Measurement Techniques

C.H. Currie
Scientific Atlanta, Inc.
Atlanta, Georgia

N.C. Currie
Georgia Institute of Technology
Atlanta, Georgia

17.1 INTRODUCTION

17.1.1 Overview

The ever-increasing interest in the MMW bands for radar applications logically leads to the need for reflectivity measurements of clutter and high-value targets at corresponding wavelengths. Such measurements are necessary because it is often difficult or impossible to predict the behavior of the reflective properties of scatterers at millimeter wavelengths from lower frequency data because of the importance of smaller scatterers and resonance effects as the frequency increases. Thus, measurements must be repeated for the MMW band, even though experiments have previously been performed at microwave frequencies. In addition, there are differences in the problems encountered in performing measurements at millimeter waves as opposed to microwaves, and different techniques must be utilized to resolve the problems. This chapter will discuss the basic techniques for radar reflectivity measurements, while emphasizing the problems and solutions unique to MMW measurements.

17.1.2 Applications of Millimeter-Wave Measurements

Currently, MMW reflectivity measurements of interest can be divided into several categories including: (1) *phenomenological* (including measurements of the backscatter from rain, snow, land clutter, and sea clutter),

(2) *basic radar cross section* (RCS) *measurements* (of aircraft, military land vehicles, missile nose cones, *et cetera*), (3) *diagnostic* (high resolution measurements in range or cross-range to support modeling and radar cross section reduction (RCSR) efforts), and (4) *polarimetric measurements* (to support target identification and classification and clutter suppression work). Many types of experimental platforms are used to perform these various classes of measurements, including tower-based and airborne systems, indoor and outdoor ranges, and real and synthetic aperture systems. Table 17.1 summarizes the types of measurements and the corresponding applications that are currently of interest to the radar community. As can be seen from Table 17.1, the majority of measurement activities are concentrated in the "window" frequencies centered around 35 and 95 GHz. This is primarily due to hardware limitations in producing systems at the higher frequencies such as low transmitting powers, high losses, and poor noise figures in the receivers, which limit the use of higher frequencies in practical systems. However, reflectivity measurements are performed at higher MMW frequencies, although these are generally limited to phenominological measurments (to verify theoretical calculations) and short-range system applications (such as fuzing). Advances in hardware at the higher MMW frequencies are continuing, so more practical applications will become feasible in the future.

17.2 BASIC MEASUREMENT CONCEPTS

17.2.1 Definitions

Some of the fundamental concepts required to understand the principles of radar reflectivity measurements will be defined in this section.

Radar Cross Section

Radar cross section (RCS) is the portion of the radiated power that is incident on a target, which is radiated toward the receiving antenna of a radar. That is,

$$\sigma = \lim_{R \to \infty} 4\pi R^2 \left| \frac{E_r}{E_i} \right|^2 \tag{17.1}$$

This definition implies that the incident wave on the target is a plane wave, since σ is defined for $R = \infty$. If the transmitting antenna is collocated

Table 17.1 Summary of Millimeter-Wave Radar Reflectivity Measurements of Current Interest.

Type	Platform	Frequency Range	Results	Application	Comments
Phenomenology	Tower, Airborne	35, 95, 140, 220	Average Values, Distributions Spectra	Radar Detection, Modeling	Primarily Real Aperture
Target RCS	Tower, Outdoor or Indoor Range	35, 95, 140	Polar Plots, Average Values, Distributions	Radar Detection, Modeling, RCS	Primarily Real Aperture
Diagnostic	Tower, Outdoor or Indoor Range	35, 95	High Resolution 2D or 3D Images	Modeling, RCSR, Target Identification	Inverse Synthetic Aperture, Includes Full-Scale and Model Ranges and Airborne SAR
Polarimetric	Tower, Ground Ranges, Airborne	35, 95	Full Polarization Matrix	Target Identification	May Be Combined with High Resolution

with the receiving antenna, then σ is known as the *monostatic* radar cross section, whereas if the transmitting and receiving antennas are separated in space, σ is known as the *bistatic* radar cross section. In general, σ monostatic $\neq$ σ bistatic. We can show that for small bistatic angles (the angle between the line of sight from the transmitting antenna to the target and that of the receiving antenna to the target), the bistatic and monostatic radar cross sections are related by the equation: [1]

$$\sigma \text{ bistatic} = \sigma \text{ monostatic (at frequency } F_m) \tag{17.2}$$

where

$F_m = F_i \cos(\beta/2) =$ measurement frequency;
$F_i =$ frequency of interest;
$\beta =$ bistatic angle.

Area Reflectivity Coefficient (Sigma Zero)

Sigma zero (σ^0) is the radar cross section of a surface per unit area. Thus, σ^0 is defined as

$$\sigma^0 = \text{RCS of Surface/Area of Surface} \tag{17.3}$$

Sigma Zero is thus a unitless quantity, which is often expressed in terms of dB where σ^0 (dB) $= 10 \log(\text{RCS/Area})$. See Chapter 5 for information on computing the area subtended by a radar beam illuminating a plane surface.

Volume Reflectivity Coefficient (σ^V)

The volume reflectivity coefficient is defined to be the radar cross section per unit volume of volumetric scattering objects such as airborne hydrometeors, σ^V, and is thus defined as

$$\sigma^V = \text{RCS/Volume of Radar Cell} \tag{17.4}$$

Chapter 5 discusses the equation for calculating the volume of the radar cell for a volume scattering medium.

Polarization Scattering Matrix

The polarization scattering matrix is a two-by-two complex matrix, which contains all of the information about the scattering properties of target for a specific geometry. Each of the complex elements of the matrix represents the amplitude and phase of the reflection from a target for one of four orthogonal polarization states. That is,

$$S = \begin{bmatrix} \sqrt{\sigma_{11}}\ e^{j\phi 11} & \sqrt{\sigma_{12}}\ e^{j\phi 12} \\ \sqrt{\sigma_{21}}\ e^{j\phi 21} & \sqrt{\sigma_{22}}\ e^{j\phi 22} \end{bmatrix} \tag{17.5}$$

where σ_{mn} is the amplitude of the reflectivity for the transmitted polarization m and received polarization n, and note that ϕ_{mn} is the corresponding phase. Typically, the polarizations used in measurements are vertical and horizontal or left and right circular, although any two orthogonal polarizations can be used, and the polarization matrix expressed in terms of one polarization set can be theoretically transformed to any other set.

For a monostatic radar, generally, $\sigma_{12} = \sigma_{21}$ and $\phi_{12} = \phi_{21}$. This gives rise to the relative polarization scattering matrix defined by

$$S_{\text{rel}} = \begin{bmatrix} \sqrt{\sigma_{11}} & \sqrt{\sigma_{12}}\ e^{j\phi 12} \\ \sqrt{\sigma_{12}}\ e^{j\phi 12} & \sqrt{\sigma_{22}}\ e^{j\phi 22} \end{bmatrix} \tag{17.6}$$

where σ_{mn} is the same as above, and note that ϕ_{12} is the relative phase between σ_{11} and σ_{12}, and ϕ_{22} is the relative phase between σ_{11} and σ_{22}; ϕ_{22} is sometimes called the *polarimetric phase*, and it has been used as a descriminant to identify targets in the presence of natural clutter.

17.2.2 The Radar Range Equation for RCS Measurements

The form of the radar range equation commonly used for radar scattering measurements allows the radar cross section of a target to be related to the received power from the target as a function of the measurable properties of the radar, the range to the radar, and the path propagation factor. The equation thus takes the form:

$$P_r = \frac{P_t G_t G_r \lambda^2 F_t^2 F_r^2}{(4\pi)^3 R_t^2 R_r^2 L_t L_r}\ \sigma \tag{17.7}$$

where

P_r = received power from the target;
G_t = transmitting antenna gain;
G_r = receiving antenna gain;
λ = transmitted wavelength;
σ = RCS;
F_t = propagation factor for the transmitted wave path;
F_r = propagation factor for the received wave path;
R_t = range from the transmitter to the target;
R_r = range from the receiver to the target;
L_t = losses between the transmitter and the antenna;
L_r = losses between the receiver and the antenna.

For the monostatic radar case, the radar range equation takes the familiar form:

$$P_r = \frac{P_t G^2 \lambda^2 F^4}{(4\pi)^3 R^4 L_t L_r} \sigma \tag{17.8}$$

17.2.3 Calibration Techniques

The most important phase of a measurement program is the calibration process, and this is even more important at millimeter wavelengths than at lower frequencies because of the greater effects of variations in the measurement equipment and in the environment on the measured results. Calibration procedures can be generally divided into three groups: *amplitude calibration*, *phase calibration*, and *polarization calibration*. All three calibration procedures are not always needed, since phase calibration is not needed for a noncoherent radar system, and polarization calibration is not required for a singly polarized radar. Currently, however, there is interest from many quarters in measuring the full polarization scattering matrix of targets and clutter. In such a case, all three techniques must be used together to calibrate the measurement. In the following sections, each of the techniques will be covered, followed by discussion of a technique for calibrating the entire polarization scattering matrix in a single process.

17.2.3.1 Amplitude Calibration

There are two methods of amplitude calibration that are generally employed, one involving comparison of the target to be measured with a standard of known RCS properties, and the other involving the use of the radar equation with careful measurements of the radar and propagation path parameters to predict the RCS of a target from its received power. Since the errors in each method are approximately independent, then "closure" of the results obtained through each technique ensures the accuracy of the measurement process (Ewell [2]).

Absolute Calibration

This technique involves the replacement of the target under test with a known calibration target at the exact location as the target under test, and direct comparison of the target returns is performed. This technique is quite often used on indoor or outdoor RCS ranges, where a sphere is typically used as the reference target. One advantage of this technique for higher frequencies is the elimination of the need to know the propagation factor as long as it is the same for the reference target as for the test target. The disadvantage is that any multipath reflections present will probably change when the target under test is replaced with the calibration target. Also, if a sphere is used for calibration, it has a relatively small RCS for a given physical size, making comparison more difficult.

Relative Calibration

A relative calibration serves to define the relationship between receiver input power and output voltage or analog-to-digital (A/D) output value. This is known as the *receiver transfer curve*. Typically, a relative calibration is performed by injecting a signal at RF or IF and varying the power in known, fixed steps, or by illuminating a calibration target at a fixed range and varying an RF or IF attenuator in known, fixed steps. Figure 17.1 gives a relative calibration performed with a linear, 35 GHz instrumentation radar, using a corner reflector as a target and varying an RF attenuator in 5 dB steps. For the case where a variable gain is used, the calibration must be performed for all gain settings that will be used, or conversely for an automatic gain control (AGC) system, the AGC voltage must be calibrated for a fixed input.

Mixed Calibration

A mixed calibration is a combination of the absolute and relative calibration techniques, which uses the absolute method to establish the value of one point on the relative receiver transfer curve. With the addition

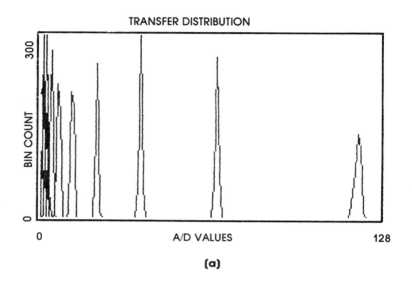

(a)

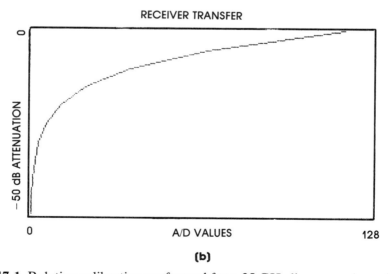

(b)

Fig. 17.1 Relative calibration performed for a 35 GHz linear receiver: (a) amplitude densities of the 5 dB calibration steps; (b) piecewise linear curve fit to the peak values of the densities in (a).

of a range-dependence factor, an equation can be developed to relate an unknown target at any range to a known calibration target at a known range. Ignoring propagation losses (which implies short ranges at millimeter waves) and assuming a linear receiver, then we have

$$\sigma_{unk}(dB) = 10 \log(V_{unk}) - 10 \log(V_{cal})$$
$$+ \sigma_{cal} - 40 \log(R_{cal}/R_{unk}) \qquad (17.8)$$

where

σ_{unk} = unknown RCS to be measured;

V_{unk} = voltage out of the receiver due to the return from the unknown target;

V_{cal} = voltage out of the receiver, due to the return from the calibration reflector;

σ_{cal} = RCS of the calibration reflector;

R_{cal} = range to the calibration reflector;

R_{unk} = range to the unknown target;

Indirect Calibration

An indirect calibration is performed by utilizing the measured parameters of the radar, such as transmitted power, antenna gain, system losses, and a receiver transfer curve in conjuction with the radar range equation, discussed in Section 17.2.2, to determine the RCS of an unknown target.

Closure

Calibration closure means that the RCS of a target determined by both the mixed calibration and indirect calibration methods falls within acceptable limits, where 3 dB is usually considered acceptable and 1 dB is excellent.

17.2.3.2 Phase Calibration

Phase calibration is required for coherent systems to ensure phase linearity and stability. Typically, phase calibration involves one of three processes: in-phase (I) and quadrature (Q) channel balance, absolute phase calibration, and phase stability checks. These processes will be discussed below.

I-Q Balance

There are three potential problems that can degrade the results achieved with an I-Q detector: I-Q gain imbalance, offset on either I or Q, and I and Q not being orthogonal. Section 17.3 discusses a method for correcting all of these problems by using an injected IF quadrature signal.

Absolute Phase Measurement

Actually, a radar does not measure absolute phase, but rather the relative phase between the delayed transmitting-receiving signal and the local oscillator (LO). For certain applications, it is desired to know how the phase from one target compares to another. The best way to calibrate the phase in such a case is through the use of a calibrated phase shifter in series with the IF signal. The I-Q outputs can then be determined for various settings in an analogous fashion to the attenuator step calibration, which is used for amplitude calibration.

Phase Quality Checks

Phase quality checks can be either manually or analytically performed. The manual method involves displaying the sampled I-Q channels on the X-Y inputs of an oscilloscope to form a polar pattern, as shown in Figure 17.2. If the radar boresight is on a strong and steady target, then the output should be a dot some distance from the center of the polar display. If the dot varies either radially or tangentially, this is evidence of amplitude or phase noise. When the phase of the signal is varied by the use of a phase shifter, or by varying the transmit frequency, then the dot should move smoothly in a circle around the display origin, and adding attenuation should vary the radius of the circle.

The stability of a system can be analytically verified if the frequency can be varied in known steps. Basically, the coherent return is recorded from a large point target (such as a corner reflector), while the frequency is stepped in equal units equal to 1/pulsewidth over a band (typically 64 or 128 steps). A Fourier transform is then performed on the data. If the system is highly stable, then the result will be a single impulse at a frequency representing the location of the target relative to the position of the front of the range gate, as shown in Figure 17.3(b). Phase or amplitude instabilities will result in sidelobes appearing around the impulse, as shown in Figure 17.3(a).

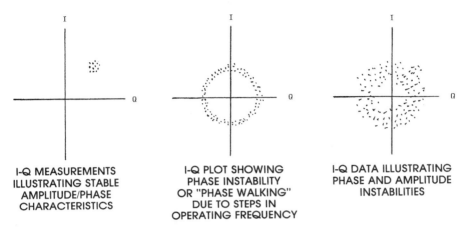

| I-Q MEASUREMENTS ILLUSTRATING STABLE AMPLITUDE/PHASE CHARACTERISTICS | I-Q PLOT SHOWING PHASE INSTABILITY OR "PHASE WALKING" DUE TO STEPS IN OPERATING FREQUENCY | I-Q DATA ILLUSTRATING PHASE AND AMPLITUDE INSTABILITIES |

Fig. 17.2 Illustration of three measurement conditions of I-Q target returns from a stationary corner reflector (from Williamson [3]).

17.2.3.3 Polarization Calibration

Polarization calibration involves the calibration of as many elements of the polarization matrix as the radar is capable of measuring. Typically, the process involves the measurement of polarization isolation of the system and the use of two of several types of calibration targets to calibrate the various components of the polarization matrix. Polarization isolation can be measured for a dual-polarized radar by transmitting one polarization and receiving the return from a nonpolarizing target with the orthogonal polarization. For example, if vertical polarization is transmitted and reflected from a trihedral reflector, then the ratio of the return in the vertical receiving channel to that of the horizontal receiving channel is the polarization isolation.

Linear Polarization Calibration

Calibration of orthogonal linear polarizations is based on a unique property of the diplane. Figure 17.4 shows a diplane with with a linearly polarized wave incident at an angle Θ relative to the seam between the faces. The reflected wave is linearly polarized, but it is oriented at an equal, but opposite, angle to the seam. Thus, if the dihedral is rotated about an axis parallel to the radar line of sight, the reflected polarization will rotate in the opposite direction at twice the rate. Thus, a diplane with

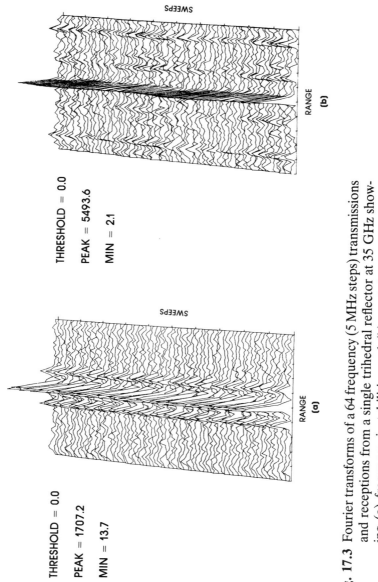

THRESHOLD = 0.0

PEAK = 5493.6

MIN = 2.1

THRESHOLD = 0.0

PEAK = 1707.2

MIN = 13.7

Fig. 17.3 Fourier transforms of a 64 frequency (5 MHz steps) transmissions and receptions from a single trihedral reflector at 35 GHz showing (a) frequency instabilities and (b) much better system frequency stability (figures courtesy of John Trostel, Georgia Institute of Technology).

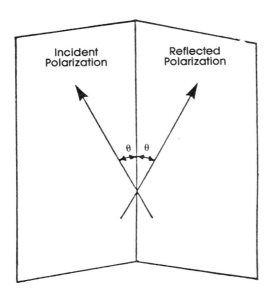

Fig. 17.4 Polarization sensitivity of a dihedral corner reflector (from Trebits [4], © 1984 Artech House).

a seam rotated at 22.5° to an incident linear polarization will return a 45° polarization relative to the incident polarization (and thus will provide equal power at the incident and orthogonal polarizations), while a rotation of 45° to an incident linear polarized wave will return an orthogonal wave. The diplane can therefore be used to calibrate orthogonal linear polarizations.

Circular Polarization Calibration

Circular polarization requires the use of two different types of targets for calibration, an odd-bounce target and an even-bounce target. Examples of odd-bounce targets include the flat plate, the trihedral reflector, and the cylinder. Each exhibits an odd number of bounces for incident radiation. Examples of even-bounce targets include the diplane, the top hat, and the Bruderhedral. An odd-bounce reflector always returns the opposite sense circular polarization because of an odd number of 180° phase reversals at each bounce. Thus, when a left-hand circularly polarized wave is reflected from a trihedral, it becomes a right-hand circularly polarized wave and *vice versa*. An even-bounce reflector always returns the same sense circular polarization as the incident wave, a left-hand circularly polarized wave is reflected as a left-hand circular wave and *vice versa*. A trihedral and a dihedral are commonly used to calibrate odd- and even-bounce circular polarized signals. Table 17.2 summarizes the commonly used calibration targets and the incident *versus* reflected polarizations for each target.

Table 17.2   Reflected Polarization for Various Reflectors (from Scheer [5], © 1984 Artech House).

Reflector	*Illuminated Polarization*					
	V	H	RHC	LHC	+45°	−45°
Flat Plate	V	H	LHC	RHC	+45°	−45°
Sphere	V	H	LHC	RHC	+45°	−45°
Trihedral	V	H	LHC	RHC	+45°	−45°
Diplane (Vertical 0° Seam*)	V	H	RHC	LHC	−45°	+45°
Diplane (22.5° Seam*)	+45°	−45°	RHC	LHC	V	H
Diplane (45° Seam*)	H	V	RHC	LHC	+45°	−45°

Polarization Distortion Matrix

The polarization distortion matrix has been defined by Barnes [6] to be a four-by-four complex matrix that contains all of the distortion terms injected onto the measured target polarization matrix by the measurement system. Thus, the product of the distortion matrix with the actual target matrix yields the matrix actually measured by the radar. The distortion matrix is inferred by performing measurements on targets for which the theoretical polarization matrix is known, such as a trihedral, a dihedral with horizontal seam, and a dihedral with a 45° seam. This approach is discussed in more detail in [6].

17.2.3.4   Properties of Calibration Targets

Figure 17.5 gives the equations for the maximum RCS values of targets most often utilized for calibration. The cylinder, sphere, trihedral, and flat plate are odd-bounce targets, while the diplane, top hat, and Bruderhedral are even-bounce targets. The properties of each will be discussed below.

---

*The top hat and Bruderhedral reflectors have the same characteristics as the diplane.

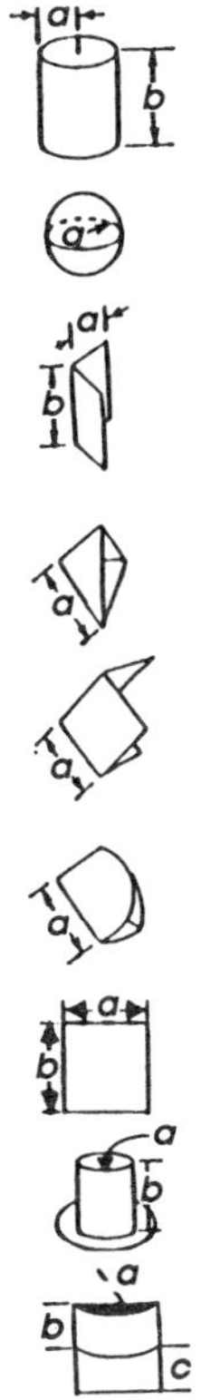

Target		Maximum Cross Section	Advantages	Disadvantages
	Cylinder	$\sigma = \dfrac{2\pi ab^2}{\lambda}$	Nonspecular along radial axis	Low RCS for site specular along axis
	Sphere	$\sigma = \dfrac{\pi D^2}{4}$ or $\dfrac{\pi D}{\lambda} > 10$	Nonspecular	Lowest RCS for size, radiates in all directions
	Diplane	$\sigma = \dfrac{8\pi a^2 b^2}{\lambda^2}$	Range RCS for size, nonspecular along one axis	Specular along one axis
	Triangular Trihedral	$\sigma = \dfrac{4\pi a^4}{3\lambda^2}$	Nonspecular	Cannot be used for cross polarized measurements
	Square Trihedral	$\sigma = \dfrac{12\pi a^4}{\lambda^2}$	Large RCS for size, nonspecular	Cannot be used for cross polarized measurements
	Circular Trihedral	$\sigma = \dfrac{0.507\pi^3 a^4}{\lambda^2}$	Large RCS for size, nonspecular	Cannot be used for cross polarized measurements
	Flat Rectangular Plate	$\sigma = \dfrac{4\pi a^2 b^2}{\lambda^2}$ Normal Incidence	Largest RCS for size	Specular along both axes
	Top Hat	$\sigma = \dfrac{2\pi ab^2}{\lambda\,\cos^3(\phi)}$ ϕ is the elevation angle to the cylinder ($\phi = 0°$ is perpendicular to the cylinder; $c > b$; ϕ is normally 45°)	Low RCS for size	Difficult to align rotated seam
	Bruderhedral		Large RCS, easier to align rotated scan	Moderately specular along one axis

Fig. 17.5 Maximum radar cross section of typical calibration targets (adapted from Scheer [5]).

Cylinder

The cylinder has a very narrow angle of return in the plane along its axis and a very broad region of return (essentially, 360°) in the plane along the radius. The cylinder is typically used for calibrating RCS ranges, where the it can be rotated in azimuth to find the specular return, while orienting the broad radial lobe in the vertical direction.

Sphere

The sphere is one of the easiest targets to manufacture, and so long as it is sized to be in its optical scattering region, the sphere's RCS is independent of frequency, which is a decided advantage for wideband systems. Its primary disadvantage is a very low RCS for a given size.

Diplane

The diplane has been the reflector traditionally used for calibrating orthogonal polarizations, as discussed in the previous section. The diplane has a broad beam in the plane perpendicular to the seam and a very narrow beam in the plane along the seam. The diplane's principle disadvantage is the difficulty in properly aiming it toward the radar because of its narrow plane, and the fact that it is often rotated from vertical to obtain orthogonally polarized returns.

Trihedral

There are three types of trihedrals with different RCS and lobe widths, depending on the shapes of the sides. The triangular reflector is used most often because it has the widest lobe. The advantage of the trihedral over other targets is that is has wide lobes in both planes, while also exhibiting a relatively large cross section. Figure 17.6 gives patterns in both the vertical and horizontal planes for a triangular trihedral.

Flat Plate

The flat plate has the largest RCS for its area of any of the targets, but this shape has a narrow lobe in both the vertical and horizontal planes

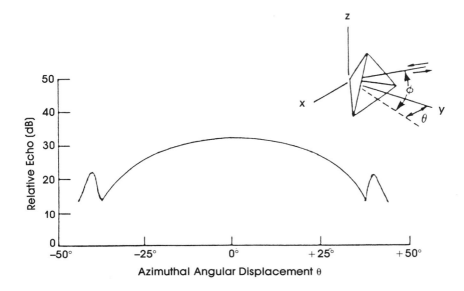

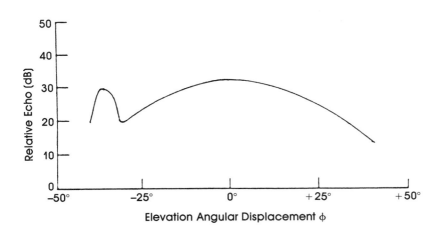

Fig. 17.6 Backscatter pattern for a trihedral corner reflector with triangular sides for the azimuth and elevation planes (from Trebits [4], © 1984 Artech House).

as is shown in Figure 17.7. Although the plate is hard to align, its properties can be put to advantage when calibration must be performed near the ground, since the narrow vertical lobe rejects multipath signals.

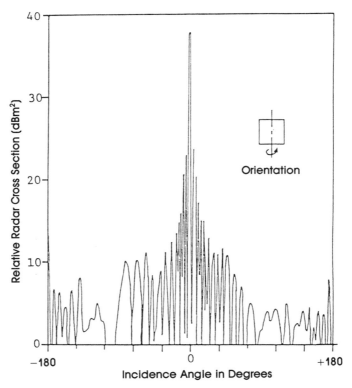

Fig. 17.7 Measured RCS of a square, conducting, flat plate as a function of incidence angle to the surface (from Trebits [4], © 1984 Artech House).

Top Hat

The top hat reflector exhibits the polarization properties of a diplane, while possessing a broad lobe in both the vertical and horizontal planes. The top hat's major disadvantage is its small RCS for a given physical size.

Bruderhedral

The Bruderhedral was developed at Georgia Tech to provide a useful target for calibrating orthogonal polarizations. Named after its developer, Joseph Bruder, by the Georgia Tech staff, the Bruderhedral is a cross between a diplane and a top hat. Basically, the Bruderhedral is a diplane in which one face is a flat plate and the other is a section of a cylinder. The Bruderhedral has a relatively large RCS, while exhibiting a wider lobe in the plane parallel to the seam than does a diplane. Figure 17.8 gives the measured patterns of a Bruderhedral in the planes parallel and perpendicular to the seam, while Figure 17.9 shows a picture of a Bruderhedral.

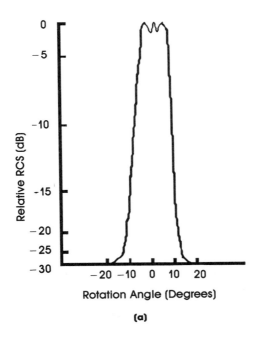

(a)

Fig. 17.8(a) RCS pattern for a Bruderhedral rotated in a plane parallel to the seam (adapted from Bruder [7]).

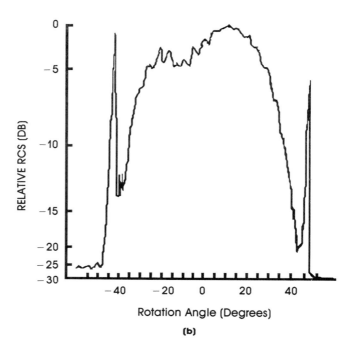

Fig. 17.8(b) RCS pattern for a Bruderhedral rotated in a plane perpendicular to the seam (adapted from Bruder [7]).

Fig. 17.9 A Bruderhedral designed for use in a MMW test program (photo courtesy of the Georgia Institute of Technology).

17.2.3.5 MMW Calibration Problems

Calibration is complicated in the MMW region because of the effects of the propagation factor, reference target errors, and degraded system performance. These problem areas are discussed below.

Propagation

As discussed in Chapter 4, clear-air attenuation can become significant for long ranges, particularly outside the window frequencies. The presence of any liquid hydrometeor in the transmission path greatly increases the loss. In such situations, the radar equation must contain a loss term in the propagation factor. Table 17.3 gives typical values for atmospheric attenuation for various conditions. As can be seen, the presence of any hydrometeor in the transmission path could significantly affect measurements, particularly at the higher frequencies.

Table 17.3 MMW Attenuation Summary.*

Medium	*Frequency Two-Way Attenuation* (dB/km)				
	1 GHz	10 GHz	35 GHz	100 GHz	200 GHz
Clear Air (Sea Level, 20°C)	—	0.04	0.1	0.6	2.0
Rain (4 mm/hr)	0.006	0.2	2.0	9.0	—
Rain (10 mm/hr)	0.16	1.0	5.2	12.0	—
Fog (0.1 gm/m^3, 100 m visibility)	—	0.02	0.4	3.0	8.0

Multipath

One of the major sources of error for most reflectivity measurements is multipath, but the problem is reduced at millimeter wavelengths. Multipath occurs because of reflections from the ground which can result in up to four paths for energy to take in traveling to and from a target, as shown in Figure 17.10. The resulting interference patterns from the four paths, each with slightly different path lengths, typically appear as shown in Figure 17.11, as a target moves relative to the radar in range or height. Since MMW antennas have narrower beamwidths for a given antenna size

---

*Adapted from Table 4.9.

than lower frequencies, the likelihood increases that energy reflected off the ground will be attenuated by the antenna. Also, the shorter wavelength results in the surface appearing rougher, which decreases the coherent scatter.

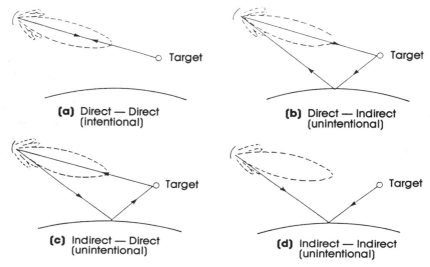

(a) Direct — Direct
(intentional)

(b) Direct — Indirect
(unintentional)

(c) Indirect — Direct
(unintentional)

(d) Indirect — Indirect
(unintentional)

Fig. 17.10 The four transmission paths between a radar antenna and a target near the earth, which leads to multipath on the received signal (from Scheer [5], © 1984 Artech House.)

Reference Reflector Errors

Errors in the manufacture of reference reflectors is more exagerated at higher frequencies. Figure 17.12 gives the effect on the RCS of an error in the flatness of one side of a diplane and a trihedral. At M band (95 GHz), a fabrication error of 1 mm yields a 3 dB error in the actual *versus* calculated RCS for a diplane.

System Instabilities

The increase in measurement system instabilities as the frequency increases also affects the accuracy of measurements. These effects can be countered by calibrating more often, and by using techniques such as biphase modulation, which is discussed in Section 17.4.3.

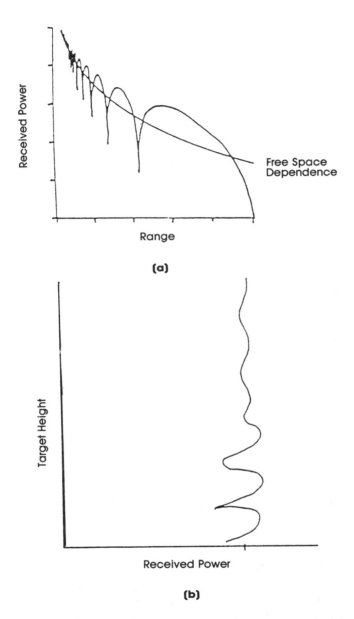

Fig. 17.11 Typical multipath lobing patterns as a target moves in (a) range and (b) elevation (from Scheer [5], © 1984 Artech House.)

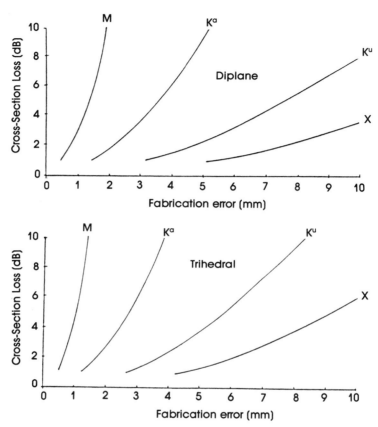

Fig. 17.12 Effects of errors in manufacture on the effective RCS of corner
reflectors (from [8]).

17.3   MM WAVE RCS MEASUREMENT SYSTEMS

17.3.1   System Considerations

Radar cross section measurements are being made from the VHF
range to the MMW frequency range. System design and performance when
operating within specific increments of this spectrum (VHF, UHF, microwave, or MMW), is dictated by the performance of system components
and environmental influences common to each frequency range. A discussion of equipment performance at MMW frequencies is provided in the
sections to follow.

17.3.1.1 System Trade-Offs

There are both advantages and limitations which must be considered when MMW systems are compared to longer wavelength microwave systems. Radar or RCS systems operating at MMW frequencies have significant advantages, which are:

- Equipment is physically smaller than microwave equipment.
- Narrow beamwidths are possible, yielding high accuracy, lower multipath or clutter, and high antenna gain.
- Large bandwidths are possible, permitting high range resolution and Doppler processing.

Limitations of MMW systems which must be considered include:

- Component cost is high.
- Component availability is low.
- Reliability of components is less than at microwave or lower frequencies.
- Test instrumentation availability is low.
- Operating range short is (10 to 20 kM maximum).
- System phase and amplitude stability *versus* temperature change or mechanical shock is poorer than at microwave frequencies due to the physically shorter millimeter wavelengths.

Some of the limitations such as cost and short operating range must be accepted. Limitations such as component availability, component reliability, and test equipment availability are being improved as more development funds are spent to satisfy the needs of the growing MMW market. System stability can be addressed by employing stabilized components and transmission paths where possible, or by including in the system design an automatic means of measuring and correcting for system phase and amplitude errors.

17.3.1.2 MMW Components for Measurement Systems

Antennas

Many MMW antennas are similar to types used at either microwave or optical wavelengths. These comprise a wide range of horn antennas (see Figure 17.13), front-fed paraboloid or Cassegrain reflector antennas (see Figure 17.14), including polarized gridded subreflector and polarization-twist versions; lens antennas (see Figure 17.15), including geodesic

lenses, waveguide lenses, and diffractive Fresnel zone plates; waveguide
arrays; dielectric rod antennas.

Scaling from microwave antennas necessitates smaller components
with closer fabrication tolerances. The relative sizes of antennas at frequencies of 10, 33, and 100 GHz, having equal gains and beamwidths, are
shown by Figure 17.16.

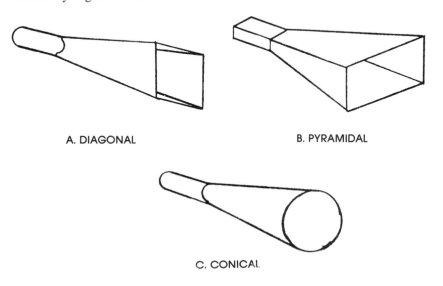

A. DIAGONAL   B. PYRAMIDAL

C. CONICAL

Fig. 17.13 Typical horn antenna shapes.

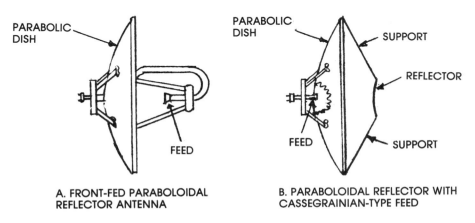

A. FRONT-FED PARABOLOIDAL
REFLECTOR ANTENNA

B. PARABOLOIDAL REFLECTOR WITH
CASSEGRAINIAN-TYPE FEED

Fig. 17.14 Reflector antennas.

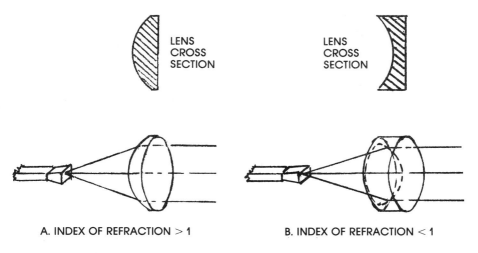

A. INDEX OF REFRACTION > 1   **B. INDEX OF REFRACTION < 1**

Fig. 17.15 Examples of lens antennas.

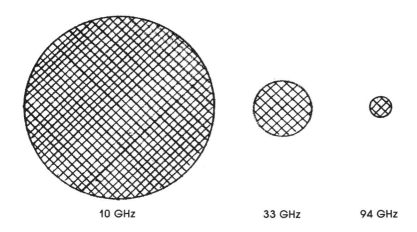

10 GHz   33 GHz   94 GHz

Fig. 17.16 Antenna size *versus* frequency for constant gain and beam-width.

Compact Ranges

Compact ranges are indoor simulations of outdoor antenna or RCS ranges, consisting of an appropriate signal source, a paraboloidal compact range reflector, and an enclosed facility properly lined with microwave absorber, as shown in Figure 17.17. This configuration results in a test zone (quiet zone) in which the characteristics of the illuminating wave are similar to far-field conditions. Compact ranges are now commercially available that are capable of operating at frequencies up to 94 GHz. Figure 17.18 shows the typical antenna pattern accuracy *versus* pattern dynamic range and frequency for standard ranges manufactured by Scientific-Atlanta, Inc., and Figure 17.19 shows a typical compact RCS range configuration. The standard models typically provide quiet zones of four feet by four feet in extent.

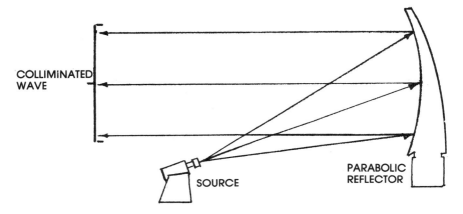

Fig. 17.17 Illustration of a compact range concept.

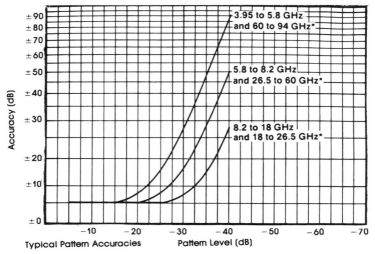

Fig. 17.18 Measurement accuracy of a compact range as a function of signal level and frequency (figure courtesy of Scientific Atlanta).

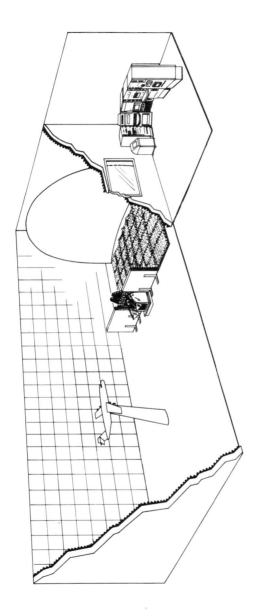

Fig. 17.19 A typical compact range configuration (drawing courtesy of Scientific Atlanta).

Improved versions of the compact range reflector now provide quiet zones which are four feet high by six feet long, and a new, larger parabolic reflector is now being developed, which will provide a quite zone that is eight feet high by 12 feet long. These new compact range improvements will make it possible to conduct RCS tests of many full-scale targets as well as models in an indoor environment with high security and away from detrimental atmospheric influences.

Waveguide Components

Waveguide components such as directional couplers, timers, attenuators, *pin* switches, and wave meters are available in sizes up to WR-3 waveguide, which has dimensions of 0.030×0.015 inches (0.75×0.38 mm) and are useful at frequencies up to 325 GHz. The calculated waveguide loss at this frequency is very high (17 dB/m). The actual loss is much greater due to the lack of precision involved in the manufacture of this small waveguide. Waveguide components at frequencies of 140 GHz and lower are practical and in common use. At frequencies above 140 GHz, many waveguide devices other than simple waveguide may perform poorly. For example, a 6 dB directional coupler in WR-5 waveguide (140 to 220 GHz) may have a 3 dB insertion loss at the upper end of its range.

Ferrite Devices

Ferrite circulators, isolators, and waveguide switches are available at frequencies up to 140 GHz. At frequencies above 90 GHz, the losses of these devices may make their usefulness marginal. In addition to losses, the components perform poorly at the higher frequencies.

Quasioptical Devices

An extension of optical techniques to the higher MMW frequencies provides one solution to the problem of increased waveguide losses. This may be accomplished by the use of lenses and mirrors to direct and focus MMW radiation. Care must be exercised in the system design to keep diffraction losses at a minimum.

The quasioptical approach is applicable to the design of transmission paths, attenuators, directional couplers, and switches. A schematic of a quasioptical diplexer is shown by Figure 17.20.

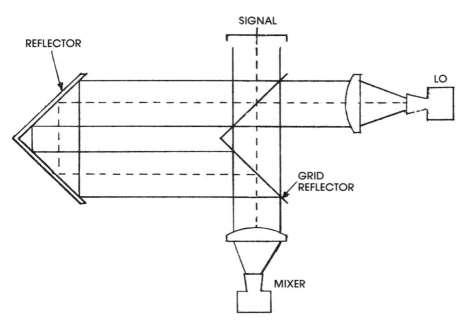

Fig. 17.20 Diagram of a Martin-Puplett quasioptical diplexer.

Signal Sources

Signal sources at MMW frequencies fall into three general categories:

1. Vacuum tube sources, including klystrons, magnetrons, and gyrotions;
2. Solid-state sources, such as Gunn and IMPATT diode sources, multipliers, and GaAs FET sources, which are now available at the lower MMW frequencies;
3. Laser sources including both discharge and optically pumped devices.

Vacuum tube technology now offers several types of MMW sources, including the extended interaction oscillator (EIO), the backward-wave oscillator (BWO), and the gyrotron, in addition to the basic klystrons and magnetrons (Ewell [9]). These devices furnish usable power at frequencies up to several hundred gigahertz. Table 17.4 gives the characteristics of typical vacuum tube oscillators, which are commercially available at popular MMW frequencies.

Table 17.4 Vacuum Tube Oscillator Characteristics.

Type of Source	Bandwidth	Pulsed Operation	Power Output Capability	General Use
Magnetron	Narrow	Yes	1 kW at 95 GHz	Pulsed[1] Transmitter
BWO	Wide	No	1 W at 300 GHz	CW source[1] Local Oscillator
Reflex Klystron	Wide[2]	Yes	500 MW at 80 GHz 10 MW at 220 GHz	Local Oscillator[3]
EIO	Narrow	Yes	5 kW at 35 GHz 40 W at 280 GHz	Pulsed[1] Transmitter

Signal source power can be boosted by the use of vacuum tube amplifiers, such as the TWT or the extended interaction amplifier (EIA). The TWT amplifiers are capable of pulsed power outputs of 100 W at 94 GHz with a 10% duty cycle. These amplifiers require high-voltage power supplies, have a high input noise figure, and have a limited operating lifetime.

The EIA is an amplifying version of the EIO. This amplifier is a narrow-band device capable of pulsed power outputs of 2 kW at 95 GHz or 100 W (CW) at 55 GHz.

Solid-state sources that are commercially available at popular frequencies are the IMPATT source (Ewell [9]), the Gunn source, the GaAs FET source, and various multiplier sources. These devices operate up to frequencies above 200 GHz, are small, efficient, long-lived, and require low power supply voltages. A summary of the characteristics of solid-state sources is given in Table 17.5.

Solid-state amplifiers that are available to increase source power are the injection-locked IMPATT amplifier capable of operation to above 200 GHz, and the new GaAs FET amplifiers now available at frequencies through 40 GHz. At present, power outputs of available GaAs FET amplifiers are limited to approximately 20 mW.

[1]May be injection-locked.
[2]Employs electromechanical tuning.
[3]May be phase-locked.

Table 17.5 Solid-State Source Characteristics.

Type of Source	Bandwidth	Pulsed Operation	Power Output Capability	General Use
GaAs FET Varactor or mechanically tuned	Wide	Yes (pulsed)	13 W at 95 GHz 500 mW at 40 GHz, CW 10 mW at 230 GHz, CW	Medium Power Pulsed or CW Source
YIG-Tuned	Wide[1]	No	20 mW, CW at 40 GHz 5 mW, CW at 50 GHz	Low Power CW Source Local Oscillator
Gunn	Wide[1]	No	200 mW, CW at 40 GHz 10 mW, CW at 100 GHz	Low Power CW Local Oscillator
Varactor Multiplier	Narrow	Yes	Doubler 45 mW at 140 GHz Doubler 100 mW at 180 GHz	Low Power CW Source or Local Oscillator

Laser sources may be either discharge pumped or optically pumped devices. Discharge pumped CW output sources, such as HCN (890 GHz) and H_2O (2500 GHz), supply outputs of only a few milliwatts. In general, the generated frequencies are useful only for short-range scale model applications.

Optically pumped laser sources have yielded outputs at a frequency as low as 153 GHz, although most useful frequencies lie above 300 GHz. One disadvantage of these sources is their extreme inefficiency (typically, 0.1%).

Mixers

Mixers employing Schottky-barrier diodes, which have either whisker contacts or beam-lead construction, are commercially available up to above

---

[1]Mechanically tuned-narrow band varactor tuning. May be phase-locked.

220 GHz. The performance of beam-lead diodes is now comparable to that of whisker-contact diodes at frequencies up to 140 GHz. Diodes using whisker contacts have been in common use for MMW applications because the whisker geometry minimizes fringing capacitance. The principal drawback has been the mechanical difficulty of assembling the diode and whisker into a package or RF circuit. In recent years, a wafer mount devised by W.M. Sharpless has overcome this disadvantage by providing easy diode interchangeability. A wafer-mounted diode package is shown in Figure 17.21.

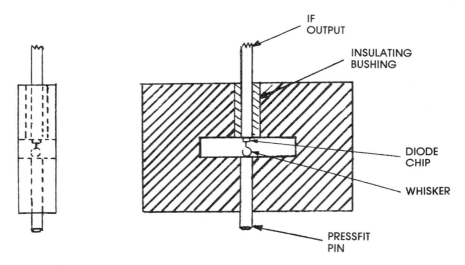

Fig. 17.21 A MMW diode wafer package.

Fundamental single-ended mixers operate to above 220 GHz, require a minimum of LO power (typically, 1 mW), provide low conversion loss, and are broadband devices.

Balanced mixers requiring approximately 5 mW of LO power are available up to 140 GHz. These mixers utilize matched diodes, provide lower conversion loss than single-ended mixers, and provide cancellation of LO noise.

Subharmonic mixers employ a LO frequency which is one half or one fourth of the signal frequency. These mixers utilize very wide RF and IF bandwidths, providing instantaneous signal bandwidths of up to 5%. Mixers driven by a LO frequency that is approximately one half the input signal frequency can exhibit DSB noise figures as low as 5 dB over the 35 to 350 GHz frequency range.

A mixer gives optimum performance only when the following conditions are met:

- The RF source impedance is matched to the diode impedance.
- The mixer back-short is optimized to minimize reactions.
- The diode whisker (if used) is kept short to minimize inductance.
- The mixer output impedance is matched to the IF load.

17.3.2   System Descriptions

17.3.2.1   Noncoherent Systems

A noncoherent RCS measurement system is able to measure and process only amplitude data. In such a system, phase data are either not available or not measured. Such a system is only capable of the measurement and processing of RCS *versus* target range, RCS *versus* target position, RCS *versus* frequency, and RCS *versus* polarization. A noncoherent RCS measurement system may have coherent subsystems for the purpose of achieving system objectives (e.g., frequency stability, frequency offset) without measuring phase. A block diagram of a basic MMW measurement system is shown by Figure 17.22. The system is general enough to represent either a pulsed system or a swept (FMCW) system.

The microwave-to-millimeter-wave up-converter and millimeter-wave-to-microwave down-converter, shown in dotted form, illustrate one method to obtain MMW performance by using a microwave RCS measurement system. A system which does not include the up-down converters uses MMW sources, amplifiers, and mixers, as well as basic waveguide components.

The basic MMW RCS system shown includes a low RCS target support (pylon) mounted on a positioning system driven by a power control unit and a position controller. Transmitted signals (pulsed or swept) are supplied by the signal source modulator, controlled by the output of the timing unit.

The receiver consists of a down-converter followed by detection circuits and a data processor. A central computer controls the overall system and directs data to displays, plotters, printers, and storage disks or magnetic tape units. The system shown can be modified to represent a coherent RCS measurement by supplying a coherent phase reference to the detection system to permit both phase and amplitude to be measured. System calibration is accomplished following the substitution of a reference target of known RCS in place of the target to be tested.

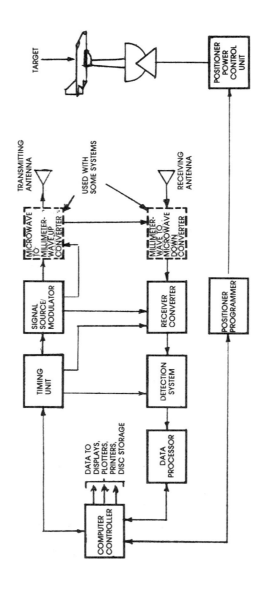

Fig. 17.22 Block diagram of a basic MMW RCS system.

The system controller and processor are designed to reduce operator interaction with the controller, and allows the subsystems to perform their respective functions. As a minimum, the system should be capable of acquiring and analyzing data as well as formatting the data for output.

Figure 17.23 is a block diagram of a coherent frequency converter, which is often used in coherent and noncoherent receiving systems that average received data by using narrow-band IF filters. This technique results in significant improvement of receiver sensitivity due to the narrow measurement bandwidth.

The system shown by Figure 17.23 includes a 300.01 MHz voltage-controlled oscillator (VCO), which is phase-locked to a 300.00 MHz crystal stabilized oscillator. A CW sample of the transmitted signal is supplied to the reference channel, and is converted to 300.00 MHz. This frequency is maintained by the reference channel phase-locked loop, which controls the frequency of the LO. Pulsed inputs at the signal channel's input are also converted to 300 MHz by the phase-locked LO. This 300 MHz signal is amplified, gated, and converted to a 10 kHz IF by the signal channel's second mixer. The IF sample gate permits only desired pulse returns to be converted to 10 kHz. The 10 kHz second IF is passed through a narrow-band bandpass filter, amplified, and supplied as an output for further processing.

Average Detection Systems

A narrow-bandwidth average detection RCS pulsed system is shown by Figure 17.24. A stable MMW source is modulated by a *pin* switch modulator supplying pulses to the MMW amplifier. The pulsed output of the amplifier is gated off, following the modulator pulse interval, to prevent the amplifier noise from coupling into the receiver during pulse reception. The pulsed output is sampled and supplied to the transmitting antenna.

The stable source may be a GaAs FET source (33 to 50 GHz), a pulse-locked Gunn-diode source, a phase-locked IMPATT source, or a phase-locked reflex-klystron source. Below 50 GHz, low-power YIG-tuned sources cover standard waveguide bands (26.5 to 40 GHz and 33 to 50 GHz). Narrow-band Gunn sources are available to supply powers of 50 or more at frequencies up to 90 GHz, narrow-band phase-locked IMPATT sources, operating at frequencies up to 96 GHz, are available with power outputs of 200 mW or more, and CW reflex klystrons having output powers of several hundred milliwatts are available to above 100 GHz.

A practical MMW pulsed amplifier must be either an EIA or a TWT. At present, solid-state amplifiers can only deliver a few milliwatts at frequencies in the 33 to 40 GHz range.

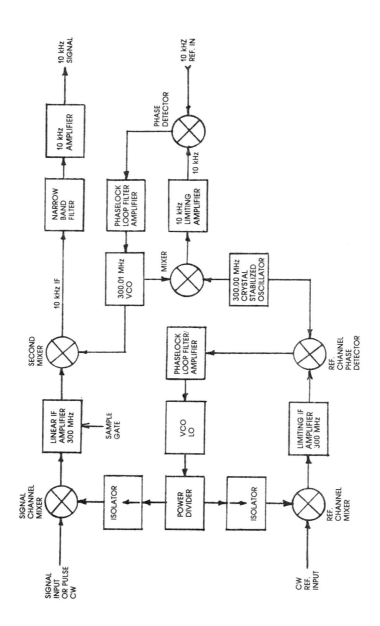

Fig. 17.23 Block diagram of a coherent frequency converter.

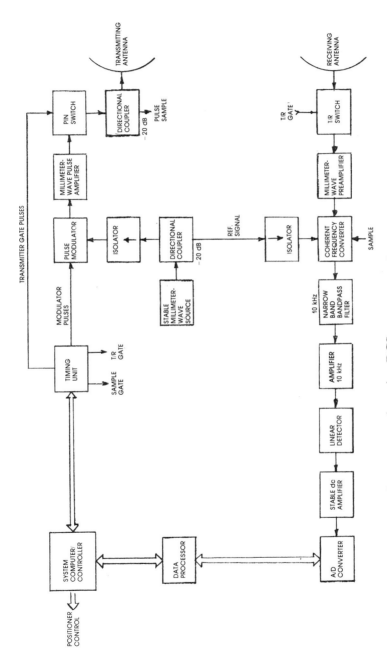

Fig. 17.24 Block diagram of an average detection RCS measurement system.

The receiver includes an input *pin*-diode T/R switch to protect the input circuits from burnout. Low-noise-figure (4 dB) preamplifiers are now available to 40 GHz. Above 40 GHz, a low-conversion-loss balanced mixer followed by a low-noise-figure IF preamplifier can yield a mixer input noise figure of 10 dB or less. The frequency converter is internally gated (see Figure 17.25) to pass only the desired received pulse. The 10 kHz output of the system is filtered, amplified, and supplied to a linear detector. The detector output is A/D converted and supplied to the data processor.

The sensitivity of the average detection receiver is dependent on the following factors:

- Input noise figure (N);
- Effective noise bandwidth (B, in Hz);
- The signal duty cycle DCS = (pulsewidth × PRF);
- The noise duty cycle (DCN = average receiver range-gate-on-interval × PRF). This assumes that the receiver noise generated after the range-gate switch is negligible.

The CW sensitivity is given by

$$S_{CW} = (-174 + N + 10 \log B) \text{ dBm} \qquad (17.10)$$

The pulse sensitivity of the system is

$$\begin{aligned} S_p &= S_{CW} + 20 \log \text{DCS} \\ &\quad - 10 \log \text{DCN (for DCS = DCN)} \end{aligned} \qquad (17.11)$$

If the signal duty cycle (DCS) and noise duty cycle (DCN) are equal then the pulse sensitivity for $S = N$ is

$$S_p (S = N) = S_{CW} + 10 \log \text{DCS} \qquad (17.12)$$

A typical system having a duty cycle of .01 (signal and noise), a noise bandwidth of 100 Hz, and an input noise figure of 10 dB will have a noise floor of -124 dBm.

Peak Detection System

An example of a peak detection RCS system is shown by Figure 17.25. The 300 MHz COHO is coherent with a stable, programmable, MMW signal source. A sample of the COHO is pulse-modulated and supplied together with a sample of the MMW source to the SSB up-converter. The frequency up-converted pulse is amplified and coupled

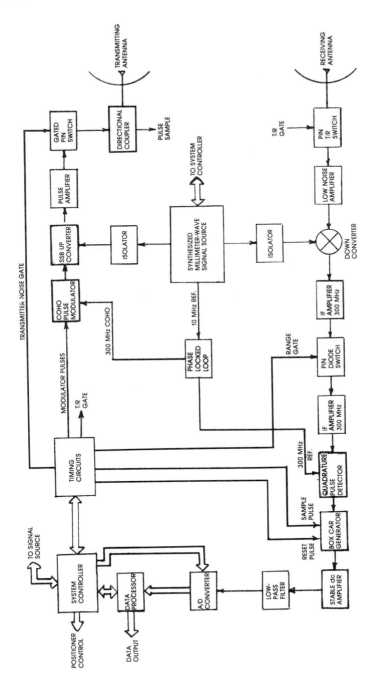

Fig. 17.25 Block diagram of a peak detector millimeter-wave RCS measurement system.

through a *pin*-diode switch and directional coupler to the transmitting antenna. Available MMW components for the transmitter are the same as for the average detection system.

A T/R switch in the receiver input enables the receiver following pulse transmission. Received pulses are converted to 300 MHz in the mixer, which is supplied a portion of the synthesized MMW signal source output as a LO signal. The *pin*-diode switch in the IF system allows a specific RCS range to be processed by the quadrature pulse detector, boxcar generator, A/D converter, and digital circuits.

Millimeter-wave components for the receiver (mixers and isolators) are available through 40 GHz. Above 40 GHz, a low-noise mixer IF system is recommended for the receiver front end.

17.3.2.2   Coherent Systems

The coherent RCS measurement system has the capability to measure both amplitude and phase of RCS data. A coherent system can be configured to measure RCS *versus* range, RCS *versus* frequency, RCS *versus* aspect, RCS *versus* cross range, and the polarization scattering matrix.

Single-Channel Pulsed Systems

A single-channel coherent RCS system using a narrow-bandwidth receiving system is shown by Figure 17.26. This system includes the following:

- A synthesized MMW signal source;
- A phase-locked COHO-second LO unit, supplying coherent 300.000 MHz COHO and 300.005 MHz second LO outputs;
- A transmitter SSB up-converter with output filtering;
- A MMW transmitter pulse amplifier with output noise gating provided by a *pin* switch;
- A directional coupler located at the input to the transmitting antenna to permit sampling of the transmitted pulse;
- A T/R *pin* switch at the receiver input to isolate the transmitter pulse from receiver input circuits;
- A low-noise MMW preamplifier, when available, at the operating frequency;
- A low-noise mixer-IF system (broadband 300 MHz);
- A second mixer to convert the 300 MHz IF to a 5 kHz IF;
- A narrow-band 5 kHz amplifier and I-Q detector;

- Output data filtering, amplification, A/D conversion, and processing for the I and Q channel signals;
- System timing circuits;
- System controller.

The received data, which are converted to 5 kHz, filtered, and detected in the I-Q detector, are coherent with the COHO signal and synthesized MMW signal source used to generate the transmitted signal. Both phase and amplitude data are present in the I-Q detector outputs, and these data are extracted by the data processor.

Receiver sensitivity is related to the following:

- Input noise figure (N);
- Effective noise bandwidth (B, in Hz);
- The signal duty cycle (pulsewidth times PRF), DC;
- The noise duty cycle (average receiver IF range-gate-on interval times PRF), DCN, which assumes that the receiver noise generated following the range gate switch is negligible.

The receiver CW sensitivity is

$$S_{CW} = -(174 + N + 10 \log B) \text{ dBm} \qquad (17.13)$$

The pulse sensitivity of the system is

$$S_p = S_{CW} + 20 \log DCS - 10 \log DCN \qquad (17.14)$$

When the signal duty cycle, DCS, and the noise duty cycle, DCN, are equal, then the pulse sensitivity for $S = N$ is

$$S_p = S_{CW} + 10 \log DCS \qquad (17.14)$$

A typical system having a duty cycle of .01 (signal and noise), a noise bandwidth of 100 Hz, and an input noise figure of 10 dB will have a noise floor of -124 dBm.

Most of the MMW components associated with the transmitter (SSB modulator, *pin* switch, directional coupler, waveguide, and antenna) are readily available from commercial sources.

For broadband systems that use a broadband synthesized signal source, the transmitter pulse amplifier must of necessity be a TWT. Narrow-band systems can use either a TWT amplifier or an EIA.

Receiver components such as the antenna, T/R switch first downconverter, and MMW isolators are available from commerical sources.

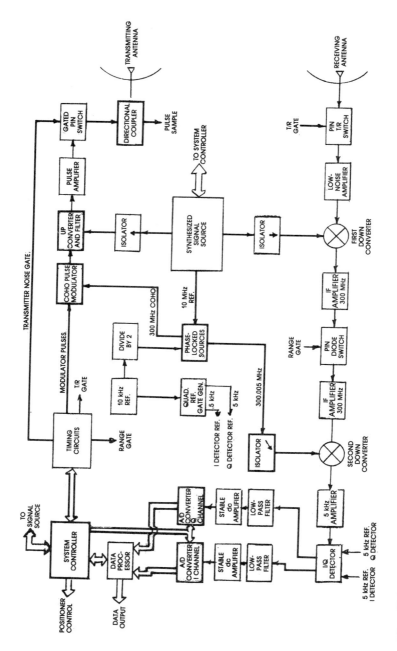

Fig. 17.26 Block diagram of a coherent average detection RCS measurement system.

Low-noise solid-state preamplifiers up to 40 GHz are now available at a price. Above 80 GHz, low-noise mixer-IF amplifier combinations are recommended.

Dual-Channel Systems

A dual-channel, peak-detection, RCS measurement system is shown by Figure 17.27. The system is coherent and capable of measurement of the polarization scattering matrix parameters, which require the pulsed RCS transmitter to transmit alternate orthogonal polarizations toward a target and the receiver to measure amplitude and phase data for the orthogonal polarizations associated with each transmission. The system shown by Figure 17.27 includes the following:

- A synthesized MMW signal source;
- A phase-locked COHO-second LO unit;
- A COHO biphase modulator, which shifts the phase of the COHO supplied to the pulse modulator by 180° on alternate pulses;
- A transmitter SSB up-converter with MMW output filtering;
- A MMW pulse amplifier with output noise gating provided by a *pin* switch;
- A directional coupler to permit the transmitted MMW pulse to be sampled;
- A transmitted polarization switch to permit orthogonally polarized pulses to be transmitted;
- A dual-polarized receiving antenna with each of two orthogonally polarized outputs fed to a separate receiving channel;
- T/R *pin* switches—one located at the input for each of the two receiving channels;
- Down-converters and IF amplifiers for each receiving channel;
- I-Q detectors for each receiving channel;
- A/D converters—one for each I and Q output;
- A data processor;
- A system controller.

Most of the MMW components associated with the transmitter (SSB modulator, *pin* switch, directional coupler, waveguide polarization switch, and antenna) are commercially available. For broadband systems that use a broadband synthesized source, the pulse amplifier must be either a solid-state amplifier (F is less than 40 GHz) or a TWT. Narrow-band systems may also use a EIA.

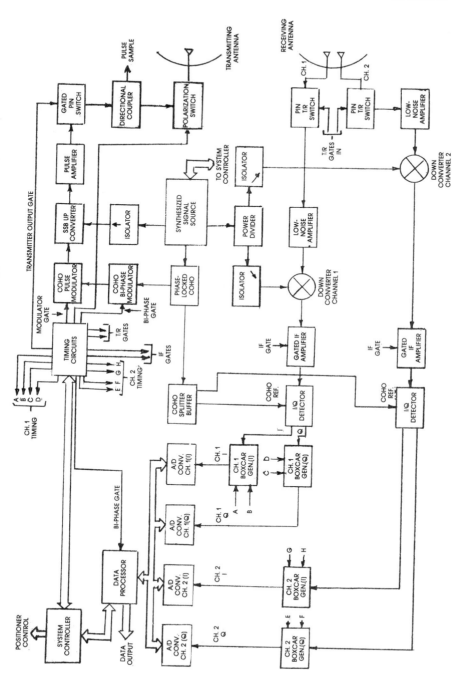

Fig. 17.27 Block diagram of a dual-channel peak detection RCS measurement system.

Receiver components such as the antenna, T/R *pin* switches, down-converters, and isolators are also commercially available. I-Q detectors can be constructed or purchased for IFs of 1.0 GHz or lower. Boxcar generators and A/D converters also are commercially available.

The system's timing unit supplies a biphase gate to the data processor to cause the sign of the processed data to be inversely synchronous with the biphase modulator. This caused data to add when averaged and system dc offsets to cancel.

The sensitivity of the receiving system is related to the following:

- Input noise figure (N);
- Effective noise bandwidth $(B$, in Hz$)$.

The CW sensitivity $(S = N)$ and single-pulse sensitivity of the receiver are equal to $S = -(174 + N + 10 \log B)$.

For a receiver noise figure (N) of 12 dB and a system noise bandwidth of 100 MHz, the single-pulse sensitivity $(S = N) = -82$. Since a minimum of two pulses must be averaged to operate with biphase modulation, the actual sensitivity must be calculated for two pulses and is equal to -85 dBm.

High Resolution FM CW Systems

The principle of operation of a linear FM CW RCS system is shown by Figure 17.28. The output of a linear swept oscillator is transmitted toward a target. The received energy that is reflected from the target is mixed with a sample of the swept oscillator, and the difference frequency is detected.

The difference frequency is proportional to the swept bandwidth and the target range (see inset of Figure 17.28) [11]:

$$F = \frac{2 \, BR}{Tc} \tag{17.15}$$

where

$F = $ difference frequency,
$B = $ swept bandwidth,
$R = $ range,
$c = $ velocity of light,
$T = $ sweep period.

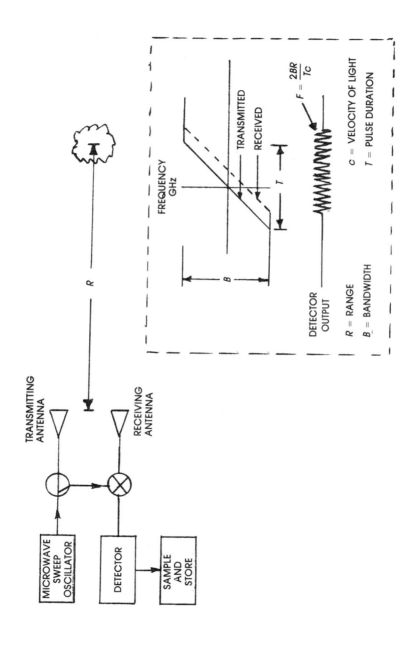

Fig. 17.28 Simplified linear FMCW RCS measurement system. (*Note*: inset shows frequency as a function of bandwidth and range.)

Figure 17.29 is a block diagram of a linear FMCW RCS system. The sweep oscillator's output is transmitted. A sample of the sweep oscillator is supplied as the local oscillator to the homodyne converter. The resulting IF is amplified, detected, sampled, and processed to provide high range resolution phase and amplitude data. The data are supplied together with target positioner data to complete the final data processing.

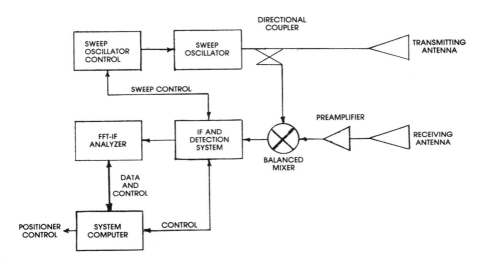

Fig. 17.29 Block diagram of a linear FMCW RCS measurement system.

A typical system includes a reference FM CW channel that employs a stable delay line and homodyne converter to monitor and to maintain the signal-source sweep linearity.

This type of RCS measurement system has several advantages:

- The system is relatively simple.
- The system is relatively inexpensive.
- The system offers high resolution.
- The system provides data over wide bandwidths.

A linear FM CW RCS measurement system must be carefully designed to overcome or minimize system limitations:

- A balanced mixer and a high gain preamplifier are required to overcome the effects of high, $1/F$ noise associated with the homodyne mixing process.

- The level of transmitted power must be kept low to avoid receiver saturation.
- The transmitting-receiving RF systems must include sufficient isolators to minimize reflections, which will appear as spurious responses.
- Operation much below 2.0 GHz is impractical due to the lack of bandwidth.
- Operation at MMW frequencies is best accomplished by an up-down converter to translate a swept microwave band to a MMW band.

Figure 17.30 is a photograph of the Model 2084 linear FM CW RCS system, which is commercially available from Scientific-Atlanta. The system currently performs over the 2.0 to 18.0 GHz frequency range with an upgrade involving frequency coverage to 40 GHz in development as of this writing. The linear sweep bandwidth is 3.0 GHz, sweep period = 0.1 s and sweep rate = 1 Hz. The sampled received data (1024 samples) are processed between sweeps. The system shown is capable of measuring the following:

- RCS *versus* range for a selected aspect angle;
- RCS *versus* range and aspect angle;
- RCS *versus* frequency for selected aspect angle;
- RCS *versus* aspect angle for any frequency within the swept band;
- RCS *versus* range and cross-range for any selected angle (isometric or contour display).

The system performance at microwave frequencies can be achieved over MMW bands by operating a system at microwaves and translating the frequency to a MMW band by the use of an up-down converter, as shown by Figure 17.31. An X-Band swept transmitter output is up-converted, amplified, and transmitted at MMW frequencies. Receiver MMW signals are down-converted to X Band and supplied to the receiver input. The crystal-stabilized LO, common to the transmitting and receiving channels, is selected to provide the required MMW frequency.

17.3.3 Techniques for System Performance Improvement

Achieving Wide Dynamic Range

The dynamic range of an IF, detection, and processing system is limited by detector nonlinearity, low-level instability, and the resolution

Fig. 17.30 The Scientific Atlanta Model 2084 RCS measurement system (photo courtesy of Scientific Atlanta).

of A/D converters. A 14-bit A/D converter will provide a -78 dB least significant bit plus a sign bit. It is possible to configure an IF system to achieve a linear dynamic range of 80 dB more, while reducing the dynamic range requirement on detectors, samplers, and A/D converters (two methods will be described below).

Figure 17.32 is a block diagram of an autoranging wide-dynamic-range IF and detection system. The system employs a programmable stable IF attenuator to adjust the IF signal level and reduce the dynamic range of operation of the detector and associated circuits. A typical system may operate over an 80 dB dynamic range, where the dynamic range presented to the detector may be as small as 6 dB. The automated system includes digital level-sensing and decision-making and digital attenuator phase and level calibration and control. One disadvantage of this system is the poor start-up accuracy, since the programmed value of attenuation is generally based on past data. This is a problem when very little or no data averaging is used, and data are not continuous, such as in the case in a frequency-stepped system.

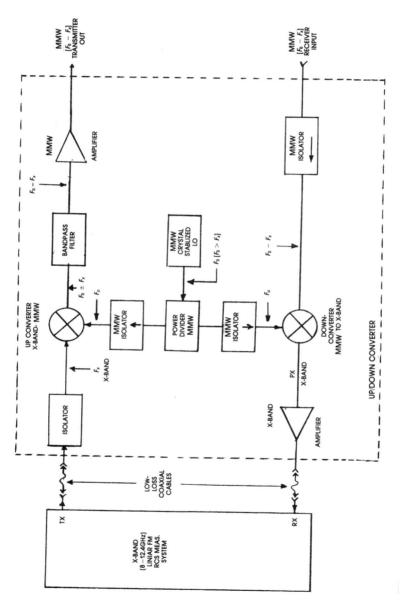

Fig. 17.31 Block diagram of a MMW linear FM RCS measurement system using an up-down converter.

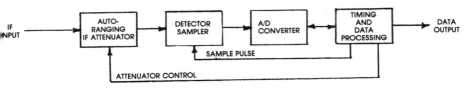

Fig. 17.32 An autoranging wide dynamic range channel.

Figure 17.33 is a block diagram of a wide-dynamic-range IF system that uses dual-detection channels. The IF signal is divided into two outputs. One IF channel is passed through a unity-gain buffer amplifier and supplied to a low-gain detector-sampler. The second IF channel is passed through a limiter and a 30 dB amplifier, and supplied to a high-gain detector-sampler. The two channels are always operational. At high signal levels, the input to the high-gain channel is limited, and data in the low-gain channel are valid. At low signal levels, data in the high gain channel are the most accurate.

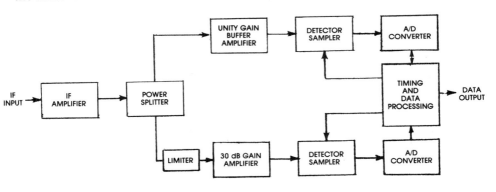

Fig. 17.33 A wide dynamic range IF channel using dual-level detectors.

The circuit shown in the figure divides an 80 dB dynamic range into 30 dB and 50 dB dynamic ranges. The automated system requires digital level-sensing, gain and phase calibration, and output selection. The system selects the most accurate data on a pulse-by-pulse basis.

Similar systems that employ more than two detectors can be used for wider dynamic ranges. The system is similar to the autoranging system, except that all possible outputs are retained to permit the selection of the most accurate data.

Minimizing Drift and Interference

Figure 17.34 gives a block diagram of a biphase-coded, pulsed, RCS measurement system. The system causes dc offsets and spurious signals, which are not phase coded to cancel. The timing unit supplies biphase timing gates to the biphase modulator and the controller-data processor. The phase of successive transmitted pulses is shifted between 0° and 180°. This causes the polarity of the I and Q outputs of the I-Q detector to shift from pulse to pulse. This polarity reversal is corrected in the data processor by changing the sign of data related to alternate pulses. When one or more data pairs are averaged the uncoded data (dc offset, spurious signals, or near-CW interference) will cancel. The arrangement permits ac coupling to be employed in the receiver channel.

Minimizing I-Q Errors

A perfect I-Q detector will have equal I and Q amplitudes, without dc offset, and will have exact phase I to Q quadrature. Typical I-Q detectors exhibit both amplitude imbalance and phase quadrature errors. Errors encountered in practical I-Q detectors are typically 1.0° amplitude and 5° phase. Correction may be accomplished by the careful adjustment of gain and phase trimming networks to balance amplitude and phase. The process, however, is time consuming and requires periodic readjustment if the I-Q detector and amplifier characteristics are not stable over the operating temperature range of the instrument.

Errors in the I and Q paths of a coherent RCS processor can be simply corrected by applying corrections derived from a test signal (Churchill, Ogar, and Thompson [12]). The correction can be simply performed by using a 2×2 matrix. In the following equation, I and Q are the desired corrected values, I_x and Q_x are the measured data.

$$\begin{bmatrix} I \\ Q \end{bmatrix} = \begin{bmatrix} C_1 & C_2 \\ C_3 & C_4 \end{bmatrix} \begin{bmatrix} I_x \\ Q_x \end{bmatrix} \tag{17.16}$$

The constants C_1, C_2, C_3, and C_4 are obtained by measuring a special quadrature test signal and computing Fourier coefficients. Correction of the data requires multiplying and summing of terms. A block diagram of an I-Q detector calibration system is shown by Figure 17.35. For the purpose of explanation, let us assume that the IF center frequency and the COHO frequency (F_o) are 400 MHz. A synchronous calibration signal $(F_o + F_c)$ 400.020 MHz if F_c is 20 kHz, is generated by phase-locking the

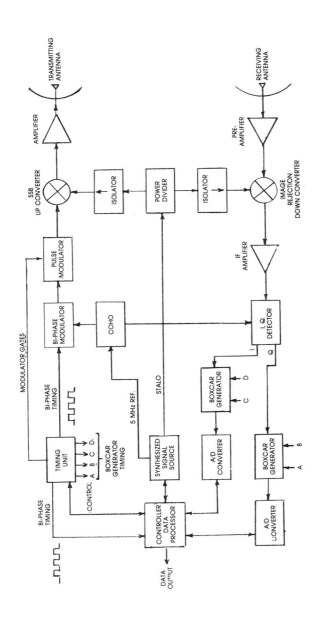

Fig. 17.34 A biphase coded, pulsed, RCS measurement system.

VCO to be 20 kHz above the COHO frequency. This signal is coupled as the IF input for I-Q calibration. The 20 kHz frequency offset of the calibration signal generates a linear phase scan (0° to 360°), which occurs at a 20 kHz rate at the input of the I-Q detector. Sample pulses at an 80 kHz rate are supplied to the sample-and-hold circuit for synchronously sampling the signal at exactly 90° intervals. This digitized data is processed to yield the necessary constants to be applied to the I-Q correction matrix. The correction process requires very little time, and can be periodically performed.

17.4 DATA ANALYSIS AND RESULTS

17.4.1 Data Requirements

Current MMW data requirements run the gamut from simple averaged RCS numbers to high resolution images of targets and clutter for their complete polarization matrix. In general, the required data vary with their intended use. A set of average values of the reflectivity of targets and the clutter background has traditionally served to define the scenario against which a radar must operate. However, in recent times, and particularly at millimeter wavelengths, it has been necessary to define the target and clutter amplitude distributions, the correlation coefficients, and possibly the clutter spectra in order to obtain more realistic radar designs.

As interest has centered on the discrimination of targets from clutter by using techniques other than just amplitude, polarization has attracted more attention. Parameters such as the polarization ratio (the ratio of the received polarization that is like that transmitted to the orthogonally received polarization) and the polarimetric phase (the phase between the two orthogonally received polarizations) have become important.

Finally, as the need to recognize and classify various types of targets has increased, interest has centered on measuring the full polarization scattering matrix and obtaining extremely high resolution. Thus, the need to process reflectivity data to obtain these results has arisen in recent years. Other uses for such data include correcting degraded data to obtain high-quality data of more standard format. For example, if the radar return from a target at low resolution is contaminated by the return from another object (such as a wall of an indoor chamber), the data can be processed to obtain high resolution, the return from the unwanted object is zeroed out, and the data are reconverted to the low resolution state without the influence of the unwanted object.

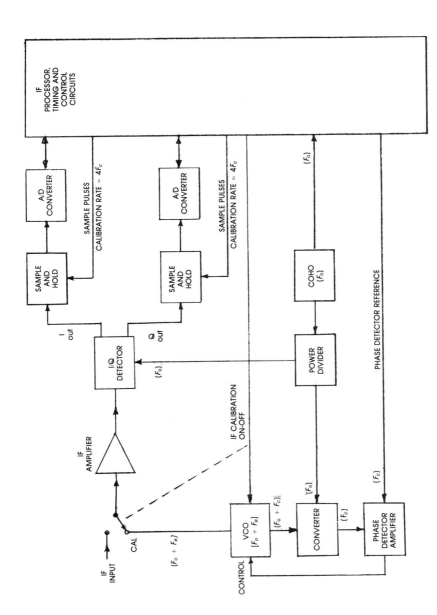

Fig. 17.35 Block diagram of an I-Q detector calibration system.

17.4.2 Analysis Techniques

As discussed in Section 16.1, MMW reflectivity measurements are performed to support specific problems. Thus, the form of data analysis being performed must also be tailored to the specific problem being addressed by the test program. Generally, data analysis activities can be considered to fall into one of two catagories: standard or advanced. Standard techniques usually involve the measurement of RCS ($\sigma°$) for a single polarization and for low resolution, usually determined by the standard radar antenna and pulse length. Advanced techniques include multiple polarizations (or the entire polarization matrix), high range resolution (usually using a swept-frequency technique), and high cross-range resolution (typically using synthetic aperture radar (SAR) techniques). Such measurements normally require coherent as well as polarimetric instrumentation radar systems. These two classes of analysis will be discussed below.

Standard Techniques

Standard techniques of data analysis involve minimum processing of the received reflectivity data (usually, only simple amplitude calibration) before the data are displayed or statistics are computed. The form of such data are typically polar plots as shown in Figure 17.36, or plots of $\sigma°$ as a function of depression angle as shown in Figure 17.37. Also, statistics of the data are often computed, including mean and median values, standard deviations, and amplitude distributions. Such parameters are quite useful in determining the amplitude performance of a radar system. For the determination of MTI performance, clutter spectra are often measured. Such spectra are often measured with noncoherent radars, and thus represent the amplitude spectra, but do not include phase effects.

Advanced Techniques

Advanced analysis techniques involve the use of multiple polarizations, a wide transmitted frequency spectrum, or target rotation to achieve more information about a specific target than standard techniques would yield. Multiple polarizations can be used to determine certain physical characteristics of a target, such as odd bounce or even bounce. Such characteristic differences can be used to detect the presence of targets in clutter as shown by Figure 17.38, which gives a grey-scale false color map of the return from snow at 95 GHz on a flat field. A 100 m^2 trihedral reflector is visible on the right side of the field. Note the absence of the reflector

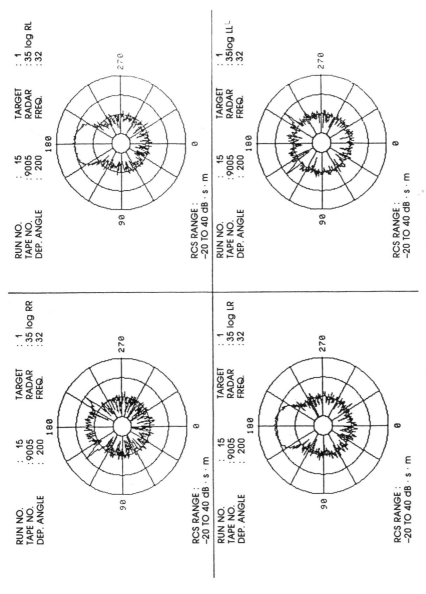

Fig. 17.36 Polar plot of the RCS of a 100 m² trihedral reflector as function of aspect angle for LL, LR, RL, and RR polarizations.

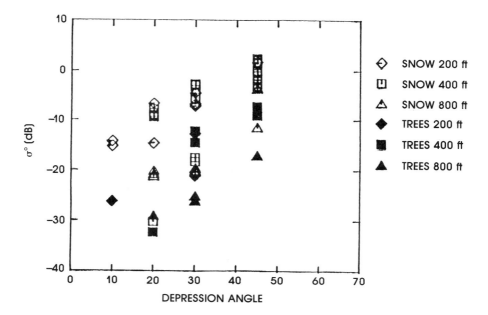

Fig. 17.37 RCS per unit area ($\sigma°$) of the return from the ground as a function of sensor altitude and depression angle at 35 GHz.

for the even-bounce polarizations, while the reflector shows up in the odd-bounce channels. The snow is similar in both polarization channels.

A wide transmitted frequency spectrum can be utilized to obtain high range resolution on a specific target. The process involves transmitting a range of frequencies in a short time, and recording the received amplitude and phase from a target for each frequency. These returns are equivalent to the spectrum of the pulse return from the target with the same bandwidth as the transmitted frequency range. (A 500 MHz bandwidth is approximately equivalent to a 2 ns pulse.) The inverse Fourier transform is computed for the "equivalent spectrum" to yield the high resolution time-domain response. This method of obtaining high resolution has the advantage over the real pulse technique, since narrow-band receivers, moderate-speed samplers, and A/D converters can be used, which greatly improves the data quality and reduces the cost.

High resolution in the cross-range dimension can be obtained if the target being measured is rotating relative to the radar. Scatterers that are present at different distances from the center of target rotation will have different relative velocities with respect to the radar, and thus different Doppler shifts. This difference in frequency can be separated by using a Fourier transform of the received signal for a number of PRF periods to yield a cross-range profile of target scatterers (called *inverse SAR*). If high

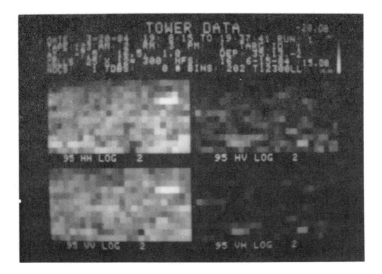

Fig. 17.38 Grey scale radar reflectivity map of the RCS per unit area ($\sigma°$) of snow-covered ground at 95 GHz.

down-range and cross-range processing are combined, then high resolution plots such as the contour plot given in Figure 17.39 can be achieved. Such data begin to approach the resolution of optical systems so that similar pattern recognition techniques can be put to use.

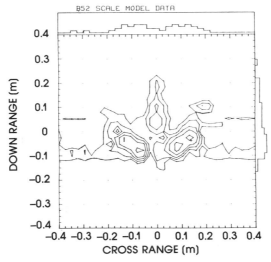

Fig. 17.39 High resolution contour plot of a B-52 22-inch scale model (figure courtesy of Scientific Atlanta).

REFERENCES

1. R.E. Kell, "On the Derivation of Bistatic RCS from Monostatic Measurements," *Proceedings of the IEEE*, August 1965, pp. 983–988.
2. G.W. Ewell, "Basic RCS Measurement Concepts and Systems," Chapter 3 of *Techniques of Radar Reflectivity Measurement*, N.C. Currie, ed., Artech House, Dedham, MA, 1984, pp. 61–63.
3. F.R. Williamson, "Private Communications," Georgia Institute of Technology, Atlanta, GA, May 1985.
4. R.N. Trebits, "Radar Cross Section," Chapter 2 of *Techniques of Radar Reflectivity Measurement*, N.C. Currie, ed., Artech House, Dedham, MA, 1984, pp. 27–53.
5. J.A. Scheer, "Radar Reflectivity Calibration Procedures," Chapter 4 of *Techniques of Radar Reflectivity Measurement*, N.C. Currie, ed., Artech House, Dedham, MA, pp. 109–134.
6. R.M. Barnes, "Calibration of a Coherent Dual-Polarized Radar Using IN-Scene Targets," ADT Project Memorandum No. 47-PM-ADT-0014, Massachusetts Institute of Technology, Lincoln Laboratories, Lexington, MA, May 1984.
7. J.A. Bruder, "Bruderhedral," Internal Technical Memorandum, Georgia Institute of Technology, Atlanta, GA, August 1986.
8. "Optical Theory of the Corner Reflector," MIT Radiation Laboratory Report No. 433, March 1944.
9. G.W. Ewell, *Radar Transmitters*, McGraw-Hill, New York, 1981, pp. 71-75.
10. J.A. Scheer and N.C. Currie, "Advanced Millimeter Wave RF Technology," *Proceedings of the Western Region Electronic Warfare Technical Meeting*, Southwestern Crow Club, White Sands Missile Range, NM, 1980.
11. D.L. Mensa, *High Resolution Radar Imaging*, Artech House, Dedham, MA, pp. 40–41.
12. F.E. Churchill, G.W. Ogar, and B.J. Thompson, "The Correction of I and Q Errors in a Coherent Processor," *IEEE Transactions on Aerospace and Electronic Systems*, vol. AES-17, No. 1, January 1981, pp. 131–137.

Chapter 18
MMW Radiometry

J.A. Gagliano

Georgia Institute of Technology
Atlanta, Georgia

18.1 INTRODUCTION TO MMW RADIOMETRY

Radiometry has been used successfully in the infrared, millimeter-wave, and the microwave bands for the detection of signals of thermal origin. The MMW region from 10 to 300 GHz is reviewed in this chapter, with particular attention to radiometric receivers. Advances in hardware components and improvement in the understanding of a need for operation in a variety of weather and climate situations have combined to provide an impetus for the development of new and advanced MMW radiometer systems.

Radiometers find applications ranging from radio astronomy to satellite-based earth surface meteorology, to air-to-surface and air-to-air missile guidance, to battlefield target detection through smoke, fog, and other obscurants, and to thermographic medical diagnostics. In each of these applications, the radiometric background may be considerably different. It is therefore necessary to radiometrically characterize each background and operating environment in order to have optimal design of the radiometer system.

18.1.1 Radiometric Principles

Radiometry is the passive reception and detection of natural radiation from an object. A significant contributor to the received radiation is the thermal emission, which is characterized by object emissivity and temperature. The passive signature is generally complex because of dependence on uncontrolled variables, such as moisture content, surface wetness, weather, cloud cover, and object temperature. The noncoherence of the passive signature is due to the fact that the emissivity of various materials is affected differently by each of the above variables. However, the reception of relatively weak signals can generally be compensated with very broad bandwidths and long integration times for MMW receivers.

Millimeter-wave radiometric signals are referred to as *blackbody radiation* in the MMW band of the electromagnetic spectrum. This blackbody radiation is primarily due to the thermal mechanisms, but it is significantly different in characteristic from blackbody radiation in the IR band. Specifically, the emitted energy in the IR band is dependent on T^4, where T is the object's physical temperature in K. This is in contrast with the MMW band by a dependence of T, which implies that object temperature is generally not the dominant factor. Object emissivity is a much more important consideration for MMW receivers.

As compared with a MMW radar system, the lack of a transmitter for passive detection provides obvious advantages for the use of radiometric sensors. The identification of the passive sensor is difficult because no transmitted signal is present. In addition, power consumption and sensor size are significantly reduced due to the absence of a transmitter.

The apparent radiometric temperature T_A of an object consists of both a thermally emitted component and a reflected component. The emissivity is defined as the ratio of that object's brightness temperature to the brightness temperature of a blackbody at the same physical temperature $(0 < \epsilon < 1)$. For an object that allows no incident RF to pass through (i.e., its transmissivity $t = 0$), $\epsilon + \rho = 1$, where ϵ is the emissivity and ρ is the reflectivity of the object. It is known from physical experiments that the reflectivity is a function of incident angle as well as material type and wavelength. The apparent radiometric temperature of an object also depends on aspect angle (grazing angle or depression angle). In a situation where an object fills the antenna beam, $T_A = \epsilon T_{object} + \rho T_{source}$, where T_{object} is the object's physical temperature and T_{source} is the reflected source's illumination temperature.

18.1.2 Antenna Temperature

The thermal source brightness (B) is a basic quantity useful for the derivation of the radiometer's antenna temperature. Assuming a blackbody source, that is a perfect absorber and a perfect emitter, leads to the following expression for B in watts/meters^2/Hz/steradian (King [1]):

$$B = \frac{2kT}{\lambda^2} \qquad (18.1)$$

where

k = Boltzmann's constant = 1.3804×10^{-23} J/K;
T = physical temperature of source, in K;
λ = signal wavelength, in meters.

Equation (18.1) is an approximation of Planck's law of radiation, which holds strictly for blackbodies. This approximation is referred to as the Rayleigh-Jeans law, which states that for observed frequencies in the MMW region, $hf \ll kT$, where h is Planck's constant ($h = 6.55 \times 10^{-34}$ J-s) and k is Boltzmann's constant. For gray bodies, which are not perfect emitters or absorbers, the physical temperature T is replaced by ϵT, where ϵ is the source emissivity, which is less than unity.

The received power W in W/Hz for a radiometric antenna with effective area A_E as shown in Figure 18.1 is given by [1]

$$W = \frac{1}{2} \iint_\Omega A_E (\theta, \phi) B(\theta, \phi) \; d\Omega \qquad (18.2)$$

The received power is reduced by one-half, as shown, because of signal polarization; that is, the direct radiation from the source is randomly polarized, resulting in only half the radiation being received at the singularly polarized antenna terminals.

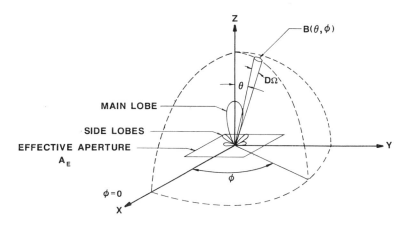

Fig. 18.1 Power received by antenna with effective aperture (A_E) from thermal source (B).

For extended thermal sources, where the source's solid angle exceeds the antenna beam angle, the power received is determined by the antenna beam angle Ω_A as given by [1]

$$\Omega_A = \frac{\lambda^2}{A_E} = \frac{4\pi}{G} \tag{18.3}$$

where G is the normalized antenna gain and A_E is the effective antenna aperture.

For radiometry, the antenna's apparent temperature T_A replaces the received power W as a measure of signal strength. T_A is defined as the temperature of a matched resistor with noise power output equal to W, that is $W = kT_A$. Utilizing the above expressions for B, W, and Ω_A leads to the following equation for the antenna temperature [1]:

$$T_A = \frac{1}{\Omega_A} \iint_\Omega T_s\,(\theta,\,\phi)G(\theta,\,\phi)\ \ d\Omega \tag{18.4}$$

where $T_s(\theta,\,\phi)$ is defined to be the brightness temperature of the thermal source B.

The basis for radiometric object detection is provided by the above expression for T_A. The brightness temperature T_s, modified by atmospheric propagation, is related to T_A as an equivalent matched load resistor in the receiver. The radiometer is used most effectively in the continuous comparison of the antenna output with that of a matched load. This basic principle leads to thermal source identification and interpretation, as well as to the determination of antenna and receiver configurations.

18.1.3 Surface Emissivity and Reflections

For a uniform thermal source, the antenna contrast temperature ΔT_A is proportional to the difference in the source and background brightness temperatures; that is,

$$\Delta T_A = T_s - T_B \tag{18.5}$$

The contrast ΔT_A is the received radiometric signal which depends on the physical temperature difference between the source and background, the atmospheric attenuation between the source and the radiometer, and the emissivity and reflections from other sources. Natural scenes often have little physical temperature differences compared to the differences in emissivity and reflectivity among various components. Except for extremely hot sources, such as rocket plumes, emissivity and reflections are the significant contributions to ΔT_A.

Electromagnetic properties of a surface are related to a set of scattering coefficients which, when measured, provide an important input to the radiometric emission process. The geometry depicted in Figure 18.2 shows an element of surface area S with incident and scattered radiation intensity I_0 and I_s, respectively. The scattering coefficient $\gamma(0, S)$ is given by [1]

$$\gamma (0, S) = \frac{4\pi R^2 I_s}{I_0 S} \cos\theta_0 \tag{18.6}$$

where R is the range and θ_0 is the incident angle.

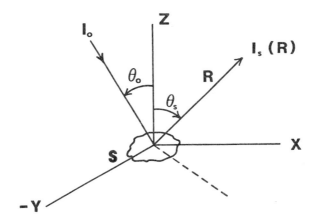

Fig. 18.2 Surface geometry for incident (I_0) and scattered (I_S) radiation intensity parameters.

Both emission and absorption coefficients are determined by surface roughness and material properties. For "smooth" surfaces, the emission is proportional to the projected area (Lambert's law) and, thus,

$$\epsilon = \epsilon_0 \cos\theta_0 \tag{18.7}$$

For "rough" surfaces, scattering properties are used with the expression for emissivity, given by

$$\epsilon = \left(1 - \frac{\gamma_0}{4}\right)\left(1 + \frac{1}{2}\sec\theta_0\right) \tag{18.8}$$

where γ_0 is obtained from backscattering data ($\theta_0 = \theta_S = 0$).

Most surfaces will exhibit a combination of smooth and rough characteristics, and thus are given by a combination of expressions (18.7) and (18.8) for surface emissivity.

18.1.4 Atmospheric Emission and Absorption

Since the early 1960s, the science of microwave radiometry has established itself as an integral part of remote sensing from spaceborne, airborne, and ground-based platforms. Figure 18.3 is a schematic representation of the contributions to energy received by an airborne radiometric antenna.

The influence of the atmosphere on the radiometric antenna temperature is represented by the loss factor L_{atm}. T_{up} is the atmospheric upwardly emitted radiation, and T_{sc} is the scattered radiation due to the downward atmospheric radiation T_{dn}. The target's self-emission temperature is given by T_B, which is affected by the atmospheric loss factor L_{atm}, as shown.

The antenna temperature, T_{Ai}, as measured in the main lobe of the antenna is, given by [2]:

$$T_{Ai} = \frac{1}{L_{\text{atm}}}(T_{Bi} + T_{Si}) + T_{\text{up}} \tag{18.9}$$

where the subscript i stands for horizontal or vertical polarization. An EM field is defined to be horizontally polarized with respect to a reference surface if its electric field is parallel to the surface. The EM field is vertically polarized if its electric field is perpendicular to the surface. For a nonlinearly polarized (elliptical or circular) antenna, the measured temperature is a linear combination of the horizontal and vertical components.

The radiometer's received temperature, T_r, present at the antenna's output, is given by

$$T_r = \eta\,\alpha_m T_A + \eta(1 - \alpha_m)T_{\text{SL}} + (1 - \eta)T_0 \tag{18.10}$$

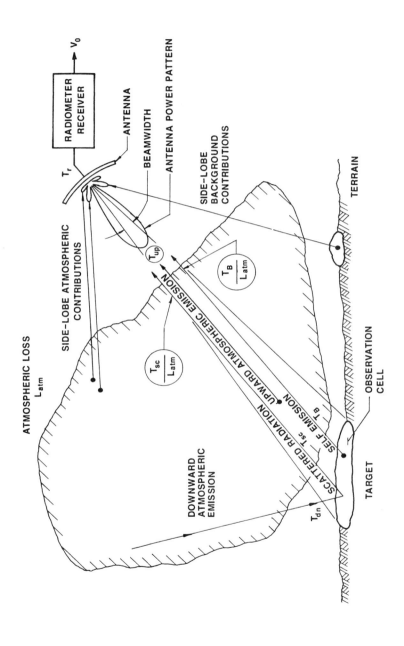

Fig. 18.3 Energy contributions to radiometric receiver including atmospheric effects (from Janza [2]).

where

  η = antenna's radiation efficiency,
  α_m = antenna's main-beam efficiency,
  T_{SL} = sidelobe average antenna temperature,
  T_0 = ambient temperature of antenna.

The factors shown in the expression for T_r represent the antenna's main-beam contribution $\eta\,\alpha_m T_A$, the sidelobe contribution $\eta\,(1 - \alpha_m) T_{SL}$, and the antenna's self-loss $(1 - \eta) T_0$. Of all these quantities, T_A and T_0 are measured directly, whereas η, α_m, and T_{SL} are either independently measured or calculated parameters.

18.1.5 System Noise Temperature

If we imagine that a receiver's ambient temperature is reduced to absolute zero and the temperature of a resistor connected to the input is adjusted until the system's output equals that of a receiver with no input signal, then the resulting output is referred to as the *equivalent input noise temperature* (T_{sys}). The noise temperature is representative of the equivalent available *noise power* (P_N) from a resistor maintained at ambient temperature. The output (in watts) of a receiver with noise temperature T_{sys}, due to the receiver's internal noise alone, is expressed as

$$P_N = kT_{sys}\beta G \tag{18.11}$$

where

  k = Boltzmann's constant = 1.38×10^{-23} J/K;
  β = system bandwidth, in Hz;
  G = system power gain.

The receiver's *noise figure* (NF) is defined as

$$\text{NF} = \left.\frac{\text{Input SNR}}{\text{Output SNR}}\right\} \text{when the input is terminated at 290 K}$$

$$= \frac{S_i/N_i}{S_0/N_0}$$

$$= \frac{N_0}{kT_0\beta G} \tag{18.12}$$

for $N_i = kT_0\beta$ (with T_0 = reference temperature), and $S_0 = GS_i$. For N_0 = receiver's output noise = $GN_i + P_N$, then,

$$NF = \frac{kT_0\beta G + kT_{sys}\beta G}{kT_0\beta G}$$

$$= 1 + \frac{T_{sys}}{T_0} \tag{18.13}$$

or, in terms of the system noise temperature:

$$T_{sys} = T_0 (NF - 1) \tag{18.14}$$

As a final note, when several stages of a receiver are cascaded, then the *total noise power* (P_T) at the last stage is given by

$$P_T = k\beta(T_1 G_1 G_2 G_3 \ldots + T_2 G_2 G_3 \ldots$$
$$+ T_3 G_3 \ldots + \ldots) \tag{18.15}$$

The overall system noise temperature becomes

$$T_{sys} = T_1 + \frac{T_2}{G_1} + \frac{T_3}{G_1 G_2} + \ldots \tag{18.16}$$

or, in terms of system noise figure:

$$NF_{sys} = NF_1 + \frac{NF_2 - 1}{G_1} + \frac{NF_3 - 1}{G_1 G_2} + \ldots \tag{18.17}$$

Consequently, the contribution of each successive stage of the receiver to the system noise figure is reduced due to the $1/G_1 G_2 \ldots$ terms.

18.1.6 System Bandwidth

The system bandwidth (β), as used in the noise temperature calculations, represents the radiometer's reception bandwidth, which affects the receiver's sensitivity. The reception bandwidth is defined by [3]:

$$\beta = \frac{\left[\int_0^\infty G(f)\,df\right]^2}{\int_0^\infty G^2(f)\,df} \tag{18.18}$$

where $G(f)$ is the radiometer's power gain response, as depicted in Figure 18.4.

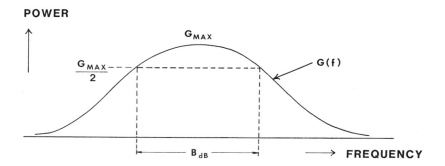

Fig. 18.4 Power gain *versus* frequency for radiometer.

A more common term for the frequency response of a circuit is the *half-power bandwidth* (β_{3dB}), as shown in Figure 18.4. There is a direct relation between the reception bandwidth (β) and the half-power bandwidth, which depends on the type of circuit (filter) used. For example, the tabular data which follow give the relationship for low-pass filters having one to three identical cascaded stages (Evans and McLeish [3]):

Type	β_{3dB}	β
Single RC	$1/2\pi RC$	$3.14\ \beta_{3dB}$
Dual RC	$0.643/2\pi RC$	$1.96\ \beta_{3dB}$
Triple RC	$0.510/2\pi RC$	$1.76\ \beta_{3dB}$

18.1.7 Radiometer Temperature Sensitivity

The sensitivity ΔT_{min} is a significant, measurable quantity for a radiometer's performance. ΔT_{min} describes the receiver's ability to distinguish input temperature changes from detected dc output changes. The sensitivity is the minimum detectable signal that produces a receiver dc output power which is equal to the noise output power. Figure 18.5 depicts a block diagram for the basic radiometer.

Fig. 18.5 Basic radiometer detection process.

It is generally assumed that amplifiers used in the predetection stage are linear devices which have constant gain and rectangular bandpass characteristics. The detector is assumed to be of the square-law type; that is, the output voltage is proportional to the input power. The derivation of ΔT_{min} for any radiometer configuration is actually the analysis of the detector output when its input signal is band-limited white noise. The section to follow describes the more common receiver configurations in terms of the sensitivity ΔT_{min}.

18.2 RADIOMETER SYSTEMS

18.2.1 Ideal Radiometer

A block diagram of the basic radiometer is represented in Figure 18.6, with each stage characterized by its power gain (G) and its reception bandwidth (B).

Fig. 18.6 Radiometric receiver with predetection gain (G_{IF}) and postdetection gain (G_{LF}).

As we previously mentioned, the detector input is band-limited white noise. Consequently, the power in the detector output's dc component is equal to the noise power output, which is commonly referred to as the *ac component*. The low-frequency components of the ac output are due to the difference frequencies generated in the IF amplifier portion of the predetection stage. The frequency range of the ac components extends from $F_{IF} - B_{IF}/2$ to $F_{IF} + B_{IF}/2$, where F_{IF} is the center frequency of the IF amplifier.

Figure 18.7 depicts the spectral power (W/Hz) for the square-law detector and the low-pass filter outputs.

The spectrum of the ac component contains frequencies that are close to zero (due to adjacent frequency components in the input) and frequencies that are close to the input bandwidth B_{IF} (due to the difference frequencies at either end of the input passband). Closely spaced pairs of frequencies across the entire input passband result in maximum power P_D near zero, while power is minimum at frequency B_{IF}. As a result, the

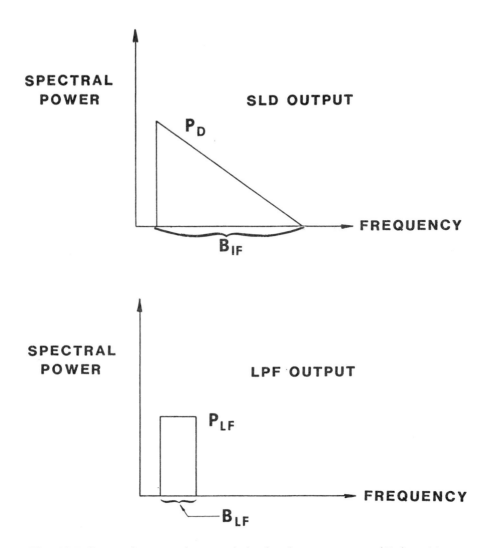

Fig. 18.7 Spectral power characteristics for detector output (P_D) and low-pass filter output (P_{LF}) (from Kraus [4] © 1966, McGraw-Hill Book Company).

square-law detector's spectral power output decreases linearly from the low-frequency end to the high-frequency end.

The low-pass filter is used at the detector's output to measure changes in the dc output component. The filter's bandwidth, B_{LF}, is set as narrow as possible, and it is determined by the frequency of the changes in the dc component to be measured. Since B_{LF} is much smaller than B_{IF}, then the spectral power at the *LPF* output is almost flat across the filter bandwidth.

The total dc power output of the detector is the area of the triangular spectrum, that is $P_{dc} = 1/2\, P_D B_{IF}$. The *LPF* ac output power is the area of the rectangular spectrum; that is, $P_{ac} = P_{LF} B_{LF}$. The ratio of the ac power to dc power results in

$$\frac{P_{ac}}{P_{dc}} = \frac{2 B_{LF}}{B_{IF}} \tag{18.19}$$

for $P_D = P_{LF}$, or in terms of voltages,

$$\frac{V_{ac}}{V_{dc}} = \sqrt{\frac{2 B_{LF}}{B_{IF}}} \tag{18.20}$$

This holds true for a band-limited white noise input to the detector and a rectangular passband low-pass filter.

Since the input temperature change ΔT can be measured by V_{ac} and the antenna temperature T_A plus the system noise temperature T_{sys} is determined by V_{dc}, then,

$$\frac{\Delta T}{T_{sys} + T_A} = \sqrt{\frac{2 B_{LF}}{B_{IF}}} \quad \text{or} \quad \Delta T = \left(T_{sys} + T_A\right)\sqrt{\frac{2 B_{LF}}{B_{IF}}} \tag{18.21}$$

If we assume that an ideal integrator, with an integration time of τ, is used as a low-pass filter, then the postdetection bandwidth (B_{LF}) is given by $1/(2\tau)$. The temperature sensitivity for the ideal radiometer is given by

$$\Delta T = \left(T_A + T_{sys}\right)\sqrt{\frac{2}{2\tau B_{IF}}} \tag{18.22}$$

$$= \frac{T_A + T_{sys}}{\sqrt{B_{IF}\, \tau}}$$

We should remember that B_{IF} is equivalent to the reception bandwidth previously given in (18.18), rather than the half-power bandwidth (B_{3dB}). For a two-pole filter, $B_{IF} = 1.96\ B_{3dB}$, according to Section 18.1.6.

18.2.2 Total-Power Radiometer

The ideal radiometer case assumes that the receiver's system gain remains constant. In practice, gain variations are unavoidable, due to supply voltage changes and ambient temperature fluctuations. Normally, the square-law detector cannot distinguish an increase in signal power from an increase due to higher predetection gain. Generally, random gain variations will effectively increase the system noise temperature.

The rms measurement uncertainty (ΔT_G) due to system gain variations is $(T_A + T_{sys})(\Delta G/G)$, where ΔG is the gain variation and G is the average power gain of the radiometer's predetection stage. For ΔT_s equal to the rms measurement uncertainty of an ideal radiometer (according to (18.22)), and since ΔT_s and ΔT_G are considered to be statistically independent, then both terms can be combined to determine the total-power radiometer's temperature sensitivity (ΔT_{min}):

$$\Delta T_{min} = [(\Delta T_s)^2 + (\Delta T_G)^2]^{1/2}$$

$$= \left(T_A + T_{sys}\right)\left[\frac{1}{\beta\tau} + \left(\frac{\Delta G}{G}\right)^2\right]^{1/2} \tag{18.23}$$

Automatic gain control (AGC) techniques have been used to stabilize the gain in the total power receiver. Care must be taken in using such techniques because AGC is likely to vary the receiver's noise figure and bandwidth, resulting in spurious output fluctuations.

18.2.3 Dicke Switched Radiometer

A technique for reducing the effects of gain fluctuations on the radiometer's temperature sensitivity was developed by R.H. Dicke in 1946 [4]. The scheme involves switching the receiver input at a constant rate between the antenna port and a constant temperature reference load. The switched frequency component of the square-law detector is then synchronously detected such that the final output is proportional to the temperature difference between the antenna and the reference load.

If we assume that V is equal to the synchronous detector voltage output, T_A is the antenna temperature, T_{sys} is the receiver noise temperature, T_D is the Dicke reference temperature, and G is the system gain factor, then, as shown in Figure 18.8, we have

$$V = G(T_A + T_{sys}) - G(T_D + T_{sys}) = G(T_s - T_0), \quad (18.24)$$

for $T_s = T_A + T_{sys}$ and $T_0 = T_D + T_{sys}$.

If we define $T_{out} = V/G$, then a radiometric measurement consists of measuring changes in V and determining the output temperature T_{out} for a known value of G, calculated by a calibration procedure. A determination of uncertainties in the output temperature T_{out} leads us to the temperature sensitivity ΔT_{min} of the Dicke radiometer. From (18.24), changes in the output voltages (ΔV) are given by

$$\begin{aligned} \Delta V &= G(\Delta T_s - \Delta T_0) + \Delta G(T_s - T_0) \\ &= G(\Delta T_s - \Delta T_0) + \Delta G T_{out}, \end{aligned} \quad (18.25)$$

for $T_{out} = T_s - T_0$.

Since ΔT_{out} (change in output temperature) is calculated from ΔV (change in output voltage) divided by the system gain, then from (18.25) we have

$$\Delta T_{out} = \frac{\Delta V}{G} = \Delta T_s - \Delta T_0 + \frac{\Delta G}{G} T_{out} \quad (18.26)$$

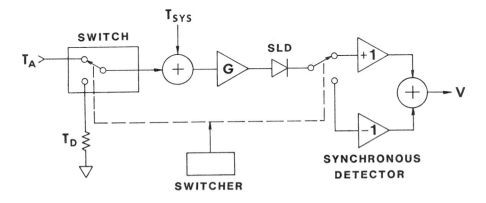

Fig. 18.8 Synchronous detection technique for Dicke-switched radiometer.

Since the radiometer's temperature sensitivity (ΔT_{min}) is defined to be the rms value of T_{out}, then,

$$\Delta T_{min} = \Delta T_{out,rms} = \left\{ (\Delta T_s)^2 + (\Delta T_0)^2 + \left[\left(\frac{\Delta G}{G} \right) T_{out} \right]^2 \right\}^{1/2}$$

$$= \left\{ 2(\Delta T_s)^2 + \left[\left(\frac{\Delta G}{G} \right) T_{out} \right]^2 \right\}^{1/2} \tag{18.27}$$

for $T_s \approx T_0$ (Dicke design) and where all terms are uncorrelated. In the ideal radiometer case, ΔT_s is given by

$$\Delta T_{s,ideal} = \frac{T_A + T_{sys}}{\sqrt{\beta\tau}} \tag{18.28}$$

whereas in the Dicke design:

$$\Delta T_{s,Dicke} = \frac{\left(\dfrac{T_A + T_D}{2} \right) + T_{sys}}{\sqrt{\beta(\tau/2)}}$$

$$= \frac{\sqrt{2} \left[\left(\dfrac{T_A + T_D}{2} \right) + T_{sys} \right]}{\sqrt{\beta\tau}} \tag{18.29}$$

The Dicke expression is based on the premise that only half the switching time (τ) is used to view the antenna, while the other half is used to view the Dicke reference. Substitution of ΔT_s into the ΔT_{min} expression for the Dicke radiometer leads to

$$\Delta T_{min} = \left\{ 4 \left[\frac{\left(\dfrac{T_A + T_D}{2} \right) + T_{sys}}{\sqrt{\beta\tau}} \right]^2 + \left[\left(\frac{\Delta G}{G} \right) T_{out} \right]^2 \right\}^{1/2}$$

$$= 2 \left(\frac{T_A + T_D}{2} + T_{sys} \right) \left[\frac{1}{\beta\tau} + \frac{\left(\dfrac{\Delta G}{G} \right)^2 (T_A - T_D)^2}{4 \left(\dfrac{T_A + T_D}{2} + T_{sys} \right)^2} \right]^{1/2}$$

$$= (T_A + T_D + 2T_{sys}) \left\{ \frac{1}{\beta\tau} + \left[\left(\frac{\Delta G}{G} \right) \left(\frac{T_A - T_D}{T_A + T_D + 2T_{sys}} \right) \right]^2 \right\}^{1/2} \tag{18.30}$$

A comparative example of the temperature sensitivity for the total power *versus* the Dicke radiometer is worth presenting. We will do so next.

If $T_A - T_D = 10$ K, $T_D = 300$ K, $T_{sys} = 1000$ K, $\beta = 1$ GHz, and $\tau = 0.1$ s,

$$\Delta T_{min(total\ power)} = \begin{cases} 0.185 \text{ K, for } \dfrac{\Delta G}{G} = 0.01\% & (18.31) \\[2em] 1.87 \text{ K, for } \dfrac{\Delta G}{G} = 0.10\% & (18.32) \end{cases}$$

and

$$\Delta T_{min, Dicke} = 0.241 \text{ K, for } 0.01\% < \frac{\Delta G}{G} < 0.10\% \qquad (18.33)$$

The Dicke radiometer's temperature sensitivity is independent over the $\Delta G/G$ range because the $1/B\tau$ term dominates in the expression for ΔT_{min}. However, the above example illustrates the effect that system gain fluctuations greater than 0.01% have on the total-power radiometer's sensitivity. From the above example, $\Delta G/G$ greater than 0.0172% suggests that the Dicke design is preferable over the total power design.

18.2.4 Correlation Radiometric Techniques

As system noise temperature limitations are approached, and if the Dicke switch adds significant loss to the overall system noise figure, then correlation techniques offer viable design options for radiometric receivers. However, a correlation radiometer requires two low-noise amplifiers, which add to the complexity of the system. Figure 18.9 is a block diagram of a single antenna correlation receiver. The output of the 3 dB hybrid is such that the phases of the output voltages produced by the input voltages cause the addition of inputs in the upper port and the subtraction of inputs in the lower port.

Given that V_A is the noise voltage due to input antenna temperature T_A, V_{ref} is the noise voltage due to input reference temperature T_{ref}, V_1 is the equivalent random noise input of amplifier with power gain G_1, and V_2 is the equivalent random noise input of amplifier G_2, then the output noise voltage from each amplifier is given by [4]:

$$V_{m1} = \sqrt{G_1} \left[\frac{(V_A + V_{ref})}{\sqrt{2}} + V_1 \right] \qquad (18.34)$$

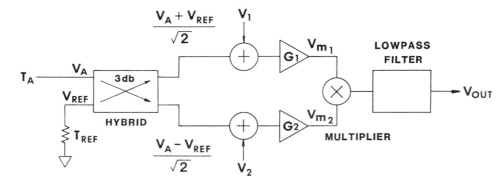

Fig. 18.9 Correlation technique for single antenna radiometric receiver.

$$V_{m2} = \sqrt{G_2} \left[\frac{(V_A - V_{ref})}{\sqrt{2}} + V_2 \right] \qquad (18.35)$$

V_{m1} and V_{m2} are multiplied to produce a voltage output of

$$V_{out} = \sqrt{G_1 G_2} \left[\frac{(V_A^2 - V_{ref}^2)}{2} + \frac{V_1(V_A - V_{ref})}{\sqrt{2}} \right.$$
$$\left. + \frac{V_2(V_A + V_{ref})}{\sqrt{2}} + V_1 V_2 \right] \qquad (18.36)$$

Since all voltages in the multiplier output are uncorrelated and random functions of time, then the $(V_A^2 - V_{ref}^2)/2$ term is the only one which produces a dc voltage at the low-pass filter output.

V_A is the noise voltage input signal due to the antenna temperature T_A, and so the output of the correlation receiver is directly proportional to the input signal (T_A) and the square root of the power gain product of amplifiers G_1 and G_2.

The derivation of the temperature sensitivity of the correlation receiver leads to the following [3]:

$$\Delta T_{min,corr} = \frac{\sqrt{2\,(x^2 + 1)}\,(T_A + T_{sys})}{\sqrt{B\tau}} \qquad (18.37)$$

for

$$x = \sqrt{\frac{G_2}{G_1}} - 1$$

If we assume $\Delta G/G = 0.05\%$ for total-power and Dicke radiometers, and $G_1 = G_2$ correlation receiver, then $\Delta T_{min} = 0.668$ K, 0.241 K, and 0.185 K respectively, for total-power, Dicke, and correlation receivers.

18.3 HARDWARE CONSIDERATIONS

18.3.1 Input Line Attenuation

A radiometer system's antenna will generally be connected to a low-noise RF amplifier or a mixer (superheterodyne receiver) by input lines that have losses. The noise effects of the input lines can be appreciable in the MMW region. Figure 18.10 depicts a system that uses an input line at ambient temperature T_0 and attenuation loss L.

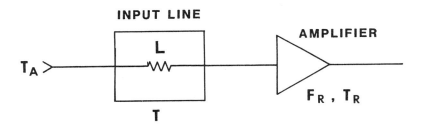

Fig. 18.10 Input line attenuation (L) for antenna-to-amplifier connection.

Assuming the amplifier noise figure and noise temperature are given by F_R and T_R, respectively, and that the gain of the input line is $1/L$, then the noise figure of the system is given by

$$
\begin{aligned}
F_s &= F_1 + \frac{(F_2 - 1)}{G_1} \\
&= L + \frac{(F_R - 1)}{1/L} \\
&= L + L(F_R - 1) \\
&= LF_R
\end{aligned}
\tag{18.38}
$$

The noise figure is expressed in decibels as

$$F_s \text{ (dB)} = L\text{(dB)} + F_R \text{ (dB)} \tag{18.39}$$

The system noise temperature (T_{sys}) is given by

$$
\begin{aligned}
T_{sys} &= T_0 \, (F_s - 1) \\
&= T_0 \, [L + L(F_R - 1) - 1] \\
&= T_0 \, [(L - 1) + L(F_R - 1)] \\
&= T_0 \left[(L - 1) + L\!\left(\frac{T_R}{T_0} + 1 - 1\right) \right] \\
&= T_0 \left[(L - 1) + L\!\left(\frac{T_R}{T_0}\right) \right] \\
&= T_0 \, (L - 1) + L T_R
\end{aligned}
\tag{18.40}
$$

Including the antenna temperature (T_A), the overall system noise temperature is given by

$$T_S = T_A + (L - 1) \, T_0 + L T_R \tag{18.41}$$

18.3.2 Mixers Used in Radiometer Front Ends

As we mentioned earlier, radiometer system front ends frequently use mixers fed directly from the antenna port. This is especially true in MMW systems that utilize wide RF bandwidths (10 to 20 GHz) for better temperature sensitivity. If both the signal and the image frequencies from the source are accepted by the mixer, then the double-sideband (DSB) noise figure is used to predict the system ΔT_{min}. However, if the image band is rejected, then only half the antenna power contributes to the system output, and the system ΔT_{min} is hence determined by the single-sideband (SSB) noise figure, which is 3 dB greater than the double-sideband figure.

Figure 18.11 depicts a radiometer system with antenna port fed by input lines having loss L to the mixer. In this configuration, both the signal and image frequencies are accepted by the mixer.

For the DSB receiver, the double-sideband noise figure (F_{DSB}) is dependent on the input line losses, the conversion loss (L_m) of the mixer, and the noise figure of the IF amplifier:

$$F_{DSB} = F_1 + \frac{(F_2 - 1)}{G_1} + \frac{(F_3 - 1)}{G_1 G_2} \tag{18.42}$$

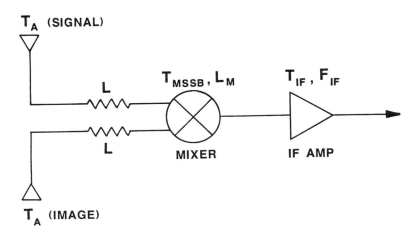

Fig. 18.11 Superheterodyne receiver configuration for double-sideband receiver.

where

$$\left.\begin{array}{ll} F_1 = L, & F_2 = L_m/2 \\ G_1 = 1/L, & G_2 = 2/L_m \end{array}\right\} \quad \text{for DSB operations,}$$

and

$$F_3 = F_{IF}$$

Therefore,

$$
\begin{aligned}
F_{DSB} &= L + \frac{\left(\dfrac{L_m}{2} - 1\right)}{1/L} + \frac{(F_{IF} - 1)}{(1/L)\,(2/L_m)} \\
&= L + L\left(\frac{L_m}{2} - 1\right) + \frac{LL_m}{2}\,(F_{IF} - 1) \\
&= \frac{LL_m}{2} + \frac{LL_m}{2}\,(F_{IF} - 1) \\
&= \frac{LL_m F_{IF}}{2}
\end{aligned}
\tag{18.43}
$$

For SSB operation where the image band is rejected, we have $F_{SSB} = LL_m F_{IF}$, which is twice the F_{DSB} factor, or 3 dB greater.

Likewise, the system noise temperature for the double-sideband receiver, including the antenna temperature, is determined by

$$T_{DSB} = T_A + T_0(F_{DSB} - 1)$$

$$= T_A + T_0 \left[L + L\left(\frac{L_m}{2} - 1\right) + \frac{LL_m}{2}(F_{IF} - 1) - 1 \right]$$

$$= T_A + T_0 \left[(L - 1) + L\left(\frac{T_m}{2T_0} + \frac{1}{2} - 1\right) \right.$$

$$+ \left. \frac{LL_m}{2}\left(\frac{T_{IF}}{T_0} + 1 - 1\right) \right]$$

$$= T_A + T_0 \left[(L - 1) + \frac{L}{2}\left(\frac{T_m}{T_0} - 1\right) + \frac{LL_m}{2}\left(\frac{T_{IF}}{T_0}\right) \right]$$

$$= T_A + T_0(L - 1) + \frac{LT_m}{2} + \frac{LL_m T_{IF}}{2} - \frac{T_0 L}{2}$$

$$= T_A + T_0\left(\frac{L}{2} - 1\right) + \frac{L}{2}(T_m + L_m T_{IF}) \qquad (18.44)$$

In a similar manner, the single-sideband noise temperature is given by

$$T_{SSB} = T_A + T_0(L - 1) + L(T_m + L_m T_{IF}) \qquad (18.45)$$

18.3.3   RF Detection

For any radiometer configuration, the goal of detecting changes in the dc output of the RF detector is a major objective. The detector that follows the predetection stage has a dc output voltage which is proportional to the change in input power (i.e., the change in system noise temperature). This only holds true for sufficiently small changes in input power, where the slope of the detector input-output characteristic is constant. Consequently, the determination of the change in input power due to a measurable change in detector output voltage requires a calibration procedure.

Generally, a known calibration signal is injected at the radiometer's input and the corresponding output change is measured to deduce the system calibrated gain. To avoid inaccuracies due to changing detector characteristics with large calibrated signals (on the order of 10% of T_{sys}), it is common practice to operate on the square-law portion of the detector's characteristic curve. Detector module manufacturers generally state the proper region of operation for each particular square-law device.

18.3.4 Synchronous Detection and Postdetection

Recall that the Dicke radiometer modulates the RF detector's dc output, and the amplitude modulation is proportional to the temperature difference between the antenna and the Dicke reference load. Synchronous detection is a method of recovering the modulation signal from the low-frequency noise in the system.

Figure 18.12 depicts the synchronous detection process in the post-detection portion of the radiometer. The bandpass filter precedes the synchronous detector, and is used to limit the RF bandwidth's dynamic range in order to reduce broadband noise. As shown in the figure, the synchronous detector effectively rectifies the Dicke modulation waveform. The square-wave modulation is maintained up to the synchronous detector input. The output is a dc offset voltage with random noise that is not affected by the synchronous detector.

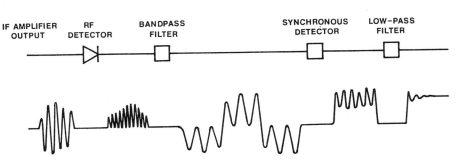

Fig. 18.12 Postdetection processing of radiometric signal.

Postdetection filtering is provided by the low-pass filter. This filter must pass the data with minimum distortion and still eliminate the noise above the data bands. A Bessel type of filter design is commonly used for the low-pass filter because it minimizes both amplitude and phase (time delay) distortion.

The Bessel filter provides constant time delay across the passband, and an amplitude response which falls steadily to the half-power point. If the low-pass filter output is digitized for further data processing, then the data sampling rate is chosen to satisfy the Nyquist criteria. A general rule is to sample the filter output at a rate of at least twice the frequency of the -20 dB point on the filter response curve.

18.4 RADIOMETER APPLICATIONS

18.4.1 MMW Target Seeker

Radiometry has been used for the remote sensing of point targets from air to ground as part of passive seeker applications. Millimeter waves offer four unique features, which make them extremely useful for target seekers. First, the shorter wavelengths relative to microwaves allow for smaller RF components, and thus reduced size and weight for missile seekers. Second, extremely large RF bandwidths are available at the MMW transmission windows. Consequently, many individual frequencies (within a specified transmission window) can be used to provide increased immunity to interference from multiple users, and thus achieve high levels of electromagnetic compatibility (EMC). For the radiometer system, this translates into improved target detection sensitivities.

The third reason for MMW usefulness in target seeker applications lies in the smaller radiated beamwidth for a given antenna size as compared with microwaves. Narrow effective beamwidths lead to improved target resolution, and therefore improved accuracy. This is important for seekers where precision target tracking is required to achieve sufficient target detail. For example, a six-inch diameter antenna provides a 1.5° beamwidth at 94 GHz as compared with 15° at 9.4 GHz.

Finally, the fourth reason is that atmospheric absorption and attenuation losses are relatively low in the MMW transmission windows. Consequently, MMW sensors are more effective than electro-optic devices, such as lasers and IR sensors, either in adverse weather or under battlefield conditions of smoke or dust. In addition, the high attenuation encountered in the MMW absorption bands can be effectively used in covert applications.

For radiometric seeker applications, passive cross section of metallic ground-based targets will depend on the radiometric temperature contrast between the background and the target. As the antenna beam passes from the background to the target, a temperature differential is observed, which depends on the ratio of target to beam interception area. The observed temperature change ΔT_A is given by (Hayes [5]):

$$\Delta T_A = \Delta T_T \frac{A_T}{A_B} E \tag{18.46}$$

where

E = antenna main-beam efficiency within the 3 dB points,
ΔT_T = target-to-background temperature contrast,
A_T = target area projected normal to the line of sight,
A_B = 3 dB beam area normal to the line of sight at the target range.

Figure 18.13 depicts the use of a 95 GHz passive radiometer in the final stage of target tracking. For this scenario, the average sky temperature at 95 GHz is assumed to be 60 K over a 140° angular sweep, 150 K over 20° sweep, and 300 K up to 10° above the surface on either side of the target. This allows us to determine the average target temperature, for a reflectivity of 1, as follows:

$$T_T = 60 \text{ K} \frac{140°}{180°} + 150 \text{ K} \frac{20°}{180°} + 300 \text{ K} \frac{20°}{180°} = 96.7 \text{ K} \tag{18.47}$$

For a target range designated by R, the radiometric beam area is given by

$$A_B = \pi R^2 \tan^2 \frac{\Theta_{3dB}}{2}$$
$$= (5.03 \times 10^{-4})R^2 \tag{18.48}$$

for a six-inch antenna at 95 GHz:

$$\left(\Theta_{3dB} = \frac{70\lambda}{D} \right)$$

If we assume that a target with a cross-sectional area of 10 m² is placed in the antenna beam area and there is an antenna main-beam efficiency of 60% (typical for a feedhorn design), then the observed temperature change for the radiometer is

$$\Delta T_A = (300 \text{ K} - 96.7 \text{ K}) \left[\frac{10 \text{ m}^2}{(5.03 \times 10^{-4})R^2} \right] (0.6)$$
$$= \frac{2.43 \times 10^6}{R^2} \tag{18.49}$$

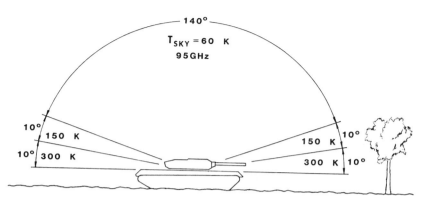

Fig. 18.13 Passive target detection scenario at 95 GHz.

for R = target range in meters. For a target range of 100 m, the temperature contrast is 243 K, whereas at R = 1 km, the contrast temperature is reduced to 2.43 K. Since the signal-to-noise ratio of the measured contrast is defined to be $\Delta T_A/\Delta T_{min}$, where ΔT_{min} is the radiometer's temperature sensitivity, then a 10 dB signal-to-noise level can be achieved at 95 GHz when ΔT_{min} approaches 0.23 K.

The effect of adverse weather during the final phase of passive target homing should be considered. Figure 18.14 depicts the scenario for this measurement under an overcast, cloudy sky at 95 GHz.

For the above scenario:

$$T_1 = \frac{3 \text{ K}}{10^{ac/10}} + \left(1 - \frac{1}{10^{ac/10}}\right) T_{atm}$$

$$= \frac{3 \text{ K}}{10^{0.1}} + \left(1 - \frac{1}{10^{0.1}}\right) 300 \text{ K}$$

$$= 64 \text{ K}$$

(18.50)

for T_{atm} = mean atmospheric temperature. Likewise,

$$T_2 = \frac{T_1}{10^{ac/10}} + \left(1 - \frac{1}{10^{ac/10}}\right) T_{atm}$$

$$= \frac{64 \text{ K}}{10^{0.4}} + \left(1 - \frac{1}{10^{0.4}}\right) 300 \text{ K}$$

$$= 206 \text{ K}$$

(18.51)

Therefore,

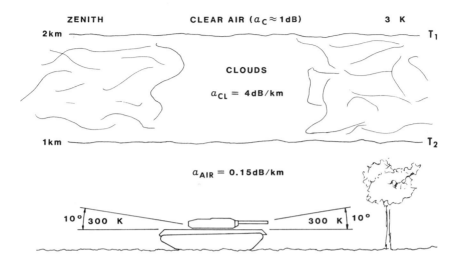

Fig. 18.14 Passive target detection at 95 GHz in adverse weather.

$$T_T = \frac{T_2}{10^{a_{air}/10}} \left(\frac{160°}{180°}\right) + \left(1 - \frac{1}{10^{a_{air}/10}}\right)$$

$$\times\ T_{atm}\left(\frac{160°}{180°}\right) + T_{gnd}\left(\frac{20°}{180°}\right)$$

$$= \frac{206\ \text{K}}{10^{0.015}}\left(\frac{160°}{180°}\right) + \left(1 - \frac{1}{10^{0.015}}\right)$$

$$\times\ 300\ \text{K}\left(\frac{160°}{180°}\right) + 300\ \text{K}\left(\frac{20°}{180°}\right)$$

$$= 219\ \text{K} \tag{18.52}$$

The observed contrast temperature ΔT_A is given by

$$\Delta T_A = \frac{\Delta T_T}{10^{a_{air}/10}}\left(\frac{A_T}{A_B}\right) E$$

$$= \left(\frac{300\ \text{K} - 219\ \text{K}}{10^{0.015}}\right)\left(\frac{10\ \text{m}^2}{A_B}\right)(0.6)$$

$$= \frac{469.5\ \text{K}}{A_B} \tag{18.53}$$

for A_B = radiometric beam area in m^2.

If we assume the same six-inch aperture antenna at 95 GHz, then the target contrast temperature is reduced to 93.3 K for a seeker-to-target range of 100 m. To achieve a minimum signal-to-noise level of 10 dB with the same radiometer requires a target range that is no greater than 635 m. The influence of the cloudy sky conditions is reflected in the effectiveness of the passive target seeker over range R as shown in Table 18.1.

Table 18.1 S/N (dB) *versus* Target Range Under Clear or Adverse Weather*.

R (meters)	Clear Weather	Cloudy Sky
1000	10.0	6.1
800	11.9	7.9
600	14.4	10.5
400	17.9	14.0
200	23.9	20.1
100	30.0	26.1

18.4.2   Airborne Imaging Radiometric System

Advances in MMW radiometric technology, particularly in the development of RF components, have led to the use of radiometric imagers for target identification and guidance. The characteristics of a passive imager are ideally suited for adverse weather, real-time, reconnaissance operations.

A variety of sensor platforms are available for carrying highly sensitive passive imaging systems. For instance, overhead drones and other remotely piloted vehicles (RPVs) can sense from low levels, and thus achieve very high spatial resolution. At higher altitudes, reconnaissance aircraft can provide imaged geographic and weather data. Tower-mounted sensors, located at fortified positions, can provide sharp imagery at close range, even in adverse weather.

The most attractive features of the radiometric imager are passive and adverse-weather operation, battlefield operation through smoke and dust, low cost, and small size. The low cost and small size of the sensor are particularly important for small, expendable RPV applications. For example, a sensor with a six-inch aperture antenna can achieve a 0.46° half-power beamwidth at 300 GHz. This sensor could spatially resolve tactical targets at ranges approaching 1000 feet.

---

*For 95 GHz radiometer using six-inch aperture antenna with ΔT_{min} of 0.23 K.

On the basis of today's technology, 95 GHz has proved to be a good compromise frequency in terms of bad-weather capability *versus* high angular resolution when using small antenna apertures. If 35 and 95 GHz systems with identical apertures are compared, the 95 GHz sensor performs best in medium fog (visibility of 120 m), clouds, or light rain for the resolution of small targets. A degradation of target contrast for the 95 GHz sensor by a factor of about four under adverse weather is compensated by a beam-filling factor that is nine times greater than that achieved by the 35 GHz sensor.

Highly sensitive MMW imaging radiometers are desirable for the detection of small ground target vehicles (2 m or less) from a flying platform. Several system constraints need to be considered in the determination of the radiometer's performance:

1. The minimum flight speed of the aircraft used for the measurements is 50 m/s;
2. The minimum ground swath width for reproducible target measurements is 40 to 50 m;
3. Beam-spot overlapping on the ground of 50% is required to generate a complete picture (image) of the target vehicle;
4. A geometrical resolution of better than 2 m at the maximum scan angle is required to detect small targets;
5. A temperature resolution of better than 1 K is required to obtain well structured maps, especially when the surface consists of different types of vegetation.

Figure 18.15 depicts an airborne scanning radiometer configuration for ground target imaging. To define a system within the operating constraints, a 90 GHz radiometer with a 25 cm antenna aperture is installed aboard the aircraft at an altitude of 90 m and a ground speed of 50 m/s. Consequently, the beam-spot diameter on the surface is given by

$$
\begin{aligned}
d &= 2h\,\tan(\theta_{3dB}/2) \\
&= 2h\,\tan(70\lambda/2D) \\
&= 1.46\ \text{m}
\end{aligned}
\tag{18.54}
$$

Since the swath width is taken to be 50 m and beam-spot overlapping of 50% is desirable, then the number of data samples per swath is given by 1.5 (50/1.46), or 51 data points per scan. An aircraft ground speed of 50 m/s and a beam-spot diameter of 1.46 m results in (1.46/50), or 29 ms time for the aircraft to move forward one beam spot. During that time, 51 data points must be sampled, which results in a dwell (integration) time of 29 ms/51 = 0.57 ms per beam spot. This integration time (τ) is the

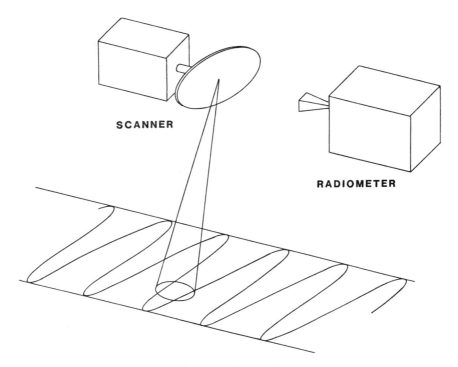

Fig. 18.15 Surface imaging for scanning radiometer.

value that would be used in the temperature sensitivity (ΔT_{min}) equation for the airborne radiometer.

For this application, a total-power radiometer is assumed, and for a negligible system gain variation ($\Delta G/G \approx 0$), we then have

$$\Delta T_{min} = \frac{T_A + T_r}{\sqrt{B\tau}} \tag{18.55}$$

where

T_A = antenna temperature = 300 K;
T_r = receiver noise temperature, in K;
B = system bandwidth = 1 GHz, typical;
τ = integration time = 0.57 ms.

Solving for the receiver noise temperature T_r yields

$$\begin{aligned} T_r &= \Delta T_{min} \sqrt{B\tau} - T_A \\ &= (1 \text{ K}) \sqrt{(10^9) (0.57 \times 10^{-3})} - 300 \text{ K} \\ &= 455 \text{ K} \end{aligned} \tag{18.56}$$

for $\Delta T_{min} = 1$ K (design goal).

To achieve such a low noise temperature at 90 GHz with today's technology requires a cooled device. Temperature stabilization of all components is necessary for low drift and gain fluctuations. The radiometer's front end must be mounted in an evacuated compartment and cooled to about 23 K by means of a temperature-controlled helium refrigerator to achieve noise temperature levels below 500 K.

Data sampling rates of $1/\tau = 1750$ samples per second place a tremendous burden on the scan mechanism required to generate a complete picture of the imaged target. Consequently, the antenna system for the radiometer requires a high-speed rotating parabolic mirror with high-precision gear and mass-balance weights. In addition, vibration-shock mounts and speed-regulated motor drive circuits are imperative for the desirable imaging system. However, a mechanical scanning system at 90 GHz is the preferred choice, as compared with electronic (phased array) scanners, because of excessive RF losses in the phase shifters, higher cost, and excessive weight, all of which are significant concerns for airborne measurements. Figure 18.16 is a photograph of an imaging radiometer, which uses a mechanical scanner, controlled by a precision stepper motor.

Fig. 18.16 Imaging MMW radiometer with calibration loads and scanning reflector.

18.4.3   Radiometric Area Correlation System

Target and background characteristics are key features in the design
of a MMW passive seeker. The target may take the form of a large man-
made object such as a building, or smaller objects such as armored vehicles
or trucks. The background has been extensively characterized in the literature for a number of materials that might form a background scenario
[6]. Table 18.2 summarizes material emissivities (ϵ_T) assuming a normal
incident angle between the target and the radiometer. The table also summarizes the radiometric target temperature, $T_T = \epsilon_T T_M + (1 - \epsilon_T) T_{sky}$,
where T_M is the target physical temperature and T_{sky} is the radiometric
sky temperature.

Table 18.2   Material Emissivities at Normal Incidence Angle (from
Button [6], © 1981 Academic Press).

Material	*Emissivity* (ϵ_T)	T_T for T_M = 300 K and T_{sky} = 100 K
Heavy Vegetation	0.93	286 K
Dry Grass	0.91	282 K
Dry Snow	0.88	276 K
Asphalt	0.83	266 K
Concrete	0.76	252 K
Metal	0	100 K

The target is characterized for a radiometric sensor by its effective
reflecting area and a series of radiometric temperature contours, which
are dependent on the antenna beamwidth, the target depression (incident)
angle, the target range, and the radiometer frequency. The passive radiometer target range equation is given by [6]:

$$R = \left[\frac{E\pi D^2 \, (A_T \Delta T_T)}{4\lambda^2 \Delta T_{min} \sqrt{S/N}} \right]^{1/2} \qquad (18.57)$$

where

$E\pi D^2/4\lambda^2$ = antenna factor;
$A_T \Delta T_T$ = target passive cross section, in m²K;
ΔT_{min} = radiometer sensitivity, in K;
S/N is the power signal-to-noise ratio.

For radiometric area-correlation applications, it is useful to characterize the radiometer's sensitivity (ΔT_{min}) with respect to the target passive cross section ($A_T \Delta T_T$). Therefore, solving for the target passive cross section yields

$$A_T \Delta T_T = \frac{\Delta T_{min} \, 4\lambda^2 R^2 \, \sqrt{S/N}}{E\pi D^2} \qquad (18.58)$$

If we assume an antenna main-beam efficiency (E) of 60%, an antenna diameter (D) of six inches, and a power signal-to-noise ratio of 10 dB, then,

$$A_T \Delta T_T = 288.93 \lambda^2 R^2 \Delta T_{min} \ (\text{m}^2\text{K})$$

If we assume a 35 GHz sensor with a sensitivity of 0.1 K, then the detectable passive cross section will vary from 21 m² K to 2122 m² K as the target range increases from 100 to 1000 m. Recall that this is based on a power signal-to-noise level of 10 dB. However, if the sensor's sensitivity degrades to 0.5 K, then the passive cross section increases from 106 m² K to 10,610 m² K, given that all other conditions remain the same.

Radiometric correlation is affected by the frequency of operation, since the target's spatial resolution improves (reduces) at higher frequency with the small beamwidth of the antenna. Table 18.3 compares the passive cross section *versus* target range for a 35 GHz sensor and a 95 GHz sensor, assuming a 0.1 K temperature sensitivity.

Table 18.3 Passive Cross Section *versus* Target Range for Temperature Sensitivity of 0.1 K and Power Signal-to-Noise Ratio of 10 dB.

R(meters)	$A_T \Delta T$ (m² K) *at Frequency*	
	35 GHz	95 GHz
100	21.2	2.9
500	530.5	72.1
1000	2122	288.5

In conclusion, a target-correlation seeker operating at MMW frequencies can provide the desired high resolution with antenna diameters of six inches or less. This is achieved by reducing the size of the beam interception pattern on the terrain, and hence the amount of clutter competing with the target signal.

18.4.4   Atmospheric Radiometer Sounder

The vertical and horizontal distribution of water vapor in the earth's atmosphere is of great importance in determining the future state of the atmosphere. For instance, the release of latent heat due to condensation of the vapor is the dominant heat source in tropical cyclones and other convective storms. Weather forecasting, particularly precipitation forecasting, benefits from improved knowledge of the water vapor distribution.

Water vapor is also an interfering phenomenon in many remote sensing activities, such as temperature sounding, altimetry, and surface temperature measurements. A MMW radiometer operating near 183.3 GHz (a strong water vapor absorption line) can provide atmospheric humidity data down to about 2 km altitude, even in the presence of cirrus clouds over land or sea. Aircraft observations with the *Advanced Microwave Moisture Sounder* (AMMS) have demonstrated that multiple channels near the 183.3 GHz water vapor absorption line are strongly affected by large ice particles associated with convective precipitation [7]. In addition to providing an indication of the area of convective precipitation, AMMS measurements have given information on the height to which the ice is lifted, and therefore an indication of the intensity of the convection.

Figure 18.17 is a block diagram of the AMMS, which is a 94/183 GHz multichannel imaging radiometer, previously flown on several high-altitude research aircraft for the US National Aeronautics and Space Administration (NASA).

The AMMS, shown in Figures 18.18 and 18.19, is a Dicke radiometer design, which uses a high-speed mechanical chopper for alternately switching, the scene between the 94 GHz and 183 GHz antennas. The radiometer's dual front end is completely solid-state, with low noise and wide RF bandwidth (91 to 97 GHz and 173 to 193 GHz). The instrument utilizes three IF channels about the 183.3 GHz water vapor absorption line and a single channel at the 94 GHz atmospheric window. The 183 GHz front-end design includes a subharmonic balanced mixer, pumped at 91.65 GHz, using a solid-state Gunn-diode oscillator (GDO). The 94 GHz system has a separate GDO for independent frequency tuning.

The antenna system for the AMMS includes a five-inch lens aperture, feeding two conical corrugated feedhorns, designed for operation at 94 and 183 GHz. The lens antenna provides 1° and 2° half-power beamwidths at 183 and 94 GHz, respectively. Imaging is provided by using a microprocessor-controlled rotating mirror across the antenna in a raster scan configuration. This same mirror is periodically directed to view precise calibration loads to provide on-line radiometric calibration. Synchronous detectors at each of the four data channels provides rectified dc outputs of the Dicke modulated (chopped) waveform. These outputs are stored on the AMMS on-board cassette recorder to provide calibrated radiometric brightness temperatures at 94 GHz (single channel) and 183 GHz (three channels). A separate ground-support computer system is used to provide radiometric images and hardcopy printouts of brightness temperatures, to transfer flight data from cartridge to computer compatible tapes, and to develop flight software support for programming modifications on a flight-to-flight basis.

Extensive analysis of the AMMS data has been performed since the initial measurement program in 1979. Aircraft data missions which have used the AMMS include: Severe Environmental Storms and Mesoscale Experiment (SESAME); Florida Thunderstorm Mission; Florida Area Cumulus Experiment (FACE); Cooperative Convective Precipitation Experiment (CCOPE); Marginal Ice Zone Experiment (MIZEX) West in the Bering Sea and MIZEX East in the North Sea (Cavalieri and Gloersen [9]). Detailed data analysis of AMMS results are documented in reports prepared for the NASA-Goddard Space Flight Center (GSFC) (Nieman and Krupp [10]).

The AMMS has provided atmospheric water vapor data that will prove very useful in the development of future radiometric sensors aboard meteorological satellites. One such sensor is the Advanced Microwave Sounder Unit (AMSU) which is planned for development in the early 1990s [11]. The AMSU will contain frequencies from 18 GHz to 183.3 GHz for temperature, tropospheric water vapor, precipitation, and sea ice measurements. In order to detect small convection rain cells, the AMSU will need resolutions on the order of 10 km, which can be achieved by using a paraboloid reflector antenna with 0.25 m diameter. Twice daily coverage of the entire earth requires a swath width of 2400 km, resulting in about 5 ms integration time. Since the precipitation effect is strong, ΔT_{min} values of 1 to 5 K should provide adequate temperature sensitivity. Assuming a total-power radiometer design for the AMSU and a minimum system bandwidth of 200 MHz (the channel that is nearest the water vapor absorption line) will require system noise temperatures of 1000 to 1500 K.

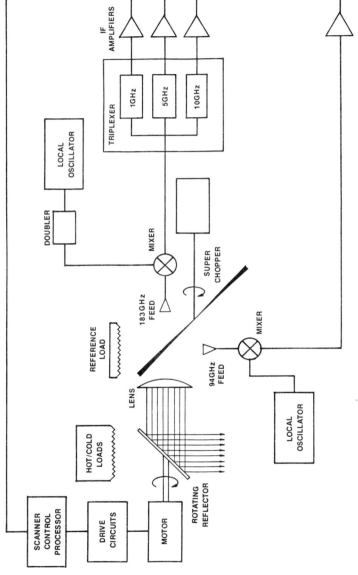

Fig. 18.17(a) Advanced microwave-moisture sounder (AMMS) predetection stage including antenna, calibration loads, and scanner (from Gagliano [8]).

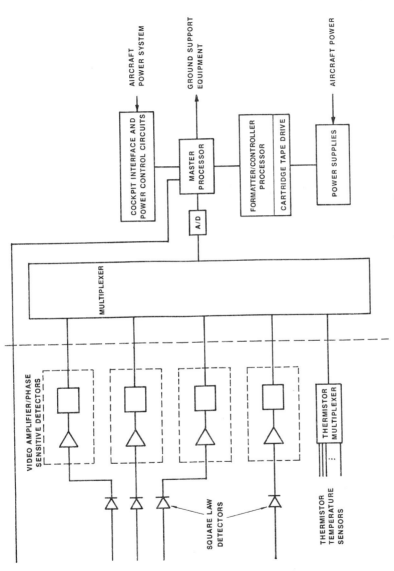

Fig. 18.17(b) Advanced microwave moisture sounder (AMMS) postde-tection stage including data multiplexer and processor (from Gagliano [8]).

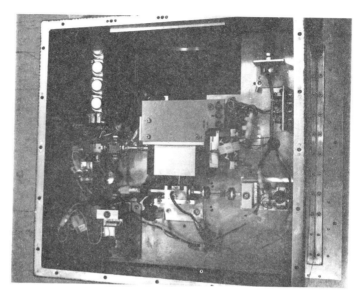

Fig. 18.18 Advanced microwave moisture sounder (AMMS) radiometer front end with RF components and five-inch diameter lens.

Figure 18.19 AMMS/ER-2 Instrumentation Pallet In Preparation For Data Flights

Fig. 18.19 AMMS/ER-2 instrumentation pallet in preparation for data flights.

18.4.5 Future Development Areas

For all types of radiometers, the achievable temperature sensitivity is directly proportional to $T_A + T_R$, where T_A is the antenna temperature and T_R is the receiver noise temperature. As long as T_R is considerably larger than T_A, then sensitivity can be improved simply by pushing the technology toward lower noise figures. However, as T_R approaches T_A, only marginal improvements in sensitivity at the expense of costly technological refinements are expected. Since, for most earth oriented measurements, T_A is in the range of 200 to 300 K, then MMW radiometric technology still has the capacity for great achievements in temperature sensitivity. Some areas expected to be investigated over the next several years include:

1. Further reduction in noise temperatures for devices above 60 GHz, where T_R is significantly above T_A;
2. Reduction in LO power requirements by improved LO matching and using better semiconductor material;
3. Development of broadband mixers to permit LO tuning (matching of diode to RF circuit), or to obtain channel splitting at IF level (matching of IF amplifier);
4. Improved reproducibility of MMW component electrical specifications, such as tighter mechanical tolerances and better whisker placement;
5. Improved stability of MMW components with respect to temperature changes and mechanical shock;
6. Better design of definitive transfer characteristics in the receiver band (i.e., reducing device-matching problems);
7. Advancing the miniaturization of the entire RF front end, which implies solving problem areas such as lossy switches (*pin* diodes, FET switches, and microstrip circulators);
8. Reduction in the reflections between receiver components, which result in oscillation in the output, by the use of baseline ripple reducers (e.g., dielectric chopper, circular polarizers, *et cetera*).

Another area of future investigation, especially for very high frequencies, is the use of quasioptical techniques. Direct applications of optical methods in radiometric systems include: the use of dicroic mirrors to split the antenna beam into different frequency ranges; quasioptical filters to reduce the noise effects of *IMPATT* diode oscillators; the calibration of radiometers. The use of optical methods for calibration avoids the use of lossy and bulky calibration switches and bulky sky horns (for cold space calibration). Calibration accuracy is increased because the procedure does not rely on a lossy signal path from the space horn through the calibration switch.

REFERENCES

1. D.D. King, "Passive Detection," Chapter 39 of *Radar Handbook*, M.I. Skolnik, ed., McGraw-Hill, New York, 1970.
2. F.J. Janza, ed., "Manual of Remote Sensing," Vol I, American Society of Photogrammetry, Falls Church, VA, 1975.
3. G. Evans, and C.W. Mcleish, *RF Radiometry Handbook*, Artech House, Dedham, MA, 1977.
4. "Radio-Telescope Receivers," Chapter 7 of *Radio Astronomy*, J.D. Kraus, ed., McGraw-Hill Book Company, NY, 1966.
5. R.D. Hayes, "Radiometric Measurements," Chapter 15 of *Techniques of Radar Reflectivity Measurement*, N.C. Currie, ed., Artech House, Dedham, MA, 1984.
6. C.R. Seashore, "Missile Guidance," Chapter 3 in *Infrared and Millimeter Waves,* Volume 4, *Millimeter Systems,* K.J. Button and J.W. Wiltse, eds., New York: Academic Press, 1981.
7. J.R. Wang, "Profiling Atmospheric Water Vapor By Microwave Radiometry," *Journal of Climate and Applied Meteorology*, Vol. 22, 1983.
8. J.A. Gagliano, "AMMS/WB-57F CCOPE Mission," Final Technical Report for NASA Contract NAS5-26528, Georgia Institute of Technology, Atlanta, GA, 1981.
9. D.J. Cavalieri, and P. Gloersen, "MIZEX-West NASA CV-990 Flight Report," NASA Technical Memorandum 85020, 1983.
10. R.A. Nieman and B.M. Krupp, "Simultaneous Active and Passive Observations of Raindrops and Ice Particles in Florida Thunderstorms," September 1979, Computer Sciences Corporation, 1980.
11. AMSU Design Study, prepared for NASA-Goddard Space Flight Center, July 1980, prepared by Aerojet Electrosystems Co., Azusa, CA.

Glossary

A	Area Projected by Antenna Beam on Ground, Antenna Aperture Area
a	Antenna Efficiency, Antenna Beam Shape Factor
A_c	Illuminated Area Within a Radar Resolution Cell
A/D	Analog-to-Digital
ADU	Air Defense Unit
A_e	Effective Area of Antenna Aperture
A_e	Effective Earth Radius
A_i	Interference Factor
A_u	Upwind-Downwind Factor
AGC	Automatic Gain Control
AI	Artificial Intelligence
AIFS	Advanced Indirect Fire Seeker
α	Alpha—Attenuation Coefficient, Antenna Beam Shape Factor, Incidence Angle
AM	Amplitude Modulation
AMMS	Advanced Microwave Moisture Sounder
AMMWS	Advanced Millimeter Wave Seeker
APC	Armored Personnel Carrier
ARM	Antiradiation Missile
B or B_n	Receiver Bandwidth
BWO	Backward-Wave Oscillator
°C	Degrees Centigrade
c	RF Propagation Velocity (3×10^8 m/s nom.)
CB	Mismatch Filter Correction Factor
CFA	Crossed-Field Amplifier
CFAR	Constant False Alarm Rate
COHO	Coherent Oscillation
CW	Continuous Wave
d	Distance
dB	Decibel

dBHz	Decibel Relative to 1 Hertz
dBi	Decibel Relative to Isotropic
dBK	Decibel Relative to 1 Kelvin
dBm	Decibel Relative to 1 Milliwatt
dBmeter	Decibel Relative to 1 Meter
dBm^4	Decibel Relative to 1 Meter to the Fourth Power
dBsm	Decibel Relative to 1 Square Meter
dBsm/m^3	Decibel Relative to 1 Square Meter per Cubic Meter
ΔF	Frequency deviation
ΔR	Range Resolution
$\Delta T_{\min}$	Minimum Discernible Temperature by a Radiometric Receiver
$\Delta \Theta$	Antenna Scan Angle
DLC	Delay Line Canceller
DSB	Double Sideband
E	Electric Field
ECCM	Electronic Counter-Countermeasures
ECM	Electronic Countermeasures
EIA	Extended Interaction Amplifier
EIKA	Extended Interaction Klystron Amplifier
EIKO	Extended Interaction Klystron Oscillator
EIO	Extended Interaction Oscillator
EMI	Electromagnetic Interference
EO	Electro-optical
ϵ	Epsilon—Dielectric Constant, Emissivity
ERAM	Extended Range Anti-Armor Munition
ERT	Energy Reflected by a Target
ESM	Electronic Support Measures
η	Eta—Radar Cross Section per Unit Volume (1/m), Combining efficiency for a two-Source Hybrid Circuit, Antenna Aperture Efficiency
F_n	Noise Figure
F_b	Beat Frequency
F_d	Doppler Frequency Shift
FEL	Free Electron Laser
FET	Field-Effect Transistor
F_s	Scan Frequency
f	Frequency
f_c	Cut-off Frequency
FM	Frequency Modulation
FMCW	Frequency Modulated Continuous Wave
G	Antenna Power Gain

GaAs	Gallium Arsenide
γ	Gamma–Reflection Coefficient $= \sigma°/\sin\theta$
GHz	Gigahertz–10^9 Hz
H	Horizontal Polarization
$H(F)$	Filter Frequency Response
$H(f)$	Output of a Matched Filter
h	rms Surface Roughness
h_a	Antenna Height
h_{av}	Average Wave Height
h_d	Duct Height Constant
h_t	Target Height
hr	Hour
HWIL	Hardware-in-the-Loop Simulation
Hz	Hertz
I	In-phase Coherent Receiver Channel
IF	Intermediate Frequency
IFF	Identification Friend or Foe
IFMCW	Interrupted FMCW
ILO	Injection-Locked Oscillator
IMPATT	Impact Avalanche Transmit Time
INS	Inertial Navigation System
IR	Infrared
K	Kelvin
k	Boltzmann's Constant—1.38×10^{-20} mW/Hz/K
kG	Kilogauss (Magnetic Field)
kHz	Kilohertz-1000 Hz
km	Kilometer-1000 m
kV	Kilovolt-1000 V
kW	Kilowatt-1000 W
l	Aperture Dimension
λ	Lambda-RF wavelength
LHC	Left-Hand Circular Polarization
LO	Local Oscillator
LOAL	Lock-On After Launch
LOBL	Lock-On Before Launch
log	Logarithmic
LOS	Line of Sight
L_{CFAR}	CFAR Loss
L_m	Miscellaneous Losses
L_R	Receiver Loss
L_s	System Losses

L_T	Transmission Loss
L_t	Total Loss Factor
m	Meter
mi	Mile (Statute)
mm	Millimeter (10^{-3} m)
mr	Milliradian
ms	Millisecond (10^{-3} s)
m/s	Meter per Second
mW	Milliwatt (10^{-3} W)
m_x	Median of x
MCGD	Millimeter Wave Contrast Guidance Demonstration
MHz	Megahertz (10^6 Hz)
MLRS/TGW	Multiple Launch Rocket System/Terminally Guided Weapon
MMIC	Millimeter-Wave Integrated Circuit
MMW	Millimeter Wave
MOPA	Master Oscillator Power Amplifier
MRSS	Microwave Radiometer Seeker Subsystem
MSGP	Millimeter-Wave Semiactive Guidance Program
MTI	Moving Target Indication
μ	Mu—Permeability
μs	Microsecond (10^{-6} s)
N	Integration Gain, Index of Refraction
N_0	Noise Spectral Density
NF	Receiver Noise Figure
$n(a)$	Drop Size Distribution
ns	Nanosecond—10^{-9} seconds
ω	Omega—Radian Frequency, Coverage (Solid Angle)
P_D	Probability of Detection
P_{FA}	Probability of False Alarm
P_t	Transmitting Power
$p(x)$	Probability Density Function
pdf	Probability Density Function
PFN	Pulse Forming Network
PGM	Precision Guided Munition
PGW	Precision Guided Weapon
PRF	Pulse Repetition Frequency
Φ	Angle between Boresight and Upwind
ϕ	Azimuth Aspect Angle
PM	Phase Modulation
POL	Petroleum Oil Lubricant
PPI	Plan Position Indicator

PPM	Periodic Permanent Magnet, Parts per million
PRF	Pulse Repetition Frequency
PRI	Pulse Repetition Interval
π	Pi (3.1417 . . .)
ψ	Psi—Three-Dimensional Free-Space Green's Function, Grazing Angle
Q	Quadrature Coherent Receiver Channel
R	Range, Rain Rate
RAM	Radar Absorbing Material
RATSCAT	Radar Target Scattering Facility (US Air Force)
RCS	Radar Cross Section
RF	Radio Frequency
RHC	Right-Hand Circular Polarization
ρ	Rho—Scattering Coefficient,
ρ_i	Rho sub i—Spherical Wave Radii
ρ_s	Rho sub s—Specular Scattering Factor
ρ_0	Rho sub zero—Reflection Coefficient for a Smooth Uniform Dielectric Surface
R_m	Maximum Range
$R_n(\tau)$	Autocorrelation Function of delay time (τ)
s	Second
SADARM	Search and Destroy Armor
SAR	Synthetic Aperture Radar
SAW	Surface Acoustic Wave
SD	Standard Deviation
SCR	Signal-to-Clutter Ratio
SDI	Strategic Defense Initiative
σ	Sigma—Radar Cross Section (m^2)
σ_c	Clutter Radar Cross Section, Conductivity
σ_h	Root-Mean-Square (rms) Surface Roughness
σ_i	Intrinsic Clutter Spectrum Standard Deviation
σ_n	Standard Deviation of Noise Power
σ_p	Platform Motion Clutter Spectrum Standard Deviation
σ_s	Antenna Scanning Clutter Spectrum Standard Deviation
σ_t	Sigma sub t—rms Angle Tracking Error
σ°	Sigma Zero—Radar Cross Section per Unit Area for a Radar Beam Mapped onto a Surface
S/N, SNR	Signal-to-Noise Ratio
STAFF	Smart Target Activated Fire and Forget
STALO	Stable Local Oscillator
T	Reference Temperature 290 K
T_a	Antenna Temperature

t	Round-Trip Propagation Time
τ	Tau—Pulse Length in Seconds, Integration Time, Correlation Lag Time
TGSM	Terminally Guided Submunition
THz	Terahertz—1000 GHz
TV	Television
T/R	Transmit-Receive
TWT	Traveling Wave Tube
TWTA	Traveling Wave Tube Amplifier
Θ_{AZ}	Azimuth 3 dB Beamwidth, One Way
Θ_{EL}	Elevation 3 dB Beamwidth, One Way
Θ, θ	Theta—Depression or grazing angle
θ_a	3 dB Azimuth Beamwidth (One-Way)
θ_x	Theta—sub x—3 dB beamwidth at frequency x
$\theta(t)$	Theta of t—Phase Modulation
UAT	Uniform Asymptotic Theory
V	Volt, Vertical Polarization
V_r	Radial Velocity
v	Velocity
VCO	Voltage Controlled Oscillator
VSWR	Voltage Standing Wave Ratio
W	Watt
$W(f)$	Power Density Spectrum

Index

Traveling wave tube (TWT), 33–34, 58, 64,
448, 456–464, 467, 470–471, 477, 740,
746, 786, 791, 797
Tree clutter, 356
Triangular trihedral reflector, 770
Triangular waveform, 681
Trihedral calibration reflector, 765, 768,
770, 812
Tube efficiency, 461
Twist reflector antenna, 554
Two-dimensional electron gas (2DEG), 411

Unambiguous Doppler, 627
Unambiguous range, 295, 626–627, 631,
645, 742
Unambiguous velocity, 627, 645
Uniform asymptotic theory (UAT), 331
Uniform illumination, 529
Unipolar video, 256
Unitary warhead, 648
Up-conversion, 487

VSWR, 474, 542, 596, 598, 636
Vacuum tube sources, 785
Varactor modulator, 432
Varactor-tuned oscillator, 432
Varactor tuning, 412, 426, 432
Vector discrimination techniques, 280
Velocity resynchronization, 463
Voltage-controlled oscillator (VCO), 423,
791, 810
Voltage standing wave ratio (*see* VSWR)
Volume clutter, Section 358, 373
Volume reflectivity coefficient, 758
Volume scattering, 192
Volumetric clutter, 623
Von Karmen radome, 569
Vulnerability, 25

WX-50, 705
Walls, 8
Warhead, 648, 661
Wasp (missile), 649
Water vapor, 132
Waveform, 625, 636
Waveguide array antenna, 780
Waveguide components, 784
Waveguide lens antenna, 779
Waveguide losses, 623
Waveguide slotted array antenna, 548
Weibull distribution, 109
Weibull target model, 96
Whisker-contact diode, 787

Windows, 8, 756
Y-junction circulator, 509
Y-junction isolator, 509